48th American Association of Swine Veterinarians Annual Meeting 2017

One World, One Health,
One Passion for Pigs

Denver, Colorado, USA
25 – 28 February 2017

ISBN: 978-1-5108-3798-0

Printed from e-media with permission by:

Curran Associates, Inc.
57 Morehouse Lane
Red Hook, NY 12571

Some format issues inherent in the e-media version may also appear in this print version.

Printed by Curran Associates, Inc. (2017)

For permission requests, please contact American Association of Swine Veterinarians at the address below.

American Association of Swine Veterinarians
830 26th Street
Perry, IA 50220-2328
USA

Phone: 515-465-5255
Fax: 515-465-3832

aasv@aasv.org

Additional copies of this publication are available from:

Curran Associates, Inc.
57 Morehouse Lane
Red Hook, NY 12571 USA
Phone: 845-758-0400
Fax: 845-758-2633
Email: curran@proceedings.com
Web: www.proceedings.com

One World, One Health, One Passion for Pigs

48th AASV Annual Meeting

February 25 - 28, 2017
Denver, Colorado

Table of Contents

Howard W. Dunne Memorial Lecture

Swine medicine in the 21st century: Immovable object meets unstoppable force 12
Jeff Zimmerman, DVM, PhD, DACVPM

Alex Hogg Memorial Lecture

One Health: Roles, responsibilities, and opportunities for swine veterinarians 20
Matthew Turner, DVM

Sunday: Research Topics

Land coverage and elevation as risk factors for PRRS outbreaks 26
A.G. Arruda, DVM, MS, PhD; C. Vilalta, DVM, PhD; A. Perez, DVM, MS, PhD; et al.

Broadly neutralizing antibodies to recent, virulent type 2 PRRSV isolates 27
M.P. Murtaugh, PhD; D.K. Gray, BS; S.R. Robinson, BVSc (Hons), MVSc, PhD; et al.

Characterization of the memory immune response to PRRSV infection 30
Michael C. Rahe; Michael P. Murtaugh

Proof of concept: PRRSV IgM/IgA ELISA detects infection in the face of circulating maternal IgG antibody 31
M. Rotolo; J.Ji; L. Gimenez-Lirola; et al.

Spatial autocorrelation and implications for oral fluid-based PRRSV surveillance 32
M. Rotolo; M. Haddad; Y. Sun; et al.

Comparative pathogenesis and characterization of recent 1-7-4 PRRSV isolates in weanling age piglets 33
A.G.M. Van Geelen, DVM, PhD; K.S. Faaberg, PhD; P. Das, DVM, PhD; et al.

Application of next-generation sequencing technology to whole genome sequencing of PRRSV under diagnostic setting........ 35
Jianqiang Zhang, PhD; Ying Zheng, MS; Qi Chen, DVM, PhD; et al.

Effects of pain mitigation during piglet castration 39
R. Park; M. Pairis-Garcia, DVM, PhD; M. Campler, MS, PhD

Effect of influenza prevalence at weaning on transmission, clinical signs and performance after weaning........ 42
F. Chamba, DVM, MS; S. Wayne, DVM, PhD; M. Culhane, DVM, PhD; et al.

Senecavirus A: Overview of experimental studies 43
A. Buckley, DVM; V. Kulshreshtha, DVM, PhD; A. van Geelen, DVM, PhD; et al.

Mycoplasma hyorhinis associated with conjunctivitis in pigs........ 45
T.P. Resende, DVM, MS; M. Pieters, DVM, PhD; F.A. Vannucci, DVM, MS, PhD; et al

Mycoplasma hyorhinis and *Mycoplasma hyosynoviae* dual colonization of dams and piglets prior to weaning 47
Luiza R. Roos, DVM, MS; Maria Pieters, DVM, PhD

Efficacy of a commercial G2b-based and two experimental G1b-based porcine epidemic diarrhea virus (PEDV) vaccines against G2b challenge 48
Tanja Opriessnig, Dr med vet, PhD; Huigang Shen, DVM, PhD; Alessandra Castro, DVM, PhD; et al.

Serum and mammary secretion antibody responses in PEDV-exposed gilts following PEDV vaccination........ 50
J.B. Kraft; K. Woodard; L. Giménez-Lirola; et al.

Modeling the transboundary survival of foreign animal disease pathogens in contaminated feed ingredients 51
Scott Dee; Gordon Spronk; Eric Nelson; et al.

Sunday Concurrent Session #1: Student Seminar

Evaluation of a peroxygen-based disinfectant to inactivate porcine epidemic diarrhea virus in swine feces on metal surfaces under freezing conditions 53
K. Baker, BS; C. Mowrer, BS; J. Zhang, MD, PhD, MS; et al.

Calculating and evaluating antibiotic usage in a United States swine production system 54
Megan Pieters; Ross Kiehne, DVM; Eric Christianson, DVM

Assessment of knowledge, compliance, and attitudes of English and Spanish speaking farm employees towards biosecurity practices 57
Michael Mardesen, BS; Jose Angulo, DVM; Clayton Johnson, DVM; et al.

Comparison of standard and bench entry protocols for prevention of environmental contamination due to personnel entry in a commercial swine facility 59
C. Fitzgerald, BS; K. Baker, BS; A. Anderson; et al.

Use of molecular characterization tools to investigate *M. hyopneumoniae* outbreaks 60
A. Anderson, BS; A. Sponheim, DVM; L. Dalquist, DVM; et al.

Development and validation of ante-mortem urine collection techniques for gilts and sows 63
M. Nickel, BS; K. Skoland, BS; A. Forseth, DVM; et al.

An evaluation of oropharyngeal swabbing as a diagnostic technique for the detection of porcine reproductive and respiratory syndrome virus (PRRSV) in weaned pigs 65
Kylie Glisson; Adam Schelkopf, DVM; Brett O'Brien, VMD, MS; et al.

A challenge study to determine whether in-feed flavophospholipol can reduce *Salmonella* shedding and colonization in nursery pigs 67
Jane Newman, BSc; Vahab Farzan, DVM, MSc, PhD; Robert Friendship, DVM, MSc

Correlating sampling methods to porcine circovirus type 2 (PCV2) viremia status of sows and piglets in a sow herd with or without PCV2 revaccination during gestation 69
Allison Knox, BS; Luis Ochoa, PhD; Laura Greiner, PhD

Sample comparison for detecting porcine circovirus type 2 in breeding herds 72
Rachel Schulte; Darrell Neuberger, DVM; Abigail Redalen, DVM; et al.

Evaluation of positive pressure filtration to reduce aerosol transmission of PRRSV during an experimental challenge of farm access points 74
H. Baldry; K. Wedel, DVM; N. Schaefer, DVM; et al.

Evaluation of methods for determining the porcine circovirus type 2 status of sow herds 77
Chelsea Ruston, BS; Jeremy S. Pittman, DVM, MS, DABVP

Dynamics of *Mycoplasma hyopneumoniae* colonization, seroconversion and onset of clinical signs in a population of gilts under field conditions 79
Chris Deegan; Daniel Linhares, DVM, MBA, PhD; Alex Ramirez, DVM, MPH, PhD, DACVPM; et al.

Investigating porcine circovirus associated disease (PCVAD) in commercial swine herds by next generation sequencing 81
Zhen Yang; Lucina Galina Pantoja, DVM, PhD; Todd Knutson, PhD; et al.

Effect of antimicrobials on non-toxigenic *E. coli* oral vaccine 85
Olivia Myers, BS; Jeremy S. Pittman, DVM, MS, DABVP; Megan Jacob, MS, PhD; et al.

Sunday Concurrent Session #2: Industrial Partners

Multiple case study reports of *Mycoplasma hyopneumoniae* elimination projects utilizing tylvalosin (Aivlosin® WSG)........................90
D. Rosener, DVM; R. Kaptur, DVM; A. Mueller, DVM; et al.

Feed truck movement and ingredient traceability using FeedTrackur™ 93
Thomas Stein, DVM, MS, PhD

Comparison of efficacy between two PCV2 vaccination protocols under PCV2d field exposure 95
Eduardo Fano, DVM MC PhD; Nathan Schaefer, DVM; Ethan Schmaling, BS; et al.

Efficacy of Ingelvac PRRS® MLV against a heterologous PRRSV 1-7-4 RFLP challenge........................98
A. Patterson, DVM, PhD; G. Haiwick, BS; J. Hermann, DVM, PhD; et al.

Comparison of intestinal lesions between Enterisol® Ileitis and Porcilis® Ileitis using a *Lawsonia intracellularis* mucosal homogenate seeder pig challenge model........................102
S. Dee, DVM, PhD; J. Nerem, DVM; D. Hanson, BS; et al.

Biosecurity controls for the industrial manufacturing of spray-dried plasma........................104
Joe Crenshaw, PhD; Joy Campbell, PhD; Javier Polo, PhD

Solution for in-barn, hand-held data collection 107
Thomas Stein, DVM, MS, PhD

The effect of feeding MCFA on performance of nursery pigs raised without antibiotics........................109
D. McManus, DVM; M. Lachmann, PhD; T. Weeden, PhD; et al.

Water in oil adjuvant selection for the formulation of one-shot safe bacterial vaccines for swine 113
Sophia Bingling Xu; Juliette Ben Arous; Nicolas Versillé; et al.

Diagnosing edema disease: the first step in prevention 115
V. Gotter, DVM, PhD, Dipl. ECPHM

Seneca Valley virus: Porcine-derived biological risk assessment 116
M. Madsen, PhD; J. Clampitt, BS; C. White, BS; et al.

Specialized VaxLiant® adjuvant for DNA vaccine development using influenza A virus-swine hemagglutination (HA) gene..........119
M. Inman, PhD; M. Pfannenstiel, PhD; D. Petrik, PhD; et al.

Feed hygiene: A layered approach 125
J.A. Jendza, PhD

Effect of combined supplementation of 25-hydroxycholecalciferol and cholecalciferol on sow and piglet Vitamin D status and progeny muscle cell parameters........................128
J. Hahn, PhD

Effect of Tonisity Px™ administration on pre-weaning mortality and weight gain........................131
Ava M. Firth, BS, DVM, MVS, MANZCVS, DACVECC, DECVECC; René Mozo Martín, DVM MSc PhD; Guillermo López Cano, DVM MSc; et al.

Effect of gruel and Tonisity Px™ on feed intake and weight gain at weaning........................134
Ava M. Firth, BS, DVM, MVS, MANZCVS, DACVECC, DECVECC; Guillermo Lopez Cano, DVM MSc; Alberto Morillo Alujas, BS DVM MS PhD MBA

Sunday Concurrent Session #3: Industrial Partners

Relationship between LP5™ and litter uniformity and birth weight 138
T. Rathje, PhD; C. Abell, PhD

Reducing antibiotic use: Educational materials for swine veterinarians and producers 140
Mike Tokach, PhD; Jose Soto, MS; H. Morgan Scott, DVM, PhD; et al.

Use of an autogenous PRRSV vaccine in conjunction with MLV elicits broadly neutralizing antibodies 141
Paulraj Lawrence, PhD; Eric Bumgardner, MS

Antibiotic-free production in the face of an acute health challenge 144
Jon A. De Jong, MS, PhD; Casey R. Neill, MS; Barry R. Kerkaert, DVM; et al.

Managing enterotoxigenic *Escherichia coli* post-weaning diarrhea with Avilamycin 146
Robert Evelsizer, DVM; J. Mark Hammer, DVM; Marcos Rostagno, DVM, PhD

Pursuing American Board of Veterinary Practitioners (ABVP) certification in Swine Health Management 150
Jeffrey A. Husa, DVM, Diplomate ABVP

Impact of piglet birthweight on post-weaning performance and system profitability 152
D.W. Newcom; C. Witte; M. Brubaker; et al.

Just how soluble are the water solubles? 156
D. Nolan, DVM; J. McCoy, CWS; R. Campbell

Masked trichothecene mycotoxins in US feed and corn samples from 2014 to 2016 161
E.G. Hendel, VMD, PhD; T. Jenkins, PhD; S.M. Mendoza, PhD; et al.

Senecavirus A, a genetic supplier perspective 164
J. Geiger, DVM, MS; P. Graham, DVM, MS; JP. Cano, DVM, PhD

The impact of trained and untrained personnel on product implementation and subsequent pig health 165
Sarah Probst Miller, DVM; Ron Ketchem

Effect of direct fed microbial, Calsporin® (*Bacillus subtilis* C-3102), on enteric health in nursery pigs after challenge with porcine epidemic diarrhea virus (PEDV) 167
P. Canning; C. Ruston; D. Madson; et al.

Impact of spray-dried animal plasma species of origin on post-weaning pig performance 168
Chad Pilcher, PhD; Sabrina Williams, MS; Sara Ebarb, MS; et al.

Validation of an ileitis mucosal homogenate seeder pig model to study the natural transmission of *Lawsonia intracellularis* in grow-finish pigs 170
Nathan L. Winkelman, DVM; Adam D. Mueller, DVM; Robin L. Winscher, CVT

A new PRRSV ELISA, highly correlated with IPMA 175
Eric van Esch, DVM, Technical Director

Sunday Concurrent Session #4: Industrial Partners

Study and field evaluation of an artificial intelligence system's ability to predict the good moment to breed sows: PigWatch® from Ro-Main 177
J. Labrecque, B. Eng.; J. Rivest, PhD; et al.

Fat-soluble vitamins: To esterify or not to esterify? 182
Robert L. Stuart, PhD

Safe-guard Aquasol® 20% suspension, a novel method of delivering fenbendazole through swine drinking water systems 184
R. Fleck, DVM; J. Slightom

Experimental PCV2d challenge of pigs vaccinated with prototype vaccines containing ORF2 from PCV2a, PCV2b or PCV2d 189
E. Strait, DVM, PhD; A. Suddith, MS; M. Allen, MS; et al.

Field experiences with Porcilis™ Ileitis: The newest tool for controlling *Lawsonia intracellularis* 191
Brad Thacker, DVM, PhD; Greg Armbruster, DVM; Angela Baysinger, DVM, MS; et al.

Electronic sow feeding (ESF) vs. the gestation stall: Three head to head comparisons of performance 195
Jeff Schoening, Technical Sales Manager

Comparative study of two vaccines against neonatal diarrhea on a Canadian commercial farm 198
W. Ariza, DVM; D. Robles, BSA; J. Badia, DVM; et al.

The impacts of in-feed antibiotics, a unique organic acid blend, or an in-feed acidifier on nursery pig performance 203
K.M. Hogan, PhD; S. Gould, BS; Y. Han, MS, PhD; et al.

Field evaluation of vaccination of piglets at processing using Fostera® PRRS 207
Keith Aljets, DVM; Jose Angulo, DVM

Determining *Mycoplasma hyopneumoniae* status in commercial breeding herds 213
E. Spronk, DVM; N. Garbes, DVM; L. Galina Pantoja, DVM, PhD

Efficacy of FluSure XP® swine influenza vaccine in pigs challenged with a cluster IV-A H3N2 swine influenza virus 216
M.C. Lenz, MS; V.J. Rapp-Gabrielson, DVM, PhD; A. Heinz, BS; et al.

The impact of activated polyphenols on performance of weaned pigs 219
D. Little, MS, DVM

The economic impact of OvuGel® 220
Dennis DiPietre, PhD; Dean Dau, DVM; Charlie Francisco, DVM, MS; et al.

Systematic implementation of VFDs using FeedLINK electronic VFD solution 225
B. Leuwerke

Pre-slaughter stunning of swine – a scientific viewpoint 227
Steve Wotton MBE, HNC, MSc

Poster Session: Veterinary Students

An interferon inducing porcine reproductive and respiratory syndrome virus vaccine candidate elicits protection against challenge with a heterologous virulent type 2 strain 230
E. Fontanella, MS; Z. Ma, PhD; Y. Zhang, PhD; et al.

Determining the efficacy of pooling different fecal swab types in the field versus the lab for detection of porcine epidemic diarrhea virus by PCR 231
Sara Hamlett, MS; Katie Woodard, DVM; Rodger Main, DVM, PhD

Evaluation of autogenous vaccine efficacy in western Canadian swine herds 232
L. Redies; J.C.S Harding, DVM, MS, Dipl. ABVP; S.E. Detmer, DVM, PhD

Trends of *Lawsonia intracellularis* polymerase chain reaction submissions at the University of Minnesota Veterinary Diagnostic Laboratory over a 10-year period 234
Donna Drebes, BS; Connie Gebhart, PhD; Julio Alvarez, DVM

Sand as a bedding for loose-housed gestating sows fed via ESF 236
J. Applebaum; M. Pierdon, VMD; T. Parsons, VMD, PhD

Evaluation of the sensitivity of Envirobooties™ and Swiffer™ kits for viral detection of PRRSV and PEDV in livestock trailers 237
T. Engle, MS; R. Jones, DVM; D. Johnson, BS; et al.

Comparison of morbidity and mortality after challenge with two North American PRRS virus isolates shows marked variation in time course and prevalence of clinical disease between isolates 239
L.A. Constance; B. Bloomberg; J.K. Lunney; et al.

Detection of influenza A virus on inanimate fair surfaces 240
Courtney M. Wright; Michele M. Zentkovich; Jacqueline M. Nolting, MS; et al.

Evaluation of the effects of flushing feed manufacturing equipment with chemically treated rice hulls on porcine epidemic diarrhea virus (PEDV) cross contamination during feed manufacturing 241
Jordan Gebhardt, BS; Steve Dritz, DVM, PhD; Jason Woodworth, PhD; et al.

Impact of sow parity on the influenza A (IAV-S) infection timing of the suckling pig in an endemic herd 242
Megan E. Bloemer, BS; Eva Jablonski, DVM, MS; Luis Ochoa, PhD; et al.

Evaluation of a novel antibiotic-free extender utilizing short-term hypothermic storage 243
R. Warner, MS; J. Brown, DVM; T. Dohlman, DVM, MS, DACT; et al.

Timing differences of *Mycoplasma hyopneumoniae* colonization in growing replacement gilts with and without tulathromycin administration at weaning 244
B. Burton, BS; K. Crawford, DVM, MS

Refinement of field anesthesia protocols for joint fluid collection in healthy swine 246
K. O'Brien, BS; P. Canning, DVM; K. Skoland, BS; et al.

Validation of snout wipes to detect influenza A virus in swine (IAV-S) 248
Joel Steckelberg; Todd Wolff, DVM, MPH; Daren Miller, DVM; et al.

Observations of sleep-related behaviors in piglets pre- and post-castration 249
Anna Martin; Kristina Horback, PhD; Thomas Parsons, VMD, PhD

Survey of enteric disease pathogens within a swine production company 251
Stephanie Betbeze; Seth Krantz, DVM; Jeffrey Luebbe, DVM; et al.

A field trial of an injectable *Lawsonia intracellularis* vaccine on an antibiotic-free farm 253
G. Chan; R. Friendship, DVM, MSc

Determining the effect of agitation of chlortetracycline stock solutions on chlortetracycline concentrations at different areas of water delivery systems 254
T. Girard, BS; A. Stricker, MS, DVM; J.Kelly, DVM; et al.

Effect of ketoprofen (Anafen®) on platelet aggregation in piglets 255
H. Golightly, HBSc; T.L. O'Sullivan, DVM, PhD; K. Reynolds, DVM; et al.

B-cell tetramer to evaluate the memory immune response to PRRSV 257
K.L. Gustafson, BS; M.C. Rahe, MPH, DVM; M.P. Murtaugh, PhD

Determining *Lawsonia intracellularis* antibody status in pigs during three early stages of life 258
Kayla Henness, BS; John Waddell, DVM, MBA; Luis Ochoa, PhD; et al.

Characterizing piglet loss from PRRSV outbreak 260
T. Homann; C. Vilalta, DVM, PhD; B. Morrison, DVM, MBA, PhD

The sow fecal microbiome across parity and hierarchy 261
John Hurst; Meghann Pierdon, VMD; Thomas Parsons, VMD, PhD

Investigating the vertical transmission of *Erysipelothrix rhusiopathiae* by real-time polymerase chain reaction (RT-PCR) and enrichment culture 262
E. Koep, BS; D. Murray, DVM; A. Sponheim, DVM

Developing urine catheterization in sows for calcium and phosphorus analysis 263
N. Lauterbach, BS; M. Hoogland, DVM, MS; G. Cline, DVM; et al.

Testing the association between multiple influenza A infections and average daily gain in nursery aged piglets under field conditions 264
Andrea Patterson, BScH; Helena Grgic, DVM, MSc, PhD; Zvonimir Poljak, DVM, MSc, PhD

Individual differences in social behavior and affective state in group-housed gestating sows 266
K. Pruett; K.M. Horback, PhD; T.D. Parsons VMD/PhD, DACAW

Determination of behavioral and physiological pain indicators associated with castration and tail docking in young piglets....268
Brent Sexton, BS; Brett O'Brien, VMD,MS; Sam Holst, DVM; et al.

Soy isoflavones and porcine respiratory and reproductive syndrome viral infection in weanling pigs....270
B.N. Smith; A. Morris; M.L. Oelschlager; et al.

Utility of antemortem joint fluid collection for diagnostic investigations of infectious lameness....272
V. Thompson; P. Canning, DVM; E. Koep, BS; et al.

We can't answer that: The importance of careful interpretation of swine production data....273
J. Tubbs, MS; M. Pieters, DVM, PhD; J. Menard, DVM; et al.

Identifying disease agents contributing to porcine respiratory disease complex (PRDC) within a large production system....274
Shannon Ure; Jana Morgan, DVM; Thomas Wetzell, DVM; et al.

Evaluating different methods of antibody detection for porcine epidemic diarrhea virus in animals exposed at various intervals....276
Kathleen Wood, BS; Brad Heins, DVM; Randy Jones, DVM; et al.

Poster Session: Research Topics

Increased frequency of isolation of multi-drug resistant *Salmonella* I 4,[5],12:i: from swine with histologic lesions consistent with salmonellosis....277
B. Arruda, DVM, PhD; A. Krull, DVM, PhD; A. Kreuder, DVM, PhD; et al.

Influenza A virus prevalence and seasonality in midwestern US breeding herds....278
F. Chamba, DVM, MS; J. Nerem, DVM; A. Alba, DVM, PhD; et al.

Identification of multiple viruses in swine fecal samples using next-generation sequencing technology coupled with Kraken algorithm-based bioinformatics analysis pipeline....279
Qi Chen; Ying Zheng; Baoqing Guo; et al.

Evaluation of pathogenesis and antibody responses of porcine deltacoronavirus in 3-week-old weaned pigs....281
Qi Chen; Phillip Gauger; Joseph Thomas; et al.

Microbiome associations in pigs with the best and worst clinical outcomes following co-infection with porcine reproductive and respiratory syndrome virus (PRRSV) and porcine circovirus type 2 (PCV2) 283
M. Niederwerder, DVM, PhD; C. Jaing, PhD; J. Thissen, BS; et al.

Evaluation of nutrient availability when using feed disinfection in nursery diets....284
L. Greiner, PhD; J. Connor, DVM; B. Knopf; et al.

Comparison of the efficacy of a commercial inactivated pandemic (p) H1N1 influenza A virus vaccine and two experimental M2e-based vaccines against pH1N1 challenge in the growing pig 286
Tanja Opriessnig, Dr med vet, PhD; Philip Gauger, DVM, MS, PhD; Xi Jiang, PhD; et al.

Mycoplasma hyopneumoniae detection *in vivo*: investigating the diagnostic sensitivity of laryngeal swabs....287
Luiza R. Roos; Nitipong Homwong; Eduardo Fano; et al.

Adaptation of porcine epidemic diarrhea virus (PEDV) to Vero cells 288
T. Sato, PhD, DVM; K. Oroku; A. Nagao; et al.

Identification of neuropathogenic *Teschovirus A* serotype 2 and serotype 11 and experimental reproduction of disease....289
F. Matias Ferreyra, DVM; B. Arruda, DVM, PhD; D. Madson, DVM, PhD; et al.

Identification of a divergent strain of Sapelovirus associated with a severe polioencephalomyelitis outbreak in the US 291
Paulo Arruda; Kent Schwartz; Bailey Arruda; et al.

Piglet feed intake during the end of the lactation period impacts post-weaning diarrhea incidence and survival rate 292
Frédéric Vangroenweghe; Hinnerk Kraneburg; Diedrich Hendriks; et al.

Mycoplasma hyopneumoniae vaccination check using quantification of IFN-gamma levels following T-lymphocyte stimulation of *M. hyopneumoniae*-vaccinated piglets 293
Frédéric Vangroenweghe; Hinnerk Kraneburg; Diedrich Hendriks; et al.

Evaluation of a vaccine prototype using a chimeric protein and a molecular enhancer produced by *Pichia pastoris* fermentation 295
A. Villamil; E. Lamazares; A. Ruíz; et al.

Poster Session: Industrial Partners

Grow-finish mortality in pigs consuming acidified water 297
J.A. Jendza, Ph.D

Development of a new swine *Mycoplasma* multiplex qPCR 299
Eric van Esch, DVM, Technical Director

Annual occurrence of mycotoxins in US corn harvested from 2012 to 2015 300
E.G. Hendel, VMD, PhD; T. Jenkins, PhD; S.M. Mendoza, PhD; et al.

Field efficacy study of RHINISENG® in Europe on a farm with problems associated with *Bordetella bronchiseptica* 301
A. Sánchez-Matamoros, PhD; A. Martos, MSc; O. Boix, MSc; et al.

Evaluation of an inactivated vaccine against virus of porcine epidemic diarrhea virus (PEDV) 303
J. Hernandez; Luis G. Giménez; A. Massa; et al.

A novel approach for using P.G. 600®: Impact on gilt performance and sow longevity 304
Derald Holtkamp, MS, DVM; Jennifer Patterson, MS

The bactericidal action of gamithromycin against *P. multocida, A. pleuropneumoniae, H. parasuis* and *B. bronchiseptica*..........306
Olivier Merdy, Agronomist, A. Richard-Mazet, DVM, PhD, Rose Huang, PhD; et al.

Comparative study to evaluate the performance in grow-finish pigs treated with Aivlosin® 17% or Denagard® 10 following an experimental challenge with *Lawsonia intracellularis* 308
R. Kaptur, DVM; D. Rosener, DVM; A. Mueller, DVM; et al

Effect of Tonisity Px™ administration on intestinal morphology 310
Ava M. Firth, BS, DVM, MVS, MANZCVS, DACVECC, DECVECC; Guillermo Lopez Cano DVM, MSc; Alberto Morillo Alujas, BS DVM MS PhD MBA

Pig Atlas - enhancing phenotypic detail to increase genetic gain in pig breeding 312
L.E. Gangsei, PhD; J. Kongsro, PhD; E. Grindflek, PhD; et al.

Disinfection in freezing temperatures: An experimental approach to identify best practices for applications in the animal health industry 314
F. Ahmadpour; S. van den Berg; G. Perez-Hernandez; L. Pantaleon, DVM, MS, DACVIM, MBA

Evaluation of reproductive performance after implementation of Fostera® PRRS in a 9,000-sow production system using continuous quality improvement tools 316
Jose Angulo, DVM; Abigail Redalen, DVM; Larry Moore, DVM; et al.

Monday General Session: One World, One Health, One Passion for Pigs

Canadian perspective on porcine epidemic diarrhea 318
Egan Brockhoff, DVM

Avian influenza: Lessons learned......321
Jill Nezworski, DVM

Consumers, pigs, vets, and zoonoses: The critical role you play in earning trust with consumers 322
J.J. Jones

Monday Concurrent Session #1: Swine Diseases

Opportunistic bacterial pathogens: Battles fought in daily practice 325
Cameron Schmitt, DVM, MS

Batch farrowing disease control benefits in a production system......326
Elise Toohill, DVM

Determining optimum PRRSV management strategies 328
Clayton Johnson, DVM

Emerging genetic strategies for disease control 329
M. Culbertson; W.O. Herring

Swine Health Information Center 2016 progress......331
P. Sundberg, DVM, PhD, DACVPM

Biosecurity disaster: Foot and mouth disease and classical swine fever outbreaks in Korea......335
W-I Kim, DVM, MS, PhD, DACVM

Monday Concurrent Session #2: Antibiotics

Antibiotic resistance mechanisms 336
Randall S. Singer

Metrics for antibiotic use in food animals 340
P.R. Davies, BVSc, PhD

Human models for reducing antibiotic use 343
Michael J. Sadowsky, PhD; Alexander Khoruts, MD

Veterinary Feed Directive: Feed industry perspective......345
Richard Sellers

Practitioner experience with implemented VFD rule......346
Paul D. Ruen, DVM

GlobalVetLINK's initial experiences following the VFD and prescription changes post-January 1, 2017......349
J.T. Holck, DVM, MS, MBA

The VFD inspectional assignment 352
Center for Veterinary Medicine, Food and Drug Administration, Department of Health and Human Services

Monday Concurrent Session #3: Managing the Reproductive Herd for High Health and Productivity

Economic impact: Fitness traits in post-weaning pigs and sows in lieu of genetic improvement in litter size and leanness 354
John Mabry, PhD, Professor Emeritus

Semen supplier contributions to sow herd performance 362
G.C. Althouse, DVM, PhD; R.L. Linardi, DVM, PhD; E.M. Woodward, PhD

A pig's early challenges 366
Maria Pieters, DVM, PhD

Considerations for batch production 368
C. Scanlon Daniels, DVM, MBA

Managing Danish nurseries in prolific sow herds with minimal antibiotic use 370
J.M.J. Agerley, DVM

Tuesday General Session: AASV's Stand: Control/Elimination

Biosecurity: The strengths and weaknesses in our industry 372
R.B. Baker, DVM, MS

Practical PED elimination and surveillance: The Quebec experience 377
Julie Ménard, DVM

US swine industry structure and disease control: A "wicked" problem 380
Mike Lemmon, DVM

Voluntary regional PRRS control: Pitfalls and progress 385
David J. Wright, DVM

Howard Dunne Memorial Lecture

Swine medicine in the 21st century: Immovable object meets unstoppable force

Jeff Zimmerman, DVM, PhD, DACVPM

In remembrance and honor of Howard W. Dunne

Dr. Howard W. Dunne received his veterinary degree from Iowa State University in 1941. He was in engaged in private practice for one year and industry for four years before entering graduate school at Michigan State University. After receiving his PhD in 1951, he spent two years as deputy chief of veterinary microbiology at the Chemical Corps Biological Laboratories in Camp Detrick, MD. In 1953, he joined the faculty of Pennsylvania State University, where he headed the veterinary science research program and the large animal diagnostic laboratory. Dr. Dunne was internationally known for his work on swine diseases, particularly hog cholera and the swine reproductive viruses then referred to as the SMEDI syndrome. Dr. Dunne authored more than 90 scientific publications, and is perhaps best known for initiating and editing the textbook *Diseases of Swine*, which is still widely considered the "bible" of swine disease.

In 1971, the American Veterinary Medical Association presented Dr. Dunne with the XII International Veterinary Congress Prize to recognize his contributions to the international understanding of veterinary medicine. At the 1974 Congress of the International Pig Veterinary Society (IPVS), Dr. Dunne was elected president of the 1976 IPVS Congress, to be held in Ames, Iowa. His sudden death at the age of 61 in September of 1974 saddened the swine veterinary community, and the 1976 IPVS Congress was dedicated to his memory.

At the suggestion of Dr. Allen Leman, a financial gift made by the 1976 IPVS Congress organizers to the fledgling American Association of Swine Practitioners was combined with additional association funds to endow this prestigious annual lecture in remembrance and honor of Dr. Howard W. Dunne.

HOWARD DUNNE MEMORIAL LECTURERS

1978	Michael Muirhead	1991	Roy Schultz	2004	Robert Desrosiers
1979	W. J. (Bill) Smith	1992	Joseph Connor	2005	Scott Dee
1980	N.C. Nielsen	1993	Camille Moore	2006	Gregg BeVier
1981	Allen Leman	1994	Robert Morrison	2007	Tom Burkgren
1982	Michael Wilson	1995	Rodney Johnson	2008	Timothy J. Loula
1983	Edward Bohl	1996	Paul Yeske	2009	John U. Thomson
1984	Richard Ross	1997	Barbara Straw	2010	James F. Lowe
1985	Kent Kislingbury	1998	Mike Daniel	2011	Locke A. Karriker
1986	Ralph Vinson	1999	Larry Rueff	2012	R. Dean Boyd
1987	Gary Dial	2000	David Reeves	2013	Craig Rowles
1988	Thomas Alexander	2001	Max Rodibaugh	2014	Daryl Olsen
1989	Steven Henry	2002	Tim Blackwell	2015	Gregory W. Stevenson
1990	Stanley Curtis	2003	Lonnie King	2016	John Harding

Swine medicine in the 21st century: Immovable object meets unstoppable force

Jeff Zimmerman, DVM, PhD, DACVPM
Iowa State University, Ames, Iowa

Introduction

What an honor to give the Howard Dunne Memorial Lecture! And what a challenge! I started preparing by reading old proceedings and watching Howard Dunne Lectures on the AASV website, but gave up pretty quickly. Wow! After eyeballing the height of the bar set by previous Howard Dunne speakers, I decided to keep it simple and stick with the things I normally think about: What we need to do to fulfill our responsibilities to swine health and welfare. What we need to do to protect and serve the industry.

"Immovable object meets unstoppable force" - why is it in the title of this talk? This confrontation of two impossibilities is one member of an ancient family of brain teasers; an antique version of "Who wins if The Hulk fights Superman?" I put it in the title because we are in a paradoxical situation today. That is, our highly efficient system of pork production (The Immovable Object) is jeopardized by our inability to deal with infectious disease (The Unstoppable Force), which is largely the result of the design of our highly efficient production system. How did we end up in this conundrum and, more importantly, how do we deal with it?

Of pigs and pork

Current market prices aside, pork production is at its zenith. More than ever before, we efficiently produce healthful, wholesome, inexpensive food for the North American consumer and pork products that compete successfully in international markets, e.g., US pork exports in 2015 (24% of production) were valued at $5.6 billion. How did we get here? Bear with me - a little history will not hurt you.

Sus scrofa was domesticated in both Asia and the Middle East around 10,000 BC.[1] Pigs were a common denominator to early agricultural societies because, as we know, they are highly resilient, mature quickly, have large litters, and (if necessary) can subsist on low-quality feedstuffs. Pigs are not native to North America, but were introduced into the New World by many of the early explorers and settlers, e.g., Christopher Columbus brought pigs on his second voyage in 1493.[13]

Pig production took off in the present day US in late 1600s when Pennsylvania farmers figured out that feeding pigs on "excess" corn production worked really well.[13] The other option was to use the excess corn to produce whisky, but that is a discussion for another time. Pig production was so successful in the US that by 1847 pigs far outnumbered people (35 million pigs vs. 20 million people) and Cincinnati was the largest pork market in the world.[27]

Initially, pigs were housed in rudimentary facilities, if housed at all. Those who object to confinement should consider Spencer's[23] comments: "*The ordinary sty with a yard attached is unhealthy for a growing or matured pig, but in the colder weather it is simply cruel for newly born pigs.*" In the early 1900's, Danish agriculturalists found that confinement housing provided for more efficient use of available land, protected animals from weather, reduced fighting, and improved feed efficiency.[22] Spencer (1919) marveled at the 2-story barn he visited near Aarhuss, Denmark -- the pigs were fed downstairs and slept upstairs, which they reached by walking up a ramp.

In the US, science started getting involved in pig health when at the turn of the last century when Bureau of Animal Industry researchers began identifying the causes of disease and devising methods of control. The BAIs first big win for pork producers was the discovery of the "hog cholera" (classical swine fever) virus in 1903 and the development of a method of immunization shortly thereafter.[27] The BAI made other practical, high-impact contributions, as well. For example, control of parasites was a tremendous problem in early swine production. The brilliant BAI parasitologist B.H. Ransom figured out parasite life cycles (some of the ones you learned in vet school) and then used his discoveries to create the "McLean County system of swine sanitation" to control them. Ransom's simple plan was to break the cycle of transmission between the dam and her piglets using basic sanitation procedures. When field tested, the McLean County system improved pig survivability and growth rates to the tune of a 20% improvement in production and was quickly adopted by producers.[3]

The trend toward confinement systems went into high gear in the 1970's-80's. Some of us remember watching new confinement facilities pop up across the horizon as producers rushed to build. The result has been highly-engineered, capital-intense facilities that reduce manual labor, improve pig welfare, and increase production efficiency. Dealing with the complexities of the environment and the production process (air handling, climate control, feeding systems, manure management, disease control) is a work-in-progress, but the goal is, and has always been, providing the best possible environment for the animal in an economically viable food production system.

The "Immovable Object"

The net result of the cumulative efforts of five generations of pork producers, geneticists, nutritionists, agricultural engineers, and veterinarians is a system with unprecedented productivity and performance. In 2015, pork production contributed $39 billion to the US gross national product.[15] 2016 measures of productivity included a new industry record of 10.58 pigs weaned per litter in the third quarter.[26] Part of this equation has been the increasing economic importance of international markets. Pork exports represented less than 2% of US pork production prior to 1992,[24] but have since become an important industry driver. In 2015, 2,132,135 metric tons of pork worth $5.6 billion were exported (www.usmef.org) - 24% of the industry's total production (www.pork.org).

So, over the course of the 20th century, pig production metamorphosed from small, outdoor, labor-dependent enterprises into capital-dependent, multi-site production systems. These changes produced safe, wholesome, inexpensive food for the US consumer and pork products that compete successfully in international markets. This is a resource worth protecting.

The "Unstoppable Force"

The one not-so-minor problem we face is infectious disease. As Robert Desrosiers pointed out at our Orlando meeting in 2014,[8] "*We have not gotten rid of any of the emerging pathogens of the last 40 years...*". And most of us would agree with Dr. Desrosiers that "*...adding diseases without eliminating any is not a good long term strategy.*"

So despite our best efforts, industry losses to endemic diseases continue to represent a massive "opportunity for improvement", the premier example being PRRSV, which weighs in at $664,000,000 in annual losses.[12] More troubling, our recent track record on the control or elimination of infectious agents provides no assurance for our ability to effectively respond to incursions of transboundary or foreign animal diseases. The detection of African swine fever virus (ASFV), classical swine fever virus (CSFV), or foot-and-mouth disease virus (FMDV) would devastate the US pork industry in a matter of hours. Rendleman and Spinelli (1999) estimated the cost of ASFV in the US at $4,500,000,000 to $5,445,000,000 in 1992 dollars. Of course, this estimate is grossly low because it does not take important factors into account: (1) changes in the value of the dollar, (2) the current importance of pork exports to the industry (not the case in 1992), and (3) the increased complexity of the industry. Notably, achieving control of a transboundary or exotic infectious agent through the classic "stop movement" or "stamping out" policies is no longer an option for economic, animal welfare, environmental, and public perception reasons. So what are we going to do? Just as Ransom could not control parasites until he understood their life cycles, we cannot control infectious diseases - endemic, transboundary, or exotic - if we do not understand how they behave in populations.

A closer look ...

There is an important distinction to draw here. Understanding a pathogen's pathogenesis, immunity, genetic diversity, diagnostics, etc. is fundamental, but constitutes only part of the picture. We also need to understand how a pathogen acts within populations. Consider the following:

Herd size

US swine inventory numbers are available as far back as 1850. For most of this period, but particularly between 1890 and the present, the total number of pigs in the US ranged from 50 to 60 million animals. Higher than historic trends, the total inventory of the US swine herd on September 1 2016 was 70.9 million head, including 6 million breeding animals and a market hog inventory of 64.8 million.[26]

In contrast to the relatively stable number of pigs in the US, the number of farms with pigs has been on a slow, steady decline since about the year 1900. Of course, the consequence is that the farms that remain must hold more pigs. Numbers are not easy to obtain and sources do not necessarily agree, but according to one source,[10] farms in Iowa averaged a total inventory of 250 animals in 1980 vs 1,500 in 2002 - a 6-fold increase. Herd size is important to this discussion because of its impact on disease. Very simply, increasing the number of individuals in a population leads to an increase in the size of outbreaks, the frequency of outbreaks, and the difficulty of achieving disease control (herd immunity). The effect of population size on disease is well-documented in humans, especially for measles,[11] but recent research has shown that the concept holds for swine, as well. In particular, influenza, PRRSV, salmonella, and probably other pathogens become endemic as herd size increases (reviewed by Pitzer et al., 2016). Of course, "herd size" is really just a proxy for other fundamental characteristics of pig production that affect the behavior of pathogens, such as the following.

Rapid population turnover

"Population change" is composed of "natural change" (total births minus total deaths) plus migration into (+) or out of (–) the population. In humans, the rate of change is typically small, i.e., populations are stable. For example, the state of Nebraska recorded a population change of +8.4% between 1990 and 2000.[7] This consisted of a natural change (births - deaths) of +5.4% and net migration of +3.1%. In contrast, a finishing barn will experience ~250% annual population change as groups of animals enter into the facility, grow, and are sent to market. Although more stable than growing pig populations, ~40% of females in sow herd populations are replaced annually.[17] Population turnover and the continual introduction of new, immunologically susceptible animals promote continual pathogen circulation and endemnicity.

Segregation of populations

Scott Dee[6] taught us the lesson of the importance of subpopulations of differing immune status in the control of PRRSV, but the concept broadly applies to all pathogens. Physical segregation by age, production stage and/or function is necessary in managing pig populations, but one consequence of the lack of interaction between these subpopulations is the creation of subpopulations on the same production site that differ greatly in disease/immune status.

Distant, but connected, populations

Relatively new to the industry (using my own timeline as a reference) is the practice of moving large numbers of young pigs from breeding farms to fill spaces on feeding operations located in the Corn Belt. It is more logical to move the pigs to the feed versus the reverse. Table 1 shows how this practice has grown over time: fewer than 5 million pigs were moved across state lines in 1990 vs nearly 50 million pigs in 2015. In addition to pigs farrowed in the US, "pig movement" includes animals imported (primarily from Canada) and shipped to grow-out facilities in the Corn Belt. Live pig imports have varied greatly over time (Figure 1), but really picked up after 1995 and reached ~10,000,000 in 2007 (fao.org/faostat). This process of physically moving pigs connects distant farms and provides an efficient mechanism for the rapid dissemination of pathogens.

So where are we?

The same infrastructure and production characteristics that led to improved efficiencies, i.e., large herds with high throughput and extensive pig movement between sites, have left the industry highly vulnerable to the rapid dissemination and perpetuation of infectious diseases. Large, physically segregated pig populations with high population turnover rates favor the persistence of pathogens because herd immunity becomes incomplete and unstable.[9] The transport of large numbers of animals between farms provides the means to spread pathogens rapidly among geographically distant populations. We have been schooled in this lesson by porcine reproductive and respiratory syndrome virus, porcine circovirus type 2, 2009 pandemic influenza A virus, porcine epidemic diarrhea virus, and others, but it has been a difficult lesson to learn.

Considering our options ...

So what to do about the control of infectious disease? As swine veterinarians, we are responsible for pig health/welfare and producer profitability. We are the ones responsible for assuring the future success of pork producers, at least in terms of the control of infectious disease. Going back to 60-sow herds is not up for discussion! Forty acres and a mule is not on the table! We need to move forward!

Of course, we continue to learn lots of new things about pathogens, disease transmission, better systems of biosecurity, better vaccines, better engineering, better diagnostics, etc., primarily through research sponsored by the National Pork Board and the Swine Health Information Center. This is all important and useful, but will it be enough? I do not think so. It will help - good information judiciously applied is always helpful - but it represents a "drift" when we actually need a "shift."

Around the time the industry started to consolidate (1970's), veterinarians began reporting "production diseases", diseases caused by the interaction of multiple infectious and non-infectious factors. This was new then, but sounds very familiar now. Calvin Schwabe, writing prophetically in 1982, argued that the simple disease models of Koch and Pasteur no longer applied and that swine veterinarians needed to re-think disease management and consider "*a form of on-going on-farm research, based upon surveillance ...* (to establish) *expected levels and patterns of diseases in particular populations, which then provide the baseline levels ... against which effects of intervention (control) efforts can be measured.*" Schwabe even expected this approach to provide "*early* (detection of) *unexpected events or deviations from expected levels and patterns, in order to appropriately and promptly respond.*"

A viable surveillance system?

In my opinion, Schwabe was on to something. Anyway, his idea never took off, primarily because "on-going on-farm surveillance" was too costly and difficult to implement. However, for the sake of conjecture, what should a viable surveillance system look like?

1. Provides a return on investment (ROI). Pork production is a business: if it is not profitable it will not survive. Data is necessary for intelligent production and/or business planning.
2. Easy to implement - all the steps from sample collection to data analysis.

Table 1: Inshipments (number of pigs) over time*.

Year	Iowa	Indiana	Minnesota	Illinois	Total US
1980	1,740,000	549,000	226,000	510,000	4,628,000
1990	1,400,000	240,000	262,000	359,000	4,317,000
2000	11,600,000	1,050,000	3,150,000	1,470,000	24,514,000
2010	21,200,000	2,632,000	7,089,000	1,898,000	39,571,000
2013	27,500,000	2,844,000	7,530,000	2,444,000	47,863,700
2014	26,800,000	2,846,000	7,500,000	2,488,000	47,184,700
2015	27,500,000	3,559,000	8,000,000	2,356,000	49,680,900

* "Inshipments" includes the number of animals entering a state for feeding or breeding purposes, but excludes animals brought in for immediate slaughter. Inshipments do not include animals transported between sites within states. (Source: USDA, NASS).

Figure 1: Number of live pigs imported into the US.

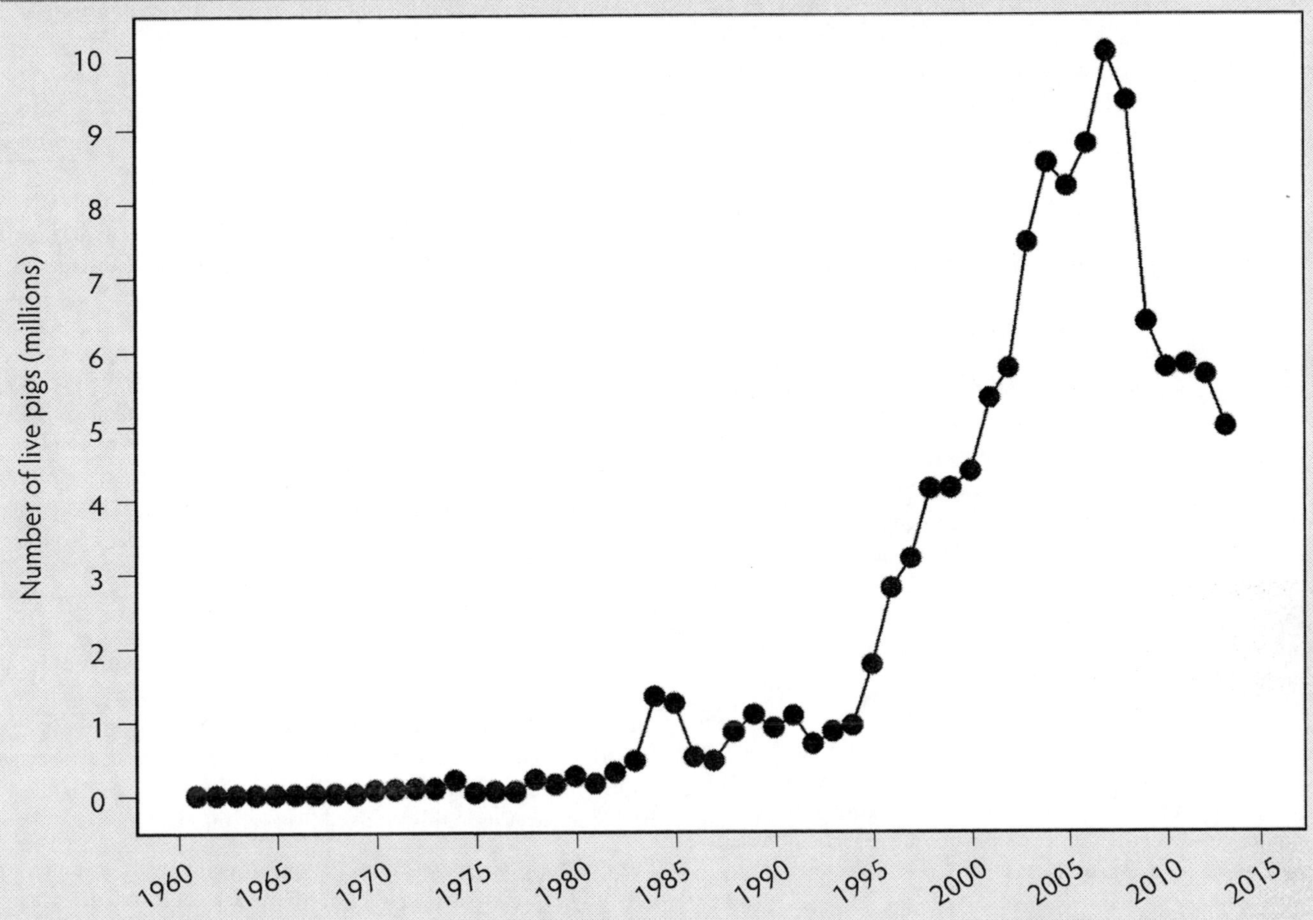

3. The people using the system believe it has value.
4. Provides timely and accurate data. In particular, false positive indicators (test results) quickly destroy confidence in the system.
5. Flexible. It must be useful for more than one purpose and able to provide data on a variety of targets (infectious and non-infectious).

We face two major barriers to achieving such a system: (1) easy, efficient on-farm sample collection and (2) a statistical valid approach to determining sample size and sampling location. How do we overcome these barriers?

On-farm collection of statistically valid numbers of individual pig specimens is too expensive and time consuming to be done routinely. Considering better options, my thoughts turn to oral fluids (no surprise!), but there are other possibilities, as well.

Technically, an oral fluid specimen is an "aggregate" sample - the accumulation of the oral fluid from all of the pigs that chewed on the rope. Other aggregate samples used in veterinary medicine include air samples, water samples, environmental samples, and bulk tank milk samples. For example, bulk tank milk is used for the detection of various diseases of cattle, e.g., bovine virus diarrhea virus,[14] *Coxiella burnetti*,[28] Schmallenberg virus,[4] and foot-and-mouth disease virus.[25] Aggregate samples differ from pooled samples in the way they are created. That is, pooled samples consist of individual samples that are combined usually, but not always, in equal proportions. In terms of ease and cost of collection, aggregate samples provide better, cheaper, and easier survey data than individual animal samples. In the case of oral fluids, we have seen that they are also more diagnostically sensitive than individual samples.[16]

A larger issue for individual pig sampling is the fact that the "hypergeometric" approach to sample size calculations - a standard that we have relied on since the 1980's - does not work at the site level in contemporary herds (at least in the way we are using it). At best, it can only be applied to certain populations that meet certain conditions. The convention of sampling 30 "randomly selected" animals for surveillance is a legacy of the PRV eradication program.[2] This number is based on a sample size providing a 95% probability of detecting at least one positive in a population with ≥ 10% prevalence.[5] The problem is that the calculation relies on assumptions (hypergeometric distribution) that were true of the small swine farms in the 1980's, but not necessarily of swine farms today. In particular, the assumption is that all individuals have the same chance of acquiring disease and that the results are independent. As discussed in Rotolo et al. (2017), these assumptions are often no longer true.

Surveillance sampling recommendations for oral fluids

Validated sample size formulas are not currently available for oral fluids or the other aggregate samples that are increasingly used in production settings. However, our research group has worked extensively to develop sample size formulas and guidelines for oral fluid samples collected from group-housed animals (Rotolo et al., 2017). The following recommendations are based on our work:

1. **Sample at the barn level (or airspace)**
   The sampling plan for group housed pigs should be designed for each barn (or air space). That is, the number of samples, location of sample collection, and frequency of sampling pertains to each barn, not the site. Again, the justification for this approach is the fact that animals in commercial production systems are segregated by age and stage, with little mixing between them. The result is that barns on a site are often of different infection status. A design based on sampling individual barns provides flexibility in tailoring surveillance to farms ranging widely in size and complexity. Furthermore, sampling across multiple barns on a site is a powerful approach for detecting infection or proving that the population on the site is truly negative.

2. **Number of samples**
   Collect 2 to 6 samples per barn (or airspace). Tables reporting the probability of detection by sample size will be available in an upcoming publication.[20] A larger sample size may provide a higher probability of detection in the barn, but may not be practical, affordable, or necessary to meet the objective. Let the purpose of sampling and the budget drive this decision.
   The number of pens in a barn is not a driver in sample size selection. If the barn is designed with many pens, samples will likely be collected from separate pens. If the barn is designed with few pens, more than one sample per pen could be collected. The key feature is a fixed spatial approach: space samples equally over the length of the barn (as described in the next paragraph).

3. **Location of sample collections within a barn**
   To collect samples, hang ropes equidistant to each other in pens on alternate sides of the center alleyway over the length of the barn. For example, if two samples are to be collected, hang each rope approximately one-third of the length of the barn from each end. Whatever the number of samples to be collected, adjust the distance between ropes so that they are evenly distributed over the space occupied by the pigs. The statistical support for this approach is presented elsewhere.[20] "Fixed spatial sampling" (as opposed to random sampling) provides a sampling design that is easily communicated to the people collecting the samples. Spatial sampling works because it reflects reality: infectious disease has a spatial component. Thus, two neighboring pens are likely to share the same disease status. Repeated sampling of the same pens over time also provides a logical picture of pathogen shedding (PCR-based testing) and/or immune responses (antibody-based testing). Spatially-based sampling is widely used in other fields, where it is considered to offer advantages in terms of cost and efficiency (Wang et al., 2013).

4. **Frequency of sampling.**
   Sampling/testing should be done at regular intervals (every 1 to 4 weeks), depending on the purpose of surveillance. This frequency will provide a clear picture of pathogen circulation OR provide strong assurance that the site is negative. If it is necessary to choose, fewer samples collected at frequent intervals is more valuable than many samples collected infrequently.

Final remarks

The goal of this commentary was to provide background on the events that led to the development of the contemporary US pork industry and to make the important point that the swine industry has changed in ways that profoundly affect infectious disease ecology. In particular, our highly productive industry has an Achilles' heel that leaves us highly vulnerable to infectious diseases.

How do we prepare to respond efficiently, effectively, and successfully to the detection of a transboundary or exotic disease in North America? Schwabe (1982) was right: we need to be able to cheaply and easily collect infectious disease data in a process of continuous analysis-and-response. And we are close to developing the methods to achieve this goal. On the farm, integration of surveillance data with production data could provide the means to: (1) identify the circulation of specific pathogens; (2) quantify their effects on pig health and productivity; (3) target interventions to the correct pathogen and population; and (4) time the intervention for maximum effect. At the regional level, aggregate samples, such as oral fluids, will make area control programs more practical, affordable, and successful.

At this point, you should be wondering about the immovable object and the unstoppable force. Where did they go in the discussion? What happens when they meet? I know it is a big letdown, but the answer is that neither an immovable object nor an unstoppable force really exist. Instead, we swine veterinarians need to find ways to make the swine industry (The Immovable Object) more agile and devise solutions that neutralize the power of infectious disease (The Irresistible Force). To paraphrase historian and philosopher Arnold J. Toynebee (1889- 975), "Civilizations ... grow by successfully responding to successive challenges." Swine veterinarians (and the pork industry) need to do the same.

Acknowledgements

I would like to thank Drs. George Beran, Howard Hill, and Hank Harris for their friendship and mentorship. I would also like to express my appreciation to the student-colleagues I have had the good fortune to work with and learn from. And I would like to thank my Iowa State colleagues for their generosity, camaraderie, and collaboration over the years. Lastly and most profoundly, I thank Silvia Zimmerman DVM MS MPH for her support, understanding, and love!

References

1. Albarella U, Dobney K, Rowley-Conwy P. 2006. The domestication of the pig (*Sus scrofa*): new challenges and approaches. In: Documenting Domestication: New Genetic and Archaeological Paradigms. University of California Press, Berkely, pp. 209-227.

2. Anderson LA, Black N, Hagerty TJ, Kluge JP, Sundberg PL. 2008. Pseudorabies (Aujeszky's Disease) and Its Eradication. A Review of the US Experience. USDA:APHIS Technical Bulletin #1923.

3. Anon. April, 1926. How sanitation improves pig yield. Meat and Livestock Digest 6(9):2.

4. Balmer S, Vogtin A, Thur B, Buchi M, Abril C, Houmard M, Danuser J, Schwermer H. 2014. Serosurveillance of Schmallenberg virus in Switzerland using bulk tank milk samples. Prev Vet Med 116:370-379.

5. Cannon RM, Roe RT. 1982. Livestock disease surveys: A field manual for veterinarians. Bureau of Rural Science, Department of Primary Industry. Australian Government Publishing Service, Canberra (35 pages).

6. Dee SA, Joo HS, Henry S, Tokach L, Park BK, Molitor T, Pijoan C. 1996. Detecting subpopulations after PRRS virus infection in large breeding herds using multiple serologic tests. Swine Health Prod 4:181-184.

7. Deichert J. June 2001. Components of population change, Nebraska Counties, 1990-2000. Focus: Nebraska 1(1):1-10. Center for Public Affairs Research, University of Nebraska at Omaha.

8. Desrosiers R. 2015. Emerging diseases: The past and the future. Proc 46th Ann Meet Am Assoc Swine Veterinarians. Orlando, Florida, pp. 519-537.

9. Evans CM, Medley GF, Creasey SJ, Green LE. 2010. A stochastic mathematical model of the within-herd transmission dynamics of porcine reproductive and respiratory syndrome virus (PRRSV): Fade-out and persistence. Prev Vet Med 93:248-257.

10. Flora JL, Chen Q, Bastian S, Hartmann R. 2007. Hog CAFOs and sustainability. The impact on local development and water quality in Iowa. A report of The Iowa Policy Project, Mount Vernon, Iowa. (34 pages).

11. Haggett, P. 2000. The Geographical Structure of Epidemics. Oxford University Press, Inc., New York.

12. Holtkamp D, Kliebenstein J, Neumann E, Zimmerman J, Rotto H, Yeske P, Yoder T, Wang C, Mowrer C, Haley CA. 2013. Assessment of the economic impact of porcine reproductive and respiratory syndrome virus on United States pork producers. J Swine Health Prod 21:72-84.

13. McGlone J, Pond W. 2003. Pig Production: Biological Principles and Applications. Thomson Learning, Inc., Clifton Park, New York.

14. Niskanen R, Alenius S, Larsson B, Jacobsson S-O. 1991. Determination of level of antibodies to bovine virus diarrhoea virus (BVDV) in bulk tank milk as a tool in the diagnosis and prophylaxis of BVDV infections in dairy herds. Arch Virol Suppl 3:245-251.

15. NPPC. 2016. A review of the US livestock and poultry sectors: Marketplace opportunities and challenges. Written testimony of the National Pork Producers Council, United House of Representative Committee on Agriculture, Livestock and Foreign Agriculture Subcommittee (May 24, 2016).

16. Olsen C, Wang C, Christopher-Hennings J, Doolittle K, Harmon K, Abate S, Kittawornrat A, Lizano S, Main R, Nelson E, Otterson T, Panyasing Y, Rademacher C, Rauh R, Shah R, Zimmerman J. 2013. Probability of detecting porcine reproductive and respiratory syndrome virus infection using pen-based swine oral fluid specimens as a function of within-pen prevalence. J Vet Diagn Invest 25:328-335.

17. PigCHAMP Benchmarking. 2012. USA 2012 - second quarter summary. Ames, Iowa.

18. Pitzer VE, Aguas R, Riley S, Loeffen WLA, Wood JLH, Grenfell BT. 2016. High turnover drives prolonged persistence of influenza in managed pig herds. J R Soc Interface 13:20160138.

19. Rendleman CM, Spinelli FJ. 1999. The costs and benefits of animal disease prevention: The case of African swine fever in the United States. Environ Impact Assess Rev 19:405-426.

20. Rotolo ML, Sun Y, Wang C, Giménez-Lirola L, Baum DH, Gauger PC, Harmon KM, Hoogland M, Main R, Zimmerman JJ. 2016. Sampling guidelines for oral fluid-based surveys of group-housed animals. Vet Microbiol (*submitted*).

21. Schwabe C. 1982. The current epidemiological revolution in veterinary medicine. Part 1. Prev Vet Med 1:5-15.

22. Shaw EB. 1938. Swine industry of Denmark. Econ Geogr 14:23-37.

23. Spencer S. 1919. The Pig: Breeding, Rearing, and Marketing. C. Arthur Pearson Ltd, London.

24. Stehle AM, Robb JG. 2014. U.S. international pork industry trade: Dimensions and perspectives on the future. Livestock Marketing Information Center, Letter #4.

25. Thurmond MC, Perez AM. 2006. Modeled detection time for surveillance for foot-and-mouth disease virus in bulk tank milk. Am J Vet Res 67:2017-2024.

26. USDA, NASS. 2016. Quarterly hogs and pigs. National Agricultural Statistics Service (NASS), Agricultural Statistics Board, United States Department of Agriculture (USDA). ISSN: 1949-1921.

27. USDA. 1981. Hog Cholera and Its Eradication. United States Department of Agriculture (USDA), Animal and Plant Health Inspection Service (APHIS), 91-55.

28. van Engeen E, Schotten N, Schimmer B, Hautvast JLA, van Schalk G, van Duijinhoven YTHP. 2014. Prevalence and risk factors for *Coxiella burnetii* (Q fever) in Dutch dairy cattle herds based on bulk tank milk testing. Prev Vet Med 117:103-109.

Alex Hogg Memorial Lecture

One Health: Roles, responsibilities, and opportunities for swine veterinarians

Matthew Turner, DVM

In remembrance and honor of Alexander Hogg

Alexander Hogg
1920-2006

A native of Scotland, Alex (pronounced "Alec") Hogg immigrated to the United States with his family when he was six years old. After reaching adulthood, he served in the US Navy during World War II before entering veterinary school at Kansas State University. He received his DVM degree in 1950, and was a mixed animal practitioner in Iowa for twenty years. Dr. Hogg returned to college – this time, Iowa State University – to complete a Masters Degree in 1972. He spent the following years serving and educating swine veterinarians and producers in his roles as Nebraska Swine Extension veterinarian and professor at the University of Lincoln-Nebraska. Upon "retirement" from the university in 1990, Alex worked as a technical services consultant for MVP Laboratories. Ever an enthusiastic learner, he graduated from the University of Illinois Executive Veterinary Program in 1995.

An active and longstanding member of the AASV, Dr. Hogg joined what was then the American Association of Swine Practitioners shortly after it was formed. He served as secretary/treasurer of the association from 1974 to 1976, and was elected president in 1979.

The many awards and honors bestowed upon him over the years included several presented by the AASV: The Howard Dunne Memorial Award in 1985, the Meritorious Service Award in 1996, and in 2005, the first-ever AASV Foundation Heritage Award for a lifetime of achievement in veterinary medicine.

ALEX HOGG MEMORIAL LECTURERS

2011 Joseph F. Connor

2012 Lisa Tokach

2013 John T. Waddell

2014 Mark J. Engle

2015 C. Scanlon Daniels

2016 Peggy Anne Hawkins

One Health: Roles, responsibilities, and opportunities for swine veterinarians

Dr. Matthew Turner, DVM
JBS USA Food Company, Greeley, Colorado

The history of One Health

The concept of One Health is not new, or even recent. The first documented recognition of the health of people being related to their environment is found in the writings of Hippocrates in 460-370 BC. In his *On Airs, Waters, and Places,* he related that the health of people could be impacted by the environment, and specifically that contact with swampy, stagnant water could make people sick. It was nearly 2000 years before it was understood that mosquitoes transmitted malaria, which literally is translated as "bad air."[1]

The relationship of animal disease and human disease has been repeatedly linked throughout history. Dr. Louis Pasteur founded the basis of vaccinology as we know it today while working with fowl cholera. Pasteur recognized that cholera broth that had been allowed to age for a month was not infectious to chickens and those chickens were then resistant to cholera when given fresh cholera broth.[1] Dr. Edward Jenner used information derived from animal agriculture to develop a vaccine for smallpox, which was eradicated from the globe in 1980.[2] Milkmaids in England in the 1800's that had been infected with cowpox never became infected with smallpox. This revelation led Jenner to use dried pus from cowpox to develop a vaccine against smallpox. Interestingly, the name vaccine is derived from *vacca*, the Latin for cow.

In the 1800's, Dr. Rudolf Virchow realized the relationship of human to animal health while studying a parasite of swine, *Trichinella spiralis*. He coined the term zoonosis as the infectious link between animal and human health and is recognized as a father of the One Health movement, as expressed in his quote, "Between animal and human medicine there are no dividing lines- nor should there be. The object is different but the experience obtained constitutes the basis of all medicine."[3]

Dr. Calvin Schwabe, is credited with coining the term "One Medicine." Dr. Schwabe is considered the father of veterinary epidemiology, as he recognized the relationship between animal and public health long before the current One Health movement. Schwabe was a veterinary graduate of Auburn University (1954) and went on to study public health and tropical medicine at Harvard University. Schwabe established the Department of Epidemiology and Preventive Medicine at UC Davis in 1966 and authored *Veterinary Medicine and Human Health,* in addition to many other works.[4] Schwabe was a true visionary in realizing the interrelationships between animal and human health.

The One Health movement fundamentally recognizes that pathogens are shared between humans and animals. In one review of human pathogens, 1,415 species of infectious organisms known to be pathogenic to humans, including 217 viruses and prions, 538 bacteria and rickettsia, 307 fungi, 66 protozoa and 287 helminths were evaluated. Out of these, 868 (61%) can be transmitted between humans and animals, and 175 pathogenic species are associated with diseases considered to be "emerging."[5]

The recent history of One Health- collaboration at the human animal interface

Dr. Roger K Mahr, president of the AVMA from 2006-2007, dedicated his presidency to championing the One Health concept. Through his efforts, the American Medical Association adopted a resolution on June 24, 2007, advocating for closer ties between human and veterinary medicine. Within the AMA resolution there was recognition of zoonotic disease, veterinary and human medical synergies, promotion of collaboration between human and veterinary medicine, support of joint educational efforts, cross-species disease surveillance, cross-species medicines and vaccines, and a need to enhance collaboration between the medical and veterinary medical professions.[4]

The growing One Health concept expanded to global recognition when the Food and Agriculture Organization of the United Nations (FAO), the World Organization for Animal Health (OIE), and the World Health Organization (WHO) developed a tripartite concept note in April of 2010. This concept note embraced the concept of One Health and provided a vision of a world capable of preventing, detecting, containing, eliminating and responding to public health risks attributable to zoonosis and animal diseases with an impact on food security. The FAO-OIE-WHO are committed to working closely together to align activities related to the animal-human-ecosystem interface to support member countries...including the impact of environmental changes and globalization, as well as new societal demands related to food security, food safety, public health and animal welfare.[7]

Antimicrobial Resistance (AMR) on an international level- important historical events

In the UK, an outbreak of antimicrobial resistant *Salmonella typhimurium* type 29 in both humans and calves caused a government investigation to be commissioned. The outcome of the investigation, called the Swann Report, released in 1969, concluded that administration of antibiotics to livestock for growth promotion and other purposes posed hazards to human and animal health by promoting antibiotic-resistant strains in enteric bacteria of animal origin and that these resistant factors could be spread to other bacteria. The report recommended that antibiotics of considerable importance as therapeutics in humans or animals not be used as feed antibiotics.[6] The Swann report was controversial to say the least, but it did set the stage for action, as parts of the Swann report's recommendations were implemented on a partial basis. Sweden was the first country to completely ban the use of antibiotic growth promoters in animal feed in 1986, followed by Denmark in 2000, and the whole of the EU in 2006. The impact of these partial or complete bans of growth-promoting uses of antimicrobials on AMR is not clear. Different groups have found positive and negative consequences of the bans, but at the end of the day there can be no reliable conclusions drawn that removing growth-promoting antibiotics will improve human health. The precautionary principle prevailed as the risk to human health is something more than zero.

AMR is a multifactorial issue

Vancomycin (a glycopeptide antibiotic) is used in human medicine to treat infections with gram-positive organisms, including *Staphylococcus aureus* and *Enterococcus faecium*. Vancomycin had been used routinely in humans for many years without resistance being detected. In 1988 patients in Paris and London were found ill with vancomycin-resistant *E. faecium* (VRE). Years later, VRE was isolated from the intestines of food animals and the use of avoparcin (also a glycopeptide) for growth promotion was empirically linked to the spread of human VRE. Avoparcin was voluntarily banned in 1995 in Denmark, and the rest of the EU followed in 1997. VRE rates in pigs and poultry decreased after the ban, but unfortunately the rates of bloodstream VRE in hospitalized patients did not. Subsequent analysis revealed the VRE clones infecting people in hospitals were genetically distinct from those found in livestock. Ironically, US hospitals have struggled with VRE also, yet avoparcin was never approved for use in the United States. Genetic analysis comparing human, livestock, and canine clones of VRE suggests that the clones of VRE in humans may have evolved from dogs.[1] This information leads us to understand that animal agriculture is not always the cause of AMR issues in humans, but it does not remove any responsibility for our profession.

One Health and the focus on AMR globally and in the USA

The FAO of the United Nations adopted a resolution in June 2015 (resolution 4/2015) to address AMR. The next year, a plan was published in support of the 2015 resolution to address AMR. Four focus areas were identified:

1. Improve awareness of AMR and related threats.

2. Develop capacity for surveillance and monitoring of AMR and antimicrobial use in food and agriculture.

3. Strengthen governance of antimicrobial use and AMR in food and agriculture.

4. Promote good practices in food and agriculture systems and prudent use of antimicrobials.

In their final report, the FAO called for a One Health and food chain approach to address AMR.[8]

In March 2015, the US Department of Health and Human Services (HHS) announced the establishment of the Presidential Advisory Council on Combating Antibiotic-Resistant Bacteria to provide advice, information, and recommendations regarding programs and policies intended to support and evaluate the implementation of the National Strategy for Combating Antibiotic-Resistant Bacteria and the National Action Plan for Combating Antibiotic-Resistant Bacteria. The five goals of the National Strategy and Action Plan are listed below with objectives listed for each goal that directly or indirectly relate to swine veterinary practice.

GOAL 1: Slow the Emergence of Resistant Bacteria and Prevent the Spread of Resistant Infections

-Eliminate the use of medically-important antibiotics for growth promotion in food-producing animals and bring other agricultural uses of antibiotics, for treatment, control, and prevention of disease, under veterinary oversight.

-Identify and implement measures to foster stewardship of antibiotics in animals.

GOAL 2: Strengthen National One Health Surveillance Efforts to Combat Resistance.

-Develop, expand, and maintain capacity in State and Federal veterinary and food safety laboratories to conduct antibiotic susceptibility testing and characterize select zoonotic and animal pathogens.

-Enhance monitoring of antibiotic-resistance patterns, as well as antibiotic sales, usage, and management practices, at multiple points in the production chain for food animals and retail meat.

GOAL 3: Advance Development and Use of Rapid and Innovative Diagnostic Tests for Identification and Characterization of Resistant Bacteria.

-Develop and validate new diagnostics—including tests that rapidly distinguish between viral and bacterial pathogens and tests that detect antibiotic-resistance—that can be implemented easily in a wide range of settings.

GOAL 4: Accelerate Basic and Applied Research and Development for New Antibiotics, Other Therapeutics, and Vaccines.

-Develop non-traditional therapeutics and innovative strategies to minimize outbreaks caused by resistant bacteria in human and animal populations.

GOAL 5: Improve International Collaboration and Capacities for Antibiotic-resistance Prevention, Surveillance, Control, and Antibiotic Research and Development.

-Collaborate with WHO, OIE, and other international efforts focused on the development of integrated, laboratory based surveillance to detect and monitor antibiotic-resistance in relevant animal and human foodborne pathogens.

-Establish and promote international collaboration and public-private partnerships to incentivize development of new therapeutics to counter antibiotic-resistance including new, next-generation, and other alternatives to antibiotics, vaccines, and affordable, rapidly deployable, point-of-need diagnostics.[9]

For food animal production, the most immediate implications of the national plan relate to antibiotic stewardship (Goal 1) and surveillance of antibiotic use (Goal 2). To date, the federal agencies (FDA, USDA) tasked with surveillance of antibiotic use in animals have not received additional funding to implement this, and the likelihood of future funding is uncertain. However, commercial pressure to document antibiotic use and practices is increasing, and all major food animal species are evaluating mechanisms and metrics for assessing antibiotic use in those industries. Therefore, transparency in antibiotic use practices has the potential to become a competitive issue both between and within food animal species. The elevated responsibilities placed on the veterinary profession by FDA Guidance 213 and the VFD regulations will bring greater scrutiny on the antibiotic use practices in food animal industries. At a bare minimum, it is incumbent on all food animal veterinarians to fully understand and comply with the new legal framework for antibiotic use. Any events of illegal use will likely be used to discredit the industry overall and the effectiveness of veterinarians as 'stewards.'[11]

The third part of One Health- the ecosystem

Global food demand is increasing and agriculture will be faced with the challenge of feeding an additional 2.5 billion people by 2050. In addition to getting food where it needs to be, there are concerns that pollution from many sources, including animal agriculture, may impact human health.

In her book *One Health and the Politics of Antimicrobial Resistance*, Dr. Laura Kahn proposes a new concept of antibiotic resistance called the global "resistome." Basically, this is a concept that all uses of antibiotics contribute to selection pressures that require microbes to adapt to survive. The resulting resistance genes survive and spread after excretion are present in manure and wastewater, and become integrated into soil microbes. Kahn proposes these resistance genes are transported globally.[1]

Producing animal protein in confinement has provided a tremendous service to society. Affordable protein is available today because of advancements in modern agriculture, including poultry, pork, and beef production agriculture systems. The pork produced in 2009 utilized 78% less land, 41% less water, and has a 35% smaller carbon footprint compared to pork produced in 1959.[13] There have also been marked improvements in pork safety due to modern practices of intensive swine production.[12] The ability to control or even eliminate the interaction of wildlife with domesticated animals has given modern man a significant advantage in terms of potential zoonotic risks in the food supply. In addition to the myriad benefits of removing livestock from interaction with wildlife, the current methods of controlling waste and utilizing nutrients for agricultural purposes adds to the sustainability of modern agriculture. With that said, there are still opportunities to improve the efficiency of animal agriculture by reducing dependence on natural resources, reducing emissions, and improving by-product utilization.

Where are the opportunities for swine veterinarians in One Health?

1. Safe food: Swine veterinarians have a long history of being focused on safe food for our customers. The continued production of food in a more efficient manner is critical to maintaining a healthy population. We do this today and must continue to advocate for practices that lead to efficient pork production.
2. Animal welfare: Veterinarians must continue to advocate for the proper care of livestock. We must consider the welfare of the animal and weigh the consequences of medical decisions, especially around antibiotic reduction policies. Veterinarians must be ready and willing to stand up for the animal to ensure its welfare is not compromised when antibiotic use is reduced. Bans on 'preventive uses' are in place in some European countries and similar political pressures are developing in the US. The ability to defend any practice of antibiotic use must first and foremost be justified by evidence of efficacy and tangible benefits to animal health and well-being.
3. Antibiotic resistance: Veterinarians are and should be concerned about the development of antibiotic resistance. Today the issues have higher stakes than ever before and every decision on antibiotic use will be scrutinized. We must manage our protocols and procedures to minimize the potentiation of antibiotic resistance, although we don't know the ideal techniques today. We must collaborate and be dedicated to continual improvement to minimize antibiotic resistance.
4. Development of antibiotic use metrics on farm for continual improvement: Antibiotic use data will be a benchmark in the future and we, as veterinarians, must lead the charge in collection, interpretation, and implementation of the knowledge gained from the data. Veterinarians have the knowledge required to accomplish efficient production in the face of this challenge.
5. Antibiotic Free Production: Affluent consumers have the right to ask for specialty products at a premium price. Free market systems will determine who is willing to pay and how much they are willing to pay. Veterinarians will be asked to help develop systems that can accomplish efficient production without utilizing antibiotics for group therapy. Health management in these systems will be more dependent on veterinarians, not less. Veterinarians are well suited to helping these systems be competitive while maintaining animal welfare standards.
6. Take advantage of GFI 209 and 213 regulations: Veterinarians have sometimes not been involved in the use of antibiotics

available over the counter, but now these drugs are prescription or VFD status. We must take responsibility and ensure these products are available when needed and used appropriately. Some feed additive antibiotics were available over the counter in the past, sometimes without significant veterinary oversight provided. The opportunity to lead, guide, and direct antibiotic use is a veterinary responsibility that cannot be taken lightly.

7. Education: Oversight of antibiotics and their use gives us an opportunity to educate producers and other stakeholders on appropriate use. The importance of understanding the "why" behind antibiotic use is just as important as the "how." There will be friction as over the counter antibiotics become restricted to veterinary prescription; we must take this opportunity to add value with our expertise. We must be able to articulate to our clients and customers that proper preventive use of antibiotics is critical to reduce total antibiotic use and maintain animal welfare.
8. Collaboration with human health: Veterinarians, especially current students, will have the opportunity to interact at levels previously not enjoyed with the human health side of the One Health movement. Opportunities to learn from and educate the human health profession could prove to be quite valuable for control of pathogens that move between pigs and people.
9. Development/utilization of animal health technology: Veterinarians working with genetic, pharma, diagnostic, and biologic companies have tremendous opportunity. There will be an increased need for tools to manage immunity/immune status and robustness. Novel techniques to administer vaccines, novel products to stimulate the innate immune system, and disease resistance genes hold remarkable promise in the future. Diagnostic tools that are available at the farm level to detect pathogens and susceptibility patterns in minutes could revolutionize medicine as we know it today- allowing targeted interventions of narrow spectrum antibiotics instead of broad spectrum antibiotics.
10. Utilization of technology: It is apparent there are not enough swine veterinarians to perform all the oversight that is considered ideal. There are opportunities to utilize new technology to perform farm visits virtually. The technology is available today but has not been adopted. Swine veterinarians must adapt to a changing environment.

It is interesting that several of the early founders of the One Health concept were involved with animals. Dr. Virchow, MD in the 1800's studied the life cycle of *Trichinella spiralis* in swine and its public health consequences. Dr. Schwabe, DVM recognized the relationships between animal and human medicine and that veterinary medicine is directly related to man's well-being. Even in our veterinary oath, the principles of One Health are included as we strive to benefit society, promote public health, and advance medical knowledge.

Being admitted to the profession of veterinary medicine, I solemnly swear to use my scientific knowledge and skills for the benefit of society through the protection of animal health and welfare, the prevention and relief of animal suffering, the conservation of animal resources, the promotion of public health, and the advancement of medical knowledge.

I will practice my profession conscientiously, with dignity, and in keeping with the principles of veterinary medical ethics.

I accept as a lifelong obligation the continual improvement of my professional knowledge and competence.[10]

Dr. Hogg would tell us that opportunity abounds and we must continue to promote our profession as veterinarians by learning more, working harder, and passing knowledge on to the next generation of professionals.

References

1. Kahn, Laura H. One Health and The Politics of Antimicrobial Resistance. Johns Hopkins University Press, 2016.

2. Centers for Disease Control. Available at CDC.gov/smallpox/index.html accessed 11-13-16.

3. Schultz M. Emerging Infectious Diseases Vol. 14, No. 9 2008. Available at www.cdc.gov/eid. Accessed 11-13-16.

4. JAVMA News. 2007. Available at www.AVMA.org/News/JAVMANews/pages/130701m.aspx. Accessed August 2016.

5. Taylor LH1, Latham SM, Woolhouse ME. Risk factors for human disease emergence. Philos Trans R Soc Lond B Biol Sci. 2001 Jul 29;356(1411):983-9.

6. Use of antibiotics in animal husbandry and veterinary medicine (Swann Report).1969. HC Deb 20 November 1969 vol 791 cc1525-31

7. FAO-OIE-WHO Collaboration.2010. Available at www.onehealthcommission.org. Accessed 10 August 2016.

8. The FAO action plan on antimicrobial resistance 2016-2020. Available at www.fao.org/3/a-i5996e.pdf. Accessed 11-12-16.

9. National action plan for combatting antibiotic resistance. 2015. Available at www.whitehouse.gov/sites/default/files/docs/national_action_plan_for_combating_antibotic-resistant_bacteria. accessed 11 November 2016.

10. Veterinary Oath 2011. Available at avma.org/KB/Policies/Pages/veterinarians-oath.aspx Accessed 13November2016.

11. Davies, PR Personal Communication 11-15-2016.

12. Davies PR. Intensive swine production and pork safety. Foodborne Pathogens and Disease 2011 8: 189-201.

13. National Pork Board 50-year sustainability study. 2013. Available at www.pork.org/pork-quick-facts/home/sustainability/todays-pork-50-year-sustainability-study. Accessed 21 November 2016.

Land coverage and elevation as risk factors for PRRS outbreaks

A.G. Arruda, DVM, MS, PhD; C. Vilalta, DVM, PhD; A. Perez, DVM, MS, PhD; R. Morrison, DVM, MBA, PhD
Veterinary Population Medicine, University of Minnesota, Saint Paul, Minnesota

Introduction

Porcine reproductive and respiratory syndrome (PRRS) is an extremely impactful disease in the North American swine industry. The Swine Health Monitoring Project (SHMP) is a national effort that aims to monitor and aid in control of PRRS and other swine diseases. This project enrolls breeding herds voluntarily and currently has swine herds representing most important swine-producing states and 42% of the sow population of the United States. There has been substantial PRRS research over the past fifteen years, but incidence has not decreased over the last 2 years.[1]

The objective of this study was to use data from the SHMP project to investigate environmental factors (e.g. land elevation and land coverage) as risk factors for PRRS outbreaks.

Materials and methods

A total of 706 sow sites participating in the SHMP and that shared weekly PRRS statuses were enrolled in this study. Information collected from the SHMP database included number of PRRS breaks, years of participation and site location. Environmental factors included land coverage (cultivated areas, shrubs and trees), land altitude (in meters above sea level) and land slope (inclination of the land in degrees compared to surrounding areas). Other factors that were included in the model were production system (company), geographical region, swine density and herd size.

The environment-related variables and pig density were captured in raster format from different sources (slope and altitude: SRTM [CGIAR-CSI 2006), GTOPO30 [USGS 2002]; land coverage: Global Land Cover 2000 Project [JRC, European Commission]) and extracted to points (swine farms) using ArcMap 10.2.2. A multilevel mixed-effect Poisson regression was built using STATA/ IC 14.1 and a backwards stepwise approach. The model included a random effect for production system, since it is known that swine sites are clustered within this level. Significant associations were declared if $P \leq 0.05$.

Results and discussion

The median number of outbreaks per farm per year was 1 (IQR: 2), and was positively skewed. Approximately 40% of the farms did not report an outbreak and the maximum number of outbreaks was 9.

Consistent with previous reports,[2,3] our final multivariable model included swine density and herd size which were positively associated with incidence of outbreaks ($P < 0.01$). Variables retained in the model included land slope, land coverage and region. Swine farms located in terrains with a slope of 9% or higher were "protected" compared to farms located in terrains with slopes of 1% or lower ($P < 0.01$). Swine farms located in an area of herbaceous cover or trees were "protected" from having PRRS outbreaks compared to farms located in cultivated/ managed areas ($P < 0.05$). Prospective follow-up studies are needed to further understand these protective factors.

References

1. Betlach et al. National Hog Farmer, 2016. Available in: http:// nationalhogfarmer.com/ health/swine-health-monitoring-project-serving-swine-disease-surveillance-system

2. Lambert et al. Epidemiological investigations in regard to porcine reproductive and respiratory syndrome (PRRS) in Quebec, Canada. Part 2: Prevalence and risk factors in breeding sites. Prev Vet Med. 2012; 104:84.

3. Velasova et al. Risk factors for porcine reproductive and respiratory syndrome virus infection and resulting challenges for effective disease surveillance. BMC Vet Res. 2012; 8:814.

Broadly neutralizing antibodies to recent, virulent type 2 PRRSV isolates

M.P. Murtaugh[1], PhD; D.K. Gray[1], BS; S.R. Robinson[1], BVSc (Hons), MVSc, PhD; M. Wagner[2], DVM; G.D. Spronk[3], DVM; P. Yeske[4], MS, DVM
[1]Department of Veterinary and Biomedical Sciences, University of Minnesota, St. Paul, Minnesota; [2]Fairmont Veterinary Clinic, Fairmont, Minnesota; [3]Pipestone Veterinary Clinic, Pipestone, Minnesota; [4]SwineVet Center, Saint Peter, Minnesota

Introduction

Indicators that predict immune protection against PRRS would facilitate herd management and vaccine development programs due to the need for efficacious cross-protection against widely divergent PRRSV isolates. Neutralization of virus infectivity by specific antibodies is a common predictor of immunological efficacy since neutralizing antibodies play a key role in anti-viral immunity in animals and humans. However, their role in protection against PRRS has been discounted since viral neutralization has not been a characteristic and consistent feature of the anti-PRRSV immune response.[1] For example, many studies in growing pigs showed that neutralizing antibody responses at weeks to months after infection were frequently absent or present at low titers of 1/8 or less. Recently, we discovered high titers of broadly neutralizing antibodies in sows from herds with a history of virulent outbreaks or repeated serum inoculation.[2]

Further investigation revealed that cross-reactive neutralization was a common feature of sows in all farms tested, was independent of parity, was present in vaccinated and field virus-exposed herds, was variable in terms of breadth of cross-protection, and varied extensively among sows in the same farm. For example, all sows (n = 20) in a routinely vaccinated herd had homologous reciprocal neutralizing titers to VR2332 ranging from 15-256; 90% had heterologous titers of 4-60 against MN184, and 20% of sera neutralized type 1 PRRSV SDEU at titers of 4-8. In the process of adapting a highly virulent 174-type PRRSV for use in neutralization assays, we found that its growth characteristics were different from all other viral isolates in the lab. At the same amount of RNA copies in cell culture media, MW174 has 1000-fold less infectious virus that other lab strains, using $TCID_{50}$. Since TCID is based on cytopathic effect, it suggests that this highly virulent family of viruses might grow better in cells by preventing or delaying cell lysis. Serum from all of 10 sows in two 174-virus outbreak herds had neutralizing activity against VR2332, MN184, MW174, NC134, and type 1 SDEU. In 22% of tests (N = 50; 5 viruses by 10 sera), neutralizing titers were >256. Cross-neutralizing activity is important since passive transfer of broadly neutralizing serum significantly reduced heterologous PRRSV infection. These findings support the concept of immune cross-protection in PRRS, and indicate that immunologically relevant conserved epitopes are present in type 1 and type 2 PRRSV.

Materials and methods

Animals

Serum samples were collected from unrelated commercial sow herds in Minnesota and North Carolina, USA. Herds had multiple exposures to PRRSV, through vaccination, natural infection or serum inoculation. Serum samples were collected from cull sows that had experienced multiple virulent PRRS virus exposures; from sows of a range of parities in a vaccinated herd not exposed to virulent viruses; and from gilts exposed to virulent viruses in a controlled exposure acclimation program. Additional sera were collected from pigs 202 days after experimental infection with PRRSV strain VR2332, and from a commercial sow having a high viral neutralizing titer.

Cell culture

Porcine alveolar macrophages (PAM) were harvested by lung lavage from pigs of various ages, but primarily from weaned pigs 5-6 weeks of age. Briefly, lavage was performed with phosphate buffered saline (PBS) containing 50 ug/ml gentamicin, and red blood cells were lysed in sterile water. Cells were passed through a 70 um filter, resuspended in 40% RPMI media, 50% fetal bovine serum (FBS) and 10% dimethyl sulfoxide (Fisher Scientific, Pittsburgh, PA) for cryopreservation. PAMs were cultured in RPMI medium supplemented with sodium pyruvate, gentamycin sulfate (Cellgro Mediatech, Manassas, VA), non-essential amino acids, HEPES buffer, and 5% FBS (Sigma, St. Louis, MO). Simian renal epithelial MARC 145 cells were cultured in MEM medium (Gibco, Grand Island NY) supplemented with sodium bicarbonate, non-essential amino acids, HEPES buffer (Sigma, St. Louis, MO), and gentamycin sulfate (Cellgro Mediatech, Manassas, VA) with 10% FBS (Sigma, St. Louis, MO).

PRRSV isolates

PRRSV isolates VR2332, MN184, SDEU, MW174, NC134 were propagated in MARC 145 cells. Infectious titers ($TCID_{50}$) were determined for each isolate by evaluation of cytopathic effect on MARC 145 cells in 96 well plates and calculated by the Reed & Muench method.[3] PRRSV open reading frame (ORF) 5 nucleotide sequences (coding for glycoprotein 5), were available from two recent virulent outbreak strains, MN0111 and MN1211, but virus isolates were not obtained. ORF5 sequences from these isolates, along with diverse PRRSV isolates including lab-adapted VR2332, vaccine strains, field isolates including MN184, Type 1 PRRSV isolate Lelystad Virus (LV), and Type 1 field isolate SDEU, were aligned in CLC Sequence Viewer version 6.8.2 (CLC bio/Qiagen, Cambridge, MA). Aligned ORF5 sequences were subjected to phylogenetic tree construction in Geneious R6 version 6.1.7 (Biomatters Ltd., Auckland, New Zealand) Tree Builder using the Tamura-Nei neighbor-joining method with 100 bootstrap resampling and no outgroup to create a consensus tree. Percent homology between strains was also compared in Geneious R6.

ELISA-based serum neutralization assay

The PRRSV serum neutralization (SN) assay was performed essentially as described.[2,4,5] Briefly, 96-well plates were treated with 25 ug/ml polyethylenimine (Sigma-Aldrich, St. Louis, MO) and PAMs were seeded at 6×10^4 cells/well and cultured for 48 hours, or MARC 145 cells were seeded at 6×10^3 cells/well. Serum samples were heat inactivated (56°C, 30 min) and serially diluted in culture media without FBS. Diluted sera were mixed with an equal volume of 5×10^4 $TCID_{50}$/ml PRRSV in non-adherent 96-well plates and incubated for 1 hour (37°C, 5% CO_2). Serum-virus mixtures containing 5000 $TCID_{50}$ were transferred to cells for adsorption for1 h at 37°C. Cells were washed in PBS, media added, and further incubated for 24-48 h. Cells were fixed in 3.7% formaldehyde (Sigma-Aldrich, St. Louis, MO) in PBS, washed, permeabilized (0.1% Triton X-100 (Sigma-Aldrich, St. Louis, MO) in PBS), and blocked (5% nonfat dry milk in PBS) for 1 hour. PRRSV monoclonal antibody SR30-A against nucleocapsid protein (Rural Technologies, Brookings, SD) was diluted 1:10,000 in 5% nonfat dry milk in PBS and added to cells at 100 ul/well for 1 hour. HRP-conjugated goat-anti-mouse IgG (H+L) (Bethyl Laboratories Inc., Montgomery, TX) at a 1:10,000 dilution was added at 100 ul/well for 1 hour. Color was developed with 100 µl of TMB peroxidase substrate (Kirkegaard & Perry Laboratories, Gaithersburg, MD) for 15 min. Reactions were stopped with 100 µl of 1 M phosphoric acid and plates were read at 450 nm in a ThermoMax Microplate Reader (Molecular Devices, Sunnyvale, CA).

Percent inhibition of viral infection at each dilution was compared with virus only controls after background subtraction of absorbance from uninfected cells. Fifty percent neutralization titer was defined as reciprocal of the dilution for which 50% inhibition of infection was reached based on curve fitting.

Results and discussion

The humoral immune response to PRRSV infection is generally considered to be weak, and neutralizing antibodies appear at low levels late in infection after acute viremia has waned, or do not appear at all. These types of studies were performed with young pigs that had been recently infected. It also was observed that neutralizing antibodies did not increase to higher levels during prolonged infection in lymphoid tissues. However, the neutralizing response had not been investigated in adult pigs in which multiple infections or antigenic exposures may have occurred and immunological changes may have taken place in the passage of time as animals aged. We hypothesized that the neutralizing response would be different in adult pigs with a history of exposure to PRRSV. Serum from sows in commercial swine herds with a history of PRRSV exposure were screened for PRRSV-neutralizing antibodies as a first step.

Multiparous sows in two herds showed uniformly high titers of antibodies that neutralized the type 2 strain, VR2332 even though there was no known exposure to the closely related vaccine strain, Ingelvac MLV. While maximal neutralization was equivalent between herds, herd 1 had a reciprocal 50% neutralization titer of 512, the highest individual titer being 2187, and herd 2 had a reciprocal 50% titer of 81. Neutralizing activity was dose-dependent and shape of the curves was similar for both herds.[2]

A subset of 5 sera from one herd were tested in the widely used fluorescent focus neutralization (FFN) by Eric Nelson at South Dakota State University against a diverse panel of type 1 and type 2 PRRSV isolates. The sera neutralized type 2 isolates at 90% titers 16 to 56, and surprisingly neutralized type 1 isolates at 90% titers >80.[2] Broadly neutralizing activity was independent of parity. High titered neutralizing antibodies also were observed in growing gilts undergoing virulent virus exposure at monthly intervals. These results indicated that broadly neutralizing antibodies can be elicited in a wide variety of conditions of age, and PRRSV exposure that includes growing pigs, adults, virulent virus exposure, and attenuated vaccine exposure.

Extensive variation has been seen among pigs on the same farm and exposure history. For example, all sows (n = 20) in a routinely vaccinated herd had homologous neutralizing titers to VR2332 ranging from 15-256; 90% of sows had heterologous titers of 4-60 against MN184, and 20% of sera neutralized type 1 PRRSV SDEU at titers of 4-8. In the process of adapting a highly virulent 174-type PRRSV for use in neutralization assays, we found that its growth characteristics were different from all other viral isolates in the lab. At the same amount of RNA copies in cell culture media, MW174 had 1000-fold less infectious virus than other lab strains, using $TCID_{50}$. Since TCID is based on cytopathic effect, it suggested that this highly virulent family of viruses might grow better in cells by preventing or delaying cell lysis. Serum from all of 10 sows in two 174-virus outbreak herds had neutralizing activity against VR2332, MN184, MW174, NC134, and type 1 SDEU. In 22% of tests (N = 50; 5 viruses by 10 sera), neutralizing titers were >256. Figure 1 shows the neutralizing titers in 5 individual sows on one farm against the indicated viruses.

Cross-neutralizing activity is important since passive transfer of broadly neutralizing serum significantly reduced heterologous PRRSV infection.[2] These findings support the concept of immune cross-protection in PRRSV, and indicate that immunologically relevant conserved neutralization epitopes are present in type 1 and type 2 PRRSV.

Figure 1: Broadly neutralizing titers of 5 randomly selected sera in a herd exposed to a 174 RFLP-type PRRSV. Legend indicates the viruses. Numbers in the stacked bar graph or 50% neutralizing titers that are ≥ 32. Pos Ctrl is the assay positive control.

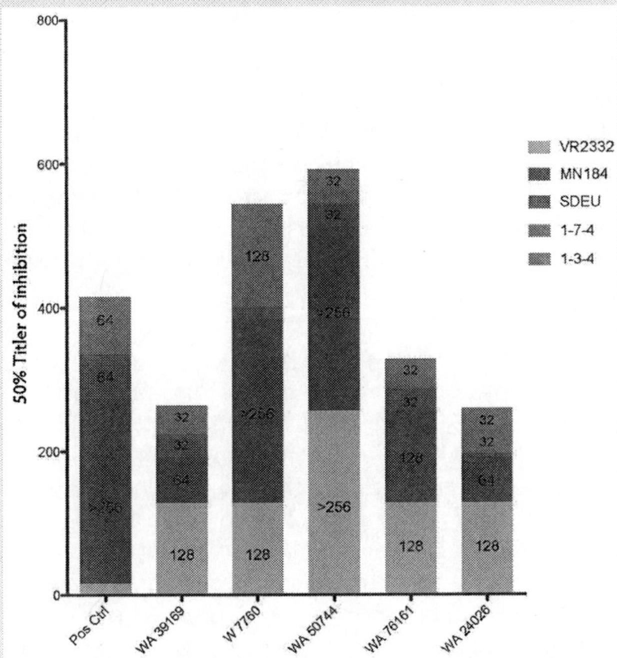

References

1. Molina, R.M., S.-H. Cha, W. Chittick, S. Lawson, M.P. Murtaugh, E.A. Nelson, J. Christopher-Hennings, K.-J. Yoon, R. Evans, R.R.R. Rowland, W.-H. Wu, and J.J. Zimmerman. 2008. Immune response against porcine reproductive and respiratory syndrome virus during acute and chronic infection. Vet. Immunol. Immunopathol. 126:283-292.

2. Robinson, S.R., J. Li, E.A. Nelson, and M.P. Murtaugh. 2015. Broadly neutralizing antibodies against the rapidly evolving porcine reproductive and respiratory syndrome virus. Virus Res. 203:56-65.

3. Reed, L.J., Muench, H., 1938, A simple method of estimating fifty per cent endpoints. American Journal of Epidemiology 27, 493-497.

4. Li, J. and M.P. Murtaugh. 2012. Dissociation of porcine reproductive and respiratory syndrome virus neutralization from antibodies specific to major envelope protein surface epitopes. Virology 433:367-376.

5. Robinson, S.R., M.C. Figueiredo, J.E. Abrahante, and M.P. Murtaugh. 2013. Immune response to ORF5a protein immunization is not protective against porcine reproductive and respiratory syndrome virus infection. Vet. Microbiol. 164:281-285.

Characterization of the memory immune response to PRRSV infection

Michael C. Rahe; Michael P. Murtaugh
Veterinary Biological Sciences, University of Minnesota, Saint Paul, Minnesota

Introduction

Immunological memory is elicited following either vaccination or natural exposure to a virus and is essential for protection against viral reinfection. The ability to characterize the porcine memory immune response has long been hindered by lack of a culture system and tools to assess the memory response. As a result, the evaluation and analysis of protective immune responses that predict immune protection in swine and facilitate vaccine development has been lacking. To fill this gap in knowledge, we created an in vitro system to efficiently and effectively interrogate the memory B cell response, a hallmark of anti-viral immunity. We then utilized this culture system to validate the specificity of a PRRSV non-structural protein 7 (nsp7) tetramer. With this novel reagent we are able to identify and characterize porcine memory B cells specific to PRRSV nsp7 as well as the immune memory response to PRRSV.

Results and discussion

Porcine splenic tissue was collected and broken down to remove structural tissue and create cellular suspensions. Splenocytes were then magnetically enriched for CD21+ positive B cells, allowing for an in vitro system free of confounding cellular heterogeneity. Various cytokines were evaluated for their ability to activate and stimulate B cells, with CD40L and IL-21 identified as inducing maximal cellular proliferation and differentiation to antibody secreting cells (ASC). Inclusion of maintenance factors, B-cell activating factor (BAFF) and a proliferation inducing ligand (APRIL), further enhanced ASC viability. Using standardized culture conditions and a PRRSV non-structural protein 7 (nsp7) enzyme-linked immunospot assay (ELISPOT), we found that .06-1.1% of CD21+ B cells from the spleen of PRRSV immune sows were nsp7-specific memory B cells while naïve sows were completely negative (Fig. 1). These findings were confirmed with a PRRSV nsp7-specific ELISA on cell culture supernatants. Immune sow #4 was negative on nsp7-specific ELISA likely due to the sensitivity of the test and the comparatively low number of nsp7 specific ASC which were identified via ELISPOT.

Future directions

Current work is focused on further refinement of antigen specific cellular identification and characterization with nsp7-specific B cell tetramers and fluorescently labeled PRRSV virions. Using these approaches, we will able to characterize the PRRSV specific memory B cell response which will be critically important for understanding variation in immune efficacy and will help guide development of improved vaccines.

Figure 1: Splenocytes from four PRRSV immune and 2 naïve animals were enriched for mature B cells. PRRSV nsp7 and IgG⁺ ELISPOTs were performed. Remaining cells were cultured for a total of 14 days. Culture supernatants were harvested and evaluated with a limiting dilution PRRSV nsp7-specific ELISA. Individual titers are displayed above the corresponding columns. Data are representative of two independent experiments with the same animals.

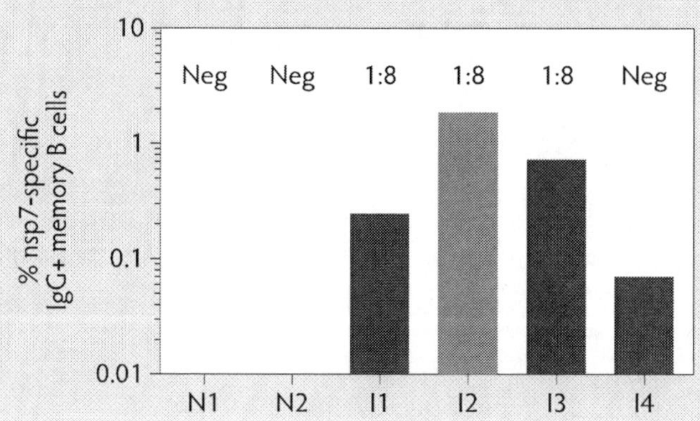

Proof of concept: PRRSV IgM/IgA ELISA detects infection in the face of circulating maternal IgG antibody

M. Rotolo; J.Ji; L. Gimenez-Lirola; R. Magtoto; C. Wang; D. Baum; M. Hoogland; R. Main; J. Zimmerman
Iowa State University, Ames, Iowa

Introduction

Oral fluids (OF) are an excellent sample for PRRSV surveillance: qRT-PCR testing detects active infection and ELISA (antibody) is useful for establishing prior viral exposure. However, in weaned pig populations originating from sow herds endemically infected and/or vaccinated against PRRSV, maternal antibody (IgG) cannot be differentiated from IgG antibody produced by the pigs in response to infection. To address this problem, we developed and evaluated IgM- and/or IgA-specific oral fluid ELISAs as a means to detect active infection in weaned pig populations with circulating maternal antibody.

Materials and methods

OF samples were collected from 3 wean-to-finish (WTF) sites (Sites A, B, C) beginning within 2 weeks of placement. Pigs originated from PRRSV vaccinated and/or exposed sow herds, but the pigs themselves were not vaccinated for PRRSV. Each site had 3 WTF barns and each barn had 36 occupied pens (~25 pigs per pen). By arrangement with the producers, pigs were not moved during the sampling period.

Samples were collected weekly from every occupied pen for 9 samplings. A total of ~2,916 OF samples were collected, i.e., 3 sites × 108 pens per site × 9 samplings. All oral fluid samples were randomized (random.org) and tested by PRRSV qRT-PCR. These results were used to establish the PRRSV status of each pen. OF samples were also tested for IgG, IgA, IgM, and the combination of IgM/IgA using ELISAs developed in our laboratory. ELISA cutoffs, diagnostic sensitivity, and diagnostic specificity were determined by ROC analyses.

Results

PRRSV qRT-PCR: On Site A, 3.7% of pens were positive at the 1st OF sampling, with all pens in all 3 barns positive ≥ 1 times by the 9th sampling. Site B was negative throughout the collection period, except for one positive pen at the 9th sampling. All samples from Site C were negative.

PRRSV Antibody: Among the 3 sites 90.6% (278/307) of the OF samples collected on the 1st sampling were IgG positive, but 273 of these were from qRT-PCR-negative pens. In contrast, IgA, IgM, and IgM/IgA positive responses were associated with qRT-PCR positivity. An ROC analysis estimated the sensitivity and specificity of the IgM, IgA, and IgM/IgA ELISAs, at ≥ 63% and ≥ 99%, respectively.

Discussion

Swine practitioners need both nucleic acid- and antibody-based tests to track PRRSV in the field. Current PRRSV IgG-based ELISAs cannot differentiate maternal antibody from the pigs' humoral response to infection. This study suggested that the pigs' IgM/IgA response could be used to detect an active response to infection even in the presence of circulating maternal antibody.

Spatial autocorrelation and implications for oral fluid-based PRRSV surveillance

M. Rotolo; M. Haddad; Y. Sun; L. Gimenez-Lirola; S. Bade; C. Wang; D. Baum; P. Gauger; M. Hoogland; R. Main; J. Zimmerman
Iowa State University, Ames, Iowa

Introduction

Oral fluids (OF) are a convenient surveillance sample because they are easily collected and can be tested for nucleic acids and/or antibodies for PRRSV and a variety of pathogens. In addition, this approach facilitates field-based research. Two questions were addressed in this research project: 1) Does PRRSV infection in the field exhibit spatial autocorrelation? 2) Is "spatially based" sampling (as opposed to random allocation of samples) a viable option for routine PRRSV surveillance?

Materials and methods

In 3 wean-to-finish barns on one finishing site, OF samples were collected weekly (9 samplings) from every occupied pen (108 pens; ~25 pigs per pen) for a total of 972 OF samples. OF samples were completely randomized and then tested for PRRSV by RT-PCR.

1. PRRSV RT-PCR results were used to test for the presence of spatial autocorrelation of PRRSV infection on the site, both globally and locally, using threshold distance as the spatial weight matrix. Moran's *I*, an indicator of global spatial autocorrelation, was used to test for clustering of PRRSV virus within the barns. LISA (Local Indicators of Spatial Association) analysis, an indicator of local spatial autocorrelation, was used to identify clusters of high PRRSV virus concentration within the barns.
2. Statistical analyses were used to compare the probability of detection based on random vs "fixed" spatial sampling.

Results

1. Moran's *I* analysis revealed positive global spatial autocorrelation in the distribution of PRRSV virus in the swine barns. LISA analysis revealed clusters of high levels of PRRSV virus at the local level. Further analyses showed that the disease status of a pen in a barn was highly influenced by the disease status of other pens in the same barn, e.g., the presence of ≥ 1 positive pens increased the probability of detecting another positive pen.
2. Analyses comparing the probability of detection ("fixed" spatial vs random) showed that "fixed" spatial sampling was as good as, or better than, random sampling for the detection of PRRSV.

Discussion

The results of the analyses showed that PRRSV (and probably other infectious diseases of swine) exhibit spatial autocorrelation. For most swine veterinarians, this is a statement of the obvious, but spatial autocorrelation has previously not been quantified at the barn level. This finding has important implications for surveillance and helps explain why fixed spatial sampling is an effective strategy. Overall, the results suggested that a simple, reliable, oral fluid-based surveillance strategy can be developed.

Comparative pathogenesis and characterization of recent 1-7-4 PRRSV isolates in weanling age piglets

A.G.M. Van Geelen[1], DVM, PhD; K.S. Faaberg[2], PhD; P. Das[1], DVM, PhD; N.A. Montiel[1], DVM, PhD; L.C. Miller[2], PhD; V. Kulshreshtha[1], DVM, PhD; A.C. Buckley[1], DVM; K.M. Lager[2], DVM, PhD
[1]Oak Ridge Institute for Science and Education, Oak Ridge, Tennessee;
[2]Virus and Prion Research Unit, ARS-USDA, Ames, Iowa

Introduction

Porcine respiratory and reproductive syndrome (PRRS) continues to be an economically important disease affecting commercial pig production in the United States and worldwide. Its considerable sequence and antigenic variation, coupled with the limited protection offered by current vaccine options impedes our abilities to control PRRS virus (PRRSV). Recently, PRRSV isolates of 1-7-4 lineage have been associated with severe clinical cases in the Mid-Western US.[1] Several publications have published genomic analyses of viruses from these outbreaks, but no animal studies have been reported. This paper reports on the comparative pathogenesis of 4 out of 22 contemporary 1-7-4 isolates we obtained, selected based on genetic diversity, and one PRRSV isolate of known moderate pathogenicity in weanling age piglets.

Materials and methods

The genomic sequences of 22 ORF5 restriction fragment length polymorphism (RFLP) 1-7-4 PRRSV isolates were analyzed by next-generation sequencing.[2] Eighty five 4-week old mixed breed piglets were split in six groups (N of 14 or 15) of which five groups were inoculated with 5 × 10E4 TCID50 intranasally. Pathogenicity was measured by analyzing average daily gain (ADG), Pyrexia, viral titers in serum and bronchiolar-alveolar lavage fluid and gross lung lesions. ADG was calculated based on starting weight, weight gain after 7 days and at necropsy. Pyrexia was measured as daily body temperature via intramuscular chip (Biothermo, Allflex, USA) viral titers were measured in serum collected at day 0, 1, 4, 7, 11 and at necropsy via TCID50 assay on MARC145 cells. Bronchio-alveolar lavage fluid (BAL) was collected at necropsy by infusing 40ml of serum-free MEM via the trachea and washing the fluid several times into the bronchiolar tree and aspirating. Thus obtained BAL fluid was cleared via centrifugation at 1000 × g for 5 minutes and filtering the supernatant via SpinX 0.45um micro filter tubes and titration on MARC145 cells. Gross pathology scores for the lungs were estimated as a percentage affected per lung lobe and a score was calculated.[3]

Results and discussion

The isolates were 87.4% to 95.1% identical for ORFs2-7 (Table 1) as determined by NGS and we selected 4 isolates based on alignment for the in vivo pathogenesis study. Although all selected strains were 1-7-4, there was considerable difference in pathogenicity between these isolates. We identified two new isolates that caused more severe disease than our reference pathogenic isolate SDSU-73. Both of these 1-7-4 isolates caused 14% mortality within 2 weeks. NADC-34 caused a decrease in ADG of 98% compared to the control group. In addition, the group of pigs infected with NADC-34 isolate had the highest average body temperatures of 41C on 5, 8 and 9 days post infection (dpi) (Figure 1) and peak serum titers of 10E2.9 at 4 dpi. Isolate ISU/2014/3 with peak viral titers at 7 dpi of 10E3.7 caused a 63.2% decrease of ADG, reached peak temperatures of 40.5 to 40.7 between 5-8dpi. On the other hand, isolate ISU/2014/2 caused no pyrexia, had a 4.6% decrease in ADG but still had peak viral titers of 10E2.6 at 7dpi, while the gross lung lesion score was less than half that of the other infected groups. This shows that the level of replication alone, as measured in serum viremia is not a good correlate for pathogenicity, but that loss in ADG, pyrexia and lung lesions are better parameters of disease severity. This study shows that there is considerable difference in disease outcome caused by PRRSV isolates grouped under the 1-7-4 lineage and that other parameters of pathogenicity remain to be identified for future evaluation of disease.

References

1. Workman AM, Smith TP, Osorio FA, et al. Complete Genome Sequence of Highly Virulent Porcine Reproductive and Respiratory Syndrome Virus Variants That Recently Emerged in the United States. *Genome Announc* 2016;4.

2. Neill JD, Bayles DO, Ridpath JF. Simultaneous rapid sequencing of multiple RNA virus genomes. *J Virol Methods* 2014;201:68-72.

3. Halbur PG, Paul PS, Frey ML, et al. Comparison of the pathogenicity of two US porcine reproductive and respiratory syndrome virus isolates with that of the Lelystad virus. *Vet Pathol* 1995;32:648-660.

Table 1: Sequence analysis of 4 selected 1-7-4 isolates and SDSU-73, our reference strain shows degree of sequence identity between the isolates and the consensus sequence after alignment. Black lines indicate a sequence difference from the consensus sequence.

	SDSU73	NADC34	2013/1	2014/2	2014/3
SDSU73		87.8	88.4	90.8	87.5
NADC34	87.8		91.0	89.0	91.2
PRRSV/ISU/2013/1	88.4	91.0		89.5	95.1
PRRSV/ISU/2014/2	90.8	89.0	89.5		88.5
PRRSV/ISU/2014/3	87.5	91.2	95.1	88.5	

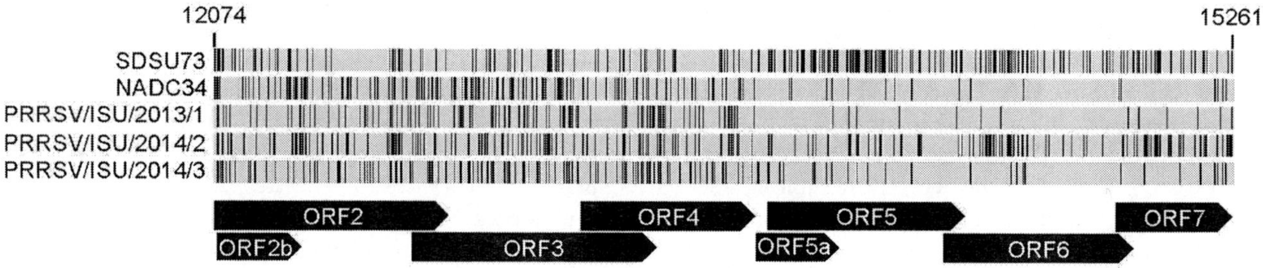

Figure 1: Temperature curves for the six groups of piglets as measured via intramuscular chip. The chips were placed three days before challenge to obtain a baseline temperature.

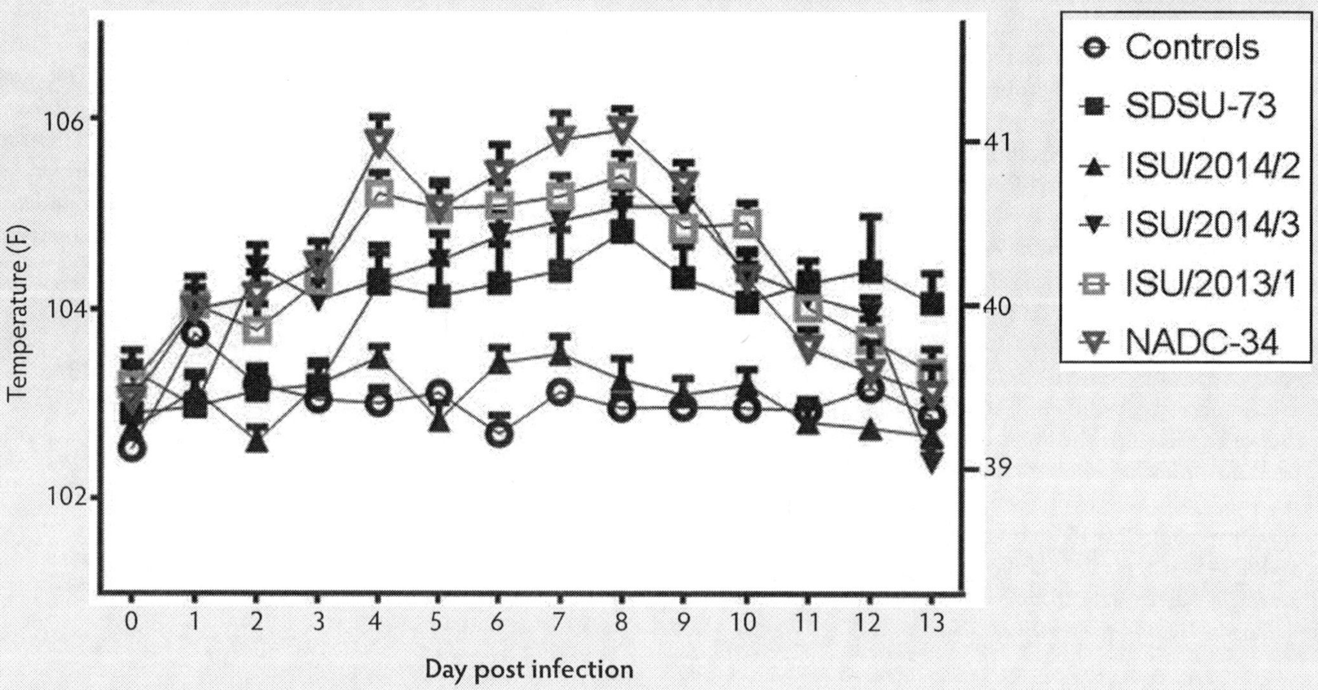

Application of next-generation sequencing technology to whole genome sequencing of PRRSV under diagnostic setting

Jianqiang Zhang, PhD; Ying Zheng, MS; Qi Chen, DVM, PhD; Sarah Bade, BS; Kyoung-Jin Yoon, DVM, PhD; Karen Harmon, PhD; Phillip Gauger, DVM, PhD; Rodger Main, DVM, PhD; Ganwu Li, DVM, PhD
Department of Veterinary Diagnostic and Production Animal Medicine, College of Veterinary Medicine, Iowa State University, Ames, Iowa

Introduction

Porcine reproductive and respiratory syndrome (PRRS), characterized by reproductive failure in breeding females and respiratory distress in pigs of all ages, is an economically important disease globally. The causative agent porcine reproductive and respiratory syndrome virus (PRRSV) includes at least 2 distinct genotypes: type 1 (European type) and type 2 (North American type), with extensive genetic variability between or even within these genotypes.[1] PRRSV has a RNA genome of ~15 kb in length and contains at least 10 open reading frames (ORF1a, 1b, 2a, 2b, 3, 4, 5a, 5, 6 and 7). Among the structural protein genes (ORF2a-ORF7), ORF5 is most genetically variable and is commonly sequenced to study molecular epidemiology and/or genetic relatedness of PRRSV.[2-4] However, ORF5 only covers ~4% of the entire genome and thus may not be able to provide the breadth of evidence needed for differentiating PRRSV strains. A rapid, high-throughput, and accurate whole genome sequencing method for PRRSV possesses enormous potential to provide more comprehensive information (e.g. more gene sequences for nucleotide and deduced amino acid identity comparison and more gene sequences for phylogenetic analysis to better assess genetic relatedness of PRRSV strains) to diagnosticians, researchers, and the swine industry.

Great progress has been made in nucleic acid sequencing technologies in the past ten years.[5] Next-generation sequencing (NGS) technologies have made it possible to determine whole genome sequences of pathogens at lower cost with short turnaround time and massive sequence output. In addition, the metagenomics-based NGS strategy (i.e. hypothesis-free) is able to detect multiple agents including the potential novel pathogens in a sample.[6] NGS technologies have increasingly played crucial role in biomedical research, but their use in time-sensitive and cost-effective veterinary diagnostics have been limited. In this study, a procedure for high-throughput whole genome sequencing for PRRSV was developed and evaluated on cell culture-derived isolates and clinical specimens for its performance and utility.

Materials and methods

Ten PRRSV isolates with known sequences (i.e. reference isolates) were used in this study which included six type 2 strains (VR-2332, VR-2385, NADC20, MN184, ISU-P, and SDSU73), one type 1 (Lelystad strain), and 3 type 2 vaccine strains (Ingelvac PRRS MLV, Ingelvac PRRS ATP, and Fostera PRRS). These PRRSV strains were propagated in MARC-145 cells and had infectious titers of 10^4 to 10^6 $TCID_{50}$/ml.

PRRSV Lelystad isolate and PRRSV VR-2385 isolate were manually mixed in the volume ratio of 1:1 to create a mixture containing a type 1 and a type 2 PRRSV isolate. Similarly, Ingelvac MLV vaccine and VR-2385 isolates were manually mixed in the volume ratio of 1:1 to create a mixture containing two type 2 strains.

Fifty-one PRRSV RNA-positive clinical specimens (19 sera, 16 lungs, and 16 oral fluids) with various PCR threshold cycle (Ct) values were used in attempts of whole genome sequencing directly on the samples.

Total RNA was extracted from aforementioned PRRSV reference isolates, mixtures of PRRSV isolates, and clinical samples, and then purified and quantified. Complementary (c)DNA libraries were constructed from total RNA using a TruSeq stranded total RNA sample preparation kit with 24 samples multiplexed. Multiplexed libraries were then normalized and pooled in an equal volume before loading to a flow cell. The pooled libraries were sequenced on an Illumina MiSeq sequencer.

Sequence reads were mapped against an in-house PRRSV reference genome library which is composed of 459 PRRSV whole genome sequences. Sequence IDs of mapped reads were extracted using a sequence alignment tool and the mapped read sets were used for *de novo* assembly. Integrative Genomics Viewer (IGV) software[7] was also used to analyze sequence data generated from mixtures of PRRSV isolates.

Results and discussion

Whole genome sequences were obtained from all 10 reference isolates with nucleotide identities ranging from 99.3-99.9% when compared to the respective sequences previously determined by Sanger method (Table 1).

For the mixture with type 1 (Lelystad strain) and type 2 (VR-2385 strain) PRRSV isolates which shared 57.8% nucleotide identity at the whole genome level, NGS was able to distinguish them and obtain their respective genome sequences. In contrast,

Table 1: Summary of NGS outcomes on PRRSV reference isolates.

PRRSV	Ct value	No. of total reads	No. of reads mapped	% Reads mapped	Full-length genome by NGS	Longest contig (bp)	GenBank accession of whole genome public sequence	Nucleotide identity between NGS and public sequences (%)
VR-2332	12.4	1,027,668	762,176	74.17	Yes	15,396	AY150564	99.9
VR-2385	11.1	1,690,980	31,702	1.87	Yes	14,956	JX044140	99.8
NADC20	11.2	165,146	6,090	3.69	Yes	15,393	Not available	
MN-184C	12.6	988,292	62,480	6.32	Yes	15,001	EF488739	99.3
ISU-P	13.5	1,596,536	774,848	48.53	Yes	15,387	EF532816	99.9
SDSU73	17.4	1,253,312	896,603	71.54	Yes	15,394	JN654458	99.9
Lelystad	14.1	1,267,240	309,634	24.43	Yes	15,083	M96262	99.8
Ingelvac MLV vaccine	16.6	91,932	54,181	58.94	Yes	15,441	AF066183	99.9
Ingelvac ATP vaccine	20.7	113,246	22,951	20.27	Yes	15,395	DQ988080	99.9
Fostera PRRS vaccine	12.9	1,726,406	394,423	22.85	Yes	15,379	AF494042*	99.4

* The complete genome sequence of the vaccine 3 is not available in GenBank. But the complete genome sequence of PRRSV P129 strain, the parental virus from which the vaccine 3 was derived, was available (AF494042), and its sequence was compared to our NGS data.

for the mixture with two type 2 isolates (i.e. Ingelvac MLV and VR-2385) which shared 92.4% nucleotide identity at the whole genome level, routine sequence analysis procedures only resulted in one consensus full-length genomic sequence; the MLV vaccine and VR-2385 genomic sequences could not be separately assembled. When the IGV software was used to align all sequence reads against the obtained consensus whole genome sequence, nucleotide variations at some positions of raw sequencing reads were observed, indicating a mixed infection. Because nucleotide variation at each single position could be clearly visualized, and identification of the PRRSV strains in the mixture was known, this enabled to determine which nucleotide corresponded to the MLV vaccine and VR-2385, respectively, at those positions with nucleotide ambiguity. As shown in Fig 1, when the sequence reads obtained from the mixture of Ingelvac MLV and VR-2385 were aligned against the Ingelvac MLV sequences using the IGV software, nucleotides among sequence reads could be visualized at each position. For example, at the positions 2480-2486, since the Ingelvac MLV had sequences 5'-CGAACTC-3', the sequences 5'-TGGGCTT-3' likely belonged to VR-2385 virus.

Full-length PRRSV genomic sequences were successfully obtained from serum samples with Ct values 15.1–23.6, lung samples with Ct values 16.4–21, and oral fluid samples with Ct values 18.7–20.6. When Ct values were ≥24.3 for serum samples, ≥21.3 for lung samples, and ≥22.9 for oral fluids, some sequence contigs could be assembled with gaps or no contigs could be assembled at all; thus full-length sequences could not be obtained under either situation under the conditions of this study.

This study revealed a few important aspects in regards to use of NGS for PRRSV whole genome sequencing:

1. NGS procedures established in this study can reliably determine the whole genome sequences of PRRSV in the form of cell culture-derived isolates.
2. NGS can detect the presence of more than one strain of PRRSV in a sample but may not always distinguish and obtain genome sequences of different strains in the same sample. NGS outcomes may depend on the degree of heterogeneity of the respective strains of PRRSV in the sample. NGS together with some sequence analysis software can be used to identify dual PRRSV infections when the sequence of one of the viruses in the sample is known (e.g. if one of the viruses is a vaccine virus). It is important for submitting veterinarians to inform the labs whether PRRSV vaccines are of use when they request NGS.
3. Specimen types (serum, lung, and oral fluid) may not markedly affect the NGS efficiency for determining the whole genome sequences of PRRSV. Instead, the PCR Ct value for PRRSV RNA in a sample is inversely related to the NGS outcomes. In addition, the sample quality and PRRSV RNA integrity could also be potential factors affecting NGS outcomes on various samples. It is critical to collect and transport samples appropriately for PRRSV NGS.
4. In the present study, 24 samples were multiplexed in a single NGS sequencing reaction. If fewer samples were to be multiplexed per run, the success rate of obtaining the PRRSV whole genome sequences from clinical specimens using MiSeq system could be higher, but the cost per sample would increase accordingly.

Figure 1: Example of alignment of sequence reads, obtained from the mixture of Ingelvac MLV and VR-2385 PRRSV strains, against the Ingelvac MLV sequences using the integrative genomic viewer (IGV) software. Only partial alignments and partial sequence reads are shown. The Ingelvac MLV PRRSV sequences are shown at the bottom as indicated.

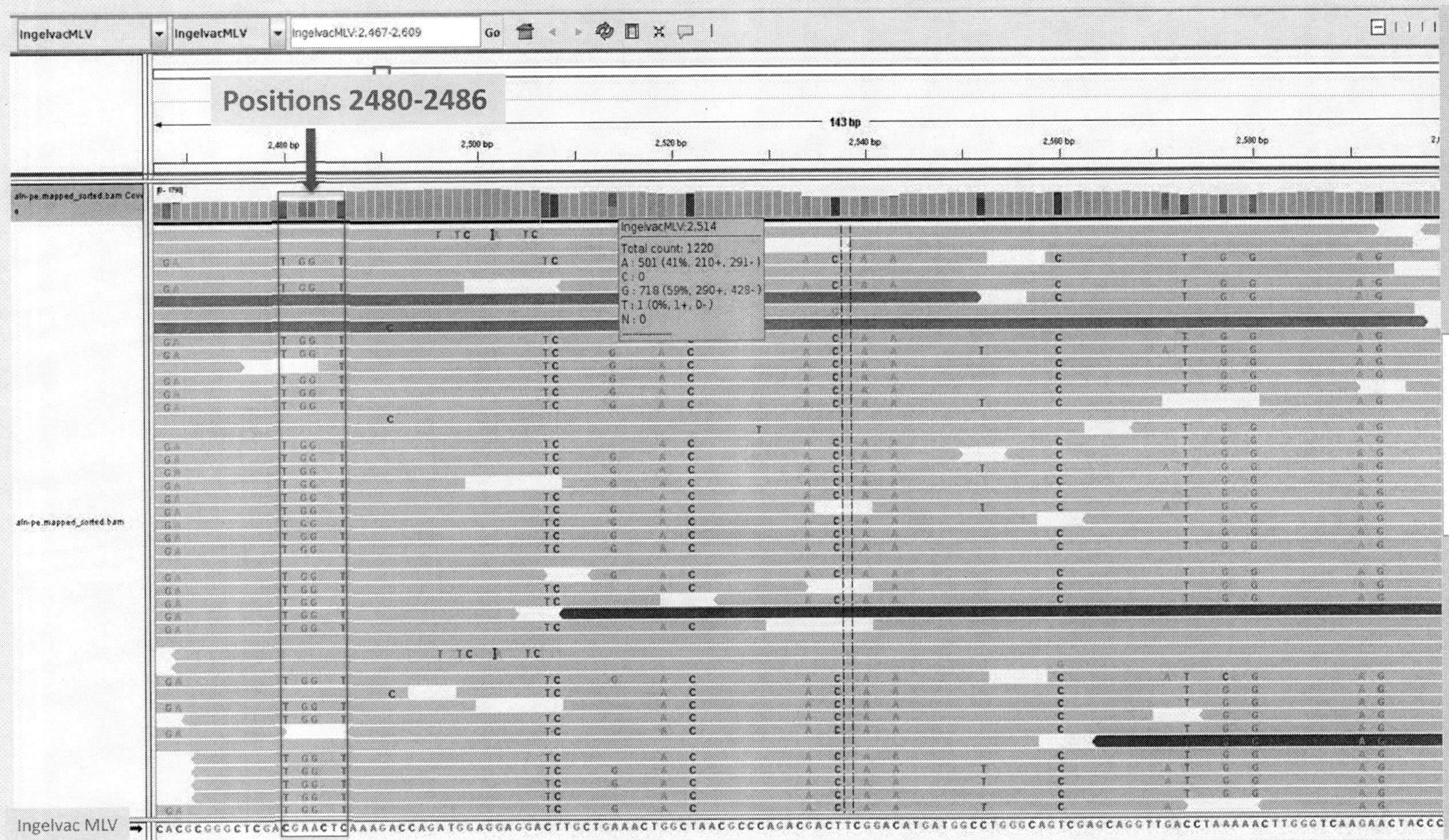

5. Reasonable cost of NGS ($300-$400 per sample, i.e. $0.02-$0.03 per base for PRRSV by NGS in contrast to $0.18 per base for PRRSV by Sanger method) offers the prospect of wider application of NGS for diagnostic investigations and/or whole genome sequencing.
6. The high-throughput NGS approach described in this study could be adapted to routine workflows in veterinary diagnostic laboratories for sequencing various pathogens, although further improvement of sequencing outcomes from clinical specimens with higher Ct values remains to be explored.

Acknowledgement

We thank Drs. Chandrasen Soans, Michael Bishop, and others from Illumina for technical training on next-generation sequencing. This study was supported in part by funds from Zoetis and the Iowa State University Veterinary Diagnostic Laboratory. Illumina provided in-kind support of some reagents.

References

1. Nelsen CJ, Murtaugh MP, and Faaberg KS. (1999). Porcine reproductive and respiratory syndrome virus comparison: divergent evolution on two continents. *J Virol.* 73: 270-280.

2. Snijder EJ, Kikkert M, and Fang Y. (2013). Arterivirus molecular biology and pathogenesis. *J Gen Virol.* 94: 2141–2163.

3. Key KF, Haqshenas G, Guenette DK, Swenson S, Toth T, and Meng XJ. (2001). Genetic variation and phylogenetic analyses of the ORF5 gene of acute porcine reproductive and respiratory syndrome virus isolates. *Vet Microbiol.* 83: 249–263.

4. Pirzadeh B, Gagnon CA, and Dea S. (1998). Genomic and antigenic variations of porcine reproductive and respiratory syndrome virus major envelope GP5 glycoprotein. *Can J Vet Res.* 62: 170–177.

5. Metzker ML. (2010). Sequencing technologies – the next generation. *Nat Rev Genet.* 11: 31-46.

6. Pallen MJ. (2014). Diagnostic metagenomics: potential applications to bacterial, viral and parasitic infections. *Parasitology*. 141: 1856-1862.

7. Thorvaldsdottir H, Robinson JT, and Mesirov JP. (2013). Integrative Genomics Viewer (IGV): high-performance genomics data visualization and exploration. *Brief Bioinform* 14: 178-192.

Effects of pain mitigation during piglet castration

R. Park; M. Pairis-Garcia, DVM, PhD; M. Campler, MS, PhD
Department of Animal Science, The Ohio State University, Columbus, Ohio

Introduction

In the United States, physical castration of male piglets is a routine practice conducted on-farm in order to eliminate boar taint; an abnormal odor associated with meat of sexually mature pigs that have not been castrated. With a high consumer sensitivity to this odor and poor meat quality associated with such cuts, much work has been done in order to minimize products with boar taint and prevent such products from reaching supermarket shelves.[1] In the United States, the primary method of controlling boar taint is castration of male piglets within the first week of life.

Castration is a surgical procedure performed on farm that results in physiological and behavioral indicators of pain. Although this procedure is recognized as painful, there are no current legislation or policies requiring the administration of pain relief for male piglets during castration. However, with a growing consumer interest in on-farm practices and concerns related to the welfare of livestock, work has begun to identify effective and realistic pain management protocols to be implemented on farm. Previous work has indicated that piglets receiving either analgesics e.g. meloxicam prior to castration (10 to 30 minutes,[2] 15 minutes)[3,4] or local anesthetics demonstrate a reduction in some pain-related behaviors and responses[5] although the level in which pain relief occurs is limited.[6,7] Recognizing these limitations, further research is needed to evaluate how to optimize analgesic efficacy by assessing optimal dosing and administration route protocols in order to identify practical management strategies to mitigate pain.

Therefore, the objective of this study was to evaluate the effect of four treatments (Meloxicam PO, Meloxicam PO and Lidocaine, Lidocaine, Control) on the behavior of piglets undergoing castration. This research has the potential to reduce or eliminate pain behavior exhibited by piglets during and after castration.

Materials and methods

One hundred eleven male DNA piglets located at The Ohio State University Don Scott Swine Research Center were enrolled in the study. Litters were selected at random prior to farrowing and individual piglets were allocated to one of four treatments: 1) No pain mitigation (control, n = 25), 2) NSAID only (Meloxidyl "meloxicam", 1.0mg/kg PO; n = 28; Ceva Animal Health, LLC, Lenexa, KS) 3) Local anesthetic (Barrier Wound Care; lidocaine, n = 30; Aurora Pharmaceutical, LLC, Northfield, MN) or 4) NSAID and local anesthetic (meloxicam 1.0 mg/kg PO and lidocaine; n= 28). Both the NSAID and local anesthetic were administered directly prior to castration. The NSAID, meloxicam, was given orally and the local anesthetic, lidocaine, was applied topically through a spray. Behavioral data was recorded using five-minute scan samples for a five-hour period on the day of castration and for two consecutive days following. The behaviors measured were grouped into four categories: active (standing and locomotion), inactive (sitting and lying), suckling and pain-related behaviors (trembling, huddling, prostrating, tail wagging and scratching); (Table 1).

Results and discussion

Data was analyzed using SAS 9.4 software. A generalized linear mixed model, PROC GLIMMIX, was used to determine differences in behavior among treatments. Treatment had no effect on total pain-related behaviors expressed over three days post-castration ($P > 0.05$) with the greatest frequency of pain related behavior occurring on day 2 ($P = 0.003$; Figure 1). Treatment had no effect on active or inactive behavior ($P > 0.05$) but had an effect on suckling behavior with piglets assigned to the NSAID treatment and NSAID + Local treatment displaying the most suckling bouts per hour (Control: 1.71 ± .07 vs. NSAID: 1.97 ± .07 vs. Local anesthetic: 1.82 ± .07 vs. NSAID + Local: 1.94 ± .07, $P < 0.05$; Figure 2).

Administration of the NSAID to castrated piglets resulted in greater suckling bouts per hour compared to other treatments. These results suggest that administering an NSAID at the time of castration resulted in a quicker resumption of natural piglet behavior and may have a positive impact on ensuring piglets continue to suckle during the post-castration period. However, no treatment group decreased total pain-related behaviors post-castration. This may be due to several factors including but not limited to time of drug administration and dosage. Treatments were administered at the time of castration. Although lidocaine will begin to work once it has been absorbed subcutaneously, NSAIDs, like meloxicam, may not achieve significant concentrations systemically until 1 to 2 hours post administration. The lack of a decrease in pain related behaviors may be a result of administrating the drug too late to effectively mitigate the peak pain response which occurs at the time castration is performed. Lastly, the choice to administer meloxicam at 1.0mg/kg were based on previous studies evaluating pain management in sows. Given differences in drug metabolism of young piglets, this dosage may not have provided a great enough concentration of drug to result in decease pain related behavior. Further research evaluating timing of drug administration and dosage of drug is needed to determine if a multimodal drug protocol could effectively mitigate pain associated with castration.

Table 1: Ethogram

Type of behavior	Definition
Standing	Weight supported by all four legs on surface.
Sitting	Weight supported by rump on surface and front legs
Locomotion	Moving using all four legs.
Lying	Any position in which the piglet's weight is supported by their body not limbs.
Sleeping	Lying down, eyes closed.
Suckling	Piglet is holding teat in mouth or making massaging movements directed at udder.
Prostrated	Head of piglet is lower than shoulders. Piglet is in an upright position.
Trembling	Body is shivering. Piglet may be lying, standing or sitting.
Huddled up	Three legs or more tucked under body.
Scratching	Rubbing rump against floor or side of farrowing stall.
Tail wagging	Tail movements in side to side or up and down motion.

Figure 1: Day effect per treatment on pain-related behaviors

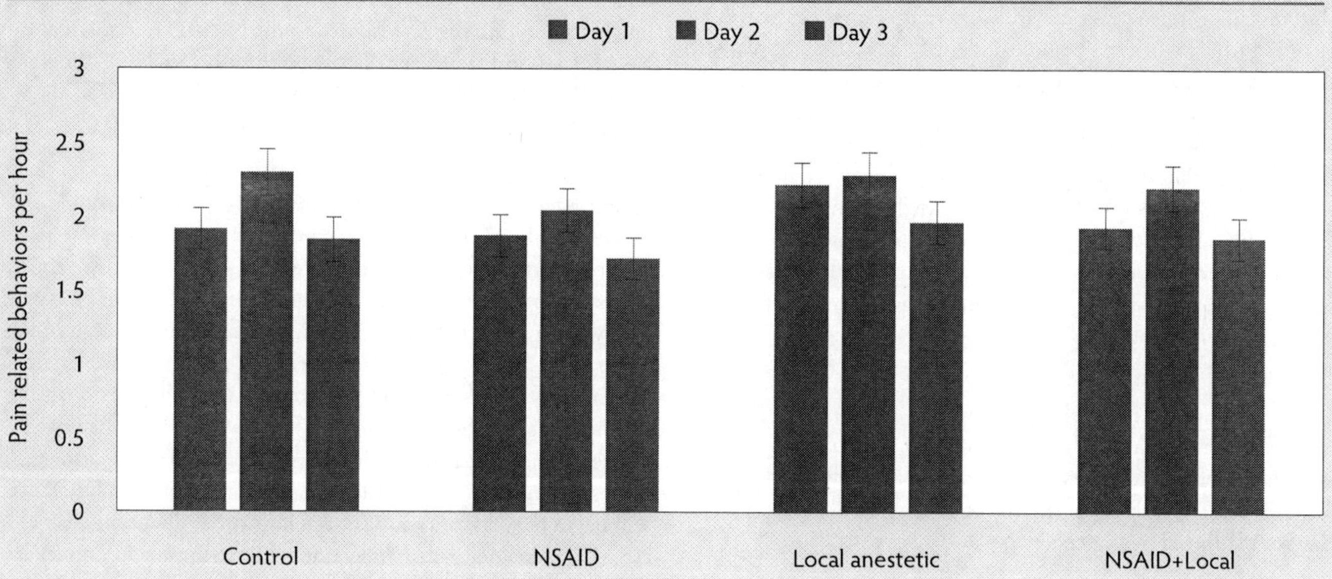

Acknowledgements

I would like to extend my gratitude to swine barn farm managers at The Ohio State University for their assistance during this project. Additionally, I would like to thank Aurora Pharmaceuticals and Vet Rx Pharmacy for their funding and continuous support of this project.

References

1. Corral, S., A. Salvador, M. Flores. 2016. Effect of the use of entire male fat in the production of reduced salt fermented sausages. *Meat Science*. 116:140-150.

2. Keita, A., E. Pagot, A. Prunier, and C. Guidarini. 2010. Pre-emptive meloxicam for postoperative analgesia in piglets undergoing surgical castration. *Veterinary Anesthesia and Analgesia*. 37:367-374.

3. Kliviers-Poodt, M., J.J. Zonderland, J. Verbraak, E. Lambooij, and L.J. Hellebrekers. 2013. Pain behavior after castration of piglets; effect of pain relief with lidocaine and/or meloxicam. *The Animal Consortium*. 7:1158-1162.

4. Von Borell, E., J. Baumgartner, M. Giersing, N. Jäggin, A. Prunier, F.A.M. Tuyttens and S. A. Edwards. 2009. Animal welfare implications of surgical castration and its alternative in pigs. *The Animal Consortium*. 3:1488-1496.

5. Hansson, M., N. Lundeheim, G. Nyman, and G. Johansson. 2011. Effect of local anesthesia and/or analgesia on pain responses induced by piglet castration. *Acta Veterinaria Scandinavica*. 53:34.

6. Herskin, M.S., P. Di Giminiani, K. Thodberg. 2016. Effects of administration of a local anesthetic and/or an NSAID and of docking length on the behavior of piglets 5 h after tail docking. *Research in Veterinary Science*. 108: 60-67.

7. Sutherland, M.A., B.L. Davis, T.A. Brooks and J.J. McGlone. 2010. Physiology and behavior of pigs before and after castration: effects of two topical anesthetics. *The Animal Consortium*. 4:2071-2079.

Figure 2: Impact of treatment group on suckling behavior

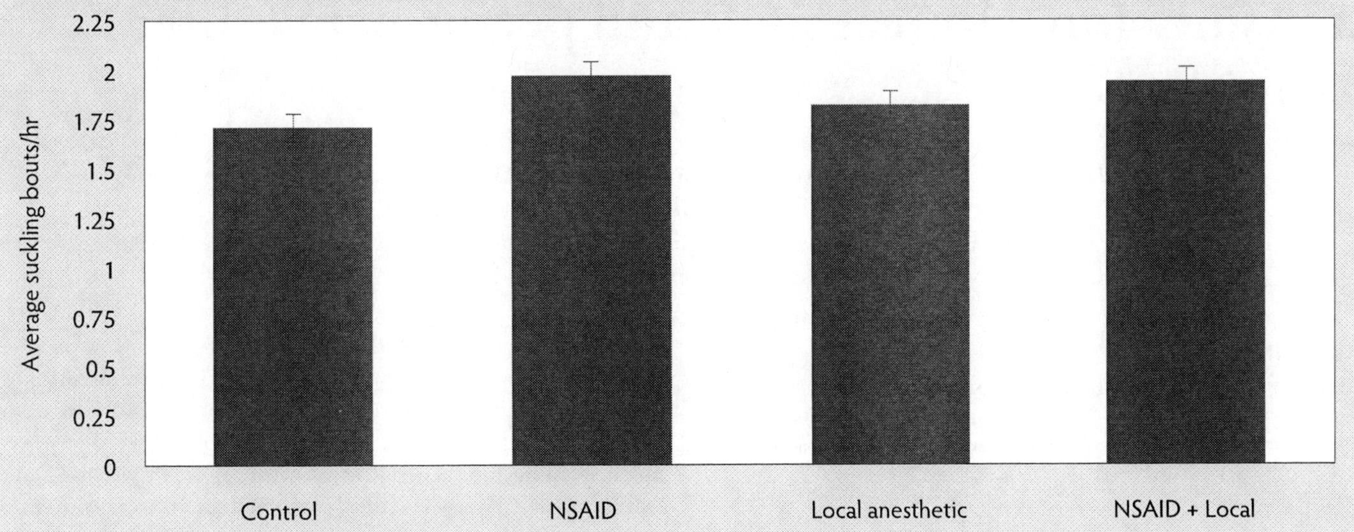

Effect of influenza prevalence at weaning on transmission, clinical signs and performance after weaning

F. Chamba[1], DVM, MS; S. Wayne[2], DVM, PhD; M. Culhane[1], DVM, PhD; A. Perez[1], DVM, PhD; M. Torremorell[1], DVM, PhD
[1]Department of Veterinary Population Medicine, College of Veterinary Medicine, University of Minnesota, Saint Paul, Minnesota; [2]Pipestone Veterinary Services, Pipestone, Minnesota

Introduction

Influenza is an important respiratory disease affecting pigs. Preweaned piglets are responsible for maintaining, diversifying and transmitting the virus to other farms.[1] Sow vaccination is the main control measure to mitigate the impact of influenza infections in pigs.

However, there is a need to understand how prevalence at weaning can impact influenza dynamics after weaning. Therefore, the aim of this study was to evaluate the effect of influenza prevalence at weaning on transmission, clinical signs and performance after weaning.

Materials and methods

A sow farm that tested positive for influenza for 2 consecutive months was selected as the source of weaned pigs. At weaning, pigs were transported to one of two nursery sites for a total of 10 cohorts of pigs weaned throughout a 5-month period. Two pens in each nursery site (1 pen per room) were randomly selected on day 0 of the study and all the pigs within the pen were individually identified. Pigs were nasal swabbed at weaning and every week during 6 weeks. Pigs were weighted at weaning and at week 6 once the study terminated. Additionally, clinical signs of sneezing and coughing were evaluated at weaning and every week in 5 randomly selected pens in each room.

Samples were tested by RT-PCR targeting the matrix gene of influenza. A sample was considered positive if the cycle threshold (ct) value was 35 or less. Moreover, proportion of positive pigs over time, and sneezing and coughing counts were compared between groups based on prevalence at weaning (high, medium and low) using a generalized linear mixed (GLM) model adjusted by farm and season. Finally, average daily weight gain (ADG) was compared between groups based on prevalence at weaning with a GLM model adjusted by farm, season, initial live body weight and days on feed. Statistical analysis was done using SAS 9.4® (SAS Foundation, Cary, NC).

Results

There were 301 pigs included in the study distributed in 10 cohorts of about 30 pigs each. There were 4 cohorts assigned to the high prevalence group, 2 to the medium, and 4 to the low prevalence group. The dynamics of influenza infection were significantly different between prevalence groups at weaning. In the high prevalence group, there were a high proportion (96%) of positive pigs at weaning and then, decreased close to zero by weeks 2 and 3 after weaning. At week 4, there was a peak of 10% of positives and then, all pigs tested negative by weeks 6 and 7 after weaning.

In contrast, in the medium prevalence group, the proportion of positive pigs was 12% at weaning, peaked one week after (60%) and decreased to levels between 5-12% during weeks 2 and 5. There was a second peak of 40% of positives at week 6 after weaning. Lastly, in the group of low prevalence, there were no positive pigs at weaning and no positives were detected throughout the study period after weaning.

Finally, we were not able to detect significant differences when comparing prevalence groups at weaning with sneezing or coughing counts per pen and with average daily weight gain.

Conclusions and discussion

In this study, we found that prevalence at weaning had an impact on the transmission dynamics of influenza after weaning. However, we were not able to detect significant differences on clinical signs and performance between prevalence groups after weaning. Based on our results so far, it is unclear whether reducing influenza prevalence at weaning is enough to mitigate the impact of influenza infection during the post weaning period. Further studies are needed to fully assess how prevalence at weaning impacts productivity in growing-finishing pigs as well as potential implications for public health concerns.

Acknowledgements

This study was funded in part by the National Pork Board (NPB).

References

1. Diaz, A. et al. Association between Influenza A Virus Infection and Pigs Subpopulations in Endemically Infected Breeding Herds. PloS one 10, e0129213, doi:10.1371/journal.pone.0129213 (2015).

Senecavirus A: Overview of experimental studies

A. Buckley[1], DVM; V. Kulshreshtha[1], DVM, PhD; A. van Geelen[1], DVM, PhD; B. Guo[2], PhD; K.J. Yoon[2], DVM, MS, PhD, DACVMP; and K. Lager[3], DVM, PhD

[1]Oak Ridge Institute for Science and Education and National Animal Disease Center, US Department of Agriculture, Ames, Iowa; [2]College of Veterinary Medicine, Iowa State University, Ames, Iowa; [3]National Animal Disease Center, US Department of Agriculture, Ames, Iowa

Introduction

Senecavirus A (SVA) is a picornavirus in the same family as Foot and Mouth Disease virus (FMDV) and swine vesicular disease virus (SVDV). SVA has been associated with rare cases of idiopathic vesicular disease (IVD) in the past. A "mini-epidemic" of IVD in US finisher and sow farms began in late summer 2015, and cases have continued to be diagnosed. SVA has been routinely detected in these cases and considered the presumptive cause, which was confirmed following experimental infections of swine with the early SVA isolates. Understanding the pathogenesis and ecology of this virus is important since SVA induced disease clinical resembles FMDV infection, and SVA may be emerging as an important primary pathogen. This manuscript summarizes SVA animal studies completed at the National Animal Disease Center.

Materials and methods

All animals used in experimental studies were inoculated intranasally with SVA. Virus was isolated from a case involving finishing pigs in South Dakota and subsequent virus was isolated from vesicles of experimentally infected swine.[1] Samples collected included serum, milk, rectal swabs, and oral swabs. Vesicle fluid aspirates as well as vesicle swabs were also collected. Real-time reverse transcriptase polymerase chain reaction (PCR) was performed on samples collected for detection of SVA. Virus neutralization was performed on serum samples for neutralizing antibody detection.

Results and discussion

Reports from a series of animal studies are summarized.

Clinical signs in nursery-age pigs to sows

Coronary band lesions began to develop around 3-5 days post inoculation (dpi) with SVA. Snout lesions were observed less commonly and typically appeared 2-3 days after the appearance of coronary band lesions. SVA was detected in the serum until 7-10 dpi. Fecal and oral swabs have continued to test positive out to 4 weeks with sporadic positive samples beyond that time period. Swabs of vesicular lesions have the lowest CT values and the most genomic copies.

Clinical signs near the time of parturition

Sows inoculated more than 10 days prior to farrowing replicated virus but their piglets were PCR negative for SVA in serum, fecal and oral swabs throughout the study. Piglets born from a sow inoculated 10 days before farrowing were PCR positive for SVA in the serum pre-suckle providing evidence for vertical transmission. There was also evidence of horizontal transmission from sows challenged within a week of farrowing to their piglets. Since virus was present in sow feces and milk samples, they may be sources of viral transmission. No signs of vesicular disease were observed in the sows challenged pre-partum or their piglets. Both sows and piglets challenged within a week after birth replicated virus in serum, fecal and oral swabs. One sow was observed with a snout vesicle, but no other signs of clinical disease were observed in the other sows and their piglets.

Protective immunity

In order to explore protective immunity, nursery pigs were given a homologous challenge of SVA 45 days after immunization and were protected with no replication of the virus or development of clinical signs. In addition, a group of gilts were immunized before breeding and challenged pre-partum. All gilts were protected and no gilt or piglet replicated virus in serum.

Isolate comparison

To study differences between historic and contemporary strains, an isolate comparison study was performed. Interestingly, all isolates produced similar clinical signs and replicated in pigs.

Transmission in sows

Studies looking at SVA transmission have been performed with mature sows. Naïve contact sows placed with infected sows at 7 and 14 dpi have developed lesions and replicated virus. Those naïve sows in contact with principle sows at 21 and 28 dpi did not develop lesions.

Diagnostic recommendations

Determine the question to be answered – Is the premise SVA positive, or is the pig SVA positive? If testing premise, then testing oral fluids by PCR may be the easiest and best use of resources. If testing pig SVA status: testing serum during the first week of infection is good, rectal swabs may be superior because less chance of contamination (like oral fluids) and swine appear to be rectal swab positive for several weeks. One "bench" testing of dust swabs collected from an isolation room demonstrated PCR-positive dust for at least 10 weeks.

Conclusion

Collectively, this body of research will help with the timing of collection of diagnostic samples and provides a better understanding of the pathogenesis and ecology of SVA which may help veterinarians and pork producers develop better control and prevention strategies to manage this virus.

References

1. Guo, B., Pineyro, P.E., Rademacher, C.J., Zheng, Y., Li, G., Yuan, J., Hoang, H., Gauger, P.C., Madson, D.M., Schwartz, K.J., Canning, P.E., Arruda, B.L., Cooper, V.L., Baum, D.H., Linhares, D.C., Main, R.G., Yoon, K.J., 2016. Novel Senecavirus A in Swine with Vesicular Disease, United States, July 2015. Emerging infectious diseases 22, 1325-1327.

Mycoplasma hyorhinis associated with conjunctivitis in pigs

T.P. Resende[1], DVM, MS; M. Pieters[2,3], DVM, PhD; F.A. Vannucci[2], DVM, MS, PhD
[1]Department of Veterinary and Biomedical Sciences, University of Minnesota, Saint Paul, Minnesota;
[2]Veterinary Diagnostic Laboratory, University of Minnesota, Saint Paul, Minnesota;
[3]Veterinary Population Medicine Department, University of Minnesota, Saint Paul, Minnesota

Introduction

Mycoplasma hyorhinis is a commensal bacterium that primarily colonizes the upper respiratory tract of swine. This bacterium has also been reported as causative agent of pneumonia, polyserositis, polyarthritis, and otitis. A variety of infectious agents have been associated with conjunctivitis in pigs, including Classical Swine Fever virus (CSF), African Swine Fever virus (ASF), Pseudorabies virus (PRV), Chlamydia, Porcine Cytomegalovirus (PCMV), Influenza A virus (IAV) and Porcine Reproductive and Respiratory Syndrome virus PRRSv. Mycoplasma-like organisms were demonstrated to be involved in sporadic outbreaks of conjunctivitis in swine.[1,2] However, the significance of Mycoplasmas in the etiology of the conjunctivitis in swine is still unclear. The aim of this study was to report outbreak investigation of conjunctivitis in three swine herds demonstrating *M. hyorhinis* as the potential causative agent of conjunctivitis by *in situ* hybridization (ISH) and PCR and comparing the results from clinically affected and non-affected animals.

Materials and methods

Conjunctivitis was clinically identified in three herds, affecting animals from 8 to 22 weeks of age (Fig.1). Anorexia and failure to thrive were the major concern of veterinarians and producers. Eye swabs and tissues were collected from pigs at early and late stages of the clinical disease. In addition, eye swabs from in-contact pen mates with no clinical signs and from a herd without history of conjunctivitis (age-matched) were also obtained. The diagnostic investigations for potential pathogens were conduct at the University of Minnesota Veterinary Diagnostic Laboratory. Histological examination of the tissues along with viral, bacterial and fungal culture, immunohistochemistry (IHC) and/or PCR were performed for CSFV, ASFV, PRV, Chlamydia, *M. hyopneumoniae*, *M. hyorhinis*, PCMV, IAV and PRRSv.

Results

At the early stage of the clinical disease (9-10 weeks of age), erosive lymphoplasmocytic conjunctivitis associated with lymphofollicular hyperplasia was observed predominantly in the palpebral conjunctiva with minimal or no extension to the ocular conjunctiva (Figure 1). Tissue samples collected when the disease was established (18-20 weeks of age) revealed ulceration associated with lymphoplasmocytic and neutrophilic infiltration distributed throughout the palpebral and ocular conjunctiva, and lymphofollicular hyperplasia. CSFV, ASFV, PRRSV, IAV, PRV and *M. hyopneumoniae* were ruled out by PCR. Samples were negative for *Chlamydia* by culture and capture ELISA. PCMV

Figure 1: *M. hyorhinis* associated conjunctivitis. Early (a) and mid stage (b) clinical signs in unilaterally or bilaterally affected pigs. Serous/mucopurulent ocular discharge was present in different severity in association with palpebral edema; c) edema and infiltration of inflammatory cells in an affected ocular conjunctiva; d) lymphoplasmocytic and neutrophilic infiltration in the palpebral conjunctiva; e) f) *M. hyorhinis* positive signals detected by in situ hybridization.

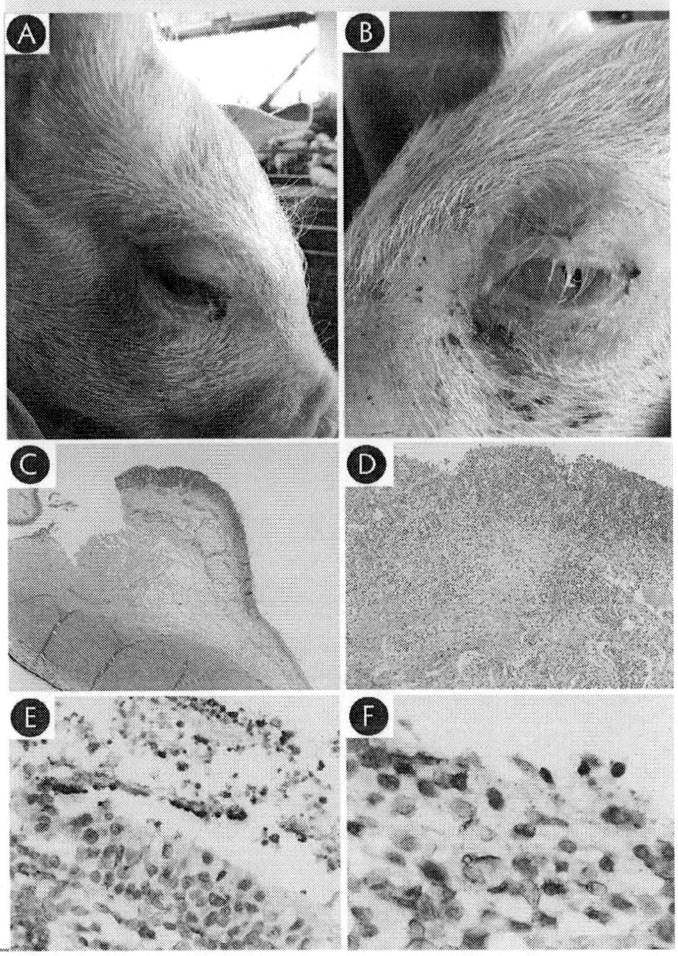

were inconsistently detected by PCR in affected and non-affected animals. Inclusion body rhinitis were not observed. *Staphylococcus chromogens* and *S. aureus* were more frequently cultured in samples from affected animals.

Eye swabs and conjunctival tissues from clinically affected animals and non-affected contact pen mates were positive for *M. hyorhinis* by PCR and culture. PCR results are shown in Table 1. ISH was performed by targeting the 16S RNA gene of *M. hyorhinis*. Positive signals (Fig.1) were observed associated with histological lesions on the conjunctival epithelium, and connective tissue where the lamina propria was exposed due to the ulcerative lesions.

Discussion and conclusion

The present study reported the identification of *M. hyorhinis* in conjunctival lesions. PCR coupled with ISH has demonstrated to be a valuable tool for outbreak investigation, especially when there are commensal bacterial organisms involved and lack of antibodies available for IHC. Friis (1976)[1] has reported the isolation of *M. hyorhinis* from conjunctiva of pigs with and without clinical disease. Based on serological evaluation, the author suggested distinct variant species of *M. hyorhinis*. Comprehensive studies involving whole genome comparison among *M. hyorhinis* strains isolated from clinical cases of conjunctivitis, polyserositis and asymptomatic carriers are warranted to clarify potential genotypic differences associated with a specific disease phenotype.

Table 1: PCR results from sampled pigs.

Pig clinical condition	Ct values
Affected - early stage (n = 11)	21.3 ± 2.4
Affected - late stage (n = 11)	23.5 ± 1.8
Non-affected - contact with affected pen mates (n = 10)	29.7 ± 3.1
Non-affected - non-contact (non-affected herd) (n = 6)	37.5 ± 1.1

References

1. Friis, N.F., 1976. A serologic variant of *Mycoplasma hyorhinis* recovered from the conjunctiva of swine. *Acta Veterinaria Scandinavica*, *17*(3), pp.343-353.

2. Rogers, D.G., Frey, M.L. and Hogg, A., 1991. Conjunctivitis associated with a Mycoplasma-like organism in swine. *Journal of the American Veterinary Medical Association*, *198*(3), pp.450-452.

Mycoplasma hyorhinis and *Mycoplasma hyosynoviae* dual colonization of dams and piglets prior to weaning

Luiza R. Roos, DVM, MS; Maria Pieters, DVM, PhD
Department of Veterinary Population Medicine, College of Veterinary Medicine, University of Minnesota, Saint Paul, Minnesota

Introduction

Mycoplasma hyorhinis and *M. hyosynoviae* are relevant microorganisms that can cause disease in swine. *Mycoplasma hyorhinis* has been established as a cause of polyserositis and arthritis in young pigs,[1] and *M. hyosynoviae* affects pigs in finishing stages also causing arthritis.[2] It has been proposed that *M. hyorhinis* is transmitted from dams to piglets soon after birth,[3] and recent studies suggest that *M. hyosynoviae* tonsillar colonization may also occur in suckling pigs in the lactation room.[4,5] Dual colonization of dams and piglets with *M. hyorhinis* and *M. hyosynoviae* has not been previously reported, although several studies have identified both microorganisms in pigs prior to weaning.[6,7] Therefore, the aim of this study was to characterize the dual colonization of dams and their offspring with *M. hyorhinis* and *M. hyosynoviae* at two points in time during the lactation period.

Materials and methods

Tonsillar swabs were collected from 29 dams of various parities and from 120 piglets randomly selected from their litters at 1 and 3 weeks after farrowing, at a farm with history of lameness in the production system. Tonsillar swabs were tested by species-specific real-time PCR to detect *M. hyorhinis* and *M. hyosynoviae* genetic material, separately. Proportions of dams and piglets with *M. hyorhinis* and *M. hyosynoviae* detection in tonsillar swabs were compared within weeks of sampling using a two-sample test for equality of proportions, and association between *M. hyorhinis* and *M. hyosynoviae* in dams and piglets colonization status was evaluated by Pearson's chi-squared. A comparison of two population proportions was also performed in order to evaluate dam parity association with detection of *M. hyorhinis* and *M. hyosynoviae* in tonsillar swabs at each week of sampling. The number of dams with *M. hyorhinis* and *M. hyosynoviae* detection, *M. hyorhinis* or *M. hyosynoviae* detection only, and no detection at each week of sampling were also evaluated by Pearson's Chi-squared test.

Results

No statistical difference was observed on dam colonization status by *M. hyorhinis* or *M. hyosynoviae*, and no association was observed between *M. hyorhinis* and *M. hyosynoviae* dam colonization status at either week of sampling. Piglet colonization with *M. hyorhinis* was significantly higher at week 3 after farrowing ($P < 0.05$). Piglet colonization with *M. hyorhinis* was also significantly higher than colonization with *M. hyosynoviae* at both weeks of sampling ($P < 0.05$). No association between *M. hyorhinis* and *M. hyosynoviae* piglet colonization status was observed at either week of sampling. Although, no association was observed between parity and *M. hyorhinis* detection one week after farrowing, *M. hyorhinis* detection was significantly higher in tonsillar swabs collected from first parity dams when compared to multiple parity dams ($P < 0.05$) at week 3 after farrowing. No association was observed between *M. hyosynoviae* detection and parity on either week of sampling.

Discussion

Dams were consistently colonized with both *M. hyorhinis* and *M. hyosynoviae*. In contrast, piglets were more colonized at week 3 after weaning with *M. hyorhinis* only. The observed colonization status of piglets might reflect the timing of disease development reported in nursery stage for *M. hyorhinis* and finishing stage for *M. hyosynoviae*. Tonsillar swabs revealed high prevalence of *M. hyorhinis* in dams and piglets, in contrast to what has been previously reported. To our knowledge, this is the first investigation to evaluate *M. hyorhinis* and *M. hyosynoviae* dual colonization of dams and their offspring. The understanding of *M. hyorhinis* and *M. hyosynoviae* dual colonization patterns can aid the development of control strategies for these agents. However, *M. hyorhinis* and *M. hyosynoviae* transmission events, and possible competitiveness between agents have yet to be unveiled.

References

1. Roberts, E.D., Switzer, S.D., Ramsey, F. K. 1963. Am J Vet Res 24, 9–18.
2. Hagedorn-Olsen, T., Basse, A., Jensen, T.K., Nielsen, N. C. 1999. APMIS 107, 201–210.
3. Friis, N. F., Feenstra, A. A. 1994. Acta Vet Scand 35, 93–98.
4. Hagedorn-Olsen, T., Nielsen, N.C., Friis, N.F., Nielsen, J. 1999c. Zentralbl Veterinarmed A 46, 555–564.
5. Schwartz, J., Bruner, L., Evelsizer, B., Konz, B., Rovira, A., Pieters, M. 2014 American Association of Swine Veterinarians (AASV) Annual Meeting (Procedings), 115–116.
6. Ross, R.F., Duncan, J.R. 1970. J Am Vet Med Assoc 157:1515–1518.
7. Clavijo, M. J. 2014. Retrieved from the University of Minnesota Digital Conservancy (`http://hdl.handle.net/11299/167036`).

Efficacy of a commercial G2b-based and two experimental G1b-based porcine epidemic diarrhea virus (PEDV) vaccines against G2b challenge

Tanja Opriessnig[1,2], Dr med vet, PhD; Huigang Shen[2], DVM, PhD; Alessandra Castro[2], DVM; PhD; Qi Chen[2], DVM; PhD; Jianqiang Zhang[2], PhD; Patrick Halbur[2], DVM, MS, PhD
[1]The Roslin Institute, University of Edinburgh, Midlothian, United Kingdom;
[2]Department of Veterinary Diagnostic and Production Animal Medicine, Iowa State University, Ames, Iowa

Introduction

Porcine epidemic diarrhea virus (PEDV) continues to be a threat to pig operations worldwide. Circulating PEDV genotype clusters today include G1a also known as CV777-like (Europe, Asia), G1b also known as INDEL (Asia, Europe, North America), G2a (Asia), and G2b also known as US prototype (North America, Asia) based on spike gene sequencing.[1] While vaccines against PEDV have become available, little is known on cross-protection between genotypes and ideal vaccination routes. In January 2014 the first conditionally licensed PEDV vaccine was introduced to the North American pig market. An RNA particle-based vaccine and an inactivated PEDV vaccine are also available to protect dams and their offspring against PEDV.[2] While the use of these vaccines is often beneficial in previously exposed herds they often fail in naïve herds.[3] The objectives of this study were to determine the ability of an experimental G1b-based live vaccine administered orally or intramuscularly and compare it to a commercial G2b-based inactivated vaccine administered intramuscularly to protect growing pigs against G2b challenge.

Materials and methods

The experimental protocol was approved by the Iowa State University Institutional Animal Care and Use Committee (Approval number: 5-14-7804-S). Thirty-nine PEDV naïve pigs were randomly divided into five groups with 7-8 pigs each: EXP-IM-PEDV (experimental live G1b vaccine intramuscularly; G2b challenge), EXP-ORAL-PEDV (experimental live G1b vaccine orally; G2b challenge), COM-IM-PEDV (commercial inactivated G2b vaccine; Zoetis; intramuscularly; G2b challenge), POS-CONTROL (sham-vaccination; G2b challenge) and NEG-CONTROL (sham-vaccination; sham-challenge). Vaccination was implemented at day post challenge (dpc) -28 and COM-IM-PEDV pigs were revaccinated at dpc -14. Pigs were challenged at dpc 0. A portion of the pigs were necropsied at dpc 3 and the reminder at dpc 14. After challenge pigs were monitored daily. Fecal swabs and serum samples were collected on a regular basis to determine PEDV shedding using a quantitative real-time RT-PCR assay[4] and IgG and IgA antibodies in serum and fecal samples using an *in-house* ELISA.[5,6]

Results and discussion

After vaccination, VAC-G1b-ORAL pigs shed virus for up to 2 weeks without showing clinical signs. These pigs showed a strong IgA response in serum which was significantly higher compared to the other two vaccinated groups. VAC-G2b-IM pigs also seroconverted and had the highest IgG response among all vaccinated pigs. There was no PEDV shedding or evidence of seroconversion before challenge in VAC-G1b-IM pigs indicating that the intramuscular route may not have been sufficient. EXP-ORAL-PEDV also had detectable IgA in feces. Anti-PEDV antibodies were not detected in any of the EXP-IM-PEDV pigs prior to dpc 7. On dpc 5-7 for COM-IM-PEDV pigs and on dpc 10 for EXP-ORAL-PEDV the amounts of fecal PEDV RNA were significantly lower compared to POS-CONTROL pigs. Under the conditions of this study a commercial inactivated G2b-based vaccine protected pigs against G2b challenge as evidenced by reduction of PEDV RNA in feces by 3-4 logs during peak shedding and a shorter duration of virus shedding. The oral, but not the intramuscular, experimental G1b-based live vaccine induced a high anti-PEDV IgA response prior to challenge which apparently did not impact PEDV shedding compared to POS-CONTROL pigs. Under the conditions of this study, a commercial inactivated G2b-based PEDV vaccine administered intramuscularly protected pigs against homologous challenge. In contrast, an experimental G1b-based live virus vaccine given intramuscularly was not protective. The same vaccine given orally induced a high IgA response but the virus shedding pattern after shedding mimicked that of the POS-CONTROL group suggesting limited protection. This perhaps could indicate that induction of a genotype specific humoral and/or cellular immune response may be important for PEDV protection.

References

1. Huang YW, Dickerman AW, Pineyro P, et al. Origin, evolution, and genotyping of emergent porcine epidemic diarrhea virus strains in the United States. MBio 2013;4:e00737-13.

2. Langel SN, Paim FC, Lager KM, Vlasova AN, Saif LJ. Lactogenic immunity and vaccines for porcine epidemic diarrhea virus (PEDV): Historical and current concepts. Virus Res 2016; pii: S0168-1702(16)30215-5.

3. Schwartz TJ, Rademacher CJ, Gimenez-Lirola LG, Sun Y, Zimmerman JJ. Evaluation of the effects of PEDV vaccine on PEDV naive and previously PEDV exposed sows in a challenge model comparing immune response and preweaning mortality. New Orleans, Louisiana, USA. Proc. Of the 47th AASV Annual Meeting 2016; 363-366.

4. Opriessnig T, Xiao CT, Gerber PF, Zhang J, Halbur PG. Porcine epidemic diarrhea virus RNA present in commercial spray-dried porcine plasma is not infectious to naive pigs. PLoS One 2014;9:e104766.

5. Gerber PF, Gong Q, Huang YW, Wang C, Holtkamp D, Opriessnig T. Detection of antibodies against porcine epidemic diarrhea virus in serum and colostrum by indirect ELISA. Vet Jl 2014;202:33-26.

6. Gerber PF, Opriessnig T. Detection of immunoglobulin (Ig) A antibodies against porcine epidemic diarrhea virus (PEDV) in fecal and serum samples. MethodsX 2015;13:368-373.

Serum and mammary secretion antibody responses in PEDV-exposed gilts following PEDV vaccination

J.B. Kraft[1]; K. Woodard[1]; L. Giménez-Lirola[1]; B. Setness[1]; J. Ji[1]; P. Lasley[2]; E. Nelson[3]; J. Zhang[1]; D. Baum[1]; P. Gauger[1]; J. Zimmerman[1]; R. Main[1]
[1]Iowa State University, Ames, Iowa; [2]Smithfield Hog Production, Princeton, Missouri;
[3]South Dakota State University, Brookings, South Dakota

Introduction

Since there is little possibility that PEDV will be eradicated in the near future, we need to identify methods to prevent and/or control its effects. In the sow herd, this implies monitoring/maintaining levels of immunity sufficient to protect neonatal pigs. The objective of this study was to compare gilt responses to vaccination using two commercially available PEDV vaccines, Harrisvaccines™ (Vaccine A) and Zoetis (Vaccine B).

Materials and methods

PEDV antibody-positive gilts (n = 36) in a commercial production system were randomly assigned to one of five vaccination protocols: 1) no vaccine (controls); 2) PEDV vaccine A (2 weeks pre-farrow); 3) PEDV vaccine A (5 and 2 weeks pre-farrow); 4) PEDV vaccine B (2 weeks pre-farrow); and 5) PEDV vaccine B (5 and 2 weeks pre-farrow).

Colostrum, serum, and fecal swab samples were collected within 12 hours of farrowing. Milk samples were collected at 3, 10, and 21 days post farrowing (DPF). Serum, colostrum, and milk samples were tested by PEDV whole virus (WV) IgG and IgA ELISAs and for neutralizing antibody by PEDV fluorescent focus neutralization assay (FFN). Thirty-three gilts completed the study, i.e., farrowed viable litters and provided a full complement of samples.

Statistical analyses were performed using SAS® (Version 9.4). A nonparametric one-way ANOVA was used to test for differences among treatment groups for IgG, IgA, and FFN by sample type (serum, colostrum, milk). A mixed-effects repeated measures model was used to analyze the difference between treatment groups for IgG, IgA, and FFN by sample type. Antibody responses in milk were evaluated by repeated measures analysis using a compound symmetry covariance structure.

Results

Gilt serum antibody levels (IgG, IgA, FFN) at 5 weeks pre-farrow, i.e., pre-vaccination, were not different. This provided evidence that any differences observed were due to vaccination. In gilts with prior exposure to PEDV, vaccination significantly ($P < 0.05$) increased IgG, IgA, and neutralizing antibody levels in serum, colostrum, and milk samples (3 DPF). A comparison of antibody responses showed gilts receiving Vaccine B had higher IgG levels in serum and colostrum compared to Vaccine A. Antibody responses in milk declined significantly between 3 and 10 DPF in vaccinated animals; however, no significant decline was evident in the control animals.

Conclusions

A strong anamnestic response to vaccination was observed in animals previously infected with PEDV. This suggested that currently available vaccines can be an important part of PEDV management in endemically-infected sow herds.

Modeling the transboundary survival of foreign animal disease pathogens in contaminated feed ingredients

Scott Dee[1]; Gordon Spronk[1]; Eric Nelson[2]; Diego Diel[2]; Travis Clement[2]; Aaron Singrey[2]; Fernando Bauermann[2]; Michele Mucciante[2]; Jane Hennings[2]; Cassandra Jones[3]; Roger Cochrane[3]; Gilbert Patterson[4]

[1]Pipestone Applied Research, Pipestone Veterinary Services, Pipestone, Minnesota; [2]Animal Disease Research and Diagnostic Laboratory, Department of Veterinary and Biomedical Sciences, South Dakota State University, Brookings, South Dakota; [3]Department of Grain Science, Kansas State University, Manhattan, Kansas; [4]Department of Veterinary Public Health & Preventative Medicine, University of Minnesota, Saint Paul, Minnesota

Project objectives

1. To model if foreign animal diseases could survive in feed ingredients shipped from Asia to the USA.
2. Evaluate whether 2 chemical mitigants could reduce risk.

Hypothesis

Pathogen survival will be influenced by ingredient and treatment.

Materials and methods

Based on the Swine Health Information Center pathogen matrix, 10 FAD viral pathogens were identified as significant risks to the US swine industry. Due to the inability to work with these actual agents, we used "surrogate viruses", which allowed us to study closely related and structurally similar viruses, but not the actual FAD pathogens. The designated FAD and the selected surrogate were as follows: FMDV (Seneca Virus A), CSFV (Bovine Virus Diarrhea Virus), PRV (Bovine HerpesVirus-1), ASFV (Vaccinia virus), Nipah virus (Canine Distemper Virus), Swine Vesicular Disease Virus (Bovine Enterovirus) and Vesicular Exanthema Virus (Feline Calici Virus). Other selected pathogens (PRRSV 174, PCV2 and Vesicular Stomatitis Virus) did not require surrogates. Using a model previously validated to study the risk of contaminated feed ingredients for the transboundary spread of PEDV,[1] we selected feed ingredients known to be imported from China to the USA based on the US Govt Harmonized Tariff Schedule (`hs.usitic.gov`). These included organic & conventional soybean meal, soy oil cake, DDGS, lysine, choline, vitamin D, pork sausage casings, and several pet foods (dry & moist). Ingredients were inoculated with representative surrogates (5g ingredient and 100 *uL* virus), with the level of viral contamination in feed based on previous observations from PEDV infection in the field.[2] Controls consisted of complete feed inoculated with surrogate or saline (negative control) as well as stock virus alone (positive control) in the absence of feed matrix. The design involved non-treated control ingredients, along with 2 mitigants: SalCURB-treated ingredients and MCFA-treated ingredients. These samples were then incubated in an environmental chamber for 37 days programmed using actual T and % RH data recorded during a journey from China to the US (Beijing to Shanghai to San Francisco to Des Moines) in December 2012 through January 2013 (SeaRates.com). Samples were tested by PCR, VI and bioassay for porcine surrogates or on primary cells for surrogates of non-porcine origin at 2 DPI (Beijing), 8 DPI (Shanghai), 25 DPI (San Francisco) and 37 DPI (Des Moines) to represent specific points in the model.

Results

As of this writing, testing of the FMDV, CSFV and PRV surrogates has been completed. Preliminary data indicate the survival of the FMDV surrogate (SVA) and the PRV surrogate (BHV-1) at all points during the 37 day shipping period from China and into the US. Both surrogates survived in conventional soybean meal and soy oil cake, while SVA also survived in lysine, pet food, Vitamin D, complete feed and casings. Both positive controls (SVA and BHV-1 stock virus) did not survive. In contrast, the CSFV surrogate (BVDV) appeared to be less stable and did not survive the 37-day journey, independent of ingredient. It did, however, survive until the samples theoretically entered the port of San Francisco (25 DPI) in conventional soybean meal and moist dog food. Of particular interest was the inability of the positive control (stock virus, no feed matrix) to survive the journey in all cases. Viruses (surrogates) currently under evaluation include African Swine Fever Virus (Vaccinia), Vesicular Stomatitis Virus, Vesicular Exanthema Virus (Feline Calicivirus) and Nipah Virus (canine distemper virus). Viruses targeted for Qt 1 2017 include PRRSV 174, PCV2 and Swine Vesicular Disease (bovine enterovirus). These data will be available for presentation at the AASV annual meeting.

Discussion

Under the conditions of this study, these preliminary results suggest that a subset of contaminated feed ingredients could serve as vehicles for FAD introduction to the US, supporting our previous results which focused on PEDV. Of particular interest is the consistency of virus survival in soy-based products as well as the inability of virus alone (stock) to survive in the absence of a feed matrix. Strengths of the study include experimental design and the consistent data across studies, while limitations include the use of surrogates in a model system, along with a lack of knowledge regarding the quantity and frequency of feed-level contamination, outside that of PEDV.

References

1. Dee S, et al Modeling the transboundary risk of feed ingredients contaminated with porcine epidemic diarrhea virus *BMC Vet Res,* 2016, 12:51.

2. Dee S, et al. An evaluation of contaminated complete feed as a vehicle for porcine epidemic diarrhea virus infection of naïve pigs following consumption via natural feeding behavior: Proof of concept. *BMC Vet Res* 2014,10:176.

Evaluation of a peroxygen-based disinfectant to inactivate porcine epidemic diarrhea virus in swine feces on metal surfaces under freezing conditions

K. Baker[1], BS; C. Mowrer[1], BS; J. Zhang[1], MD, PhD, MS; A. Ramirez[1], DVM, MPH, PhD, DACVPM; L. Karriker[2], DVM, MS, DACVPM; D. Holtkamp[1], DVM, MS

[1]Iowa State University College of Veterinary Medicine, Ames, Iowa;

[2]Swine Medicine Education Center, Iowa State University, Ames, Iowa

Introduction

Since its emergence in 2013, porcine epidemic diarrhea virus (PEDV) spread rapidly throughout the country due, in part, to contaminated livestock trailers. Disinfection is an important part of procedures for decontaminating trailers. The evaluation of the efficacy of disinfectants against PEDV under field conditions is essential for PEDV prevention. The objective of this study was to test a peroxygen-based disinfectant for the ability to inactivate PEDV in the presence of swine feces on aluminum surfaces at 4°C or -10°C.

Materials and methods

Forty 3-week-old individually housed barrows were used as a bioassay to determine the infectivity of PEDV after treatment with either a 1:100 or 1:600 dilution of a peroxygen-based disinfectant (Virkon™ S, Lanxess, Wilmington, DE). The peroxygen-based disinfectant was allowed 10 or 30 minutes of contact time at either 4°C or -10°C. The 1:600 dilution was not evaluated at -10°C for 10 minutes due to facility constraints, resulting in seven disinfectant treatment groups. A negative control and 2 positive control groups (4°C and -10°C) were evaluated. Disinfectant mixtures applied at -10°C were 10% propylene glycol by volume. One 15.24 × 15.24 × 2.54cm coupon, constructed of diamond plate aluminum flooring, randomly matched to one pig was the experimental unit. Each treatment group consisted of 4 replicates, 4 coupons and 4 pigs per treatment. Coupons were contaminated with 2 mL of PEDV-positive feces, PEDV-negative for the negative control. PEDV-positive feces were created by spiking 2 mL of PEDV-negative feces with 1 mL of PEDV isolate USA/IN19338/2013 that had an infectious titer of 3.9×10^6 $TCID_{50}$/mL. An environmental swab was obtained from the coupons immediately following contamination. The contaminated coupons were then subjected to their randomly allotted treatment group. Following treatment, the solution of feces and disinfectant remaining in the coupons post-treatment was re-collected and used to inoculate the pigs via oral gastric gavage. A sample of inoculum was collected post-treatment and tested for infectious PEDV via virus isolation at the Iowa State University Veterinary Diagnostic Laboratory (ISU VDL). Rectal swabs were collected from each pig 3 and 7 days post-inoculation (DPI) and were tested for PEDV by PCR at the ISU VDL.

Results

Environmental samples from all coupons, except the negative control, were PCR positive for PEDV both before and after treatment. All rectal swabs from the pigs in the negative control and the 7 disinfectant treatment groups were PCR negative for PEDV on both 3 and 7 DPI. All pigs in the positive control group conducted at 4°C and 3 of 4 pigs in the positive control group conducted at -10°C were PCR positive for PEDV on both 3 and 7 DPI.

Discussion

Both the 1:100 and 1:600 dilution of the peroxygen-based disinfectant inactivated PEDV in 2mL of swine feces at -10°C with short contact times providing evidence that it may be an effective option for trailer sanitation and decontamination in the winter and remains effective in the presence of low levels of feces. This study also demonstrates that PEDV positive PCR results on environmental swabs should be expected even after 30 minutes of contact time with the peroxygen-based disinfectant. A sample of the inoculum given to the positive control pig that did not test positive for PEDV, was PEDV positive by virus isolation indicating that the lack of infection was due to an ineffectual gavage.

Calculating and evaluating antibiotic usage in a United States swine production system

Megan Pieters[1]; Ross Kiehne[2], DVM; Eric Christianson[3], DVM
[1]Iowa State University, Ames, Iowa; [2]Swine Vet Center, Saint Peter, Minnesota;
[3]Elanco Animal Health, Greenfield, Indiana

Introduction

Information about antibiotic usage in production animal agriculture is on the top of mind for veterinarians, producers, consumers and most importantly retailers and food marketers. In 2015, a United States House Representative introduced the Delivering Antimicrobial Transparency in Animals Act of 2015 (H.R. 2459) to the House of Representatives, which if passed, would federally mandate tracking all antibiotics.[1] Today, the demand for knowledge on antibiotic usage continues with the release of FDA's Guidance for the Industry 152, 209, and 213.[2,3,4] Thus, the objective of this project was to quantify the antibiotic usage, based on Denmark's Danish Integrated Antimicrobial Resistance Monitoring and Research Program (DANMAP) for a swine production system in the United States and to understand the opportunities and limitations it provides.[5]

Material and methods

Data from 2015 and 2016 closeout groups for both nursery and finishing stages of production was collected from MetaFarms (with producer consent); accounting for 918,050 pigs and 21 different antibiotics for feed-grade, water soluble, and injectable. Antibiotics were categorized based on those formed in FDA Guidance 152 as shared class (medically important or not medically important) and animal use only antibiotics.[3] Due to varying concentrations between products (g/lb for feed-grade antibiotics, or g/bottle or packet for injectable and water soluble products), all antibiotics were standardized to grams of active ingredient.

Results

Of the total grams of antibiotics used, 57.3% of were used in finishing, and 75.9% of total antibiotics were feed-grade for both finishing and nursery. An antibiotic usage report was completed by antibiotic category and route of administration found grams of active ingredient per pound of live weight were 0.00071 gram/lb and 0.00081 g/lb for shared class antibiotics (0.00065 g/lb and 0.00079 g/lb for medically important and 0.000057 and .0000033 g/lb for not medically important) and 0.000063 g/lb and 0.000113 g/lb for animal use only for both 2015 and 2016 respectively (Figure 1). Grams of active ingredient per pig marketed for all categories and routes of administration were 41.16 grams/pig for 2015 and 38.39 grams/pig for 2016 (Figure 2). Mortality percentage as pigs in closeout groups show no statistical significance between grams of total active antibiotic ingredients (all categories and all routes of administration) per

Figure 1: Grams of active ingredients of antibiotics per pound of live weight for 2015 and 2016 nursery and finisher closeout groups.

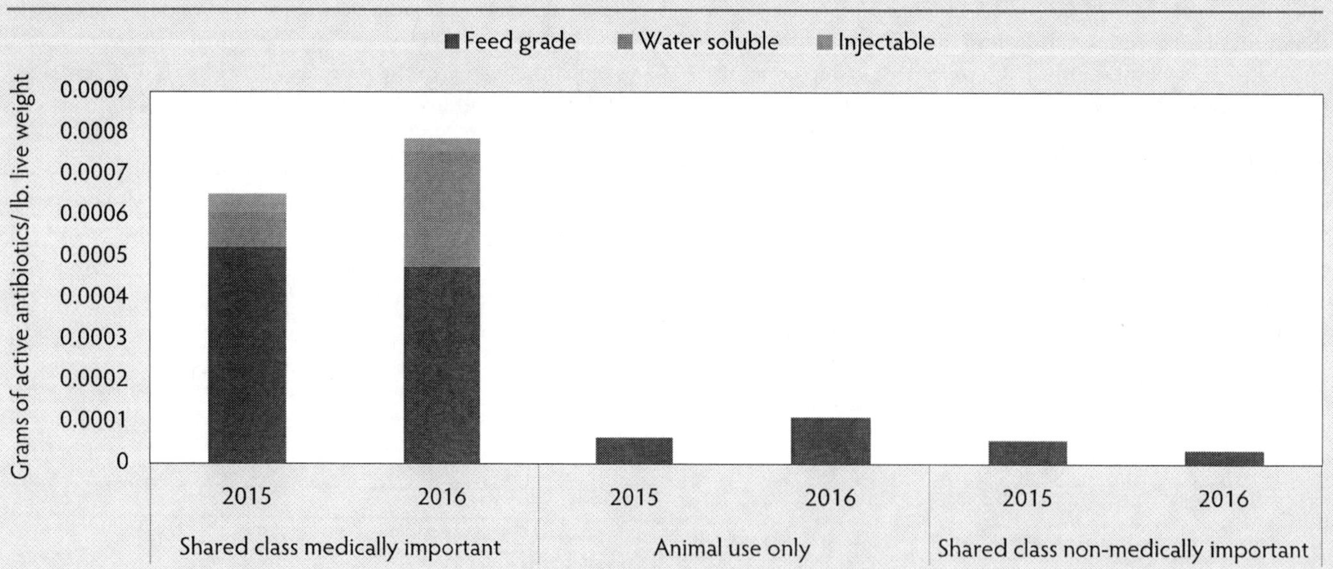

pigs for 2015 and 2016 closeout groups for nursey and finisher (Figure 3). Assigned Health status accounting for porcine reproductive and respiratory syndrome virus, porcine epidemic diarrhea virus, and *Mycoplasma* sp., was analyzed and showed no statistical difference with total grams of active ingredients of antibiotics consumed and presumed health status.

Discussion

Results from this project cannot be compared to another company's data, unless there is complete certainty each antibiotic metric is calculated the exact same. Although mortality percentage shows no statistical correlation between grams of total active antibiotic ingredients/ mortality percentage as pigs in, it can be noted there is low variation of mortality percentage in this across the production system for both 2015 and 2016. (Figure 3) It is important to consider what is reflected in the presumed health status comparison is not representative of other diseases occurring in the barn, thus further analysis of additional disease events occurring in the barns needs to be evaluated. It is also important

Figure 2: Grams of active ingredient of antibiotics per pig closed out in 2015 and 2016 nursery and finisher groups.

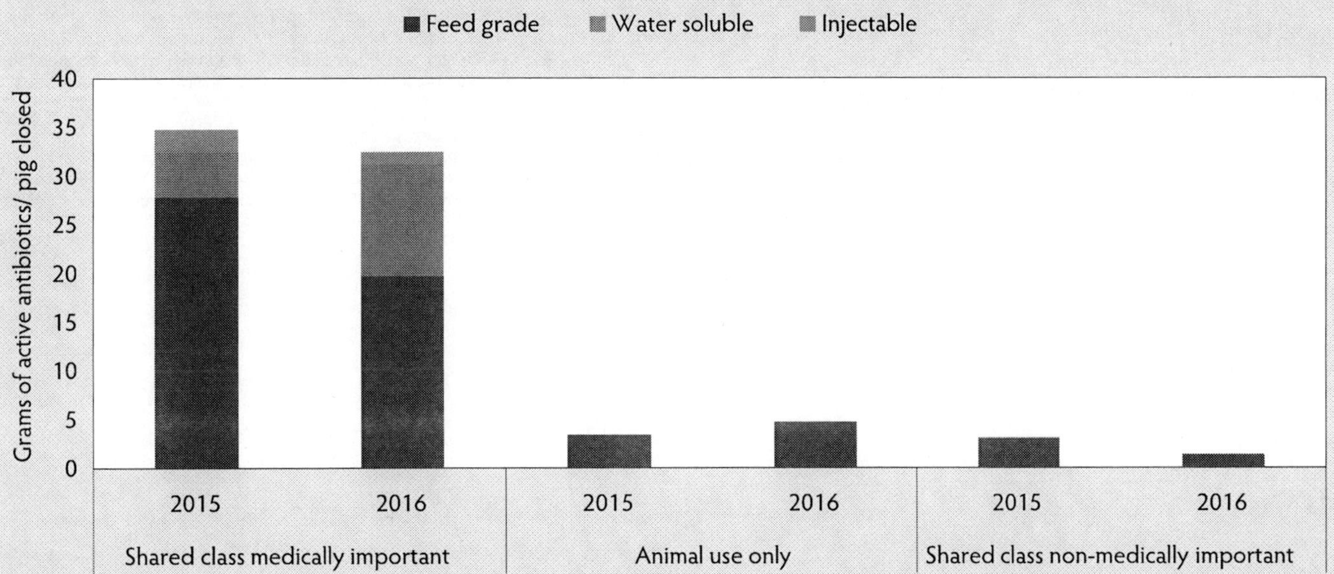

Figure 3: Grams of active ingredient of antibioitc per pig for 2015 and 2016 closeout nursery and finisher groups.

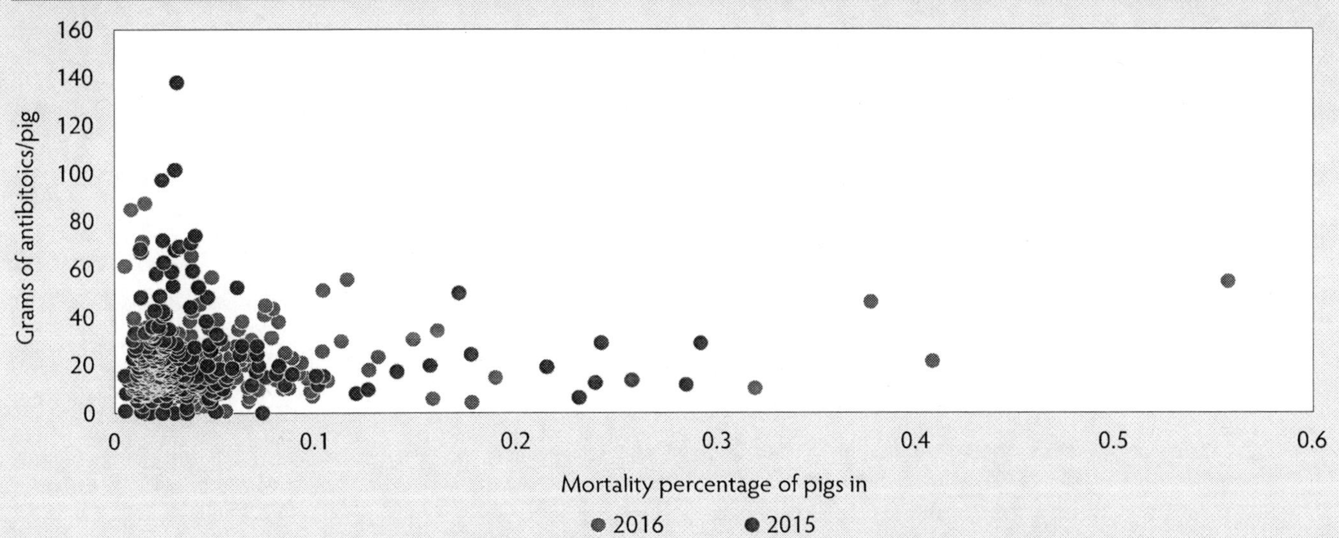

to consider animal welfare implications when making management and marketing decisions based on the results of this metric. Limitations to calculating and analyzing a system's antibiotic usage are based on accuracy and completeness of a production system's records, a lack of agreeance between production systems on how to calculate antibiotic usage, and assumptions within each calculation including each pig consuming/receiving the same amount of antibiotic, assuming on-label dosages, and factors altering potency of a drug. Along with the antibiotic usage number, written criteria needs to be established ensuring judicious use of antibiotics, selection of animal use only over medically important products when available, and veterinary oversight over all medications. Therefore, a consistent antibiotic tracking system (i.e. grams of active ingredient per pound, grams of active ingredient per pig, Defined Daily Dose, and etc.) along with written criteria needs to be established prior to comparison between production systems and public release due to the potential industry impact.

References

1. Health Subcommittee in the Energy and Commerce Committee (US). Delivering Antimicrobial Transparency in Animals Act of 2015. 2013. United States House of Representatives. Available at: https://www.congress.gov/bill/113th-congress/house-bill/820. Accessed 1 September 2016.

2. Food and Drug Administration (US). Evaluating the Safety of Antimicrobial New Animal Drugs with Regard to Their Microbiological Effects on Bacteria of Human Health Concern. Guidance for Industry 152. 2003. United States Department of Health and Human Services. Available at: http://www.fda.gov/downloads/AnimalVeterinary/GuidanceComplianceEnforcement/GuidanceforIndustry/ucm052519.pdf. Accessed 01 September 2016.

3. Food and Drug Administration (US). The Judicious use of Medically Important Antimicrobial Drugs in Food-Producing Animals. Guidance for industry 209. 2012. United States Department of Health and Human Services. Available at: http://www.fda.gov/downloads/AnimalVeterinary/GuidanceComplianceEnforcement/GuidanceforIndustry/UCM216936.pdf. Accessed 01 September 2016

4. Food and Drug Administration (US). New Animal Drugs and New Animal Drug Combination Products Administered in or on Medicated Feed or Drinking Water of Food-Producing Animals: Recommendations for Drug Sponsors for Voluntarily Aligning Product Use Conditions with GFI #209. Guidance for industry 213. 2013. Available at: http://www.fda.gov/downloads/AnimalVeterinary/GuidanceComplianceEnforcement/GuidanceforIndustry/UCM299624.pdf. Accessed 01 September 2016.

5. National Veterinary Institute. Use of Antimicrobial Agents and Occurrence of Antimicrobial Resistance in Bacteria from Food Animals, Food and Humans in Denmark. 2014. Available at: http://www.danmap.org/~/media/Projekt%20sites/Danmap/DANMAP%20reports/DANMAP%202014/Danmap_2014.ashx. Accessed 01 September 2016.

Assessment of knowledge, compliance, and attitudes of English and Spanish speaking farm employees towards biosecurity practices

Michael Mardesen[1], BS; Jose Angulo[2], DVM; Clayton Johnson[3], DVM; Daniel Linhares[1], DVM, MPH, PhD, Diplomate ACVPM; Derald Holtkamp[1], DVM, MS; Alejandro Ramirez[1], DVM, MPH, PhD, Diplomate ACVPM
[1]Iowa State University CVM; [2]Zoetis; [3]Carthage Veterinary Service

Introduction

Biosecurity is a key component to any herd health program in terms of limiting the spread and transmission of disease from animal to animal and herd to herd. For biosecurity to be effective, it is crucial that all employees follow all protocols in all stages of farm operations and for them to have good knowledge, compliance and attitudes of the protocols. Breakdowns in biosecurity can often be attributed to a failure of one of these factors. In human healthcare settings, studies have been done to assess these factors with varying results. Knowledge has been shown to be mixed, [1-2]attitudes have been high.[2] and all in all compliance is good as well, with most reporting to be as good as peers.[3] Many studies been conducted in human health settings, little has been done to assess attitudes and compliance in a farm setting. The purpose of this study was to assess knowledge, compliance, and attitudes of English and Spanish speaking farm employees towards biosecurity Standard Operating Procedures (SOPs).

Materials and methods

For this study, 14 breed-to-wean swine farms in the Carthage Professional Swine Management System having a minimum of 3 English speaking employees and 3 Spanish speaking employees were selected. A survey instrument was administered to each employee individually by either an English or Spanish speaking proctor according to the employee's first language. The survey contained 8 knowledge based questions about farm biosecurity SOPs (entering the premises, dead disposal, etc) with 2 additional questions for filtered farms. Initially an open ended knowledge question was asked that required participants to mention all key points of the practice in question to receive credit on primary attempt. If participants failed to address all points, leading questions were asked to allow participants to receive points in a secondary attempt. For selected points, questions were asked to assess self vs peer compliance and overall importance to farm biosecurity using a likert-scale ranging from 1-5. Multiple choice questions were included to address responses to failure in biosecurity and time and effort required to comply with the practice. In total, 73 employees (55 English speaking, 18 Spanish speaking) were surveyed.

Results

Overall, compliance and attitudes were similar across farms. Most biosecurity practices were considered by employees as highly important procedures, and averages for all self and peer reported compliance were over 4 (Almost Always). For many aspects there was a difference in reported compliance for self vs peer, with self-reported compliance almost always being higher than peer reported (Table 1). In the knowledge portion, both language groups received a total score (primary attempt correct + secondary attempt correct) of around 90%. Spanish speaking employees answered more questions on the primary attempt (65.9%) than English speaking employees (42.5%). Farm managers scored higher on primary attempt (60.6%) than technicians (49.8%). In addition, employees who have been at the farm for over 36 months also scored higher than employees who have been there for 12 months or less (59.3% vs 48%) (Table 2).

Discussion

Surveys such as these can serve as a valuable model to help identify gaps in protocols and training programs for employees. The results from this study identified which groups (technicians, less than 12 months on the farm) training should be focused toward based on the knowledge scores. By understanding where the gaps in biosecurity are, production systems can focus training, reduce training time and costs, and improve interventions in deficient areas. Systems can also better identify if problems are knowledge, compliance, or attitude based, and respond accordingly by adjusting training protocols, worker mentality, or management. The main limitations of this study was the small sample group and the different proctors for different language groups. Scores could have potentially been influenced by the administration of the different proctors between the different language groups. However, the survey instrument was constructed in such a manner as to limit this variability and proctors were trained as well. Having participants report compliance vs observing compliance may have also inflated actual compliance numbers. This limitation was accounted for by having participants report peer compliance as well as self-compliance.

References

1. Paudyal Priyamvada, Simkhada Padam, Bruce Julie. "Infection Control Knowledge, Attitude, and Practice among Nepalese Health Care Workers." *American Journal of Infection Control* 36.8 (2008): 595-97.

2. Stein A.D., Makarawo T.P., Ahmad M.F.R. "A Survey of Doctors' and Nurses' Knowledge, Attitudes and Compliance with Infection Control Guidelines in Birmingham Teaching Hospitals." *Journal of Hospital Infection* 54.1 (2003): 68-73.

3. Berhe Mesgebe, Edmond Michael B., Bearman Gonzalo M.L. "Practices and an Assessment of Health Care Workers' Perceptions of Compliance with Infection Control Knowledge of Nosocomial Infections." *American Journal of Infection Control* 33.1 (2005): 55-57.

Table 1: Self versus peer compliance score

Biosecurity protocol, all farms (n =146)	Mean(Std. Dev) Self-reported score[1]	Peer reported score[2]	Difference
Farm Entry Protocol: Wears plastic booties after exiting vehicle.	4.93[a](.3)	4.6[b](.66)	.33
Farm Entry Protocol: Properly uses bench/coat room.	4.9[a](.16)	4.71[b](.61)	.26
Farm Entry Protocol: Proper shower protocol.	4.96[a](.2)	4.82[b](.48)	.14
Farm Entry Protocol: Has adequate downtime if contacted pigs.	4.97[a](.12)	4.73[b](.56)	.24
Supplies Entry Protocol: Properly loads fog chamber.	4.78[a](.65)	4.44[a](.82)	.34
Supplies Entry Protocol: Runs fogger for appropriate amount of time.	4.88[a](.5)	4.7[a](.59)	.18
Lunch Entry Protocol: Proper use of clean/dirty table.	4.92[a](.36)	4.6[b](.64)	.32
Lunch Entry Protocol: Properly loads UV chamber.	4.82[a](.51)	4.41[b](.81)	.41
Semen Entry Protocol: Removes outer packaging.	5.00[a](0)	4.88[b](.41)	.12
Dead Disposal Protocol: Performs task at proper time.	4.92[a](.28)	4.9[a](.3)	.02
Dead Disposal Protocol: Does not re-enter farm.	5.00[a](0)	4.95[a](.28)	.05
Dead Disposal Protocol: Completely covers in compost.	4.97[a](.16)	4.59[b](.62)	.38
Rodent Control Protocol: Checks and fills at proper interval.	4.84[a](.62)	4.51[b](.78)	.33
Feed Spill Protocol: Cleans spill in prompt manner.	4.84[a](.53)	4.48[b](.88)	.36
Biosecurity protocols, filtered farms only (n=62)			
Farm Entry Protocol: Pressurizes entry room before entering.	4.97[a](.18)	4.61[b](.84)	.36
Supplies Entry Protocol: Has fan on when door is open.	4.81[a](.54)	4.65[b](.75)	.16
Animal Loadout Protocol: Follows Danish entry system.	4.67[a](1.01)	4.61[a](1.02)	.06
Animal Loadout Protocol: Proper use of fans to pressurize room.	5.00[a](0)	4.84[a](.45)	.16
Dead Removal Protocol: Proper use of fans to pressurize room.	4.94[a](.36)	4.55[b](.81)	.39

[a,b] Different letters indicates statistically significant ($P < .05$) difference in mean self versus peer compliance score for the biosecurity protocol
[1] Mean response to how likely are you to comply on a likert-scale ranging from 1(never) to 5 (always)
[2] Mean response to how likely are your fellow employees to comply on a likert-scale ranging from 1(never) to 5 (always)

Table 2: Percent correct responses to all biosecurity protocols on primary attempt

Variable	Response group	Least squares mean
Filtration status	Filtered	54.2%[a]
	Non-filtered	54.2%[a]
Primary language	English	42.5%[a]
	Spanish	65.9%[a]
Role on farm	Farm manager	60.6%[a]
	Section manager	52.3%[a,b]
	Technician	49.8%[b]
Time on farm	<12 months	48%[a]
	12-36 months	55.4%[a,b]
	>36 months	59.3%[b]

[a,b] Different letters indicate statistical significance ($P \leq .05$) within variable

Comparison of standard and bench entry protocols for prevention of environmental contamination due to personnel entry in a commercial swine facility

C. Fitzgerald, BS; K. Baker, BS; A. Anderson; R. Stika; D. Linhares, DVM, MBA, PhD; D. Holtkamp, DVM, MS
Iowa State University College of Veterinary Medicine, Ames, Iowa

Introduction

The swine industry continues to struggle with the introduction of emerging transboundary pathogens as well as endemic disease. Failing to prevent novel pathogen introductions into herds results in economic losses for producers. On-farm employee entry is one of the most frequent events that occurs on swine farms and can pose significant risk for pathogen entry. The addition of a bench entry system is an additional layer of biosecurity that is being used to lower the risk of pathogen transmission from contaminated footwear and ante-rooms. However, no research has been done to date to evaluate if the addition of a bench is effective at reducing environmental contamination on the clean side of showers in swine farms. The objective of this study was to determine if installing a bench entry system in swine facilities was effective at preventing entry of swine pathogen contaminated particles from on-farm personnel footwear using a fluorescent powder as a model for the contaminated particles.

Materials and methods

A control group (NoBench) and a treatment group (Bench) were evaluated in this study. Fluorescent powder (Glo Germ™, Moab, Utah) was used to simulate swine pathogen contaminated particles and was spread on the floor of the hallway leading to the shower. During Bench treatment days, a bench was installed in the hallway immediately before the shower room and after the fluorescent powder. During NoBench treatment days there was no bench installed after the fluorescent powder. Both treatments had a shoe rack placed in the same spot between the fluorescent powder and bench area. The entry of four female employees was monitored for 20 study days (10 Bench, 10 NoBench). Environmental contamination was measured each day at 4 sampling points by counting the number of squares with contamination using a grid. The sampling points were before the bench (A), after the bench (B), on the clean side of the shower (C) and dirty side of the shower (D). After each study day, all fluorescent powder was removed to ensure no residual powder contaminated the next study day's results.

Results

Sampling points were analyzed for significant differences in environmental contamination using the ANOVA procedure (SAS® 9.4; SAS® Institute Inc., Cary, NC). The mean number of contaminated squares was significantly ($P < 0.05$) lower for the Bench treatment group at sampling point B (directly after bench) but not at any of the other sampling points.

Discussion

The bench entry inconsistently reduced the level of contamination in this study, highlighting the need to focus on compliance and execution. Employees were instructed and monitored for proper use of the bench. However, the employees in this study frequently touched the bottom of their shoes, transferring the contaminant to their hands and to whatever their hands touched subsequently. Personnel clothing also needs to be considered. Pants that were long enough to drag on the ground led to high levels of contamination shown on one Bench day in this study. Lastly, this study highlights the importance of layering biosecurity practices. On two NoBench treatment days, fluorescent powder was found on the clean side of the shower, but was never found on Bench treatment days. Individual practices, that are partially effective in isolation, when layered with other practices can substantially reduce the risk of pathogen introduction.

Use of molecular characterization tools to investigate *M. hyopneumoniae* outbreaks

A. Anderson[1], BS; A. Sponheim[1,3], DVM; L. Dalquist[2], DVM; E. Fano[3], DVM, MS, PhD; M. Pieters[1], DVM, PhD
[1]College of Veterinary Medicine, University of Minnesota, Saint Paul, Minnesota;
[2]Swine Veterinary Center, Saint Peter, Minnesota; [3]Boehringer Ingelheim Vetmedica, Saint Joseph, Missouri

Introduction

Mycoplasma hyopneumoniae (*M. hyopneumoniae*) is the causative agent of enzootic pneumonia and known to economically impact herds, especially those co-infected with other respiratory pathogens such as influenza virus and porcine reproductive and respiratory syndrome virus.[1,2] *M. hyopneumoniae* elimination strategies, such as herd closure and medication, have been successfully attempted for disease eradication.[3] However, confirming the negative status of a herd post-elimination is a difficult task due to the expected low prevalence and slow transmission rate of this pathogen. In some cases, clinical disease and *M. hyopneumoniae* detection can be evidenced post-elimination, questioning the success of the program or potential lateral transmission. Therefore, this investigation assessed *M. hyopneumoniae* outbreaks in reproductive herds within one production system employing molecular characterization tools in order to determine bacterial variants origin after the implementation of disease elimination programs.

Case history and diagnostics

Two 2,500 farrow-to-wean farms initiated a *M. hyopneumoniae* elimination program in March, 2014. Both farms used a 240-day herd closure, strategic medication, and herd vaccination. The success of the elimination program was established by testing 90 sera at four consecutive sampling events approximately 8, 10, 11, and 12 months post-elimination for *M. hyopneumoniae* using ELISA (Idexx; 95% detection probability of a $\leq$ 1% prevalence).[4] Subsequently, 30 introduced replacement gilts from each gilt developing unit (GDU) lot group (n = 110-220) sourcing the farms showed no *M. hyopneumoniae* seroconversion using ELISA (Idexx) prior to farm entry (95% detection probability of a 10% prevalence). Fourteen months post-completion of the elimination program, clinical signs suggestive of *M. hyopneumoniae* infection, such as dry cough, were evidenced at both farms, as well as in four other sow farms that underwent a *M. hyopneumoniae* elimination program in April, 2014. A *M. hyopneumoniae* outbreak was confirmed at all six farms within a two month period by detecting this pathogen in bronchial or laryngeal swabs from clinically affected pigs using species-specific real-time PCR.[5] P146 sequencing was performed in DNA from positive swabs to characterize *M. hyopneumoniae* variants. Using BioPortal, a phylogenetic tree was created to analyze the sequences obtained from the seven sites as well as the original variant(s) identified prior to the implementation of sow farm elimination program.

Informative findings

Mycoplasma hyopneumoniae sequences obtained from the farms before and after the implementation of *M. hyopneumoniae* elimination strategies were compared using a phylogenetic tree (Figure 1). Sequences obtained from sow farms and/or their downstream flows prior to the implementation of *M. hyopneumoniae* elimination strategies along with a site that would have sourced all sow farms within the system in 2013 are labeled in green (Figure 1). Sequences labeled in red were obtained from all six farms approximately 14 months post-completion of the elimination program within a two month period. As shown in Figure 1, sequences obtained post-elimination clustered together but differently and independently from the original variants identified prior to elimination. This finding suggests that a different bacterial variant was identified at the time clinical signs were observed, indicating that a new *M. hyopneumoniae* outbreak occurred. By investigating GDU shipment records, a GDU was identified that sourced all six farms approximately 2.5 months prior to the detection of *M. hyopneumoniae* post-elimination. It is important to note that all gilts tested (n = 30) within each lot group at this GDU revealed no *M. hyopneumoniae* seroconversion prior to entry into the farms (Idexx; 95% detection probability of a < 10% prevalence). Despite this outcome, the possibility of the GDU becoming positive through lateral transmission and the lack of *M. hyopneumoniae* detection due to insufficient sample size, timing for seroconversion, and diagnostic method cannot be ruled out. Furthermore, the evaluation of transportation records revealed that a trailer was shared from one of the six *M. hyopneumoniae* positive sow farms to a GDU 17 days prior to *M. hyopneumoniae* detection in sera and laryngeal swabs collected from sampled gilts within the GDU. Sequences obtained from this GDU and the six farms after the elimination program clustered together (red; Figure 1). Based on these findings, it can be hypothesized that the GDU *M. hyopneumoniae* outbreak could have resulted from lateral transmission of this pathogen from a trailer used between positive and negative sites.

Recommendations

Through the use of molecular characterization tools, insights regarding new *M. hyopneumoniae* outbreaks were provided in multiple different scenarios, such as assessing a potential failed elimination attempt or lateral transmission between pig sites. To aid veterinarians in the investigation of *M. hyopneumoniae* activity post-elimination efforts, transmission, and variant origin, the

Figure 1: Comparison of p146 sequences obtained prior to and/or after *M. hyopneumoniae* outbreaks

* Green=sequences obtained prior to *M. hyopneumoniae* elimination program from the six sow farms; red=sequences obtained post completion of the *M. hyopneumoniae* elimination program; pink = *M. hyopneumoniae* 232 reference strain; black=sequences obtained from other farms within the system.

molecular characterization of variants in sow farms and GDUs should be an essential and vital component of an elimination program. From this case, we can also recommend that assaying sera samples based on a 10% *M. hyopneumoniae* prevalence within GDUs as a method for screening gilts prior to farm entry may not suffice in all scenarios. This statement is further supported by previous research demonstrating that laryngeal swabs and other sample types assayed by real-time PCR are more sensitive for *M. hyopneumoniae* detection compared to antibodies (experimentally).[6] Therefore, additional sampling guidelines for *M. hyopneumoniae* detection based on different sample types and assumed prevalence levels should be provided. These guidelines can be used to help validate the screening of gilts prior to sow farm entry and avoid future *M. hyopneumoniae* outbreaks.

References

1. Thacker EL, Minion C. Mycoplasmsosis: Zimmerman JJ, et al., eds. Diseases of Swine. 10th ed. Ames, IA. 2012; 779-797.

2. Holtkamp D, et al. Assessing production parameters and economic impact of swine influenza, PRRS, and *Mycoplasma hyopneumoniae* on finishing pigs in a large production system. Proceedings from American Association of Swine Veterinarians Conference. 2012; 75-76.

3. Holst S, et al. Elimination of *Mycoplasma* hyopneumoniae from breed-to-wean farms: A review of current protocols with emphasis on herd closure and medication. Journal Swine Health and Production. 2015; 23(6): 321-330.

4. Polson, D. Personal communication. 11 November, 2016.

5. Strait EL, et al. Real-time PCR assays to address genetic diversity among strains of *Mycoplasma hyopneumoniae*. J Clin Microbiol.2008; 46(8):2491-2498.

6. Pieters A, Rovira A. Comparison of various sample types for detection of *Mycoplasma hyopneumoniae* in recently infected pigs. Proceedings from Allen D. Leman Swine Conference. 2013; 75-76.

Development and validation of ante-mortem urine collection techniques for gilts and sows

M. Nickel[1], BS; K. Skoland[1], BS; A. Forseth[1], DVM; J. Brown[1], DVM; V. Thompson[1]; M. Breuer[1]; R. Warner[1], MS; K. O'Brien[1], BS; P. Canning[1], DVM; P. Thomas[2], DVM, MS; M. Hoogland[3], DVM; N. Lauterbach[3], BS; L. Karriker[1], DVM, MS, DACVPM

[1]Swine Medicine Education Center, Iowa State University, Ames, Iowa;
[2]Audubon-Manning Veterinary Clinic, Audubon, Iowa; [3]Smithfield Foods, Algona, Iowa

Introduction

Urine collection and evaluation is underutilized within the swine industry as a diagnostic tool. A factor restricting its use in the field is the limited ability to collect urine efficiently and reliably. A snowball trial design, applying systematic trial and error, allowed for the identification of two collection techniques for gilts and sows: the tampon method and the Whirl Pak® method. The objectives of this study were to develop ante-mortem urine collection techniques for female swine that were reliable, reproducible and economical and to determine if the alternative techniques would alter urinalysis (UA) or calcium, phosphorus and creatinine values when compared to a free catch sample from the same individual.

Materials and methods

Fourteen sows were selected for urine collection and analysis using three techniques: free catch, tampon and Whirl Pak®. A free catch urine sample was collected from each sow using a 120 milliliter (mL) urine cup. The tampon technique included the use of a super-sized, unscented tampon with a plastic applicator and umbrella absorbency. The tampon was inserted into a female's vestibule and the exterior string was taped on the side of the vulva with elastic waterproof tape. Following urination, the tampon was transferred into a 120 mL urine cup. The Whirl Pak® technique included the use of a 24-ounce Whirl Pak® bag and elastic adhesive tape. The bag was positioned around the vulva and secured at the lateral, dorsal and ventral aspects. Following urination, the bag was removed from the rear of the sow and the urine sample was transferred into a 120 mL urine cup. From each sample collected, fifteen mL of urine were used to perform a UA. The UA included a specific gravity measurement using a Reichert TS 400 refractometer, dipstick testing using VetStix 11 parameter test strips and a microscopic evaluation of sediment. The eleven parameter test included evaluation of blood, bilirubin, urobilinogen, ketones, protein, nitrite, glucose, leukocytes, and ascorbic acid as well as pH and specific gravity. Ten milliliters of each sample were then spun at 1,600 RPM (rotations per minute) for five minutes using a C3-Select centrifuge and the sediment was evaluated at 10× power under a light microscope. Urine samples from ten of the fourteen sows were submitted to the Iowa State University Veterinary Diagnostic Laboratory for evaluation of calcium, phosphorus and creatinine. The results were used to calculate the calcium to creatinine and phosphorus to creatinine ratios of each sample. A correlation test was then completed using MedCalc software to directly compare the sample ratios. In addition, timing of placement, reproducibility, reliability and cost were evaluated while in the field.

Results and discussion

The tampon was placed in the shortest amount of time with an average of nineteen seconds, while placement of the Whirl Pak® took an average of 104 seconds. Both methods were placed successfully by a person after having observed an example placement only one time. The tampon produced urine samples in a large enough volume for evaluation 82% of the time, while the Whirl Pak® method was successful 59% of the time. The cost of the materials used in each method were $0.23 per tampon sample and $1.19 for each Whirl Pak® sample. The UA results did not differ significantly between sample methods. Fibers were identified during microscopic evaluation of tampon sample sediment but did not obstruct the observer's ability to evaluate the sample. The phosphorus to creatinine ratios (Figure 1) and calcium to creatinine ratios (Figure 2) were not significantly different among the three sample methods used for individual sows, and each had a significance level of $P < 0.001$. In conclusion, this investigation has identified two, reproducible, reliable and economical urine collection techniques that can be utilized in a variety of field settings to provide practitioners with new information on sow and gilt nutritional status and physiology.

Figure 1: Urine phosphorus:creatinine ratio comparison of different collection methods from ten female swine.

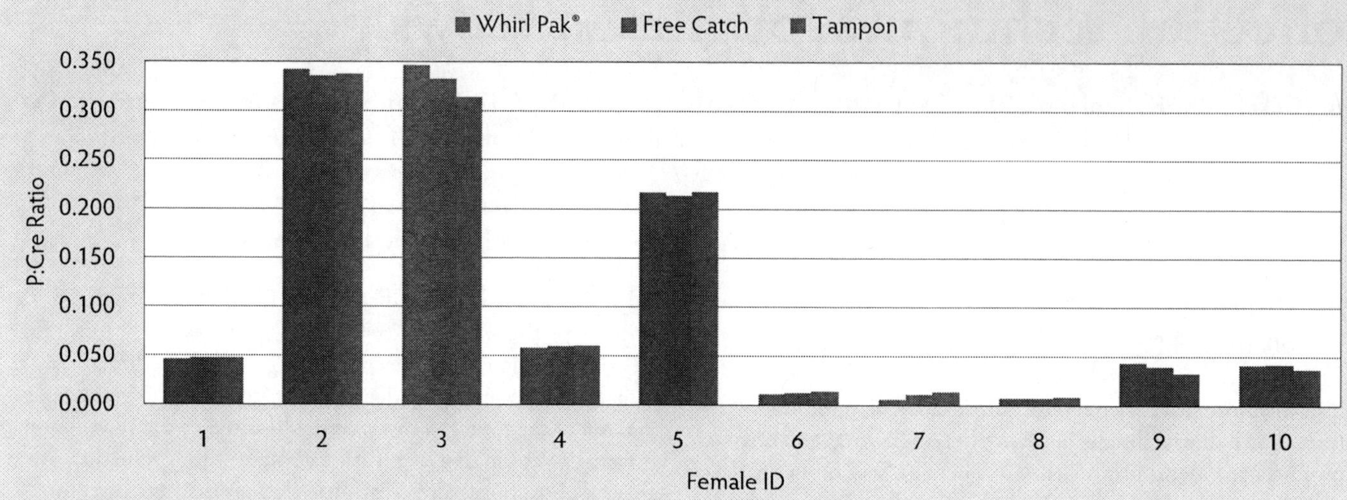

Figure 2: Urine calcium:creatinine ratio comparison of different collection methods from ten female swine.

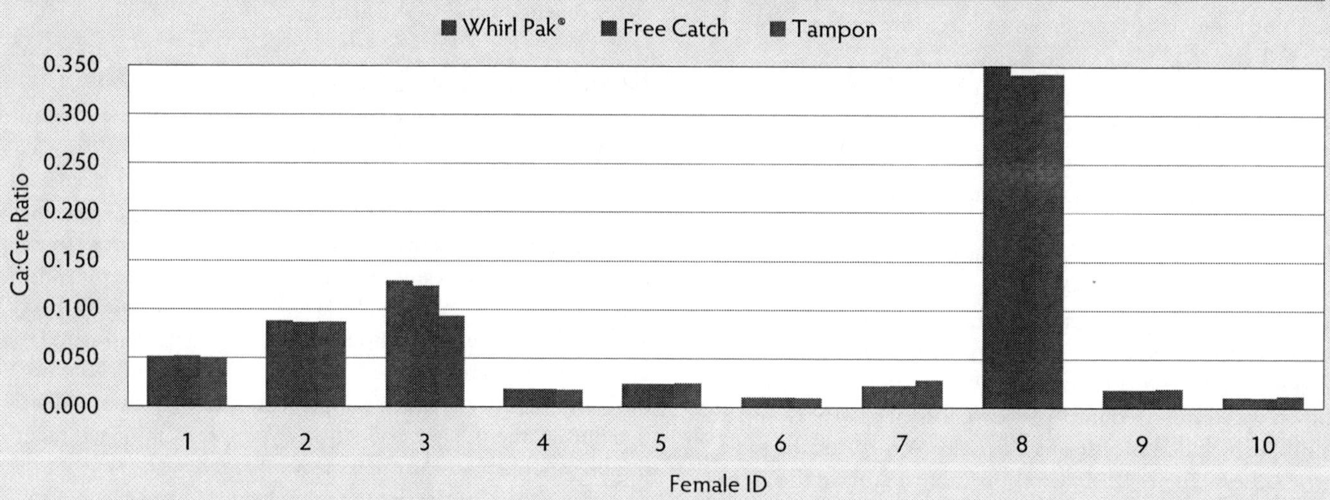

An evaluation of oropharyngeal swabbing as a diagnostic technique for the detection of porcine reproductive and respiratory syndrome virus (PRRSV) in weaned pigs

Kylie Glisson[1]; Adam Schelkopf[2], DVM; Brett O'Brien[3], VMD, MS; Julio Alvarez[4], DVM, PhD; Bob Morrison[4], DVM, MBA, PhD

[1]North Carolina State University, Raleigh, North Carolina; [2]Pipestone Veterinary Services, Pipestone, Minnesota; [3]Zoetis, Parsippany, New Jersey; [4]University of Minnesota, Saint Paul, Minnesota

Introduction

PRRS remains one of the most critically important reproductive and respiratory diseases in the swine industry today resulting in significant health and production losses.[1] Currently, sow farm monitoring for PRRSv centers around wean pig diagnostic testing of various sample types including oral fluids, tissues, and serum, with serum being the most widely accepted sample type of choice in the industry. Oropharyngeal scrapings have been shown to be the superior method for ante-mortem detection of PRRSv in sows as it samples lymphoid tissue, the site of PRRSv persistence.[2] However, this procedure is invasive, stressful, and impractical for wean pigs. The industry could benefit from a less invasive and time consuming method to detect PRRSv in wean pigs that still targets the tonsillar area. Oropharyngeal (OP) swabs have been used successfully to detect several swine pathogens including influenza A virus of swine (IAV-S),[3] *Mycoplasma hyopneumoniae*,[4] and Porcine Circovirus Type 2 (PCV-2).[5] As many veterinarians monitor their sow herds for both PRRSv and IAV-S, a diagnostic sample that could be used to detect both viruses and enhance the ability to sequence positives would be a desirable tool to use for wean pig monitoring.

Objective

The objective of this study was to determine the sensitivity of oropharyngeal swabs, relative to serum samples, as a diagnostic technique for the detection of PRRSv by PCR in wean pigs.

Materials and methods

Wean pigs from two sow farms, each with a PRRSv infection diagnosed 2 months prior to the start of the study, were selected for this study. Both farms were visited twice, two weeks apart, with 25 to be weaned pigs selected at random to be sampled at each visit (n = 100). Each pig was swabbed using a BD BBL™ CultureSwab® by inserting the swab in the back of the oral cavity and swabbing the soft palate vigorously for a minimum of 5 seconds. Individual serum samples were also collected on the same pigs. All samples were stored on ice and sent overnight to the South Dakota State University Animal Disease Research and Diagnostic Laboratory (Brookings, SD) to be tested individually for PRRSv by PCR. A correlation and sensitivity analysis was performed on the data.

Results

The overall sensitivity of OP swabs was 82.6% based on true positives as validated by the presence of virus in serum (Table 1). The sensitivity of OP swabs was highest (95.8%) when serum Ct values were < 26 but declined as viral load decreased with a sensitivity of 53.3% when serum values were ≥ 29 (Table 2). Mean Ct values of positive samples for OP swabs and serum were 30.7 and 25.7, respectfully; however, results were only moderately correlated (Pearson's coefficient = 0.26).

Discussion and conclusion

Results from this study suggest that OP swabs are an effective, alternative means for detection of PRRSv in wean pigs particularly when the prevalence of infection is high. A limitation of this study was that it was conducted in acutely infected farms; therefore, further work is needed to evaluate the sensitivity of the technique in chronically infected populations. OP swabs are a valuable tool for the swine industry and could impact sow farm monitoring, as they offer an alternative means of detection for PRRSv that is easy to teach to farm employees, highly sensitive, and less invasive.

References

1. Holtkamp D et al. Assessment of the economic impact of porcine reproductive and respiratory syndrome virus on United States pork producers. *J Swine Health and Prod.* 2013; 21:72.

2. Kleiboeker S et al. Concurrent use of reverse transcription-polymerase chain reaction testing of oropharyngeal scrapings and paired serological testing for detection of porcine reproductive and respiratory syndrome virus infection in sow. *J Swine Health and Prod.* 2002; 10:251.

3. Strobel R et al. Evaluation of diagnostic sampling techniques used in the farrowing house to increase sensitivity of detection and sequencing if influenza A virus in swine (IAV-S). AASV. 2016;77.

4. Kuecker K et al. A comparison of tracheobronchial, oropharyngeal, and nasal swabs to determine the Mycoplasma hyopneumoniae status in suckling pigs. AASV. 2012;321.

5. Shibata I et al. PCRR detection of Porcine Circovirus type 2 DNA in whole blood, serum, oropharyngeal swab, nasasl swab, and feces from experimentally infected pigs and field cases. *J Veterinary Medical Science.* 2003; 405.

Table 1: Serum and OP swab PCR (positives/total)

	Farm 1 sample		Farm 2 sample		
	Visit 1	Visit 2	Visit 1	Visit 2	Total
Serum	22/25	14/25	25/25	25/25	86/100
OP Swab	15/25	11/25	22/25	23/25	71/100

Table 2: Sensitivity of OP swabs based on viral load

	OP Swab+	Serum+	Sensitivity
< 26	46	48	95.83
26-29	18	23	78.26
> 29	8	15	53.33

A challenge study to determine whether in-feed flavophospholipol can reduce *Salmonella* shedding and colonization in nursery pigs

Jane Newman, BSc; Vahab Farzan, DVM, MSc, PhD; Robert Friendship, DVM, MSc
Department of Population Medicine, University of Guelph, Guelph Ontario, Canada

Introduction

Salmonella is an increasing public health concern, with the emergence of multi-drug resistant strains. Pigs are often asymptomatic *Salmonella* carriers and during periods of stress, such as weaning, may shed *Salmonella* in feces and colonize the bacteria in tissue. This may result in animal to human transmission through pork, contaminated produce, or groundwater contamination. On-farm control of *Salmonella* has proven difficult and there is need for a practical and inexpensive solution. Flavophospholipol is a phosphoglycolipid antimicrobial agent used in swine and poultry as an antimicrobial growth promoter (AMGP). Previous studies have suggested flavophospholipol, an antibiotic with non-transferable resistance, may alter the microflora in favour of beneficial bacteria, and lead to reduced *Salmonella* shedding and colonization.[1,2]

Objectives

The objectives of this study were i) to compare *Salmonella* shedding and colonization as well as the antibody response to *Salmonella* in pigs receiving flavophospholipol compared to control pigs, ii) to determine time course of *Salmonella* shedding and antibody response in pigs challenged with *Salmonella* Typhimurium.

Materials and methods

Pig and sample collection

Thirty-two weaned pigs (21-28 days) were housed in 4 rooms in a level 2 biosafety isolation unit, with 8 pigs per room. Half of the pigs (n = 16) were fed a diet containing 4 ppm flavophospholipol (Flavomycin®, Huvepharma) and the other half, the same feed without flavophospholipol (control). Pigs entered the facility on Day 0 and all pigs were challenged with a 2 mL oral dose of 10^8 colony forming units (CFU)/mL of *Salmonella* Typhimurium DT 104, with resistance to nalidixic acid on Days 7 and 8. Pigs were monitored daily for clinical signs of salmonellosis and rectal temperatures were taken. Feces were collected 11 times over the course of 36 days and blood samples were collected 9 times. Pigs were euthanized at the end of trial (Day 36) and tissue samples (liver, lymph node, and spleen) were collected.

Salmonella culturing

Fecal/tissue samples were enriched in tetrathionate broth (TTB) in a 1 to 9 ratio, homogenized for 30 sec and incubated at 37°C for 18 to 24 h. Then, 0.1 mL of TTB was transferred to a 9.9 mL Rappaport Vassiliadis (RV) tube and incubated at 41°C for 18 to 24 h. Lastly, a loop-full of RV culture was plated on Xylose Lactose Tergitol™ 4 (XLT-4) agar containing nalidixic acid and incubated at 37°C for 18 to 24 h.

In addition, fecal samples were diluted with 0.1% buffered peptone water (BPW) to make serial dilutions of 10^{-1}, 10^{-2}, and 10^{-3}. Each dilution was plated on XLT-4 agar, containing nalidixic acid. The plates were incubated at 37°C for 24 h and number of colonies per plate were counted.

Salmonella antibody detection

An indirect enzyme-linked immunosorbent assay (ELISA) was used to determine level of antibody to *Salmonella* and seropositivity in pigs.

Data analysis

A mixed effects linear regression method was used to compare *Salmonella* CFU in fecal and tissue samples as well as antibody response (Sp ratio) in pigs receiving the medicated diet with that in the controls. In addition, a mixed effects logistic regression method was used to compare seropositivity between pigs on the medicated diet and control diet.

Results and discussion

Two pigs (both from the control group), died prior to the assigned euthanasia date (Day 36). One pig developed septicemia on Day 13 of the trial and was euthanized. The other pig died due to bronchopneumonia and peritonitis and tested positive for *Salmonella* in lung tissue. *Salmonella* was recovered from feces collected at each sampling point post-challenge; only one pig in the treatment group tested negative on Day 14 and 28. The medicated pigs had a larger CFU of *Salmonella* in fecal samples than the control pigs ($P < 0.05$) (Figure 1). In total, 5/31 pigs were positive for *Salmonella* in lymph nodes (4), liver (1), and spleen (2) samples. There was no significant difference in antibody response among pigs in the 2 groups ($P > 0.05$). Overall, 56%, 28%, and 87% of pigs were seropositive at Day 1 (28 days of age), Day 12 (40 days of age), and Day 36 (64 days of age), respectively. The seropositivity trend indicates that piglets entering the nursery stage had maternal antibodies (21/32 seropositive), which appeared to become depleted by 33 days of age, with only 9/32 pigs being seropositive. Pigs then acquired immunity, with peak values seen at 57 days of age.

Figure 1: Log$_{10}$ of *Salmonella* colony forming unit (CFU)/g of feces in pigs fed a medicated (flavophospholipol) or conventional (control) diet. Pigs entered the facility on Day 0, and were challenged with a 2 mL oral dose of 10^8 colony forming units (CFU)/mL of *Salmonella* Typhimurium DT 104 on Days 7 and 8.

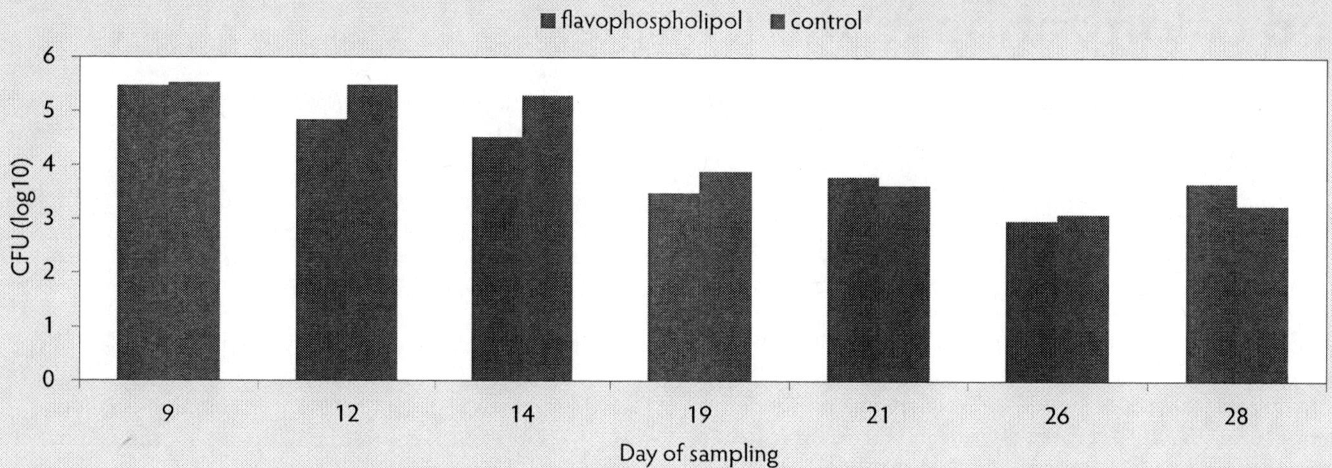

The findings indicate that 4 ppm of flavophospholipol in-feed had no impact on *Salmonella* shedding and colonization in nursery pigs. In fact, the pigs receiving flavophospholipol had a higher number of *Salmonella* in feces contradicting reports in the literature of a positive therapeutic effect. Although both premature deaths occurred in the control group, more trials are needed to evaluate impact of flavophospholipol on morbidity and mortality rates. Further, the seropositivity trend indicates that piglets entering the nursery stage may have had maternal antibodies to *Salmonella* but this protection appeared to wane over the next 2 weeks and acquired immunity began to appear. Finally, the *Salmonella* infection model developed in this trial appeared to work well and can be used for future studies.

Acknowledgements

Funding sources: Huvepharma, University of Guelph-Ontario Ministry of Agriculture, Food and Rural Affairs Research Partnership, and Swine Innovation Porc

References

1. van den Boogaard et al. Antimicrobial agents and Chemotherapy, 2002; 46(1); 110-118.
2. Dealy J & Moelle M. Journal of Animal Science. 1976; 42: 1331-1336.

Figure 2: *Salmonella* seropositivity in pigs fed a medicated (flavophospholipol) or non-medicated (control) diet determined by an indirect enzyme-linked immunosorbent assay. Pigs entered the facility on Day 0, and were challenged with a 2 mL oral dose of 10^8 colony forming units (CFU)/mL of *Salmonella* Typhimurium DT 104 on Days 7 and 8.

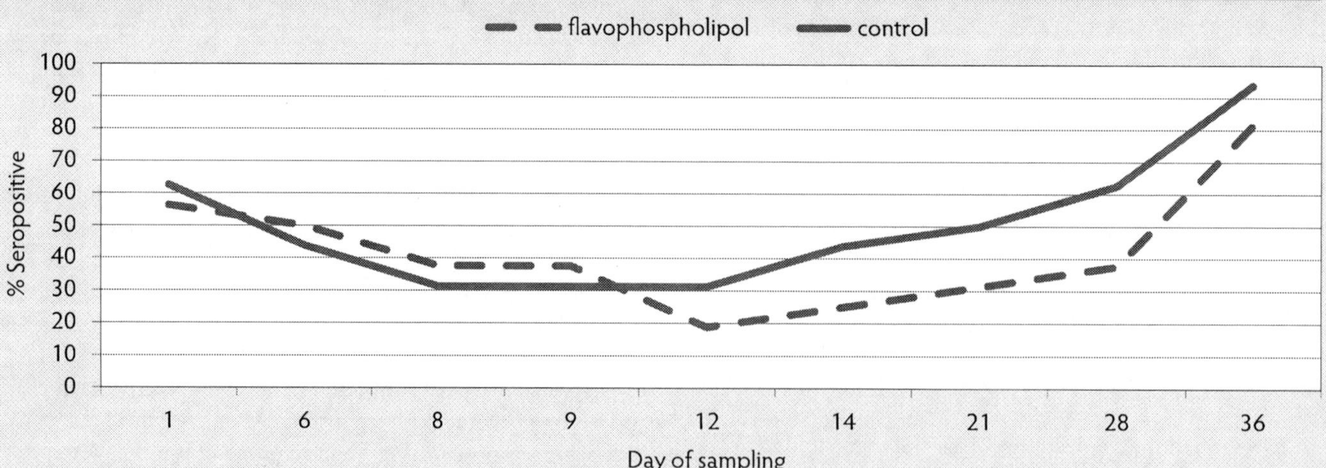

Correlating sampling methods to porcine circovirus type 2 (PCV2) viremia status of sows and piglets in a sow herd with or without PCV2 revaccination during gestation

Allison Knox[1], BS; Luis Ochoa[2], PhD; Laura Greiner[2], PhD

[1]University of Illinois, Urbana, Illinois; [2]Carthage Innovative Swine Solutions, LLC, Carthage, Illinois

Introduction

Clinical signs of porcine circovirus type 2 (PCV2) are most commonly observed in late nursery and early finishing pigs. However, clinical signs may appear in the early nursery phase, indicating exposure and infection prior to weaning. This puts pressure on sow farms to better identify and prevent piglets infected with PCV2. Methods currently utilized to determine PCV2 status via PCR virus testing rely on the sampling of various sera, tissues, secretory fluids, and the pig's environment.[1] However, it remains unclear what method better correlates to the PCV2 status of sows and piglets.

Objective

The objective of the study was to correlate different sampling methods to the PCV2 viremia status of sows and piglets in a herd with or without PCV2 revaccination during gestation.

Materials and methods

Treatment groups

Samples were collected from 60 sows and their litters in a commercial, 5,600 sow herd using PIC 1050 genetics. The herd had a history of producing pigs that exhibited downstream PCV2 signs in late nursery and early finishing. Sows were selected based on parity (P1 or P4+) and revaccination status for PCV2. The vaccination groups were defined as revaccinated or control (without revaccination). All sows and gilts were vaccinated, regardless of treatment, around 140 days of age. Revaccination of gilts and sows occurred five weeks prior to farrowing using a commercial PCV2 vaccine.

Sampling techniques

Sow and piglet sera were collected post-farrowing. Piglet serum was collected prior to the piglets nursing via anterior vena cava venipuncture and pooled within litter. Sow serum was collected via ear venipuncture or external jugular venipuncture. Placental umbilical cord serum (PUCS) samples were collected post-farrowing. Between 3 and 6 placental cords were identified per sow, and using clean gloves, the contents of the cords were stripped into a serum separator tube. Colostrum was collected within 12 hours of farrowing from teats in four quadrants of the underline after wiping with chlorhexidine soaked gauze pads.[2] Environmental samples were collected from a 2' by 2' area in the farrowing crate directly posterior to the sow. Gauze pads moistened with phosphate buffered solution (PBS) were used to swab the crate and the swab contents were transferred to 10 ml of PBS solution.

Diagnostic and data analysis

All samples were evaluated using quantitative PCR for PCV2 at the Iowa State University Veterinary Diagnostic Laboratory using standard methods. The experimental unit for the purpose of this study was the sow. Data analysis was performed using SAS.

Results

The PCV2 PCR results are presented in Table 1.

Sow and litter serum samples were negative, while PUCS, colostrum, and environmental samples had varying numbers of positive samples. Due to the lack of positive serum samples, no correlation analysis was conducted between serum and PUCS, colostrum, or environmental samples. In addition, the number of PCV2 qPCR positive environmental samples was 29.2% lower for animals revaccinated in gestation than those without revaccination. There was no significant difference in total born (13.39 ± 0.52), born alive (12.4 ± 0.59), stillborn (0.65 ± 0.17), and mummies (0.34 ± 0.16) between the revaccinated and control groups.

Discussion

This study identified a large difference in the number of PCR positive serum samples compared to PUCS, colostrum, and environmental swabs. This is likely due to the method of sampling and the increased risk of contamination with shed virus in the environment. Comparing the serum and environmental samples, environmental swabs indicate a higher frequency and level of contamination that is not proportionate to those found in the pig. Therefore, serum from piglets before consumption of colostrum appears to be the most accurate measure of individual PCV2 status. The increasing frequency and quantity of virus found in colostrum, PUCS, and environmental swabs is likely due to their respective increased risk of exposure to shed virus in the environment. This potential contamination decreases the

Table 1: Effect of PCV2 vaccination during gestation on qPCR PCV2 analysis of sow serum and litter serum at birth, colostrum, PUCS samples, and environmental swabs, and farrowing performance of sows.

Item	Gestation vaccine[1]			
	No	Yes	SEM	*P*-value
PCV2 qPCR analysis				
Sow serum at farrowing				
Number of samples	32	28	-	-
Number of PCV2 Positive samples	0	0	-	-
Litter serum at birth				
Number of samples	32	28	-	-
Number of PCV2 Positive samples	0	0	-	-
Colostrum				
Number of samples	32	28	-	-
Number of PCV2 Positive samples	9	8	-	0.97[z]
Genomic copies on positive samples, copies/ml	6,457	18,366	9,095	0.37
Average cycle time on positive samples (Ct)	35.2	34.9		
PUCS				
Number of samples	32	28	-	-
Number of PCV2 Positive samples	13	12	-	0.86[z]
Genomic copies on positive samples, copies/ml	99,779	75,827	42,803	0.70
Average cycle time on positive samples (Ct)	33.9	34.9		
Environmental swabs				
Number of samples	24	20	-	-
Number of PCV2 Positive samples	24[b]	17[a]	-	0.049[z]
Genomic copies on positive samples, copies/ml	171,832	198,496	56,657	0.49
Average cycle time on positive samples (Ct)	32.5	33.5		
Farrowing performance				
Born alive	12.78	12.04	0.585	0.37
Still born	0.72	0.57	0.172	0.55
Mummies	0.31	0.36	0.162	0.85
Total born	13.81	12.96	0.522	0.26

1 All sows received a PCV2 vaccine at approximately 140 d of age regardless of treatment.
z Chi-square analysis
ab Means with different superscript within a row differ ($P < 0.05$)

accuracy of using colostrum and PUCS to predict piglet or sow PCV2 viremia when taken at one time point. Multiple samplings of PUCS in addition to or in place of pre-suckle piglet serum may be needed to determine the true prevalence of PCV2 viremia in the sow herd.

In the herd sampled, PCV2 had been circulating for an extended period of time prior to the study. By the time samples were collected, there was little evidence in the sow herd and downstream pigs of PCV2 clinical signs. Therefore, the positive samples identified from colostrum, PUCS, and the environmental swabs likely indicated residual virus in the environment, and not the viremia in the sows or piglets. In herds that are not actively infected, as evidenced by negative PCR analyses from replacement gilts or sows, PUCS and colostral sampling may be better suited for herd screening and used over time to assess changes in viral exposure within the herd. Ongoing surveillance in a herd can be utilized to determine correlation of these sampling methods to the risk of PCV2 infection post weaning.

In regard to the effects of vaccination on PCV2 sampling, the results suggest that revaccinating due-to-farrow sows may reduce the environmental viral loads of PCV2 (Figure 1). However, further research is needed to determine the implications of this reduction in environmental PCV2 levels.

References

1. Dvorak, C.M.T. et al. (2013). Multiple routes of porcine circovirus type 2 transmission to piglets in the presence of maternal immunity. *Veterinary Microbiology* 166: 365–374.

2. Gerber, P.F., et. al. (2011). Serum antibodies and shedding of infectious porcine circovirus 2 into colostrum and milk of vaccinated and unvaccinated naturally infected sows. *Veterinary Journal*, 188 (2), pp. 240-242.

Figure 1: Effect of gestation PCV2 vaccine on environmental viral load (genome copies/ml)

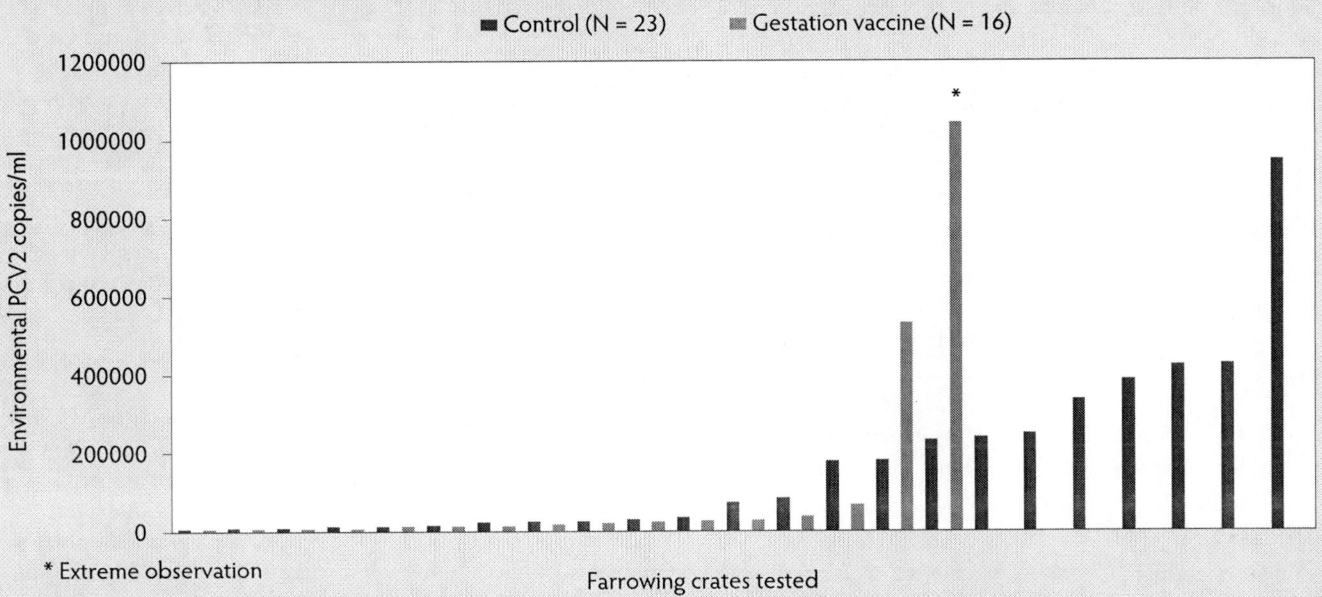

* Extreme observation

Sample comparison for detecting porcine circovirus type 2 in breeding herds

Rachel Schulte[1]; Darrell Neuberger[2], DVM; Abigail Redalen[2], DVM; Clark Huinker[3], DVM, MS; Mark Wagner[3], DVM; Daniel Linhares[1], DVM, MBA, PhD, FSVM
[1]Iowa State University, Ames, Iowa; [2]Zoetis, Parsippany, New Jersey; [3]Fairmont Veterinary Clinic, Fairmont, Minnesota

Introduction

Although the understanding of porcine circovirus 2-associated diseases (PCVAD) in growing pigs continues to progress, little research has been published on specific practices at the sow farm to minimize PCVAD manifestations downstream.[1] In this study, one pig flow had an estimated death loss of 15% in vaccinated pigs during the nursery phase. The producer estimated a loss of $7 per pig weaned due to broken contracts ($6 loss per pig) and the need for increased sick pig treatments (average $1 loss per pig).

A prior study showed that piglets from viremic sows had increased pre-wean mortality as compared to piglets from non-viremic sows in the same herd.[2] There are different biological sample types that can be used to detect and quantify the presence of PCV2 in sow farms. Most recently, placental umbilical cord serum (PUCS) sample use on sow farms has been described as a good indicator of level of PCV2 circulation of pre-suckle pigs.[3]

Objective

The objective of this study was to compare sow colostrum, sow serum, piglet serum and PUCS samples as methods for detecting PCV2 DNA by PCR in sow herds.

Materials and methods

Seventy eight sows from three non-vaccinated sow farms were sampled for PCV2 using sow serum and colostrum. Within 4 hours of farrowing, serum and PUCS samples of 3 piglets from 30, 18, and 30 sows at Farms A, B, and C were also collected. Samples were stored at -20 degrees Celsius and tested by PCR at the University of Minnesota Veterinary Diagnostic Laboratory. The piglet serum and PUCS samples were pooled by sow. The percentage of positives in each sample category was compared. When multiple sample types per litter were positive, this was also recorded.

Results

Results showed that 12.99% (10/77) of colostrum samples, 6.41% (5/78) of sow serum samples, 10.26% (8/78) of PUCS samples, and 0% (0/78) of piglet serum samples were positive (Table 1). In every case where PUCS samples from the baby pig tested positive, the sow tested negative via both sow colostrum or serum samples. Four sows tested positive for both sow serum and colostrum. Adjusted by farm, there was a statistically significant (P 0.0062) difference using a regression model (PROC GLM from SAS9.4) between Ct values of positive samples. Comparing sample types, sow serum Ct was lower than sow colostrum, PUCS, and piglet serum Cts (Figure 1).

Conclusions

Because colostrum showed the highest detection of PCV2, this may indicate that it is the most sensitive indicator of PCV2 presence in the sow herd. Sow serum samples had higher Cts than other sample types, but this result should be interpreted with caution due to the small sample size analyzed. The absence of positive piglet serum samples may indicate that this sampling method is an unreliable means of detection or that none of the piglets were viremic. Future studies should focus on establishing reliable methods for the evaluation and analyzation of sow herd PCV2 stability in order to produce marketable weaned pigs and minimize production losses during finishing.

References

1. Shen, H, Wang, C, Madson, D, Opriessnig, T. (2010). High prevalence of porcine circovirus viremia in newborn piglets in five clinically normal swine breeding herds in North America. Preventative Veterinary Medicine 97 (2010) 228-236.

2. Calsamiglia, et al. (2007) Sow porcine circovirus type 2 (PCV2) status effect on litter mortality in postweaning multisystemic wasting syndrome (PMWS). Research in Veterinary Science 82(3):299-304.

3. Fangman, T, Baumert, D, Rustvold, J. 2013. Diagnostic investigation to determine herd stability of a sow herd with downstream PCVAD in nursery pigs. Allen D Leman Swine Conference, Saint Paul, Minnesota.

Table 1: PCV2 PCR positives per total samples collected

Farm	Piglet serum	PUCS	Sow serum	Colostrum
A	0/30	7/30	4/30	7/30
B	0/18	0/18	0/18	3/18
C	0/30	1/30	1/30	0/29
Total	0/78	8/78	5/78	10/77

Figure 1: Distribution of PCV2 DNA PCR Ct values by sample type

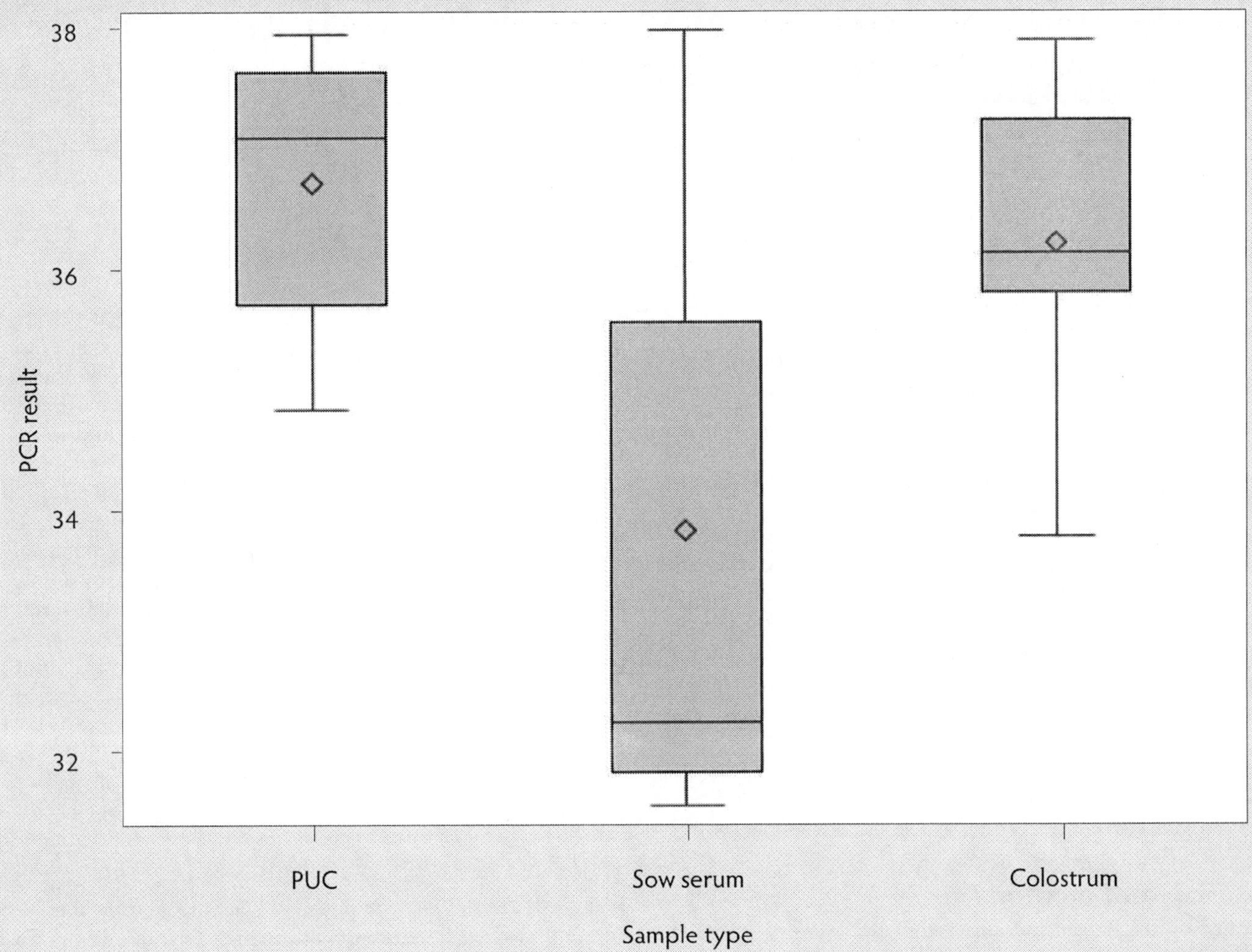

Evaluation of positive pressure filtration to reduce aerosol transmission of PRRSV during an experimental challenge of farm access points

H. Baldry[1]; K. Wedel[2], DVM; N. Schaefer[3], DVM; P. Thomas,[2] DVM
[1]University of Minnesota College of Veterinary Medicine, Saint Paul, Minnesota; [2]Iowa Select Farms, Iowa Falls, Iowa; [3]Boehringer Ingelheim Vetmedica, Saint Joseph, Missouri

Introduction

Since its discovery in the late 1980's, porcine reproductive and respiratory syndrome virus (PRRSV) continues to be one of the most economically important swine diseases in the United States.[1] The virus's ability to spread via aerosol over large distances makes it difficult to control, especially for breeding herds located in pig-dense regions.[2] Many farms have implemented air filtration as a method of reducing the risk of infection; however, filtration is expensive, requires significant maintenance to prevent intake of unfiltered air, and PRRSV transmission may still occur via retrograde air movement through open farm access points.[3,4] Some systems are investing in positive pressure filtration (PPF) to reduce the risk of aerosol PRRSV transmission via retrograde air movement, but no research currently exists on the positive pressure air speed required to prevent aerosol PRRSV entry onto PPF farms. The objectives of this study were to 1) develop a technique to test aerosol PRRSV transmission on PPF farms, 2) assess the PRRSV aerosol transmission risk via retrograde air movement at PPF farm access points, and 3) provide best practices to prevent aerosol PRRSV entry.

Materials and methods

Modified-live PRRS vaccine (Ingelvac PRRS® MLV, Boehringer Ingelheim Vetmedica, St. Joseph, Missouri) was diluted to high and low concentrations (Ct x□: 23, 29) using phosphate buffered saline in order to simulate a realistic PRRS challenge while still being detectable by polymerase chain reaction (PCR). The diluted vaccine was dispersed through a cold mist fogger (Dyna-Fog® Hurricane Model No. 2790) at a speed of 13.9 m/s using a variable flow rate setting (266 ml/min, 189 ml/min) and fog time of 10 or 20 s. Swiffer® sweeper cloths saturated with neutralization broth were used to collect environmental samples using one of two methods: the saturated cloths were swabbed over either a 1.8 by 0.6 m board covered in a plastic sheet or an individual wearing disposable Tyvek coveralls. These methods were selected in order to mimic farm personnel standing in the open doorway of farm access points during animal load out, load in, or supply entry procedures. After swabbing, the cloths were wrung into a plastic bag, and the resulting liquid pipetted into a tube and refrigerated until testing. Samples were taken before (negative control) and after fogging, and plastic gloves, Tyvek coveralls and plastic sheets were changed for each replicate. A 5 mL sample of diluted vaccine was collected as positive control. All samples were submitted to the Boehringer Ingelheim Health Management Center Testing Laboratory in Ames, Iowa and tested by PCR.

Trial 1

Preliminary testing was completed in a parking lot using the Tyvek coveralls method. Distance from the fogger, fog time, flow rate and vaccine dilution were varied to test a variety of experimental setups. Based on these results (Table 1), the 1.5-meter

Table 1: Summary of Trial 1 results

Distance (m)	Fog time (s)	Flow rate	Vaccine concentration	PCR
0	10	High	High	3/3
1.5	10	High	High	3/3
3	10	High	High	3/3
7.6	10	High	High	3/3
7.6	20	High	High	1/1
13.7	20	High	High	1/1
19.8	20	High	High	0/1
0	20	Low	Low	3/3
1.5	20	Low	Low	3/3
3	20	Low	Low	0/3
7.6	20	Low	Low	0/3

distance, 20 s fog time and low flow rate were selected for on-farm challenge in Trial 2.

Trial 2

Testing was completed at a 4,200 head PPF sow farm located in a pig-dense region of Iowa. Four farm access points (gilt/cull chute, mortality removal door, weaned pig chute and supply room) were each challenged by placing the 1.8 × 0.6-meter board in the open doorway or chute to the outside. The fogger was setup 1.5 m away from the doorway and run for 20 s per replicate on the low flow rate. Farm building attic pressure and access point inlet openings were altered to vary the pressure at each access point in order to simulate season-dependent differences in building settings. Positive pressure air speeds at each access point doorway were measured using a handheld anemometer (Kestrel 3000 Weather Meter).

Results

The results from Trial 1 and 2 are summarized in Tables 1 and 2, respectively. All samples collected at the wean chute and supply room access points using high vaccine concentration were PCR positive, with an average Ct value of 29.1. One sample obtained at the mortality door using low vaccine concentration was PCR positive (Ct: 36.94). All control samples taken before fogging were PCR negative. Positive pressure air speeds measured at each access point doorway were positively associated with access point pressure. Positive pressure air speed ranged from 0.17 to 4.39 m/s ($\bar{x}$: 1.68 m/s).

Discussion

Changing the attic pressure and inlet openings during Trial 2 allowed for alteration of the access point pressure and the speed of the air going out the door. Under these experimental conditions, all samples collected when positive pressure air speeds were > 1.85 m/s were PCR negative, suggesting that the positive pressure was sufficient to prevent retrograde movement of aerosol. As a comparison, Alonso et al determined that air speed must be at least 0.76 m/s for aerosolized PRRSV to enter a negative pressure filtered farm through an idle fan. However, given that 9/10 positive results were obtained when using high vaccine concentration at only two access points, further testing is needed to determine if

Table 2: Summary of Trial 2 results using 20 s fog time, low flow rate, and 1.5 m distance

Vaccine concentration	Location	Pressure (Pa)	Air speed (m/s)	PCR	Ct
Low	G/C chute	7.5	1.85	Negative	
			1.92	Negative	
			2.12	Negative	
Low	G/C chute	15.0	2.76	Negative	
			3.16	Negative	
			3.12	Negative	
Low	G/C chute	25.0	4.39	Negative	
			3.75	Negative	
			4.06	Negative	
Low	Mortality door	15.0	2.18	Negative	
			2.53	Negative	
			2.86	Negative	
Low	Mortality door	2.5	0.52	Positive	36.94
			0.54	Negative	
			0.65	Negative	
High	Wean chute	5.0	0.40	Positive	27.83
			0.36	Positive	29.91
			0.21	Positive	29.58
High	Wean chute	10.0	1.06	Positive	29.90
			0.69	Positive	29.40
			0.66	Positive	28.97
High	Supply room	5.0	0.17	Positive	29.93
			0.23	Positive	29.09
			0.18	Positive	27.95

these results are repeatable at different access points and farms at high vaccine concentration.These data suggest that steps taken to increase positive pressure air speeds may reduce the risk of PRRS transmission on PPF farms; for this reason, it is recommended that all access points be equipped with manometers so that farm personnel can evaluate the access point pressure prior to opening outside doors. Personnel should also fully open all access point inlets prior to opening the door in order to increase the access point pressure. Minimize the amount of time outside doors are open, and limit the doorway opening to increase the speed of the air traveling through it. Steps to further reduce risk in access points with low positive pressure air speeds include having personnel wear disposable coveralls when outside doors are open, and disinfecting the area once they are closed.

Conclusion

1) Diluted vaccine aerosolization combined with novel environmental sampling techniques allowed for testing of PRRSV aerosol entry into PPF farm access points. 2) Under the experimental conditions of this study, positive pressure air speeds > 1.85 m/s resulted in no aerosol transmission. 3) Ensuring adequate positive pressure air speed through steps taken to increase access point pressure can further reduce the risk of aerosol PRRSV transmission on PPF farms.

References

1. Holtkamp DJ et al. Assessment of the economic impact of porcine reproductive and respiratory syndrome virus on United States pork producers. *J Swine Heal and Prod.* 2013;21:72.

2. Dee S et al. Evidence of long distance airborne transport of porcine reproductive and respiratory syndrome virus and Mycoplasma hyopneumoniae. *Vet Res.* 2009;40:39.

3. Alonso C et al. Epidemiological study of air filtration systems for preventing PRRSV infection in large sow herds. *Prev Vet Med.* 2013;112:109.

4. Alonso C et al. An evaluation of interventions for reducing the risk of PRRSV introduction to filtered farms via retrograde air movement through idle fans. *Vet Microbiol.* 2012;157:304.

Evaluation of methods for determining the porcine circovirus type 2 status of sow herds

Chelsea Ruston[1], BS; Jeremy S. Pittman[2], DVM, MS, DABVP
[1]Iowa State University College of Veterinary Medicine, Ames, Iowa;
[2]Smithfield Hog Production Division – North Region Waverly, Virginia

Introduction

Cases of porcine circovirus type 2 (PCV2) associated disease in growing pigs have been traced to instability in the sow herd, defined as *in utero* infection and viremic piglets at birth. To date there have been several diagnostic methods evaluated to determine PCV2 status of sow herds such as detection via PCR in sow serum, colostrum, piglet serum up to 6 weeks of age (WOA), placental umbilical cord serum (PUCS), pre-suckle piglet serum and born dead piglet tissues (lungs, heart) or by evaluation of maternal antibody levels in 2 to 3 WOA piglets by immunofluorescent antibody (IFA) and enzyme linked immunosorbent assay (ELISA).[1-3] Another option might be to detect PCV2 via PCR in PUC tissue homogenate (PUCT). The objective of this study was to determine which diagnostic sample, colostrum, PUCS, PUCT, or piglet serum, was best for detecting PCV2.

Materials and methods

A 10,000 head commercial sow farm with a recent history of PCV2 instability (i.e., downstream grow/finish PCVAD and PCR positive PUCS) was used for this study. From 150 sow-litter pairs farrowing over 3 consecutive weeks, colostrum (1-2 ml), pooled PUCS (~5 ml) and multiple one inch pieces of intermittent PUCT (6-10 pieces) from the same 3 to 5 umbilical cords were collected. Approximately 50 sow-litter pairs were sampled each week. Three piglets from each sow were convenience enrolled, tagged and remained on the sow until weaning at 3 WOA. Serum from tagged piglets was obtained at 5 to 6 days of age (DOA). Thereafter, 52 litters were selected based on PCR findings and serum was obtained at 3 and 6 WOA. Semi-quantitative PCR for PCV2 was performed on colostrum, PUCS, PUCT and piglet serum, pooled by sow-litter. Full range and 4 dilution IFAs were performed on 3 WOA serum. Percent agreement (PA) and Kappa was calculated to compare the PCR results for colostrum, PUCS and PUCT samples, to each other and to a hypothetical "gold standard" defined as a sow-litter positive to any sample type.

Results

Less than 5% of piglet serum at all ages sampled were PCR positive for PCV2. There was no relationship between PCR results and IFA titers at 3 WOA. Total PCR percent positives out of the 150 sow-litters sample is shown in Figure 1. There was week-to-week variation in the percent positives among each sample type as shown in Figure 2. Colostrum and PUCS PCR positives had a PA (67.3%) and PUCS and PUCT PCR positives had a PA (36.7%). The Kappa score comparing PUCS, colostrum, PUCT to a "gold standard" was 0.475, 0.239 and 0.033, respectively. The average CT was highest for PUCS and lowest with PUCT with colostrum intermediate.

Discussion

In this study, PUCS appeared to be the most sensitive sample for evaluating PCV2 status of a sow herd. Colostrum, while less sensitive, was also successful in detecting PCV2 at a relatively high rate. PUCT and post-suckle pig serum at any age was not able to detect PCV2 by PCR reliably and should be discouraged as a diagnostic test. Tissue homogenization possibly dilutes virus resulting in higher or negative CT values and maternal antibodies might clear the viremia and mask infection in piglet serum. There was significant variation in percent positive PUCS across the three sampling weeks, suggesting evaluation of sow herd PCV2 status should be performed over a period of time.[4] The Kappa score results suggest that performing PCV2 PCR on PUCS or colostrum could be helpful to practitioners when trying to determine the PCV2 stability of a sow herd.

Figure 1: Total percent PCR positive results.

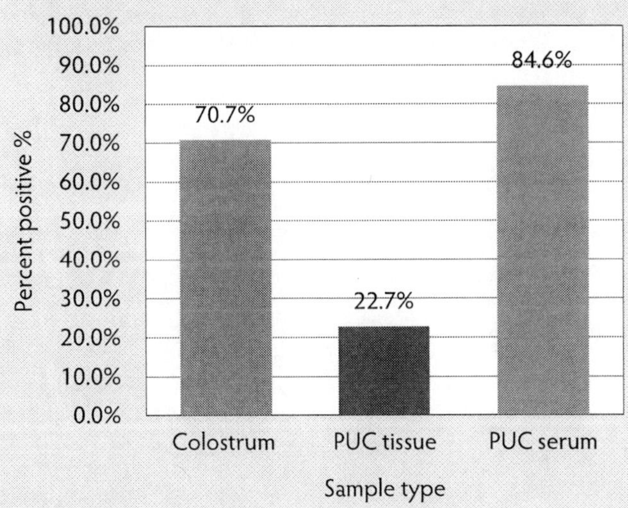

Figure 2: Percent PCR positive results across the three collection weeks.

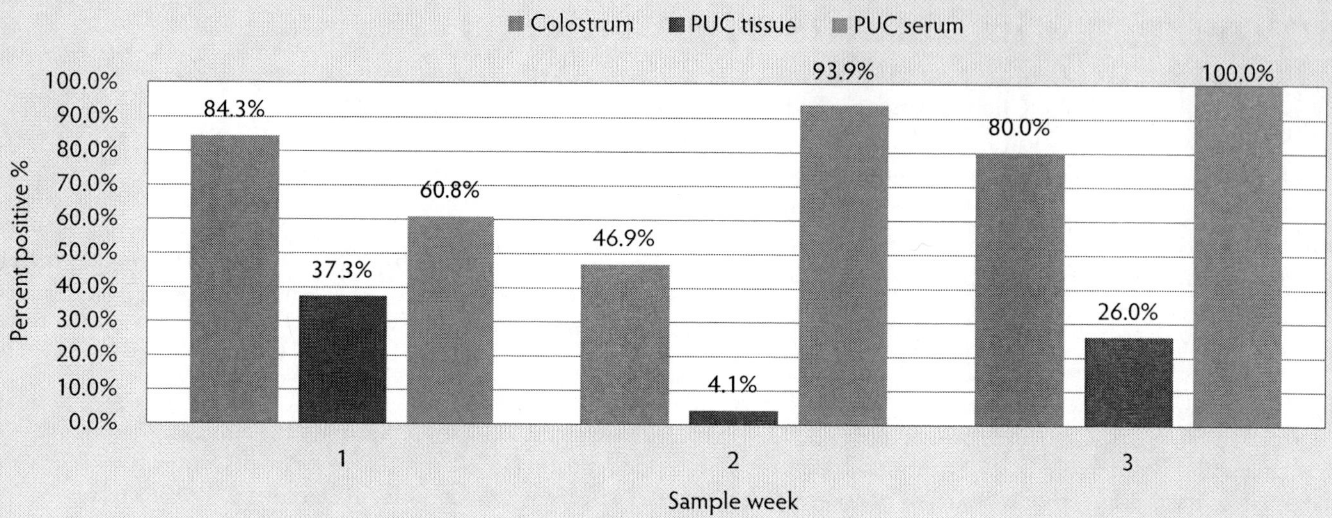

References

1. Stoffel, C., Schwartz, T., Stricker, A. Comparison of sample types for assessing porcine circovirus type 2 status of sow herds. Proc AASV. New Orleans, Louisiana. 2016; 281.

2. Madson, D. Iowa State University Veterinary Diagnostic Laboratory Seminar, speaking at the 2010 World Pork Expo.

3. Madson, D., Opressnig, T. PCV2-associated reproductive failure. Proc AASV. Phoenix, Arizona. 2011; 430.

4. Te Grotenhuis, T., Fangman, T., Baumert, D. Using placental umbilical cord serum to determine porcine circovirus type 2 status of breeding herd. Proc AASV. Dallas, Texas. 2014; 311.

Dynamics of *Mycoplasma hyopneumoniae* colonization, seroconversion and onset of clinical signs in a population of gilts under field conditions

Chris Deegan[1]; Daniel Linhares[2], DVM, MBA, PhD; Alex Ramirez[2], DVM, MPH, PhD, DACVPM; Deb Murray[3], DVM; Marcelo Almeida[2], DVM; Lucina Galina Pantoja[4], DVM, PhD

[1]College of Veterinary Medicine, University of Minnesota, St. Paul, Minnesota; [2]College of Veterinary Medicine, Iowa State University, Ames, Iowa; [3]New Fashion Pork, Jackson, Minnesota; [4]Zoetis, Parsippany, New Jersey

Introduction

Mycoplasma hyopneumoniae (Mh) infection results in a significant economic impact on pig production due to its effect on porcine respiratory disease complex (PRDC) in growing pigs as primary pathogen and/or by potentiating infection caused by other pathogens including porcine reproductive and respiratory syndrome virus (PRRSV), porcine circovirus type 2, or Influenza A virus (IAV).[2] A recent study under field conditions showed that Mh increased pig production cost by $ 1.00 in uncomplicated infections and by $ 10.00 per pig in pig flows co-infected with PRRSV or IAV.[1]

For successful control of Mh in swine production systems, effective gilt acclimation to Mh is critical.[2] When done correctly, a uniform herd immunity will be developed, which will mitigate Mh shedding by the time the gilts are entered into the breeding herd. Thus, establishing anti Mh-herd immunity is a critical step in strategies to control and eliminate Mh from breeding herds. More specifically, the ideal scenario for introducing gilts into breeding herds undergoing Mh control/elimination (low Mh pressure of infection) would be: all gilts being previously exposed to Mh (i.e. having detectable anti-Mh antibody titers), but not actively shedding the bacteria (i.e. Mh DNA not detected by PCR from the respiratory tract). Characterization of Mh shedding and exposure status in pigs has been done through PCR-based assays on sample types such as laryngeal swabs (LS), nasal swabs (NS), oral fluids (OF), and antibody response (S/P). The objective of this study was to describe the dynamics of Mh DNA and anti-Mh antibody detection over time in a naturally exposed herd to determine the method that is most sensitive at determining Mh shedding status, in comparison to timing of seroconversion.

Materials and methods

A continuous flow, gilt development unit (GDU 1) with 9 barns with a clinically active Mh infection (based on PCR testing of nasal swabs, histopathology and clinical signs compatible with Mh disease) was used for the first four weeks of this study. Between weeks four and five, the gilts were moved to another site (GDU 2). The GDUs (1 and 2) were both negative for PRRSV and IAV. Thirty, 7-week old gilts were tagged and tested on arrival at GDU 1 using LS, NS, pen OF and serology to confirm their negative status. Study pigs were tested weekly (PCR on LS, NS, OF and ELISA on blood serum) for 5 weeks. The University of Minnesota Veterinary Diagnostic Lab tested the LS, NS and OF for Mh DNA by quantitative PCR and serum for anti-Mh antibodies by IDEXX ELISA. The time from gilt entry to testing positive via LS, NS, OF and serology was described using interaction plot.

Results and discussion

For the purpose of this study, we evaluated the dynamics of Mh infection breaking down into phases of colonization (Mh DNA detected on LS), evidence of shedding (Mh DNA detected on OF or NS), seroconversion (anti-Mh antibodies detected on blood serum) and clinical disease (presence of a dry, unproductive cough consistent with Mh literature). Our data showed Mh had a quick colonization period, where 73% of pigs were positive after 12 days post natural exposure and 100% positive after 4 weeks (Figure 1). This colonization rate is faster than previously reported by Dr. Kent Schwartz where it could take longer than 5 months for every pig to test positive.[3] This may be attributed to Mh strain properties, pathogen load or sensitivity of testing method of selected biological samples. The frequency of detecting shedding by NS and OF was lower than detection of shedding pigs by LS. We interpreted and propose that colonization was able to be detected sooner and more effectively than shedding. As expected, the onset of clinical signs (coughing) was delayed compared to time to detect Mh DNA in LS samples. Our data shows the relatively low cough (incidence of 5%) at week 2 compared to the 73% positivity rate of LS qPCRs (Figure 1).

Similarly, antibody response was effective at determining exposure to Mh and, as expected, detection of Mh exposure by serology was delayed (between weeks 3-5) compared to LS testing by PCR (between weeks 1-3)(Figure 1). In conclusion, this study outlines the differences in sample types in detecting and confirming a Mh infection and demonstrates the importance of sample selection when testing for Mh based on expected time point since natural exposure. Since the data shows that detection of colonization using LS occurred much quicker than detection of shedding (using NS or OF), seroconversion and onset of clinical signs, an acute Mh infection can best be detected using the LS swab method. This is important information for practitioners because it not only helps to determine which sample type to collect when

Figure 1: Frequency of samples positive to Mh DNA by PCR over time and by sample type.

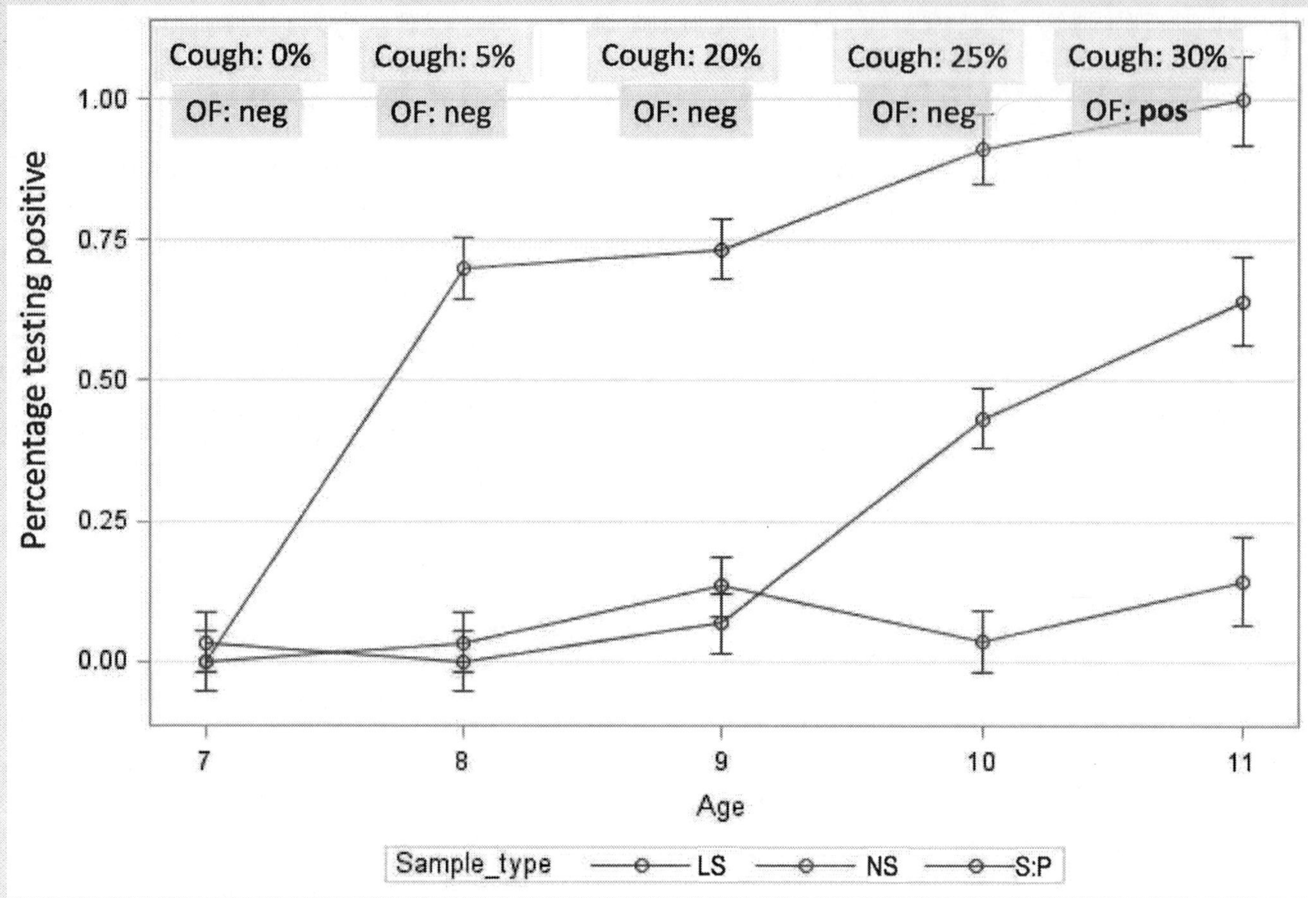

testing for Mh, but also will aid in the interpretation of the results for tests. When selecting a sample type to test for Mh, ease of collection, employee training, and the equipment available at the farm need to be considered. This is a reason that OF or NS may be selected over LS, but with training protocols and fairly cheap equipment upgrades, LS collection is feasible to be taught to employees in the field, increasing the accuracy of testing and detecting Mh.

References

1. Haden DC, Painter T, Fangman T, et al. Assessing production parameters and economic impact of swine influenza, PRRS and *Mycoplasma hyopneumoniae* on finishing pigs in a large production system. *Proceedings 43rd Annual Meeting Am Assoc Swine Veterinarians*, Denver, Colorado. 2012:75-76.

2. Pieters M, Pijoan C, Fano E, et al. An assessment of the duration of *Mycoplasma hyopneumoniae* infection in an experimentally infected population of pigs. *Vet Microbiol* 2009;134:261-266.

3. Schwartz K, et al. Choosing the right test: Zoetis Inc, *A contemporary review of Mycoplasma hyopneumoniae control strategies*, 2016; 18.

Investigating porcine circovirus associated disease (PCVAD) in commercial swine herds by next generation sequencing

Zhen Yang[1]; Lucina Galina Pantoja[2], DVM, PhD; Todd Knutson[1], PhD;
Albert Rovira[1], DVM, MS, PhD; Douglas Marthaler[1], PhD
[1]College of Veterinary Medicine, University of Minnesota, Saint Paul, Minnesota; [2]Zoetis, Parsippany, New Jersey

Introduction

Porcine circovirus type 2 (PCV2) is one of the most economically important viral diseases of pigs, causing a series of disease and syndromes such as postweaning multisystemic wasting syndrome (PMWS), porcine dermatitis and nephropathy syndrome (PDNS) and reproductive disorder, which are collectively named as porcine circovirus-associated disease (PCVAD).[1-4] Porcine circovirus type 2 is a small circular, single-stranded DNA virus with two major open reading frames (ORF1 and ORF2). Five distinct genotypes of PCV2 were identified, namely PCV2a, PCV2b, PCV2c, PCV2d and recently PCV2e.[5] Current commercial PCV2a vaccines are consistently efficacious against PCVAD via reducing clinical signs and improving productivity.[6-8] In the US, the percentage of PCR positive cases declined after the massive introduction of commercial vaccines in 2006.[9] However, clinical PCVAD cases sporadically appear in commercial herds, and the virus still circulates in production systems with a genetic shift to PCV2d.[10] In the summer of 2016, a few PCVAD cases from Midwest swine farms were investigated, and we hypothesized that unknown viral co-infections and/or potential genetic changes in the whole genome PCV2 were contributing to the complexity of PCVAD. In this retrospective study, information from diagnostic reports, vaccination protocols, next-generation sequencing (NGS) and molecular analysis were integrated to investigate recent cases of PCVAD in vaccinated herds. The objectives were to use NGS technology to 1) investigate the involvement of other viruses with PCVAD and 2) analyze the variability in the whole PCV2 genome of PCVAD cases.

Materials and methods

Between April and June of 2016, a single case from Ohio and five cases from Minnesota PCVAD were selected for this study. Historical information was retrospectively collected from each case through diagnostic reports and consultations with the case veterinarian (Table 1). Six lung specimens from a single case of PCVAD, located in Ohio (OH1 to OH6), were obtained from Iowa State University Veterinary Diagnostic Laboratory while five tissue homogenate specimens (MN1 to MN5) of five different Minnesota cases were from University of Minnesota Veterinary Diagnostic Laboratory (UMN VDL) (Figure 1A). The 11 tissue samples were homogenized, the supernatants were centrifuged at 4500 rpm for 60 min, and filtered through a 0.22-µm filter to remove bacterium-sized particles. The nucleic acid was extracted using the Direct-zol™ RNA MiniPrep. All nucleic acid samples were submitted to the University of Minnesota Genomic Center for Quality Control of the nucleic acid and library preparation, then sequenced using pair-end, 250bp reads on the Illumina MiSeq platform, yielding an average of 1.3 million reads per individual tissue specimen. FASTQ files generated by the MiSeq were input directly into UMN VDL's NGS analysis pipelines for pathogen detection (Figure 1B). The paired-end (PE) reads were trimmed, adapter sequences and low quality sequences were removed, using Trimmomatic software. Surviving PE reads were classified using Kraken software, which contains a database of known reference sequences available in GenBank. Pipeline 2 (genome assembly) utilized the specific read viral classified by Kraken for de novo viral assembly, using tree viral-specific assemblers: Integrative Virus Assembler (IVA), Paired-Read Iterative Contig Extension (PRICE), and A5. Full-length PCV2 assemblies were verified by re-mapping the short reads to the sequence using Bowtie2. The PCV2 ORF1, ORF2, and the whole genome from each strain were aligned and compared with selected reference strains and the PCV2 database from a recent PCV2 manuscript.[5]

Results

Mortality across cases ranged ranged from 3-12%, with PCV2 Ct values between 8-27 (Table 1). Histologically, all the cases had lymphoid depletion and macrophage infiltration in lymphoid tissues. Information on the PCV2 vaccination protocol was obtained from three cases, and deviations from the recommended manufacture protocols occurred in all three cases. In the Ohio case, with the exception of PCV2, no other viral pathogens were found while in four of the five Minnesota cases, porcine reproductive and respiratory syndrome virus (PRRSV) was detected. Eight PCV2 full genomes from different samples were successfully assembled, four from the Ohio case and four from the Minnesota cases. A single PCV2a (MN2) and seven PCV2d strains (OH1, OH4, OH5, OH6, MN3, MN4 and MN5) were identified (Figure 2), and the nucleotide identity between PCV2a and PCV2d was 97.8-98.1% and 90.2-90.3% in ORF1 and ORF2, respectively (Table 2). Two clusters of PCV2d strains were identified, with more nucleotide changes in ORF1 (97% identity) than ORF2 (99% identity) (Table 2).

Table 1: Diagnostic summary of five PCVAD cases *

Case	Age (wks)	Clinical signs	Mortality	Lymphoid depletion	PCV2 (Ct)	PRRSV (Ct)	Vaccination protocol
MN1	13	weight loss	NA	Yes	14.03	27.5	NA
MN2	Nursery	Not given	3-12%	Yes	8	13	First dose several days pre wean (14 days) and again 3 weeks post wean.
MN3	6	PCVAD respiratory abortion	NA	Yes	7.8	26	NA
MN4	11	Not given	NA	Yes	7.6	32	NA
MN5	8	PWMS sudden death	6.50%	Yes	8.44	Negative	1 cc at weaning followed by two 0.5 cc shots at 3 weeks and 5 weeks post-placement in the nursery.
OH (1-6)	9-18	PCVAD	3-4%	Yes	8.4-27.2	Negative	2cc two days after weaning(11-15d)

* MN = Minnesota, each MN PCVAD case had one tissue homogenate sample for next generation sequencing(NGS); OH = Ohio, Ohio case had six lung tissue samples for NGS. Histological diagnosis showed that all cases had lymphoid depletion. Bacteria results were not listed in this table.

Figure 1: 11 tissue or tissue homogenate PCVAD specimens underwent the following processes. The flow chart A was the simplified sample processing. The flow chart B was the simplified bioinformatic pipelines used in this PCVAD investigation, including pathogen identification, genome assembly, comparative genomics and phylogenetic analysis.

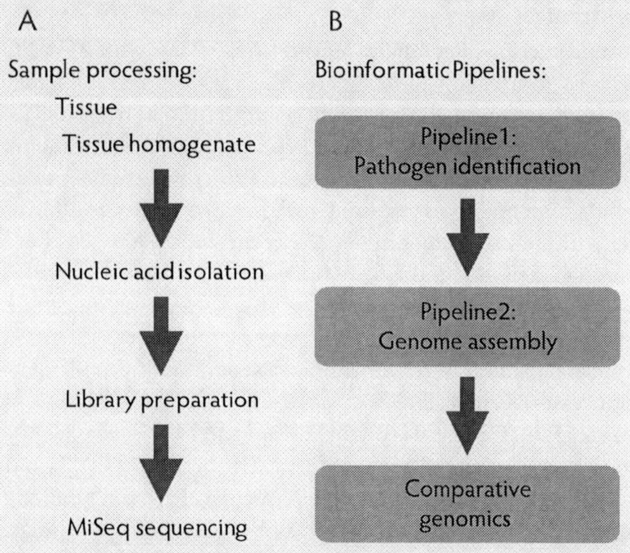

Discussion and implications

Studying pathogen coinfections as well as vaccination protocols is critical for advances in the prevention of PCVAD. Viral co-infections, like PRRSV and porcine parvovirus confection, have been proven to be one of the main triggers for the development of PCVAD.[9,11-14] On the other hand, there is a concern that the emergence of new or variants of PCV2 may cause PCVAD in vaccinated herds, although current commercial vaccines are effective against the different genotypes of PCV2 under experimental conditions.[6-8] A novel porcine circovirus (PCV3) causing similar PCVAD signs, including porcine dermatitis and nephropathy syndrome (PDNS) and reproductive disorders, was recently identified by NGS, which supports NGS was an applicable technology to discover unknown viruses in pigs.[15] In this study, we explored the feasibility of using NGS to identify potential unknown pathogens in pigs with PCVAD. While unknown viruses including newly reported PCV3 were not identified in these samples, PCV2 and PRRSV were identified, indicating that NGS did not miss any important pathogens contributing to PCVAD in our study. In addition, NGS proved very useful to generate the whole genome of PCV2 in PCVAD cases. According to the phylogenetic analysis of ORF2, PCV2 strains were similar to previously reported sequences from the Midwest region. As for whole genome analysis, the PCV2d strains had more variations in ORF1 than ORF2, indicating the variability in the ORF1 of PCV2 needs further investigation. With regards to vaccination protocols, three farms used different commercial PCV2a vaccines, and differences were observed in the number of

Figure 2: Phylogenetic analysis of the open reading frame 2 (ORF2) for PCV2 sequences. An unrooted phylogenetic tree was constructed by the neighbor-joining method from aligned sequences: 729 PCV2 ORF2 sequences from a recent manuscript[5] and 8 PCV2 ORF2 sequences in this study, including OH1, OH4, OH5, OH6, MN2, MN3, MN4, MN5. The strains OH1, OH4, OH5, OH6, MN5 and MN3, MN4 were clustered into two clades within PCV2d group while MN2 was identified as PCV2a.

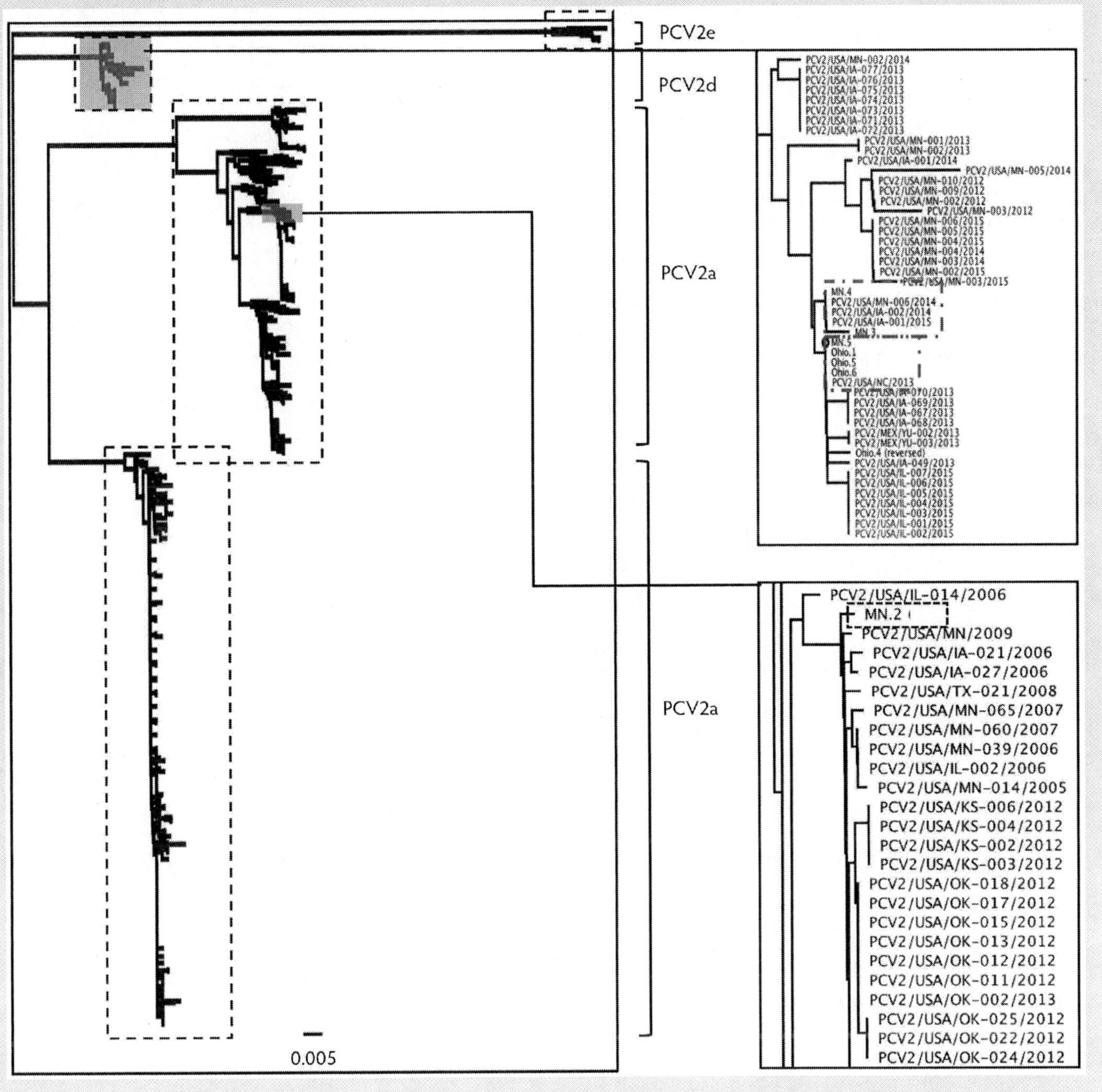

Table 2: The nucleotide identity between PCV2a and PCV2d and two clusters of PCV2d strains in ORF1, ORF2 and whole genome. PCV2a = MN2; PCV2d = OH1, OH4, OH5, OH6, MN2, MN3, MN4, MN5; PCV2d cluster 1 = OH1, OH4, OH5, OH6, MN5; PCV2d cluster 2= MN3, MN4.

Genotypes	ORF1	ORF2	Whole genome
PCV2a - PCV2d	97.8% - 98.1%	90.2% - 90.3%	95.9% - 95.1%
PCV2d cluster 1 and 2	97.3% - 97.6%	99.6% - 99.9%	98.6% - 100%

doses, age of vaccination and timing. Since no other viral co-infections (except PRRSV) or novel changes within the PCV2 genome were identified, our PCVAD cases could be the result of deviation from the vaccination protocols, different management within farms, and/or co-infections with PRRSV. Given the complexity of current PCVAD, case investigations must include vaccine information, diagnostic workups, and the complement of alternative tools to solve the associations of PCVAD for each case. Next generation sequencing was a powerful tool to identify co-infections and understand the variability of the whole genome of circoviruses.

References

1. Chae C. A review of porcine circovirus 2-associated syndromes and diseases. *Vet J.* 2005;169(3):326-336.

2. Segalés J et al. Porcine circovirus diseases. *Anim Heal Res Rev.* 2005;6(2):119-142.

3. Opriessnig T et al. Porcine Circovirus Type 2 associated disease: Update on current terminology, clinical manifestations, pathogenesis, diagnosis, and intervention strategies. *J Vet Diagn Invest.* 2007;19(6):591-615.

4. Alarcon P et al. Cost of post-weaning multi-systemic wasting syndrome and porcine circovirus type-2 subclinical infection in England - An economic disease model. *Prev Vet Med.* 2013;110(2):88-102.

5. Davies B et al. Diagnostic phylogenetics reveals a new Porcine circovirus 2 cluster. *Virus Res.* 2016;217:32-37.

6. Shen HG et al. Comparison of commercial and experimental porcine circovirus type 2 (PCV2) vaccines using a triple challenge with PCV2, porcine reproductive and respiratory syndrome virus (PRRSV), and porcine parvovirus (PPV). *Vaccine.* 2010;28(37):5960-5966.

7. Chae C. Commercial porcine circovirus type 2 vaccines: Efficacy and clinical application. *Vet J.* 2012;194(2):151-157. doi:10.1016/j.tvjl.2012.06.031.

8. Rose N et al. A commercial PCV2a-based vaccine significantly reduces PCV2b transmission in experimental conditions. *Vaccine.* 2016;34(33):3738-3745.

9. Afghah Z et al. Ten years of PCV2 vaccines and vaccination: Is eradication a possibility? *Vet Microbiol.* 2016.

10. Xiao CT et al. Global molecular genetic analysis of porcine circovirus type 2 (PCV2) sequences confirms the presence of four main PCV2 genotypes and reveals a rapid increase of PCV2d. *J Gen Virol.* 2015;96(7):1830-1841.

11. Pogranichniy RM et al. Case-control study on the association of porcine circovirus type 2 and other swine viral pathogens with postweaning multisystemic wasting syndrome. *J Vet Diagn Invest.* 2002;14(6):449-456.

12. Allan GM et al. Experimental Reproduction of Severe Wasting Disease by Co-infection of Pigs with Porcine Circovirus and Porcine Parvovirus. *J Comp Pathol.* 1999;121(1):1-11.

13. Allan GM et al. Experimental infection of colostrum deprived piglets with porcine circovirus 2 (PCV2) and porcine reproductive and respiratory syndrome virus (PRRSV) potentiates PCV2 replication. *Arch Virol.* 2000;145(11):2421-2429.

14. Gillespie J et al. Porcine circovirus type 2 and porcine circovirus-associated disease. *J Vet Intern Med.* 2009;23(6):1151-1163.

15. Palinski R et al. A novel porcine circovirus distantly related to known circoviruses is associated with porcine dermatitis and nephropathy syndrome and reproductive failure. *J Virol.* 2016;(October):JVI.01879-16.

Effect of antimicrobials on non-toxigenic *E. coli* oral vaccine

Olivia Myers[1], BS; Jeremy S. Pittman[2], DVM, MS, DABVP; Megan Jacob[1], MS, PhD; Mark Hammer[3], DVM; Greg Stevenson[4], DVM, MS, PhD, DACVP
[1]College of Veterinary Medicine, North Carolina State University, Raleigh, North Carolina;
[2]Smithfield Hog Production Division North Region, Waverly, Virginia; [3]Elanco, Greenfield, Indiana;
[4]Iowa State University, Ames, Iowa

Introduction

Escherichia coli strains expressing K88 (F4) fimbriae are one of the most important causes of post-weaning diarrhea (PWD) in pigs.[1] *E.coli* results in economic loss due to increased mortality, decreased growth rate, and an increased need for treatment. However, many systems use a live, oral non-toxigenic *E.coli* vaccines to prevent PWD around the time of weaning.[2] Also around the time of weaning, pigs are often given antimicrobials in feed, water or as injectables. Three antimicrobials used in wean age pigs are avilamycin (Kavault, Elanco, Greenfield, IN), ceftiofur crystalline free acid (Excede for swine, Zoetis, Madison, NJ) and enrofloxacin (Baytril, Bayer, Shawnee Mission, KS). Kavault is an orthosomycin antibiotic that impairs expression of K88 (F4) *E.coli* fimbriae and therefore prevents attachment.[3] Kavault is labeled for reduction in incidence and overall severity of diarrhea in the presence of pathogenic *E.coli* in groups of weaned pigs. Excede is a third generation cephalosporin that is labeled to control swine respiratory pathogens. It has been confirmed to be an effective treatment for colibacillosis and reduces mortality in neonatal piglets orally challenged with K88 *E.coli*.[4,5] Baytril is a fluoroquinolone labeled for swine respiratory pathogens and colibacillosis associated with *E. coli* in groups of weaned pigs. The objective of this study was to determine if these antimicrobials commonly used at weaning interfere with the viability of an oral, live, non-toxigenic *E.coli* vaccine, and thereby reduce immunity to *E.coli*.

Materials and methods

Ten pregnant sows on a commercial farm were identified as homozygous dominant for the K88+ receptor.[6] Eighty-five piglets from eight of the sows and five piglets from a K88 heterozygous sow were blocked by weight and gender and randomly allocated into six different treatment groups (Table 1); negative control group (NEG) that did not receive vaccine (VAC) or antibiotics but was challenged (CHAL), a positive control group (POS) that received VAC and CHAL but no antibiotics, Excede group (EXC) that received Excede, VAC and CHAL, Baytril group (BAY) that received Baytril, VAC and CHAL, Kavault partial group (KVP) that received 80 ppm (73 g/T) of Kavault d-1 through d6, VAX and CHAL, and Kavault full group (KVF) that received 80 ppm (73 g/T) of Kavault for the entire duration, VAC and CHAL. Excede and Baytril injections were given 24 hours (d-1) prior to VAC. Fifteen pigs within treatment groups were allocated to three adjacent pens, while treatment group pens were physically separated. Two diets (non-medicated and Kavault) were identical corn, soy, whey pre-starter pelleted diets, except for medication. Feed was weighed and hand-fed for 42 days. NEG, POS, EXC, BAY received the non-medicated diet d-1 through d14. On d6, KVP was switched to non-medicated feed. At completion of the clinical evaluation period (d14), all pigs received Kavault until completion (d42). On d0, all pigs (except NEG) were vaccinated with 10^9 CFU of oral, live non-toxigenic (K88+, STa-, STb-, LT-) *E.coli* sensitive to ceftiofur and enrofloxacin. On d2, one pig per pen (n = 3/treatment) was randomly selected and necropsied. Six small intestinal mucosal impression smears and adjacent fixed intestine were taken every 4 inches proximally, from the ileo-cecal junction. Impression smears were pooled by pig and tested with a K88+ *E.coli* qPCR. Fixed tissue was scored semi-qualitatively for coliform colonization by a blinded pathologist. On d9, pigs were challenged with 10^9 CFU of a pathogenic (K88+, STa+, STb+, LT+) *E.coli* isolate (CHAL). Rectal swabs were taken on d0 and d9 for qualitative culture. Fecal samples were taken on d2, d11 and d13 for quantitative culture. Packed cell volume was taken on d9 and d13 to assess dehydration. Clinical scores including lethargy, fecal consistency and gauntness were recorded daily through d14. Pigs were weighed on d-1, d9 and d42. Feed and feeders were weighed periodically to determine feed conversion.

Results

One NEG and two BAY pigs died from colibacillosis during the study. On d2, KVP, KVF and BAY had higher log of *E.coli* as compared to NEG ($P < 0.01$) (Figure 1). On d11, EXC, BAY and KVF had higher log *E.coli* as compared to KVP or NEG ($P = 0.03$) (Figure 2). Fecal consistency was less formed from d10-14 in BAY and NEG compared to POS and KVF ($P = 0.02$) (Figure 3). BAY had higher ADG d-1-14 compared to KVF, KVP and NEG ($P < 0.01$) (Figure 4). EXC and POS had lower cycle-threshold values for K88+ *E.coli* qPCR on impression smears than NEG ($P < 0.01$) (Figure 5). There was not a difference in lethargy, gauntness or PCV between treatments.

Discussion

Under the conditions of this study, EXC had an impact on colonization of *E.coli* and EXC, BAY and KVF had an impact

Table 1: Timeline of samples and procedures taken throughout the trial by treatment groups from days -4 to 14.

<table>
<tr><th>Treatment</th><th>Day -4</th><th>Day -1</th><th colspan="2">Day 0</th><th>Day 2</th><th>Day 6</th><th>Day 9</th><th>Day 11</th><th>Day 13</th><th>Day 14</th></tr>
<tr><td>Ceftiofur (Excede)-EXC=1.0 mg/kg</td><td rowspan="6">Tag, Count, Gender</td><td rowspan="6">Enrollment, Weaning, Started Treatment (gave injections and Kavault)</td><td rowspan="6">Fecal Swabs (in AM)</td><td rowspan="4">Vaccination</td><td rowspan="6">Necropsy (1 pig/pen) Fecal samples (+/-Swabs)</td><td></td><td rowspan="6">Challenge, Fecal swabs, PCV</td><td rowspan="6">Fecal Samples</td><td rowspan="6">Fecal Samples, PCV</td><td rowspan="6">All Pigs on Kavault</td></tr>
<tr><td>Avilamycin (Kavault) Partial-KVP=80ppm (73 g/T)</td><td>Switched to Non-Medicated Feed</td></tr>
<tr><td>Enrofloxacin (Baytril)-BAY= 1.55 mg/kg</td><td rowspan="4"></td></tr>
<tr><td>Avilamycin (Kavault) Full-KVF=80ppm (73 g/T)</td></tr>
<tr><td>Negative Control-NEG</td><td></td></tr>
<tr><td>Positive Control-POS</td><td></td></tr>
</table>

on immune development. Impression smear K88+ qPCR was a better indicator of *E.coli* colonization than histopathology. Treatment of wean pigs for systemic infections should not exclude prevention of enteric diseases. The benefits of treatment with Baytril, Excede or Kavault at the time of oral, non-toxigenic *E.coli* vaccine should be weighed since there appears to be interference of antimicrobials with the vaccination. Given these results, adjustment of *E.coli* vaccination, treatment protocols or timing may need to be considered when using both close together.

References

1. Fairbrother JM, Harel J, D'Allaire S and Bonneau M (1994). Characterization of *Escherichia coli* isolated from postweaning piglets with and without diarrhea. In: Proceedings of the 13th IPVS Congress, Bangkok, Thailand, p. 212.

2. Fuentes M, Pijoan C, Becton L, Morrison B and Pieters M. (2004). Inoculation of nonpathogenic *Escherichia coli* to control disease and reduce antibiotic usage. In: Proceedings of the 18th Congress IPVS, Hamburg, Germany, vol. 18, p. 258.

3. Rostagno, M and Pelger, G. (2014). Exposure to the antibiotic availmycin inhibits *E.coli* fimbriae and attachment. In: Proceedings of the 23rd IPVS Congress, Mexico, p. 246.

4. Hibbard B, Kausche FM, Lechtenberg KF, et al. Efficacy of ceftiofur sodium for the control of mortality in neonatal pigs orally inoculated with K88+ (F4+) enterotoxigenic *Escherichia coli*. J Swine Health Prod. 2003;11(1):7–11.

5.Yancey RJ (1990). Efficacy of ceftiofur hydrochloride for treatment of experimentally induced colibacillosis in neonatal swine. American journal of veterinary research 0002-9645, 51:3, p. 349.

6. Sellwood R, Gibbons RA, Jones GW and Rutter JM (1975). Adhesion of enteropathogenic *Escherichia coli* to pig intestinal brush borders: the existence of two pig phenotypes. Journal of Medical Microbiology 8: 405–411.

Figure 1: Quantitative culture results from fecal samples taken on Day 2 (2 days post-vaccination) of the trial showed increased *E. coli* colonization in the Kavault partial (KVP), Kavault full (KVF) and Baytril (BAY) groups as compared to the negative control (NEG) group ($P < 0.01$).

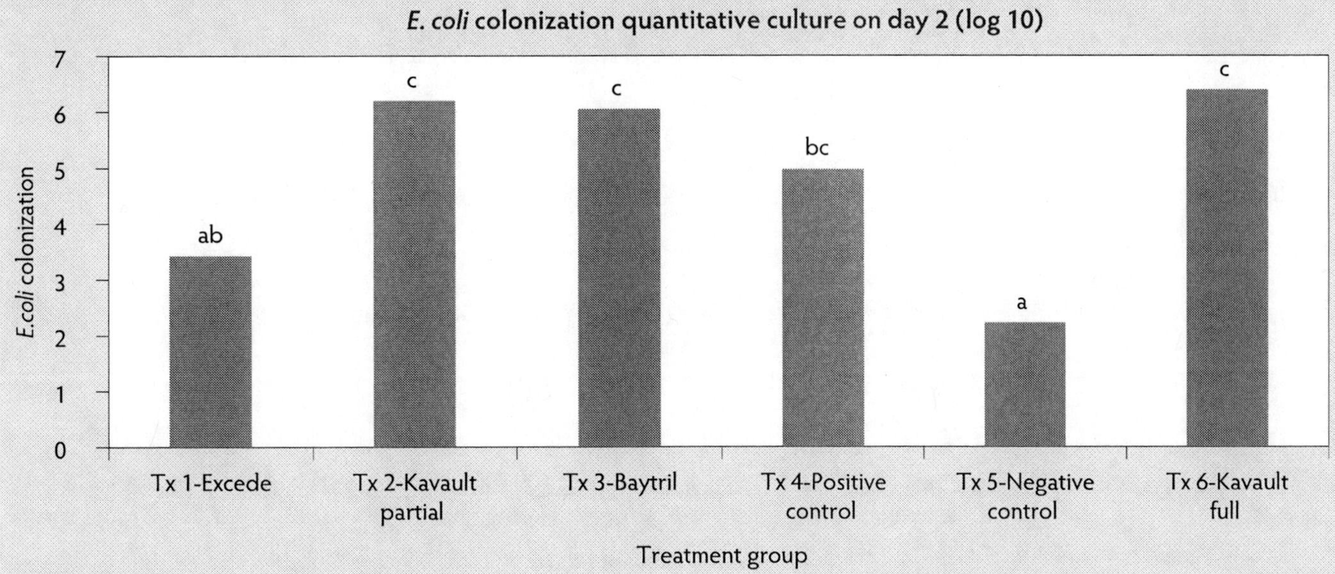

abc Indicate groups that are significantly different

Figure 2: Quantitative culture results from fecal samples taken on Day 11 (2 days post-challenge) of the trial showed increased *E. coli* colonization in the Excede (EXC), Baytril (BAY), and Kavault full (KVF) groups as compared to the Kavault partial (KVP) and negative control (NEG) groups ($P = 0.03$).

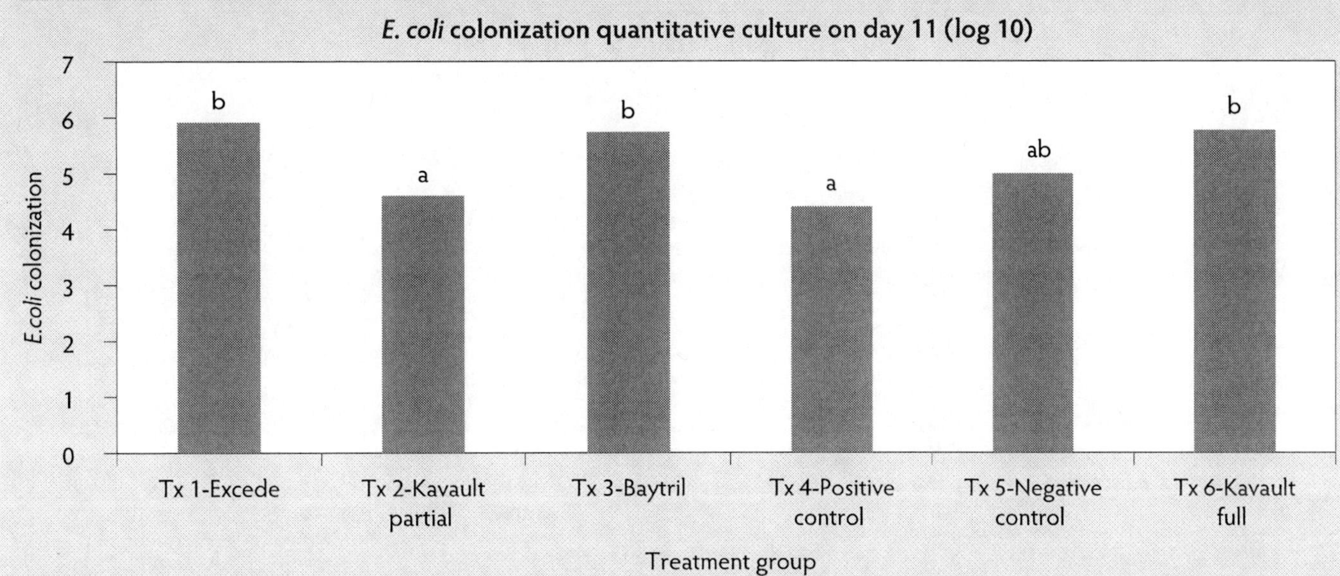

abc Indicate groups that are significantly different

Figure 3: Fecal consistency as judged from clinical scores was higher (less formed stool) on Days 10-14 of the trial (1-5 days post-challenge) in Baytril (BAY) and negative control (NEG) groups as compared to postive control (POS) and Kavault full (KVF) groups ($P = 0.02$).

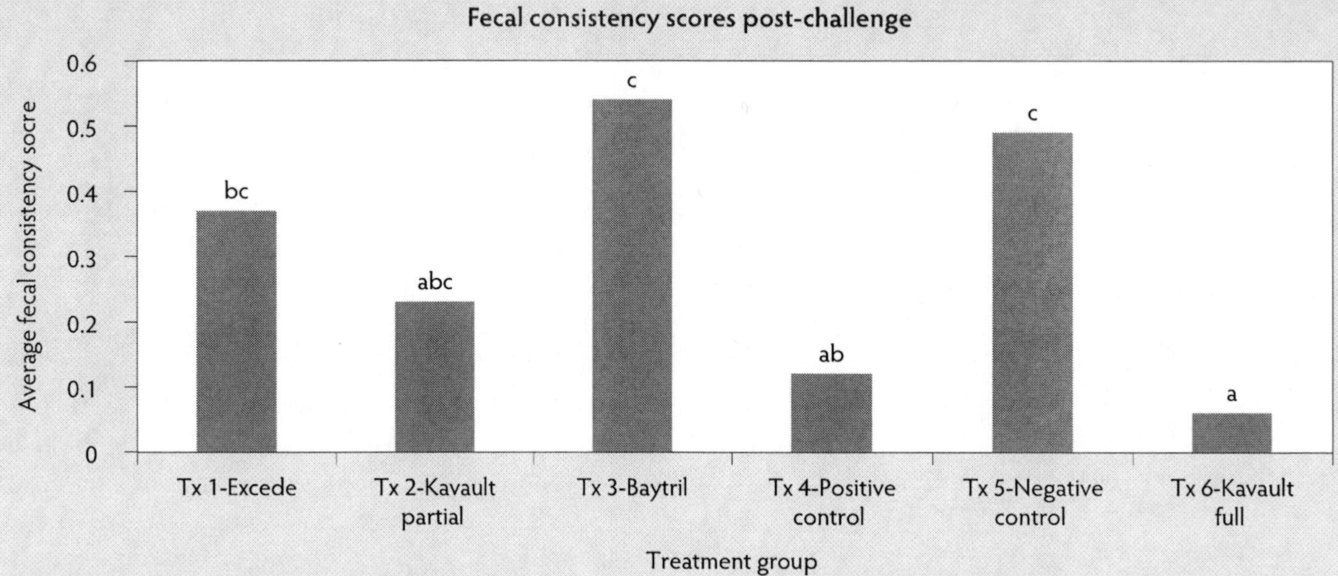

abc Indicate groups that are significantly different

Figure 4: Average daily gain of pigs from days -1 to 14 (day of weaning until 5 days post-challenge) was higher in the Baytril (BAY) group as compared to the Kavault partial (KVP), Kavault full (KVF) and negative control (NEG) groups ($P < 0.01$).

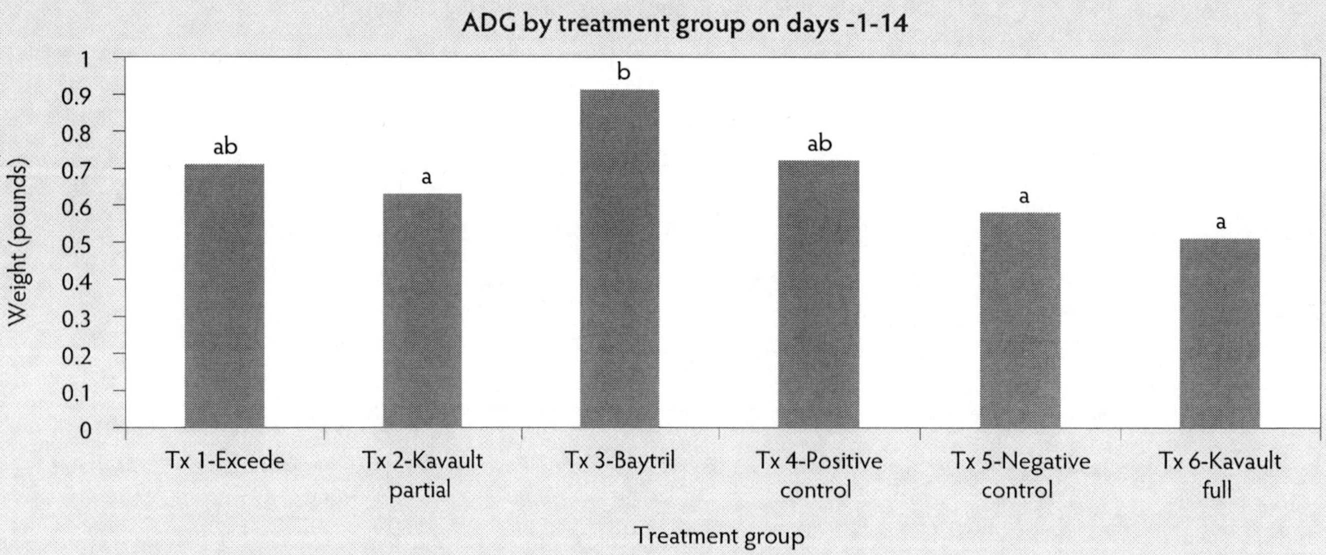

abc Indicate groups that are significantly different

Figure 5: Impression smears of intestines taken on Day 2 (2 days post-vaccination) had lower cycle-threshold values for K88+ *E. coli* qPCR in Excede (EXC) and positive control (POS) groups as compared to negative control (NEG) groups ($P < 0.01$).

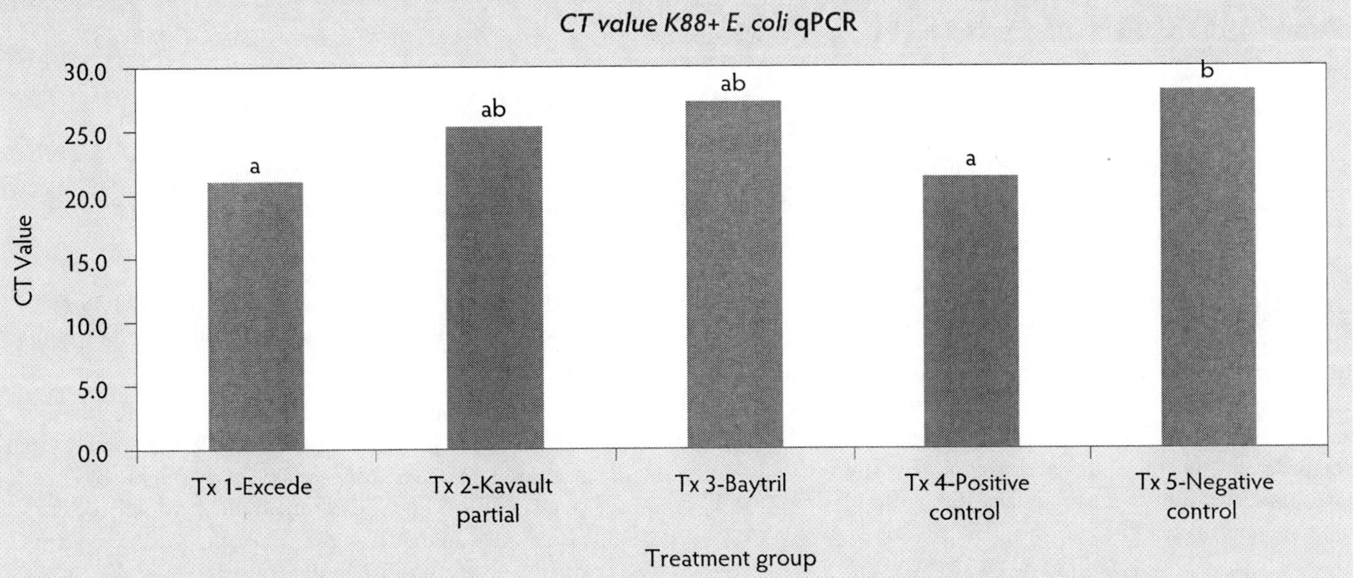

abc Indicate groups that are significantly different

Multiple case study reports of *Mycoplasma hyopneumoniae* elimination projects utilizing tylvalosin (Aivlosin® WSG)

D. Rosener[1], DVM; R. Kaptur[1], DVM; A. Mueller[2], DVM; A. Supple[3], DVM;
D. Magrath[4], BVSc, MRCVS; W. Lyons[5], DVM
[1]Pharmgate Animal Health, Wilmington, North Carolina; [2]Swine Services Unlimited, Inc., Rice, Minnesota;
[3]East Fork Swine, Hopkinton, Iowa; [4]Innovative Veterinary Services Inc., Lethbridge, Alberta, Canada;
[5]Bethany Swine Health Services, Bethany, Illinois

Introduction

Mycoplasma hyopneumoniae (Mhp), the microorganism that causes mycoplasmal (or enzootic) pneumonia in pigs continues to be a major economic pathogen despite wide spread vaccine use and various disease reduction strategies utilizing antibiotics, acclimation programs, and other approaches. Producers and veterinarians have known for years that Mhp causes problems with slow growth and tail-enders, otherwise known as the "18-week wall". Recently, Mark Schwartz, Director of Production Systems, Schwartz Farms, Inc., has demonstrated that the cost of Mhp in their system is a $3.25 per head disadvantage, with costs escalating even further (upwards of a $9.00 disadvantage) during periods of instability due to co-infections with SIV and/or PRRS.[1] In an effort to mitigate this cost, various approaches to eliminate Mhp from sow farms have been utilized, including depopulation and repopulation, partial depopulation (Zimmermann[2] or "Swiss Method"), herd closure and medication, and whole-herd medication without herd closure.[3] The herd closure and medication approach, utilizing various antibiotics with mycoplasmal activity, e.g., tylvalosin, has become widely accepted. Numerous successful eliminations utilizing tylvalosin have been reported in Europe[4] and Canada,[5] where Aivlosin® has been on the market for a longer period of time than in the US This paper is a compilation of recent case studies in the US and Canada where Mhp was successfully eliminated utilizing herd closure with tylvalosin medication of the sow herd.

Methods

A General Protocol is outlined in Table 1. Deviations from that protocol are detailed under the four case study reports. Herd closure often goes by the mantra: load, close, and expose. Loading refers to accumulating enough replacement gilts to meet the needs of the sow herd during the closure period. Closure refers to the time period required for the replacement gilts to mount an immune response and stop shedding. In the case of Mhp, shedding has been documented for up to 200 days, with total clearance being less than 254 days after initial exposure.[6] Lastly, exposure refers both to vaccine exposure, and arguably more important -- natural exposure to the resident, live organism. All four case studies used a combination of vaccination and natural exposure during acclimation.

Eight months after the initiation of the General Protocol, the entire sow herd is medicated with Aivlosin® Water Soluble Granules at a targeted dosage of 2.125 mg/kg bodyweight (BW) for a three week period. During this time frame, any piglets born should also receive an injection of tulathromycin (2.5 mg/kg BW) at birth and again at 10-14 days of age.

At the completion of the protocol, laryngeal swab PCR is typically utilized to see if Mhp is still present in the gilts or their offspring. If those tests come back negative, new replacement gilts are given the green light to enter the isolation facilities. A final confirmation frequently utilized is the testing of sentinel gilts. If the sentinel gilts are naïve and not vaccinated for Mhp, serology may be utilized to confirm that they are not exposed after entry to the sow unit. If the sentinel gilts have been vaccinated, then laryngeal swab PCR is the test of choice. In Case Studies 1 and 4, PCR testing on the sentinel gilts was performed, while in Case Studies 2 and 3, ELISA serology on unvaccinated gilts was performed.

Case study 1

This was a 1000 head sow unit located in the upper Midwest with the elimination project steered by Swine Services Unlimited, Inc. There were three exceptions to the General Protocol in this case study. One was that Aivlosin® Water Soluble Granules were administered at 5 mg/kg BW, which is the targeted dose per label indication in the US, albeit for the control of porcine proliferative enteropathy. Another deviation was that in an attempt to accomplish an APP elimination at the same time, tilmicosin in the feed was administered to the breeding herd for 3 weeks immediately after the 3 weeks of Aivlosin® in the water. The final exception was that instead of a sow herd vaccination and booster for Mhp on weeks 28 and 30, this system elected to vaccinate the entire sow herd three times with a 1-dose product 3, 5, and 7 months into the protocol.

Aivlosin® is a registered trademark of ECO Animal Health Ltd.

Table 1: *M. hyopneumoniae* herd elimination general protocol

Week	Action items
1	Load, close, and expose
	Acquire enough gilts to close the herd for at least 240 days, the clock starts after the last gilt has been exposed
	Gilts should be of various ages with the youngest preferably at least 2 months of age
	Vaccinate gilts for Mhp
	Expose replacement gilts to Mhp via infected animals (actively coughing gilts, cull sows) or lung tissue homogenate
3	Booster vaccinate for Mhp
8	Verify gilt exposure (serology or preferably laryngeal swab PCR)
28	Vaccinate whole breeding herd for Mhp
30	Booster vaccinate whole breeding herd for Mhp
33	Evaluate herd inventory and euthanize/cull any compromised animals
34	Wash and disinfect breeding and gestation barns
	Begin 3 week treatment of whole breeding herd with Aivlosin® WSG at 2.125 mg/kg BW (this is an extra-label dosage and duration in the U.S.)
	Begin treating piglets with tulathromycin (2.5 mg/kg BW) at birth for three weeks
	Colostrum intake is very important and minimal cross fostering should occur.
	Unthrifty pigs should be euthanized as soon as they appear
	Any sows off feed during the medication period should be injected with tulathromycin (2.5 mg/kg BW)
35	Begin treating all piglets that received a dose at birth with a second injection of tulathromycin at 10-14 days of age
	Weaning age may be dropped to 14 days for one turn through the crates.
	Alternatively, if batch farrowing, having a 2 week break in farrowing is beneficial.
	Serology or preferably laryngeal swab PCR on youngest gilts and piglets
39	Negative gilts enter isolation facilities
44	Can monitor new (sentinel) gilts via serology if they were not vaccinated, otherwise test sentinels and piglets via laryngeal swab PCR to confirm negative.

Case study 2

This was a 6000 head sow unit located in a hog dense area of the Midwest. This elimination project was managed by East Fork Swine with input from South West Ontario Veterinary Services, who have completed over fifty Mhp eliminations (the vast majority of which utilized Aivlosin®17% Premix) with a 90+ % success rate.[5] The exception to the General Protocol was that one dose of a 1-dose Mhp vaccine was administered at week 28.

Case study 3

This was a 400 head sow unit in Alberta under the veterinary care of Innovative Veterinary Services Inc. There were three primary exceptions to the General Protocol utilized by this herd. First, Aivlosin® 17% Premix was used in the feed in lieu of Aivlosin® Water Soluble Granules at the same 2.125 mg/kg BW dosage. (Canada, unlike the US, allows the extra-label use of feed antibiotics.) Secondly, a 4 week rather than a 3 week treatment period was used. Lastly, any sow farrowing during the 4 week medication period received an injection of tulathromycin (2.5 mg/kg BW) to compensate for sub-therapeutic dosage levels due to reduced feed intakes around farrowing.

Case study 4

This was a 2600 head sow unit located in the Midwest under the veterinary care of Bethany Swine Health Services. There were two extra safe guards utilized in this case study as compared to the General Protocol. First, half way through the closure period, all of the remaining gilts in the gilt development unit were bled and tested for Mhp antibodies via Idexx ELISA. 11 negative gilts and 13 suspects were subsequently culled from the herd prior to the medication period. Secondly, Aivlosin® was administered for a 4 week period rather than for 3 weeks.

Results and discussion

Despite the subtle differences in the approaches utilized, all of these Case Studies appear to have had successful eliminations of Mhp. They all report improved performance over their past baselines, with the only real surprise occurring in Case Study 3 (a very unstable herd) where improved performance was noted in the nursery phase rather than the more typical late finishing period. In Case Study 3, the pigs were 20 pounds heavier leaving the nursery and maintained that improvement throughout the finisher going to market 10 days earlier than in the past.

The veterinarians involved in these case studies feel medication with Aivlosin® provides an additional level of insurance to a simple herd closure protocol and may offer the ability to reduce the closure period and/or clean up those gilts which may not have been exposed on Day 0. One of the advantages of utilizing tylvalosin in a Mhp elimination is that it is mycoplasmacidal, unlike many other antibiotics which are simply bacteriostatic.[7] While it is difficult to guarantee 100% success with any eradication project, even with an elimination failure one could take consolation in the fact that the prevalence would likely be markedly reduced. Given the slow transmission rate of Mhp, pig performance would likely be improved for a number of months, thereby demonstrating a favorable return on investment.

References

1. https://www.aasv.org/library/swineinfo/Content/LEMAN/2015/2015-Schwartz.The%20Cost%20of%20M.pdf.

2. Zimmermann et al. Swine enzootic pneumonia: Partial depopulation as an alternative to total depopulation in eradication of the infection. *Schweiz. Arch. Tierheijk.* 1989;131:179-181.

3. Holst et al. Elimination of *Mycoplasma hyopneumoniae* from breed-to-wean farms: A review of current protocols with emphasis on herd closure and medication. *J Swine Health and Prod.* 2015;23(6):321-330.

4. Roozen et al. Recent experiences with tylvalosin (Aivlosin®) intervention for the eradication of *Mycoplasma hyopneumoniae* in commercial swine operations. AASV. 2014;479.

5. Personal Communication, Drs. Brent Jones and Martin Misener, South West Ontario Veterinary Services, 2016.

6. Pieters et al. An assessment of the duration of *Mycoplasma hyopneumoniae* infection in an experimentally infected population of pigs. *Veterinary Microbiology.* 2009;134:261-266.

7. Tavio et al. In vitro activity of tylvalosin against Spanish field strains of *Mycoplasma hyopneumoniae. Veterinary Record.* 2014;175:539.

Feed truck movement and ingredient traceability using FeedTrackur™

Thomas Stein, DVM, MS, PhD
DP Techlink, LLC, Johnston, Iowa

FeedTrackur™ is a hardware-software solution for automated, real-time tracking and tracing of animal feed (diet and ingredients) from the time it leaves a feed mill on a feed trailer until a driver delivers it to a feed bin on a production site. At the same time, it automates the process of managing both the driver and the feed delivery work-flow.

FeedTrackur™ was developed to meet a critical need in today's modern feed industry: the ability to track and monitor feed truck movements and feed delivery business processes, precisely tracking where, when and what feed and diet ingredients were delivered to which pigs and which producer.

As a complete solution, FeedTrackur has three main parts (Figure 1):

1. A private, cloud-based Enterprise Software Service that contains an integrated, real-time, customer-specific database that includes feed orders, load-out information, driver information, trailer information, and continuously updated GPS data. It also includes data being captured from a feed trailer's CANBUS, a standard data and communications controller area network that interconnects vehicle components. Relevant to feed deliveries, this means vehicle miles, air bag pressure sensors that can track the weight being delivered, sensors that track compartment unloading, and sensors that "know" the stinger (feed trailer augur) angle and positioning characteristics.
2. A secure Data Agent that sits behind a customer's firewall at their feed mill and does an ongoing, real-time push of both feed orders and load-out information from whatever software systems a mill is using (e.g. Feed Mill Manager, Easy Automation, WEM, Beta-Raven, i-Control, Repete, etc.)
3. An Interactive Mobile application written in Java for the Android operating system installed on a tablet computer that sits inside the cab of the tractor pulling a feed trailer. It does the heavy lifting needed to continuously monitor, read, and send GPS and trailer data to the FeedTrackur Enterprise Software Service. It receives updated information from the enterprise software and stores an updated GPS-based route map as well as preventing drivers from delivering feed to an incorrect location.

Today, feed mills are using sophisticated software to track and fulfill orders, and they use batching and load-out software to manage the process of manufacturing feed and filling feed trailers. But they lose visibility and traceability after a feed trailer leaves the mill. Typically, feed mills use a pen-and-paper manual system where drivers have printed orders and leave a hand-written document at an animal production site when they complete a delivery. That means a feed company or a pork producer cannot monitor and track drivers' operating efficiencies such as route taken, stops made, total time and fuel used, etc. And they cannot track and know with certainty when or if a delivery has been made and whether or not the correct feed was delivered to the right place.

FeedTrackur solves this problem by combining a feed mill's key information about each feed delivery with a mobile software application that tracks movement of a feed truck starting from a feed mill to unloading at an animal production site and then back to the mill.

FeedTrackur has been used in day-to-day operations for the past two years by a very large toll milling business based in northwest Iowa. This company makes and delivers feed for some of the largest pork producers in the US. It produces over 1.2 million tons a year across four mills with six pellet mills. It employs more than 50 drivers who drive a fleet of 32 feed truck-trailers. They go 24 hours a day, 7 days a week. It's being used to track more than 75,000 feed deliveries per year.

DP Techlink introduced FeedTrackur to the North American industry early in 2016. We have customers who are now tracking over 3 million tons of feed being delivered to swine and poultry production sites in the Midwest.

Based on analysis of feed mill operating performance over the past two years, we have been able to determine that FeedTrackur provides at least a 3:1 return-on-investment. The value proposition includes: (a) 50% reduction in "simple" delivery mistakes, i.e. feed delivered to the wrong production site; (b) elimination of catastrophic delivery mistakes e.g. feed with Mecadox delivered to market-ready pigs; (c) reduced driver turnover; (d) a 30% reduction in "excess feed" events (feed bin too full to receive the full amount ordered); (e) reduced labor for processing excess feed events; (f) reduced labor for regulatory compliance; (g) reduced labor for managing VFDs; and, (h) improved driver efficiency measured by a 20% drop in hours per delivery.

Figure 1: FeedTrackur™

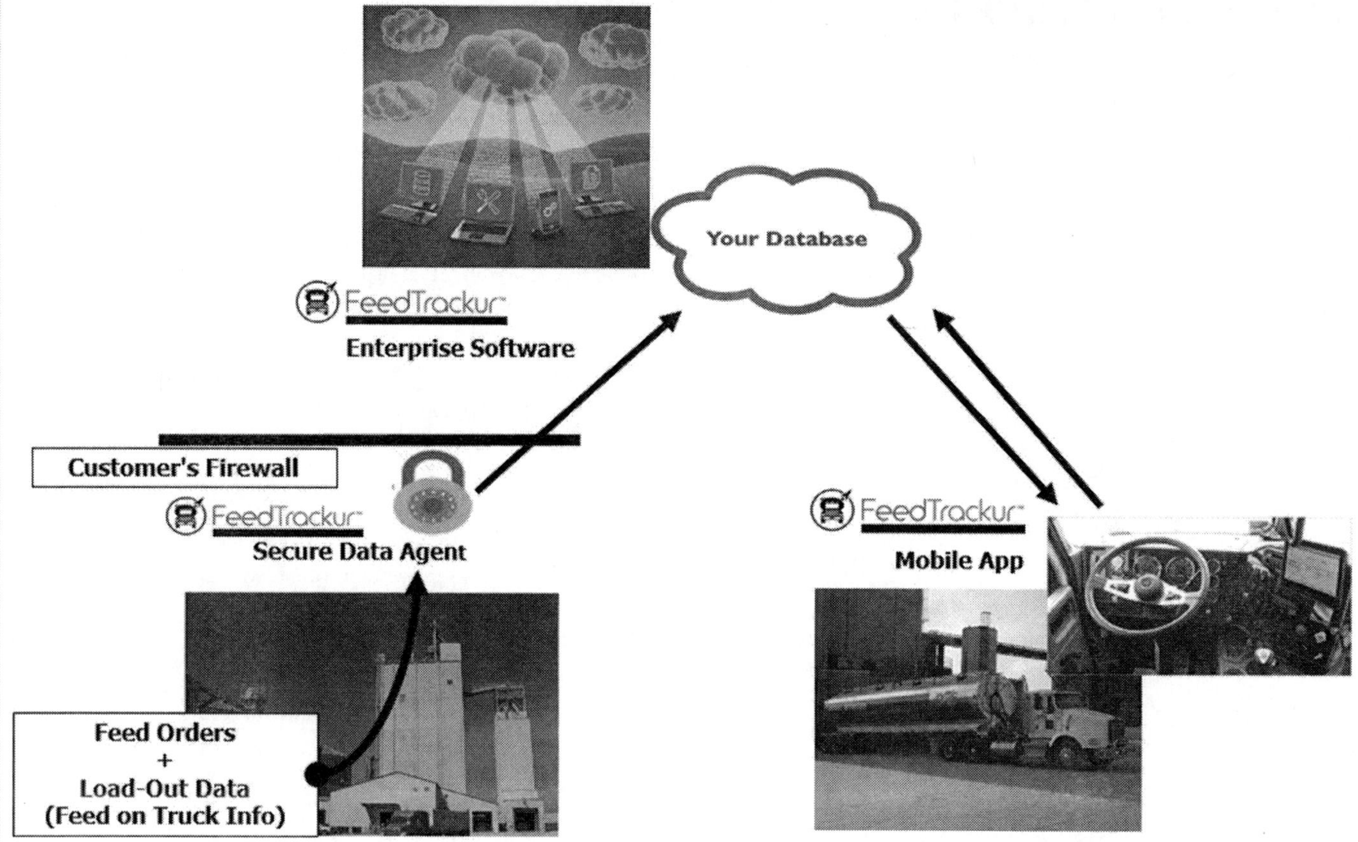

Comparison of efficacy between two PCV2 vaccination protocols under PCV2d field exposure

Eduardo Fano[1], DVM MC PhD; Nathan Schaefer[1], DVM; Ethan Schmaling[2], BS; Roy Edler[2], MS
[1]Boehringer Ingelheim Vetmedica, Inc., Saint Joseph, Missouri;
[2]Health Management Center, Boehringer Ingelheim Vetmedica, Inc., Ames, Iowa

Introduction

Porcine circovirus is a viral disease in swine that causes multi-systemic disease. The disease caused by porcine circovirus is preventable through the use of vaccines. Practitioners and producers have special interest in best vaccination protocols conferring protection against prevalent PCV2 strains.[1-4] Schwartz et al, 2015 reported that there had been a shift from PCV2b to PCV2d (new designation for the PCV2 mutation reported by Opriessnig et al., in 2012[5]) in the industry since 2015,[6] at least from clinical cases submitted to the Iowa State University Veterinary Diagnostic Laboratory.[3] Primary control for porcine circovirus type 2 (PCV2) is vaccination of pigs around the age of weaning with PCV2 vaccines, which are efficacious at controlling the disease when administered to the pigs according to the manufacturer's label. Several studies evaluating the efficacy of CircoFLEX® in clinical and field situations have shown it has been efficacious and equivalent to two dose protocols.[5,7] However, there are no comparative efficacy trials under PCV2d field exposure. The primary objective of this study was to measure the comparative efficacy of Ingelvac FLEXCombo (one dose) and Circumvent PCVM G2 (two doses) in growing pigs as measured by ADG and mortality/culling rates following field exposure to PCV2d.

Materials and methods

The experimental animals were housed in a 1200 head wean-to-finish commercial swine facility. Pigs were housed in a barn appropriate for breed and age and were kept under similar conditions of climate, ventilation, temperature, humidity and light. A total of 704 pigs were blocked by weight and randomly assigned to one of three treatments (Treatment 1 = 331 pigs, treatment 2 = 333, treatment 3 = 40 pigs selected as NVC, Table 1). The pigs were approximately 21 days of age at weaning and the barn was filled over the course of 1 week. Pigs were individually weighed and ear tagged (trial day 0). The neck area of each pig was observed and recorded to access a baseline (normal status). On trial day 1, pigs were allocated to treatment groups and administered their treatments. On trial day 7, the neck area of all pigs was observed to access vaccine reactions and signs of inflammation. On trial day 1, the T1 group received a single dose of vaccine. On trial days 1 and 22, the T2 group received their first and second doses of vaccine. All injections for trial vaccine were administered on the pig's right side of the neck. On trial day 28, the neck area of all pigs was observed to access vaccine reactions and signs of inflammation. All pigs were weighed at 13 weeks of age (trial day 71) and at first marketing cut of the growing period at 25 weeks of age (trial day 154, off test). Forty non-vaccinated sentinel pigs, one from each pen selected as the first pig encountered, were tagged for serial bleeding at and tested at trial days 0, 21, 50, 113, and 154. Serum was tested by PCV2 PCR, PRRS ELISA, and *Mycoplasma hyopneumoniae* (Mhp) IDEXX ELISA. PCV2 sequencing was done in positive samples to determine strain.

Performance data included day 0, 71, and 153 weights as well as day 0-71, 71-153, and 0-153 average daily gain. Growth performance data was analyzed using ANOVA, where treatment and gender (gilts and barrows) were included in the model along with the treatment x gender interaction. The model was subsequently optimized by removing the interaction term. Mortality and cull data was analyzed using Pearson's chi square. Cull rate was based on pigs weighing less than 225 pounds on day 153. The mean size of swelling observed as well as minimum and maximum size were calculated in mm^2 per treatment group. The percent of pigs with measurable injection site swelling was analyzed using Pearson's chi square.

Results

The effect of vaccination protocol in growth performance is shown in table 2, the effect on mortality and culling rates is shown in table 3, and the effect on injection site swelling in table 4. There were no significant differences in growth performance between the treatment groups. There were no significant differences in mortality and culling rates between the treatment groups. There was significantly greater injection site swelling in the T2 group on day 28 ($P < 0.0001$). PCV2 viremia was detected in NVC on trial days 113 and 153 and molecular characterization indicated PCV2d infection.

Discussion

Pigs vaccinated with both protocols in this study showed no clinical signs of PCVAD. Under the conditions of this field study there were no significant differences in growth performance between the two treatment groups, and there were no significant differences in mortality and culling rates between the two treatment groups. This is in agreement with previous studies where no differences have been found evaluating the efficacy of CircoFLEX® (as the PCV2 fraction in FLEXCombo®) vs two dose protocols.[5,7] The field strain involved in this study was PCV2d; therefore, we can state that under conditions of this study, no

Table 1: Vaccination protocol and treatment description

ID	Name	n	Group description	Dose	Route of administration
T1	CircoFLEX® (FLEX Combo®)	331	1-Full dose at weaning	1 × 2 ml	IM
T2	Circumvent PCVM G2 (Merck)	333	1st dose at weaning, 2nd dose at 21 days post-weaning	2 × 1 ml	IM
NVC	Non-vaccinated sentinels,	40	Sentinels used for diagnostics but not for performance conclusions	None	N/A

Table 2: Least square means and SEM for growth performance by vaccine treatment (protocol) and gender

	Vaccine protocol		Gender	
Response variable	T1	T2	Barrows	Gilts
Day 0 weight, lbs	12.80	13.06	13.00	12.90
Day 71 weight, lbs	109.14	108.63	108.97	108.43
Day 153 weight, lbs	279.15	277.81	289.82[a]	266.4[b]
Day 0-71 average daily gain, lbs/day	1.36	1.35	1.35	1.35
Day 72-153 average daily gain, lbs/day	2.07	2.06	2.21[a]	1.93[b]
Day 0-153 average daily gain, lbs/day	1.74	1.73	1.81[a]	1.66[b]

*ab Means with different superscripts indicate difference at $P \leq 0.05$ (Student's t)

Table 3: Mortality and cull rates by treatment group

	Vaccine protocol	
	T1	T2
Day 0-71 mortality rate, (%)	10/331 (3.02)	8/333 (2.40)
Day 72-153 mortality rate, (%)	7/321 (2.18)	9/325 (2.78)
Day 0-153 mortality rate, (%)	17/331 (5.14)	17/333 (5.11)
Day 153 cull rate, (%)	12/294 (4.08)	13/303 (4.29)

ab Means with different superscripts indicate difference at $P \leq 0.05$
* Cull rate base on pigs weighing less than 225 lbs on day 153.

Table 4: Percent of pigs with measurable injection site swelling

	Vaccine treatment		Mean size of swelling observed (mm2), (min, max)	
	T1	T2	T1	T2
Day 6 palpable swelling, (%)	0/127 (0.00)	0/130 (0.00)	n/a	n/a
Day 28 palpable swelling, (%)	0/127[a] (0.00)	43/130[b] (33.08)	n/a	1125.1 (216, 4123)

[1] Analyzed using Pearson chi-Square.
[ab] Means differ (Pearson chi-square).

efficacy differences were documented under PCV2d field/natural challenge. A previous study has suggested that no differences in pathological and clinical outcome should be expected between the two vaccines evaluated in this study vs PCV2d experimental challenge.[5,8] In addition and as described before, there was significantly greater injection site swelling in the two dose protocol on day 28.

References

1. Desrosiers et al, Use of a one-dose subunit vaccine to prevent losses associated with porcine circovirus type 2 *JSHAP*. 2009;148-54.

2. Cline et al, Efficacy of immunizing pigs against porcine circovirus type 2 at three or six weeks of age. *Vet Record*, 2008;1-4.

3. Haiwick et al, A single dose, PCV2a based ORF2 subunit vaccine cross-protects against challenge with a recent PCV2b strain. AASV. 2014;129-30.

4. Schwartz et al, PCV2: Tools to assess the subclinical impact, AASV. 2015;487-91.

5. Opriessnig et al, Commercial PCV2a-based vaccines are effective in protecting naturally PCV2b-infected finisher pigs against experimental challenge with a 2012 mutant PCV2, *Vaccine*. 2014; 32.4342–4348.

6. Iowa State University Veterinary Diagnostic Laboratory. 2015. Data derived from the Veterinary Diagnostic Laboratory.

7. N da Silva et al, Mixed treatment comparison meta-analysis of porcinecircovirus type 2 (PCV2) vaccines used in piglets. Preventive Veterinary Medicine 117 (2014) 413–424.

8. Jeong et al, Comparison of three commercial one-dose porcine circovirus type 2 (PCV2) vaccines in a herd with concurrent circulation of PCV2b and mutant PCV2b cross protection paper, *Vet Micro*. 2015. 177;43–52.

Efficacy of Ingelvac PRRS® MLV against a heterologous PRRSV 1-7-4 RFLP challenge

A. Patterson, DVM, PhD; G. Haiwick, BS; J. Hermann, DVM, PhD; B. Fergen, MS, PhD; W. Chittick, MS; K. Doolittle, BS; N. Goeser, BS; R. Philips, DVM
Boehringer Ingelheim Vetmedica, Inc., Saint Joseph, Missouri

Introduction

The swine industry continues to experience significant losses due to PRRS virus infections of sow farms and growing pigs.[1] The use of Ingelvac PRRS® vaccines can significantly reduce lung lesions following challenge with heterologous PRRSV isolates (86-94% ORF5 nucleotide similarity) in the three-week-old pig respiratory challenge model.[2] In addition, it has been demonstrated that the implementation of a systematic methodology for PRRS control that utilizes modified-live vaccine for the control of wild type-PRRSV infections can mitigate the consequences of infection on health and performance.[3,4] It is necessary for modified-live vaccines to be effective against heterologous challenge with current PRRSV field isolates. However, the efficacy of Ingelvac PRRS® MLV vaccine against a current virulent PRRSV RFLP 1-7-4 isolate has not been reported to date. The objective of this study was to evaluate the efficacy of Ingelvac PRRS® MLV (BIVI-Boehringer Ingelheim Vetmedica, Inc., St. Joseph, MO) and Fostera® PRRS (Zoetis Inc., Kalamazoo, MI) in a three-week-old pig respiratory challenge model, using a heterologous PRRSV RFLP 1-7-4 field isolate that was isolated from North Carolina in 2014.

Materials and methods

A randomized, blinded vaccination-challenge study was performed in 159, three week old pigs from a PRRS naïve source. Pigs were confirmed to be PCR negative for PRRS prior to the initiation of the study. The study consisted of 4 treatment groups of pigs. Group 1; n = 45, was vaccinated with Ingelvac PRRS® MLV (2 ml intramuscular (I.M.) per label recommendations) and Group 2; n = 45, was vaccinated with Fostera® PRRS (2 ml I.M. per label recommendations). Groups 1 and 2 were vaccinated on day 0. Group 3; n = 64, was a non-vaccinated challenge control treatment group that received a 2 ml placebo I.M. injection on day 0. Group 4; n = 5, was a non-vaccinated and non-challenged strict control group that served as an environmental control. Pigs were housed in rooms by group during the vaccination period. Group 4 pigs were removed and euthanized prior to challenge and examined for lung lesions and confounding bacterial pathogens. At day 28 of the study (D28), all pigs were comingled and challenged with 2.0 ml I.M. and 2.0 ml intranasally (1 ml per nostril) of inoculum containing $10^{4.6}$ $TCID_{50}$/ml of virulent PRRSV RFLP 1-7-4. Ingelvac PRRS® MLV and Zoetis' PRRS have 87.7% and 86.6% similarity respectively to the PRRSV 1-7-4 challenge isolate based on ORF 5 sequence. Blood was collected from all pigs on days 0, 7, 14, 21, 28, 29, 31, 35 and 42. Serum samples were tested by RT-PCR for the presence of viremia and by ELISA for the presence of anti-PRRSV antibody. A subset of samples from D28, 29, 31 and 35 were assayed by Fluorescent Multiplexed Bead-Based Immunoassay (FMIA) for multiple cytokines including IFN-α, performed by BIVI-Health Management Center, Ames, IA (HMC). Viremia testing for PRRSV was done by HMC using Quantitative PCR and results reported as "positive or negative" in addition to PRRSV quantity in units of genomic copies per ml serum. Pigs were weighed at day 0, day of challenge (D28), and termination of the study (D42) to assess average daily weight gain (ADWG). On D42 (14 days post-challenge), all pigs were necropsied and lungs were scored for the presence of macroscopic lesions and BALF samples were collected. Data were analyzed using Generalized and Linear Mixed Models. Pairwise comparisons between groups were conducted as appropriate using a level of confidence of 0.05 to indicate statistical significance. The study design is summarized in Table 1.

Results

Lung lesions

The day 42 (14 days post-challenge) lung lesion assessment found both vaccinated treatment groups to have significantly lower median gross lung lesion scores compared to the non-vaccinated (Placebo) challenged treatment group. The median percent lung lesion scores for Ingelvac PRRS® MLV and Fostera® PRRS were 8.4% and 12.9% respectively, compared to a median lung lesion score of 25.4% for the non-vaccinated challenged treatment group. No significant differences were noted between the vaccine groups; however, pigs vaccinated with Ingelvac PRRS® MLV had the lowest numerical median lung lesion score. Post-challenge lung lesion results are summarized in Table 2.

Average daily weight gain

ADWG was evaluated during the post-challenge period of the study (days 28-42). Both vaccinated treatment groups demonstrated a significant increase in ADWG compared to the non-vaccinated (Placebo) challenged treatment group. ADWG for Ingelvac PRRS® MLV and Fostera® PRRS was 0.61 lbs/day and 0.49 lbs/day respectively, compared to a 0.24 lbs/day ADWG for the non-vaccinated challenged group. While no significant differences were noted between the vaccine groups, pigs vaccinated with Ingelvac PRRS® MLV had a numerically higher ADWG. The ADWG results are summarized in Table 3.

Table 1: Study design

Group	N	Treatment	Day 0	Day 28	Day 42
1	45	Ingelvac PRRS® MLV	Vaccinated	PRRSV 1-7-4 Challenge	Necropsy – lung lesion evaluation
2	45	Fostera PRRS®	Vaccinated		
3	64	Placebo-Challenge Control	Non-vaccinated		
4	5	Strict Control	Non-vaccinated	NA	NA

Table 2: Day 42 lung lesions (median %) in vaccinated and non-vaccinated pigs challenged with PRRSV 1-7-4.

Group	Treatment	Median percent lung score
1	Ingelvac PRRS® MLV	8.4[a]
2	Fostera® PRRS	12.9[a]
3	Placebo - challenge control (non-vaccinated)	25.4[b]

[a] significantly different from the placebo at $P \leq 0.05$.

Table 3: Post-challenge ADWG

Group	Treatment	ADWG in lbs.
1	Ingelvac PRRS® MLV	0.61[a]
2	Fostera® PRRS	0.49[a]
3	Placebo - challenge control (non-vaccinated)	0.24[b]

[a] significantly different from the placebo at $P \leq 0.05$.

Viremia

Following challenge, the non-vaccinated challenge control pigs became viremic within 24 hrs of infection (D29), and all animals remained viremic during the post-challenge period. A decrease in viremia of approximately 1log was noted at D31 in the non-vaccinated challenge control pigs. This decrease was unexpected as use of the same challenge material in a previous study (BI-2015269) indicated a steady increase in viral load from days post-challenge (DPC) 1 through DPC7. In comparison to the challenge control pigs, pigs in both vaccinated groups had significantly lower levels of viremia at D29 ($P < 0.001$). By D42, levels of viremia were significantly lower in animals vaccinated with Ingelvac PRRS® MLV in comparison to animals in the challenge control and Fostera® PRRS groups ($P < 0.001$ and $P < 0.0017$, respectively). Post-challenge viremia results are summarized in Figure 1.

Interferon-alpha (IFN-α) response

Within the challenge control group, levels of IFN-α were greater than 667pg/mL in 95% of the animals tested by one day following infection. Levels then decreased by D31 (three days post challenge) and approached baseline parameters by D35 (seven days post challenge). In contrast, IFN-α levels in animals within the vaccinated groups remained between 100 and 400pg/mL throughout the seven day evaluation period. The portion of animals with a detectable IFN-α response was lower in vaccinated animals in comparison to the challenge control animals at D29. By D31, the portion of animals with detectable IFN-α was similar in vaccinated and challenge control animals. However, vaccinated animals had significantly lower amounts of IFN-α ($P < 0.004$).

Post-challenge IFN-α results are summarized in Figure 2.

Discussion and conclusions

The objective of this study was to evaluate the efficacy of Ingelvac PRRS® MLV and Fostera® PRRS against a current and virulent heterologous PRRSV 1-7-4 isolate in a respiratory challenge model. As evidenced in this study, PRRSV infection can negatively impact pig performance. Mitigating this negative biological impact can have a significant economic benefit. In this study, the Ingelvac PRRS® MLV group had a 20% and 61% higher ADWG when compared to the Fostera® PRRS group and the challenge control group respectively. In addition, Ingelvac PRRS® MLV demonstrated a numerical advantage compared to Fostera® PRRS for the efficacy parameters of lung lesions and showed a significant difference in the reduction of the level of viremia at D42 post-challenge.

Type I interferons (IFN-α and IFN-β) are important contributors to innate antiviral immunity. Specific to PRRSV, a study by Brockmeier *et al.* has shown that the presence of IFN-α at the time of challenge delays PRRSV viremia and reduces the quantity of detected virus.[5] However, the role of IFN-α induced following vaccination and challenge remains unclear. In this study, elevated levels of IFN-α were observed in the challenge control animals. This increase in IFN-α is hypothesized to be real and representative of the immune response following viral replication. The lower IFN-α response in the vaccinated animals was likely secondary to reduced viremia. Even though both vaccines induced significantly lower amounts of IFN-α

Figure 1: Post-challenge viremia – Group least square mean PRRSV qRT-PCR in serum.

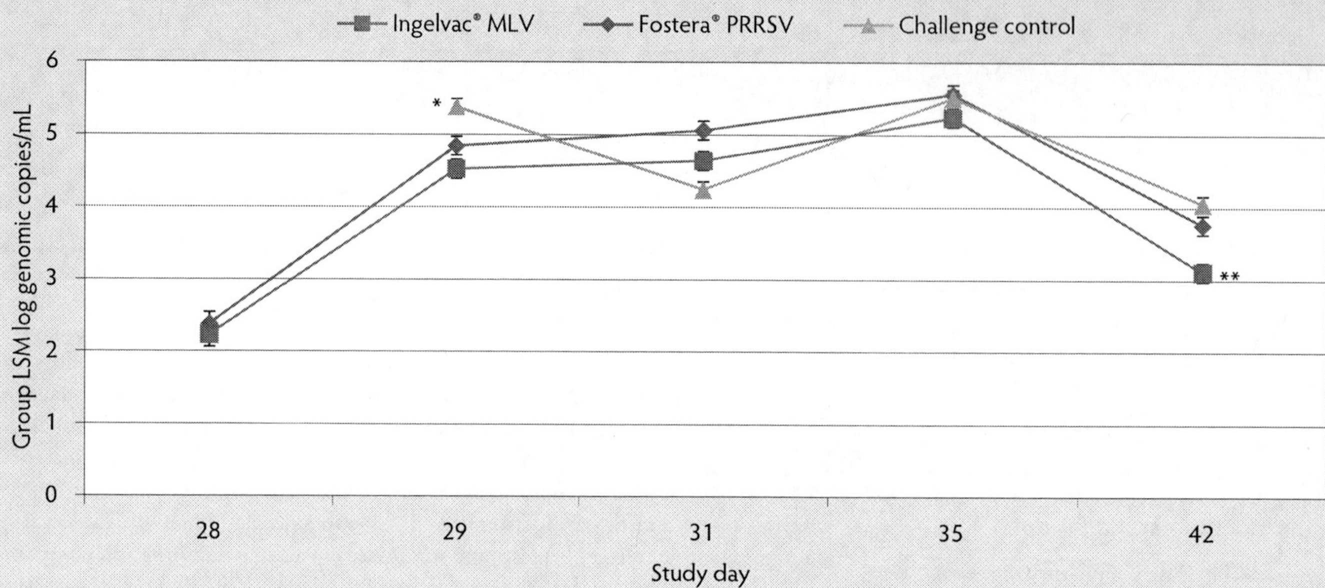

* significantly different from the vaccinates at $P < 0.001$.

** significantly different from Fostera® PRRS ($P < 0.0017$) and challenge control ($P < 0.001$).

Figure 2: Post-challenge IFN-α data – percentage positive and median amounts of IFN-α (pg/mL) present in serum by group and study day.

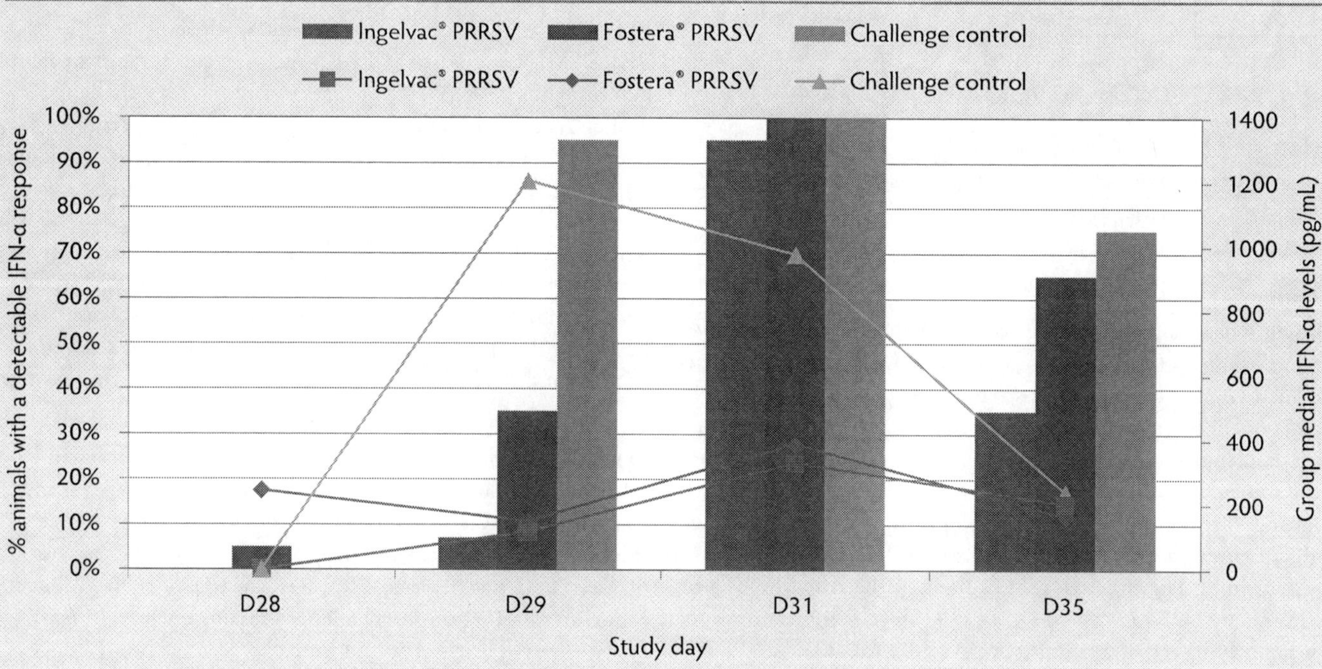

in comparison to the challenge virus, they were still able to significantly reduce lung lesions and significantly increase ADWG.

This study is another example demonstrating the ability of Ingelvac PRRSV® MLV to protect against a relevant PRRSV challenge. Implementing vaccine in a systematic methodology for PRRS control can mitigate the consequences of infection and subsequently improve the health and performance of pigs.

References

1. Holtkamp, D., et al. 2013. Assessment of the economic impact of porcine reproductive and respiratory syndrome virus on United States pork producers. *J Swine Health and Prod.* 21(2):72-84.

2. Patterson, A., et al. 2013. Modified live PRRSV vaccination is efficacious following challenge with eight genetically diverse PRRSV isolates. *A.D. Leman Swine Conference Proceedings.*

3. Oropeza, A., et al. 2012. A methodical approach to PRRS management, a success story of production improvements. *IPVS Proceedings.*

4. Kolb, J., et al. 2015. A summary of four large scale system-based PRRS control projects. *AASV Proceedings.*

5. Brockmeier, S., et al. 2012. The presence of alpha interferon at the time of infection alters the innate and adaptive immune responses to porcine reproductive and respiratory syndrome virus. *Clin Vaccine Immunol.* 19:508-514.

Comparison of intestinal lesions between Enterisol® Ileitis and Porcilis® Ileitis using a *Lawsonia intracellularis* mucosal homogenate seeder pig challenge model

S. Dee[1], DVM, PhD; J. Nerem[1], DVM; D. Hanson[1], BS; J. Seate[2], DVM; T. Wetzell[2], DVM; R. Edler[2], MS; E. Schmaling[2], BS
[1]Pipestone Veterinary Services, Pipestone, Minnesota; [2]Boehringer Ingelheim Vetmedica, Inc., Saint Joseph, Missouri

Introduction

Lawsonia intracellularis affects millions of growing pigs yearly with clinical and subclinical ileitis.[1] Enterisol® Ileitis is an orally modified live vaccine that provides protection against ileitis. Recently an injectable killed vaccine, Porcilis® Ileitis has been made available. The objective of this study was to evaluate the efficacy of Enterisol® Ileitis and Porcilis® Ileitis in growing pigs via comparison of ileocecal lesions.

Materials and methods

Weaned pigs were placed in a tunnel ventilated 2400 head, two room, 84 pen wean to finish barn managed by the Pipestone Applied Research team with Pipestone Veterinary Services. Twenty-four pigs were placed within a pen, and the pens were stratified by weight and gender. The three treatment groups (non-vaccinated controls (NVC), Enterisol® Ileitis and Porcilis® Ileitis) were then randomly assigned within block to construct a randomized block design. Additionally, three non-vaccinated, non-treatment group pigs were added to each pen to serve as seeder pigs. Pigs in the Porcilis® Ileitis group were vaccinated IM at placement (3 weeks of age) with 2 ml according to label. Pigs in the Enterisol® Ileitis pens were vaccinated orally (water line) using a peristaltic pump (Stenner Pump Company) five weeks post-weaning (approximately eight weeks of age) according to the standard on-farm protocol. The non-vaccinated seeder pigs in each pen were challenged by gastric lavage with an intestinal homogenate (provided by Gut Bugs Inc.) containing 40 ml *Lawsonia intracellularis* (targeting a concentration log 10 × 8) at nine weeks post weaning (12 weeks of age). Four weeks post-challenge (16 weeks of age) 30 pigs from each treatment group showing clinical signs of ileitis were randomly selected, euthanized and necropsied. Tissues were submitted to Iowa State University veterinary diagnostic lab. Immunohistochemistry (IHC) and Hematoxylin & Eosin (H&E) staining of ileum and cecum sections were used to detect the presence and severity of *L. intracellularis* and lesions resulting from bacterial colonization and proliferation. Observers, veterinarians, diagnosticians and statisticians were blinded to the treatment groups.

Results

Microscopic lesions, within the ileum sections, were evident in all three treatment groups, however; the Enterisol® Ileitis group had no lesions in the cecum sections (Table 1). The Enterisol® Ileitis group had significantly less IHC positives (6.67%) in ileum sections compared to the non-vaccinated controls (63.33%) and the Porcilis® Ileitis (43.33%) treated pigs (Table 1). Both the Enterisol® Ileitis group and the Porcilis® Ileitis group had significantly less percentage of pigs affected on ileum H & E sections than the non-vaccinated controls. However, the Enterisol® Ileitis group had numerically less pigs affected on ileum H&E than the Porcilis® Ileitis group (Table 1). Enterisol® Ileitis had lower numbers of Lawsonia PCR positives and lower levels from cecal content compared to Porcilis® Ileitis and non-vaccinate control groups at necropsy (Table 2).

Discussion

Previous work has shown a strong correlation between ileocecal lesions and its impact on growth performance.[2,3] It has been demonstrated that with increased prevalence and severity of ileocecal lesions there is a decrease in pig performance. In this seeder pig challenge model, pigs that were vaccinated with Enterisol® Ileitis then indirectly challenged showed significantly lower IHC scores when compared to both pigs vaccinated with Porcilis® Ileitis and non-vaccinated pigs. We evaluated cecal Lawsonia PCRs from all necropsied pigs. There were fewer Enterisol® Ileitis vaccinated pigs with positive cecal PCRs than for either the Porcilis® Ileitis vaccinated group or the non-vaccinated group. In summary this study demonstrated that vaccinating with Enterisol® Ileitis provides more effective protection against ileitis than Porcilis® Ileitis.

References

1. McOrist S, Gebhart CJ et al. Proliferative Enteropathy. *Diseases of Swine* 10th Ed:811-820.

2. Collins, A.M, Barchia, I.M. The critical threshold of Lawsonia intracellularis in pig faeces that causes reduced average daily weight gains in experimentally challenged pigs. *Vet. Microbiol.* 2013.12.003.

3. Burrough ER, Rotolo ML, Gauger PC, et al. Correlation of *Lawsonia intracellularis* semi-quantitative fecal polymerase chain reaction assay results with the presence of histologic lesions of proliferative enteropathy and positive immunohistochemical staining. *J Swine Health Prod.* 2015;23(4):204–207.

Table 1: Summary of affected pigs for microscopic lesions.

	Enterisol® Ileitis	Porcilis® Ileitis	NVC	*P*-value[1]
Ileum IHC (%)	6.67[a]	43.33[b]	63.33[b]	0.0001
Cecum IHC (%)	0.00	3.33	13.33	0.0637
Ileum H&E (%)	20.00[a]	43.33[a]	66.67[b]	0.0013
Cecum H&E (%)	0.00[a]	30.00[b]	10.00[ab]	0.0023

[1] Analyzed using Pearson's Chi-square.
[ab] Means differ < 0.05 Fisher's Exact Test.

Table 2: Summary of Lawsonia PCR+ for necropsy pigs by treatment group.

Treatment group	Frequency	CT Median	CT Min	CT Max
Enterisol® Ileitis	5/30 (16.67%)	36.75	33.12	36.91
Porcilis® Ileitis	9/30 (30.00%)	35	29.22	37.38
NVC	22/30 (73.33%)	33.25	26.47	37.69

Figure 1: Infected ileum from a non-vaccinated control pig (view 1).

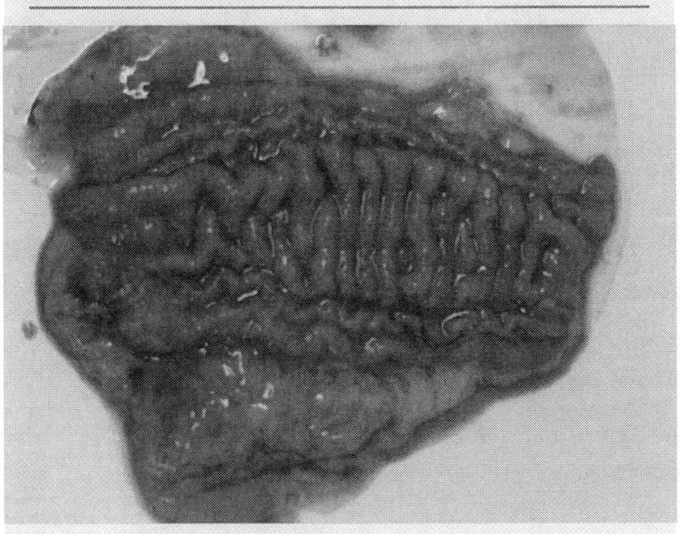

Figure 2: Infected ileum from a non-vaccinated control pig (view 2).

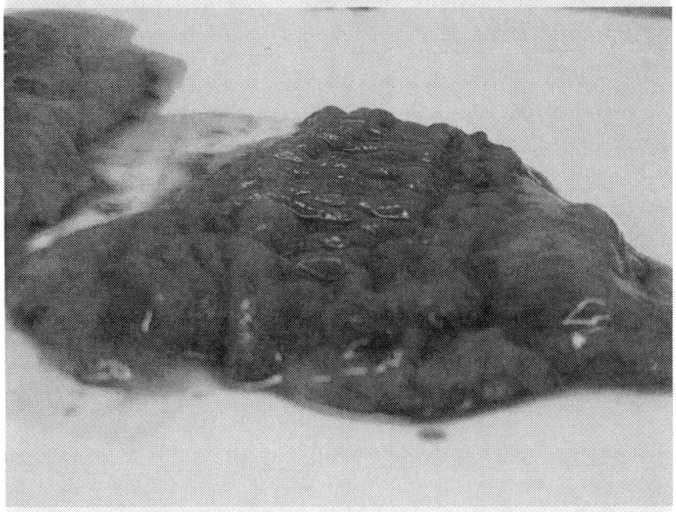

Biosecurity controls for the industrial manufacturing of spray-dried plasma

Joe Crenshaw[1], PhD; Joy Campbell[1], PhD; Javier Polo[2], PhD
[1]APC Inc., Ankeny, Iowa; [2]APC-Europe, Granollers, Spain

Introduction

APC[1] is committed to produce safe and effective products for animal nutrition that improve the life of animals worldwide. APC has manufactured and supplied spray-dried animal blood products for global use in animal nutrition for nearly three decades. Spray-dried animal plasma (SDP) is a safe ingredient used as a functional protein source in feed for weaned pigs because the nutrition it provides has proven to enhance pig performance and survival during the stressful post-weaning period of production.[2] A recent review of literature discusses the multiple modes of action of spray-dried plasma as a functional protein ingredient, and the results of numerous studies that support the biosafety of SDP in relation to pathogens of concern for the swine industry, such as porcine epidemic diarrhea virus (PEDV), porcine circovirus type 2, porcine reproductive and respiratory syndrome virus, swine vesicular disease virus, and pseudorabies virus.[3] Collectively, the studies related to biosafety indicated that the production process for manufacturing spray-dried plasma protein inactivated swine viruses, and even though the viral genome was detected in the finished product, feeding studies indicated the product was not infectious.

The intent of this manuscript is to discuss manufacturing processes and procedures including critical control points for biosecurity and quality control and assurance that APC uses to produce SDP and to discuss adaptation of best biosecurity practices at manufacturing plants, as well as APC exclusive technology that provides additional safety features to the manufacturing process.

Manufacturing process and hazard control points

A schematic of the processing steps for manufacturing of spray dried plasma used by APC is presented in Figure 1. The process begins with collection of blood at federally inspected abattoirs and finishes with storage of finished packaged product at APC manufacturing facilities.

Blood is collected only from abattoirs where animals have been federally inspected and determined fit for slaughter for human consumption. This is the initial critical control point avoiding collection of blood from clinically affected animals which reduces risks of biohazard contamination in the raw material. The next critical control point is that the blood collection system is designed to capture the blood and remove it rapidly from the area of animal processing reducing the risk of contamination with extraneous material. From this point the blood remains in a fully dedicated and enclosed stainless steel system precluding contamination with environmental air-borne biohazard. Depending upon the situation, whole blood may be transported to a remote drying facility or the blood may be centrifuged at the abattoir and only the plasma fraction may be transported to the drying facility.

The plasma or blood is rapidly cooled to 4°C through a chilling system usually within 5 minutes of collection to reduce risk of bacterial growth in the raw material. The chilled plasma or blood is stored in dedicated insulated stainless steel tanks at the abattoir until it is transported in specie-specific, dedicated, sealed tankers to a remote facility for further processing and drying. Blood collection systems and holding tanks at abattoirs are regularly cleaned and sanitized using standard operating procedures specific for each plant to further reduce risks of biohazard contamination. Another bio-safety feature associated with processing SDP is that liquid plasma or blood collected at abattoirs from numerous animals results in pooling of antibodies present in plasma which has been shown to have neutralizing capacity against a diversity of pathogens, even after the plasma has been spray-dried.[4]

Transport tankers for raw material are enclosed and unloaded into insulated tanks within the facility in an enclosed system with sealed couplings in a covered area protected from the environment. Incoming plasma or blood must arrive with the tanker seal intact and prior to unloading must meet quality standards that include temperature, color and odor (fresh smell).

The processing of the liquid plasma is a highly controlled process in which operators continuously monitor specialized software that maintains appropriate flow for membrane filtration, concentration and temperatures throughout the process. All holding tanks, filtration and concentration systems, and transport tankers are cleaned and sanitized per standard operating procedures. Typically, the time from collection to drying is less than 24 hours. Strict controls of minimum inlet and outlet temperatures for spray dryers are used and must be adhered to for product release for sale by the Quality Assurance staff. During spray drying, the product is exposed to temperatures over 80°C throughout substance and this is one of the most important critical control points in the manufacturing process.

Post-drying staging and packaging procedures are defined per standard operating procedures specific for each plant and only new packaging materials, totes or bags, are used. All products must undergo quality control and assurance testing to be released for sale. Finished products are assigned a lot number which allows traceability through distribution to customers.

Figure 1: Manufacturing process used by APC for production of spray-dried animal plasma.

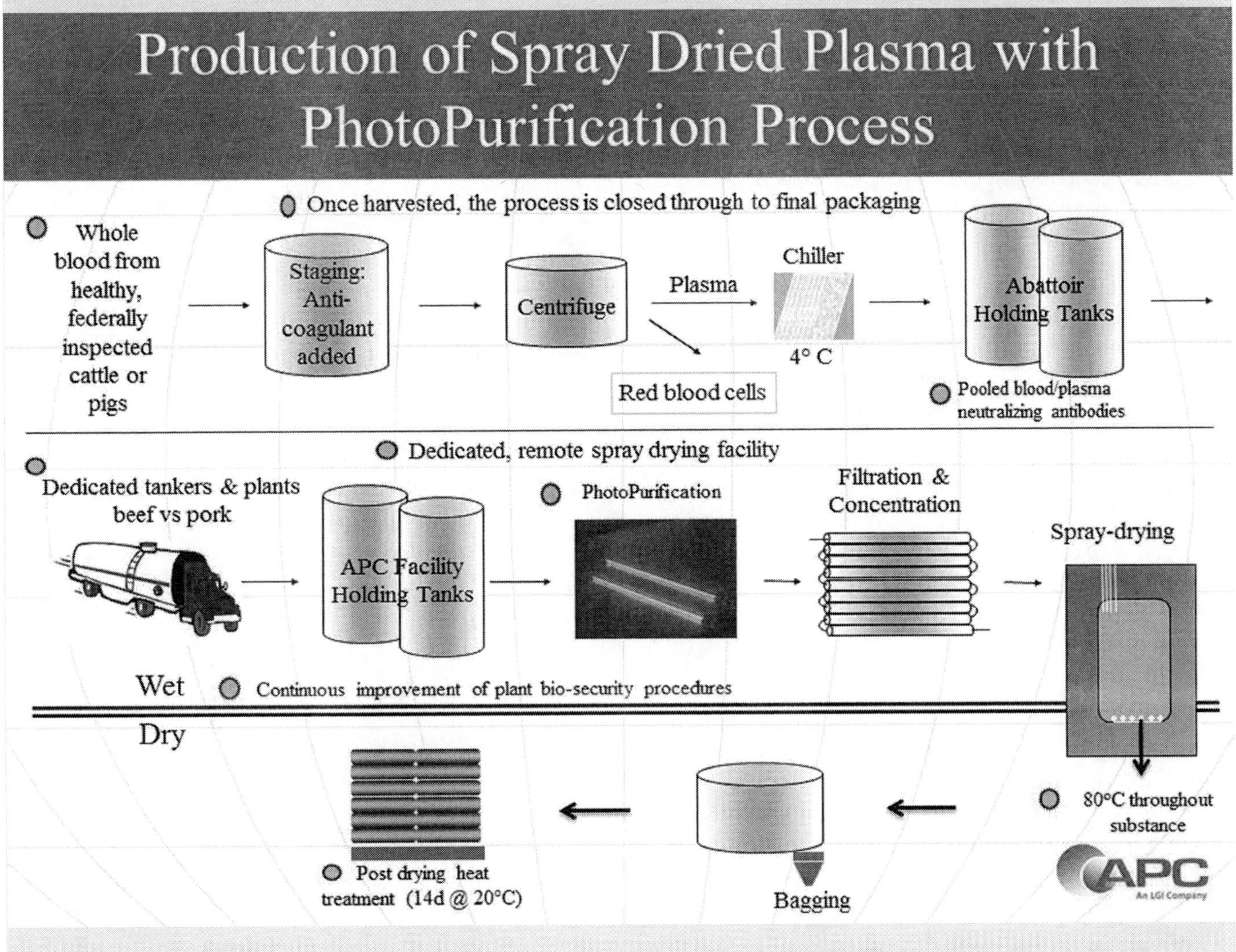

APC facilities are dedicated to only processing blood products, thus eliminating the potential from cross-contamination with biohazards from other products. In addition, most APC plants are dedicated to process only porcine or only bovine blood products eliminating the potential cross-contamination of specie origin. In the case of a mixed plant in Canada which receives both bovine and porcine raw material, the facility has specie-dedicated areas for receiving, and the facility has designated areas separated with bio-secure barriers to process only bovine plasma or only porcine whole blood. Specific to this plant, specie-specific finished products are tested for presence of either bovine or porcine genome by a third party laboratory to assure absence of potential cross-contamination of specie origin. Entry and exit areas at APC plants are controlled with designated pathways identified for truck traffic and have dedicated areas for raw material and finished products.

Biosecurity controls for people

Employees are dedicated to either wet or dry processing areas and wear distinctive uniforms in each area. Tools and equipment are dedicated to each processing area. Strict wet/dry separation is in place with clearly marked pathways as well as physical barriers. Sanitary barriers with hand and boot washing stations are located strategically throughout the facility. Furthermore, employees and contractors are trained using biosafety standard operating procedures. Truck drivers are not permitted to enter the facility. Visitors permitted to enter the plant for audits or tours are always accompanied by plant staff, provided personal protective equipment, and required to sign in and declare that they have not been in contact with animals in the past 24 hours.

Additional bio-safety features

Since the emergence of porcine epidemic diarrhea (PED) in North America, APC has implemented insulated temperature controlled warehouse storage at certain plants such that finished products can be certified for post-drying heat treatment at 20°C for a minimum of 14 days. Past research demonstrated that PEDV inoculated on spray-dried plasma survived less than 7 days when stored at room temperature,[5,6] and less than 1 day when stored in ambient winter time conditions.[7]

Also over the past 10 years APC has worked to develop an industrial scale exclusive photopurification process using UV-C exposure of the liquid plasma prior to spray drying. Research has demonstrated the photopurification process to be effective in completely inactivating porcine parvovirus inoculated in the liquid plasma and subsequently demonstrated no detrimental effects on the functional properties of the finished spray dried product when fed in diets for weaned pigs.[8] In addition, an assessment of the commercial scale process has indicated > 3 log reduction of aerobic plate count and elimination of Salmonella and E. Coli in liquid plasma with no effect on finished product physical or chemical properties.[9] Also feeding pigs diets containing commercially produced spray-dried porcine plasma which had been subjected to the commercial scale photopurification process, although positive for PEDV genome, did not transmit PEDV to naïve weaned pigs and performance of these pigs was similar to that of pigs fed a diet with spray-dried bovine plasma.[10] Currently the photopurification process has been implemented at two APC manufacturing plants.

Summary

APC has collaborated with swine industry stakeholders in a biosecurity risk assessment[11] and in the development of feed and ingredient biosecurity guidelines[12] and committed resources to assure customers that our products are safe. Furthermore, APC provides several product options of either bovine or porcine origin for our customers based on their individual preferences (Table 1).

References

1. APC Inc., Ankeny, IA. 2016; http://www.functionalproteins.com. Accessed 1 November 2016.

2. Torrallardona, D. Spray dried animal plasma as an alternative to antibiotics in weanling pigs – A review. *Asian-Australian J Anim Sci.* 2010;23:131.

3. Pérez-Bosque A et al. Spray dried plasma as an alternative to antibiotics in piglet feeds, mode of action and biosafety. *Porcine Health Management.* 2016;2:16.

4. Polo J et al. Neutralizing antibodies against porcine circovirus type 2 in liquid pooled plasma contribute to the biosafety of commercially manufactured spray-dried porcine plasma. *J Anim Sci.* 2013;91:2192.

5. Pujols J et al. Survivability of porcine epidemic diarrhea virus (PEDV) in bovine plasma submitted to spray drying processing and held at different time by temperature storage conditions. *Vet Microbiol.* 2014;174:427.

6. Trudeau M et al. Are there differences in survival among swine coronaviruses in feed? National Hog Farmer. 2016; http://nationalhogfarmer.com/nutrition/are-there-differences-survival-among-swine-coronaviruses-feed. Accessed 28 October 2016.

7. Dee S et al. An evaluation of porcine epidemic diarrhea virus survival in individual feed ingredients in the presence or absence of a liquid antimicrobial. *Porcine Health Management.* 2015;1:9.

8. Polo J et al. Ultraviolet light (UV) inactivation of porcine parvovirus in liquid plasma and effect of UV irradiated spray dried porcine plasma on performance of weaned pigs. *PLoS One.* 2015;10(7):e0133008.

9. Campbell J et al. Ultraviolet treated liquid porcine plasma and spray dried to powder form maintained product quality. *IPVS.* 2016;Poster #1245.

10. Campbell J et al. Ultraviolet treated liquid plasma PCR+ for PEDV spray-dried and fed to naïve pigs was not infective and performed equal to pigs fed untreated plasma. *IPVS.* 2016;Poster #551.

11. Sampredo F et al. Risk assessment of feed ingredients of porcine origin as vehicles for transmission of porcine epidemic diarrhea virus. 2015; http://research.pork.org/FileLibrary/ResearchDocuments.14-164-DAVIES-UofMN.pdf. Accessed 4 October 2016.

12. American Feed Industry Association. Developing biosecurity practices for feed & ingredient manufacturing. 2015; http://www.afia.org/rc_files/645/guidance_for_developing_biosecurity_practices_2015.pdf. Accessed 4 October 2016.

Table 1: Current APC spray-dried plasma product options for use in swine feed in USA or Canada.

Product	Plant locations	Plant status	Comments
Bovine plasma	Dubuque, IA, USA	Dedicated bovine	Not approved in Canada
Bovine plasma	Sublette, KS, USA	Dedicated bovine	Not approved in Canada
Bovine plasma	Calgary, AB, Canada	Mixed, bovine plasma or whole porcine blood	Approved in Canada and US. Bovine plasma certified as negative for porcine DNA and as negative for PEDV RNA. Product certified 14 day hold at 20°C
Porcine plasma	Vercheres, QC, Canada	Dedicated porcine	Approved in Canada and US. Certified negative for PEDV RNA. Product certified 14 day hold at 20°C
Porcine plasma	Arion, IA, USA	Dedicated porcine	Approved in Canada and US. Product certified 14 day hold at 20°C
UV-treated porcine plasma	Arion, IA, USA	Dedicated porcine	Approved in US. Product certified 14 day hold at 20°C

Solution for in-barn, hand-held data collection

Thomas Stein, DVM, MS, PhD
i-Novarum, Inc, & Maximum Ag Technologies, Jenison, Michigan

In response to consumer demand and retail customer requirements, pork producers are converting sow farms from stall to pen gestation systems. As producers either remodel or build new barns, they are also installing static or dynamic ESF (electronic sow feeding) systems. Sow identification with RFID tags is a key part of making an ESF system work. It won't be long before RFID tags are present on almost all sows and gilts in commercial sow farms. Omnipresent RFID tags provide a real opportunity to advance in-barn, sow-side data collection technology which will improve the speed and accuracy of farm data collection as well as reduce the overall cost of data collection and data management.

There are several good, solid sow data management software systems in use throughout the world: MetaFarms, PigKnows, Porcitec, and PigCHAMP, for example, just to name the ones being used in North America. The biggest challenge in using software effectively has been data acquisition and data entry. Many pork producers continue to send hand-written barn sheets to a central data office where clerical staff enter data by hand. Proprietary hand-held data collection devices have not gained much market share because the economics and form factor haven't worked (small screen, small keyboards, proprietary and expense devices leading to vendor lock-in).

i-Novarum in collaboration with Maximum Ag Technologies have introduced a new in-barn data collection system called WANDA™, which uses an RFID tag reader on a wand-like device linked via Bluetooth to a hand-held computer tablet (Figure 1). WANDA runs on either Android or iOS (Apple) tablet computers.

The tablet computer contains a simple and straightforward sow record software application which synchronizes with a cloud database. With WANDA, barn workers can read a sow's RFID tag, bring up her history on the tablet, review her sow card, see her lifetime performance, and view actions needed or alerts for that sow. They can also enter basic sow data such as breeding, pregnancy check, farrowing, weaning, movements, etc. These data immediately update a sow's record on the tablet computer

Figure 1:

Figure 2:

as well as updating the cloud database (Figure 2). Importantly, WANDA's cloud database can immediately synchronize with whichever sow management software they use, i.e. MetaFarms, PigKnows, Porcitec, PigCHAMP, and others.

It's easy to see the advantages the WANDA system provides: larger screen, larger keyboard, full sow histories, easier data entry, real-time data collection, and better ergonomics by using a wand-like device to read RFID ear tags. More importantly, there is no vendor lock-in because WANDA can send data to any sow software system.

WANDA also links to and synchronizes with the Maximus barn controller system and the Maximus ESF system from Maximum Ag Technologies, which means that each sow's daily feed intake in gestation automatically connects to their history. This also gives barn workers the ability to manage or change a sow's intake on-the-fly and in real-time as they do their daily check on sow body condition in gestation.

The effect of feeding MCFA on performance of nursery pigs raised without antibiotics

D. McManus[1], DVM; M. Lachmann[1], PhD; T. Weeden[1], PhD; B. de Rodas[1], PhD; S. Crowder[1], MS; S. Petersen[2], PhD
[1]Purina Animal Nutrition LLC; [2] PMI Nutritional Additives

Introduction

In North America, public concern regarding the use of antibiotics in pork production has grown over the last years. The human medicine segment is looking into animal production practices that can contribute to reducing the risk of proliferation of resistant bacteria in the environment. As result, starting Jan 1, 2017, all medically-important antimicrobials intended for use in swine feed that are currently available "over-the-counter" will require a Veterinary Feed Directive (VFD). Consequently, the swine industry is working on the design and evaluation of feed additives, feeding programs and management practices to support animal growth and health without the use of antibiotics (AB).

Feed additives are not designed to provide nutrient density to the diet; they provide functional components to support production efficiency and pig performance. To be considered an effective feed additive product, it must provide a performance improvement equivalent to the added cost[2] or to maintain animal performance at a lower cost (improve pig production and profitability).

Medium chain fatty acids (MCFA), acidifiers, prebiotics, probiotics and botanicals in swine diets have been considered as options to support the gut health, the immune system and subsequently, performance. Prior trials (unpublished Purina Animal Nutrition research) have indicated the effectiveness of a commercial blend of MCFA with other functional components (OutPace® Feed Additive (OutPace® FA) from PMI Nutritional Additives) in improving performance of nursery pigs. Two experiments were conducted to evaluate the effect of feeding OutPace® FA on nursery performance of pigs raised with or without AB.

Material and methods

Experiment 1

Experiment 1 (Exp 1) was conducted with the objective to evaluate the effect of feeding OutPace® FA at 3 different levels (0, 2.5 and 5 lbs/ton) within the complete feed on pig performance when pigs were fed diets containing AB during the nursery period. To achieve this objective, a total of 660 pigs were allotted to 1 of 3 dietary treatments (20 pigs per pen with 11 replicates). The AB were used as follows for all Trt groups: a combination of Denagard® (Novartis Animal Health, Greensboro, NC) at 35g/ton and chlortetracycline at 400g/ton (Denagard®/CTC) in phase 1 and 2, 50g/ton Mecadox® (Carbadox – Phibro Animal Health Corporation, Teaneck, NJ) in phase 3 and CTC at 500g/ton in phase 4. All feeding programs were similar in nutrient composition and functional components with the exception of: Treatment (Trt) 1: consisted of a control diet without OutPace® FA, Trt 2: consisted of a similar feeding program as Trt 1 with the exception that OutPace® FA was added at a rate of 2.5 lbs/ton of complete feed and Trt 3: was similar to treatment 1 with the exception OutPace® FA was added at a rate of 5 lbs/ton of complete feed. See Table 1.

Experiment 2

Experiment 2 (Exp 2) was conducted to compare the performance of the pigs fed feeding programs containing AB in combination with OutPace® FA at 2.5 lbs per ton vs. the performance of nursery pigs raised without AB in combination with 5 lbs per ton of OutPace® FA (as an alternative AB free program). To achieve this objective, a total of 442 pigs were allotted to 1 of 2 dietary treatments (22 pigs per pen with 11 replicates). Trt 1: consisted of a 4 phase feeding program with the inclusion of 2.5 lbs/ton of OutPace® FA in addition to antibiotics in phase 1, 2 and 4 (based on the best performing Trt from Exp 1). Trt 2: consisted of a similar feeding program as Trt 1 with the exception that OutPace® FA was added at a rate of 5 lbs/ton without the use of AB in the feed (alternative AB free feeding program). The AB protocol use in Trt 1 was Denagard (Novartis Animal Health, Greensboro, NC) at 200g/ton in phase 1, a combination of Denagard at 35g/ton and chlortetracycline at 400g/ton (Denagard/CTC) in phase 2, Non-Medication in phase 3 and repeated the Denagard/CTC combination in phase 4. Both feeding programs were similar in nutrient composition and included other functional components currently used in the Purina Animal Nutrition lactose-free nursery feed program.

The two dietary treatments were designed to be fed in 4 phases during a 43-days period, following a feeding budget of 10 lbs of feed/pig in phase 1 and phase 2, and 20 lbs of feed/pig in phases 3 and 4, respectively. However, not all the feed was consumed by day 43. Therefore, pigs were fed until day 50. The 2 dietary treatments are described in Table 2.

In Exp 1 and 2, Pigs had free access to feed and water. Production performance parameters including initial and final pig body weight, pen feed consumption, pen weight, and mortality were measured and recorded weekly. Weight gain and Feed:Gain (F:G) ratio was calculated. All the collected data was analyzed using the mixed procedure of SAS.[3]

Table 1: Exp 1: Dietary treatments

Treatment	OutPace® FA	Antibiotics	Feed budget, lbs. of feed/pig
1	0.0 lbs./ton	Phase 1: Denagard® 35 g/ton + CTC 400 g/ton	Phase 1: 2.5
2	2.5 lbs./ton	Phase 2: Denagard® 35 g/ton + CTC 400 g/ton	Phase 2: 5.0
3	5.0 lbs./ton	Phase 3: 50g/ton Mecadox®	Phase 3: 12.0
		Phase 4: CTC 500 g/ton	Phase 4: 34.0

Table 2: Exp 2: Dietary treatments

Treatment	OutPace® FA	Antibiotics	Feed budget, lbs of feed/pig
1	2.5 lbs/ton	Phase 1: Denagard® 200 g/ton Phase 2: CTC 400 g/ton /Denagard® 35 g/ton Phase 3: No Meds Phase 4: CTC 400 g/ton -Denagard® 35 g/ton	Phase 1: 10.0 Phase 2: 10.0 Phase 3: 20.0 Phase 4: 20.0
2	5 lbs/ton	None	Phase 1: 10.0 Phase 2: 10.0 Phase 3: 20.0 Phase 4: 20.0

Results and discussion

Experiment 1

In Exp. 1, the best performance and economic return was obtained (Tables 3 and 4) when pigs were fed diets containing 2.5 lbs/ton of OutPace® FA with AB (Trt 2). These pigs were 1.98 lbs heavier ($P = 0.09$) when compared to pigs raised with AB and not fed OutPace® FA and 0.98 lbs heavier ($P = 0.09$) when compared to pigs fed diets containing AB and OutPace® FA at 5 lbs/ton of complete feed. Additionally, this group of pigs used 0.08 lbs less feed ($P < 0.01$) per each pound of body weight gain when compared to the control pigs (7.8 point). These results indicate that the most efficient level of OutPace® FA is 2.5 lbs/ton of feed in combination with AB to improve pig performance.

The next step to identify an effective alternative feeding program was to compare the nursery performance of pigs fed a feeding program with AB and 2.5 lbs of OutPace® FA/ton of complete feed across all nursery diets vs. an alternative feeding program for pigs raise without AB (using OutPace® FA at 5lbs/ton of feed). This was the approach used in Exp 2.

Experiment 2

There were no significant differences in pig performance between pigs fed OutPace® FA at 5 lbs/ton in non-medicated diets compared to feeding a program with AB and 2.5 lbs of Outpace® FA (Tables 5 and 6). Although, these results indicated a strong numerical trend ($P = 0.06$) for improved feed efficiency (F:G) between days 0 to 43 when pigs were fed AB in combination with 2.5 lbs/ton of OutPace® FA.

Conclusions

The results of these studies indicate that adding OutPace® FA at a rate of 2.5 lbs/ton of complete feed improves performance of nursery pigs fed diets containing antibiotic, and that pigs fed diets containing 5 lb/ton OutPace® FA and no antibiotic had similar gains and feed conversion to those fed diets containing antibiotic and 2.5 lb/ton of OutPace® FA during day 0 to 50 postweaning. OutPace® FA should be consider as an efficient tool to improve nursery pig performance when pigs are fed diets with or without antibiotics.

References

1. OutPace® is a registered trademark of Land O'Lakes Inc.

2. Jacela JY, DeRouchey JM, Tokach MD, et al. Feed additives for swine: Fact sheets – acidifiers and antibiotics. J Swine Health Prod. 2009;17(5):270–275.

3. SAS Institute Inc., Cary, NC.

Table 3: Exp 1: Effect of feeding 3 level of OutPace® FA in nursery pig diets on the performance of nursery pigs raised with antibiotics during a 41-day period.

Variable	Trt 1: Control No OutPace® FA + AB	Trt 2: OutPace® FA 2.5 lbs./ton + AB	Trt 3: OutPace® FA 5 lbs./ton + AB	*P*-value
Start weight, lbs	14.31	14.44	14.44	> 0.1
D 14 Wt, lbs	18.39	18.32	18.32	> 0.1
D 21 Wt, lbs	23.79	24.22	24.00	> 0.1
D 27 Wt, lbs	29.88[d]	30.92[e]	30.48[de]	= 0.1
D 41 Wt, lbs	49.63[d]	51.61[e]	50.69[de]	= 0.09
Day 0-14				
ADG, lb	0.472	0.469	0.471	> 0.1
ADFI, lb	0.603	0.596	0.601	> 0.1
F:G	1.29	1.28	1.28	> 0.1
Day 0-21				
ADG, lb	0.578[d]	0.604[e]	0.600[ed]	= 0.1
ADFI, lb	0.798	0.788	0.802	> 0.1
F:G	1.39	1.34	1.31	
Day 0-27				
ADG, lb	0.656	0.692	0.683	> 0.1
ADFI, lb	0.932	0.930	0.931	> 0.1
F:G	1.43[f]	1.35[d]	1.37[e]	= 0.1
Day 0-41				
ADG, lb	0.893[f]	0.944[d]	0.930[e]	= 0.06
ADFI, lb	1.135	1.320	1.326	> 0.1
F:G	1.48[f]	1.40[d]	1.43[e]	< 0.01
Mortality, %	2.727[a]	0.455[b]	1.818[a]	< 0.05

[a,b,c] means within a row with different superscripts are significantly different, $P < 0.05$.
[d,e,f] means with different superscripts indicate a trend to be different, $P \leq 0.10$.

Table 4: Exp 1: Economics from the experimental period (Summer 2016).

Variable	Trt 1: Control No OutPace® FA + AB	Trt 2: OutPace® FA 2.5 lbs./ton + AB	Trt 3: OutPace® FA 5 lbs./ton + AB
Total feed cost	$8.6324	$8.6393	$8.7686
Difference		$0.0069	$0.1362
Cost per/lb gain	$0.2291	$0.2168	$0.2252
Cost of 40 lbs of gain	$9.164	$8.672	$9.008
Savings/pig/40 lbs gain		-$0.492	-$0.1560

Table 5: Exp 2: Performance of nursery pigs fed diets containing AB and 2.5 lbs/ton OutPace® FA vs. pigs fed non-medicated diets containing 5.0 lbs/ton of OutPace® FA during a 43-day period.

Variable	Trt 1: 2.5 lbs/ton of OutPace® with AB phase 1,2 and 4	Trt 2: 5.0 lbs/ton of OutPace® without AB	*P* value
Start Wt, lb	11.992	11.958	0.92
D 22 Wt, lb	23.89	23.49	0.58
D 43 W, lb	49.84	48.05	0.40
Day 0-22			
ADG	0.528	0.524	0.81
ADFI	0.628	0.640	0.43
F/G	1.19	1.22	0.37
Day 0-43			
ADG	0.87	0.84	0.07
ADFI	1.20	1.18	0.40
F/G	1.38c	1.41d	0.06

Table 6: Exp 2: Overall performance of nursery pigs fed diets containing AB and 2.5 lbs/ton OutPace® FA vs. pigs fed non-medicated diets containing 5.0 lbs/ton of OutPace® FA (day 0-50).

Variable	Trt 1: 2.5 lbs/ton of OutPace® with AB phase 1,2 and 4	Trt 2: 5.0 lbs/ton of OutPace® without AB	*P* value
Start wt, lb	11.992	11.958	0.92
D 50 wt, lb	60.12	59.04	0.40
Day 0-50			
ADG	0.95	0.94	0.47
ADFI	1.39	1.39	0.74
F/G	1.47	1.47	0.48
Mortality, %*	2.90	0.83	0.13
Savings**, $/pig		0.45	

* The majority of the death loss (77.8%) was from strep in the heavier pigs.
** Economics from the experimental period (Summer, 2016).

Water in oil adjuvant selection for the formulation of one-shot safe bacterial vaccines for swine

Sophia Bingling Xu; Juliette Ben Arous; Nicolas Versillé; Laurent Dupuis
SEPPIC, Puteaux, Paris, France

Introduction

Oil emulsion adjuvants are extensively used in swine inactivated vaccines. Classical oil based swine vaccines consists in oil-in-water emulsion formulations, or water-in-oil-in-water double emulsions.[1] These adjuvants induce a strong short term response, but usually require two injections for a long term protection. The formulation of one-shot vaccines can be performed with the use of water-in-oil adjuvants. Water-in-oil vaccines for swine must be optimized carefully to avoid problems of injectability and risks of local reactions at slaughter, especially with reactive antigens. Here we show that the selection of an adapted water-in-oil adjuvant allows the formulation of a one-shot vaccine against *Actinobacillus pleuropneumoniae* (APP) with a strong antigenic load reduction compared to oil in water formulation.

Materials and methods

APP S2 vaccines were formulated with 2 different water-in-oil adjuvants (WO1 and WO2). 3 vaccines were formulated for each adjuvant with a range of APP concentration: 100% (2.25 10^8 CFU/ml), 10%, 1%. Injectability of all vaccines was assessed (ease to administer the vaccine). 80 6-week old pigs were randomly separated in 7 test groups and 1 non-adjuvanted control group. At D0, 3 groups received 1 ml of WO1 based vaccines respectively at 100%, 10% and 1%, 3 groups received 1 ml of WO2 based vaccines at 100%, 10% and 1%, 1 group received 2 ml of oil-in-water (OW, Montanide™ ISA 15A VG) classical vaccine at 100% of antigen concentration. Body temperature, local (manual palpation) and general (such as anaphylactoid shocks) reactions were measured in all groups at 4 and 24h after injection, and carcass quality was also assessed at slaughter at D120. Blood samples were taken at D0, 28, 80 and 120, and specific IgG1 and IgG2 antibody titers against APP were measured by ELISA titration.

Results

Administration of the vaccines was done without injectability issue in all groups. Water-in-oil vaccines did not induce any body temperature increase (Figure 1), local or general reaction after the injection, whereas the classical vaccine induced a transient increase of temperature (< 1°C). With 100% antigenic load, both WO1 and WO2 induced highly stronger humoral responses to APP than classical OW vaccine antibody response, but they also induced non acceptable local reactions at slaughter such as abscesses (< 1 cm). With 10% of antigenic load, 1 ml injection of WO2 based vaccine did not induce injection site lesions at slaughter, and induced a humoral response (IgG1 and IgG2) similar to the injection of 2ml of OW based vaccine containing 100% antigenic load (Figure 2).

Figure 1: Body temperature evolution before injection, at 4 and 24h p.i.

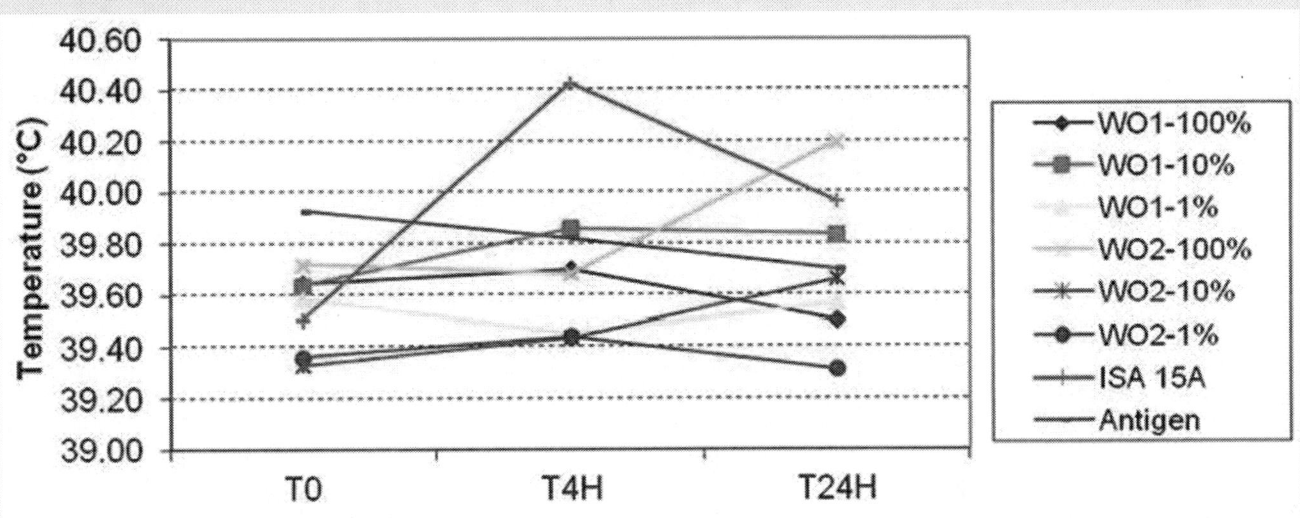

Figure 2: Humoral response titration

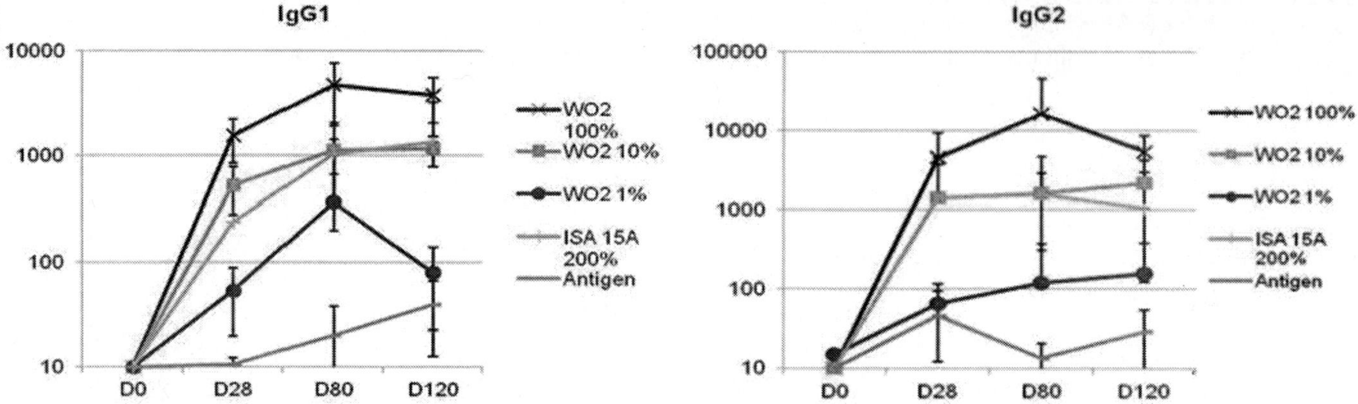

Conclusions

These results show that water-in-oil adjuvants are effective formulations for swine vaccination. However, the selection of an adapted adjuvant and antigenic load and vaccine dose adjustment is critical to avoid lesions at slaughter. Water-in-oil adjuvants can allow the formulation of one-shot vaccines for pigs with limited amount of antigen, with low pyrogenicity and low risks of anaphylactoid shocks.

References

1. Heegaard PMH, Dedieu L, Johnson N, Le Potier MF, Mockey M, Mutinelli F, Vahlenkamp T, Vascellari M, Sørensen NS. Adjuvants and delivery systems in veterinary vaccinology: current state and future developments. *Arch Virol.* 2011;156:183–202.

Diagnosing edema disease: the first step in prevention

V. Gotter, DVM, PhD, Dipl. ECPHM
IDT Biologika, Dessau-Rosslau, Germany

Introduction

Edema disease (ED) is one of the primary reasons of death and reduced performance in weaned pigs worldwide.[1] The disease is caused by the shigatoxin type 2e (Stx2e) of shigatoxin producing *Escherichia coli* (STEC), sometimes also called edema disease *Escherichia coli* (EDEC).[2] Typical clinical signs include edemas of the eyelids and snout (which gave the disease its name). However, other, more unspecific clinical signs can occur as well, such as lying on the side with paddling legs, "squeaking" vocalizations and stunted growth. Even sudden death without any prior clinical signs has been reported.[2] Therefore, there are several differential diagnoses should be considered. These include (but are not limited to): infections with *Streptococcus suis, Haemophilus parasuis,* classical swine fever virus (CSFV), Aujeszsky's Disease virus (ADV) and salt intoxication.[2] The preliminary on-farm diagnosis must therefore be confirmed or refuted by laboratory diagnostics. For many years it has been considered state-of-the art to serotype hemolytic *E.coli* colonies that have been previously isolated via culture. The objective of our study was to show that by using only this method, some cases of ED might not be identified for what they are.

Material and methods

In the year 2015, 462 samples (naïve fecal samples and intestinal swabs) from 167 farms were sent in for analysis to a diagnostic laboratory in Germany (Table 1). All samples came from pigs with the preliminary diagnosis of ED on-farm. The samples were each submitted to bacterial culture. Afterwards colony pools of up to six *E. coli* with similar morphologic characteristics were analyzed for virulence factors via multiplex PCR.

Table 1: No. of samples, farms and PCRs in 2015.

No. of samples	462
No. of farms	167
No. of PCRs	821

Results and discussion

In total, 821 PCRs were performed of which 122 were positive for Stx2e (Figure 2). Thirteen of these Stx2e positive PCRs were performed on non-hemolytic colonies (Table 2). If these PCRs had not been performed, eleven herds would falsely have been classified as not positive for Stx2e and therefore ED as cause of the clinical signs would have been ruled out. Our study demonstrates that not only is it important to preform additional laboratory diagnostics, but also how they are performed. Veterinarians should be aware of the newest possibilities for diagnostic analyses and should preferably send samples to laboratories that are able to perform these. Only if the diagnosis of ED is based on solid evidence, a preventive strategy can be designed that will not only help to improve animal welfare, but also reduce the use of antimicrobial substances and provide an economic benefit for the farmer.

Table 2: Results of the analysis from 2015.

No. of positive samples	108	% of positive samples	23.3
No. of positive farms	50	% of positive farms	29.9
No. of positive PCRs	122	% of positive PCRs	14.8
No. of Stx2e positive PCRs from non-hemolytic *E. coli*	13	% of Stx2e positive PCRs from non-hemolytic *E. coli*	10.6
No. of herds misclassified	11	% of herds misclassified	22.0

References

1. Frydendahl K (2002): Prevalence of serogroups and virulence genes in *Escherichia coli* associated with post-weaning diarrhea and edema disease in pigs and comparison of diagnostic approaches. Vet Micro 85: 169-182

2. Fairbrother JM, Gyles CL (2012): Colibacillosis. In: Zimmermann, J.J., Karriker, L.A., Ramirez, A., Schwartz, K.J., Stevenson, G.W. (Eds.), Diseases of Swine 10th Edition, Wiley-Blackwell, Oxford, pp. 723-749.

Seneca Valley virus: Porcine-derived biological risk assessment

M. Madsen, PhD; J. Clampitt, BS; C. White, BS; S. Brimer, BS; E. Brimer, BS; Y. Gardin, DVM; J. El-Attrache, PhD
Scientific Services and Investigative Unit, Ceva Biomune, Lenexa, Kansas

Introduction

Senecavirus A (SVA) is a swine viral pathogen causing clinical lameness, coronary band separation and vesicular lesions on the snout and hooves, a naturally occurring (+) ssRNA Picornavirus. Of particular concern is the similarity of symptoms to foot-and-mouth disease and the possibility of overlooking symptoms because they are believed to be an SVA infection and not the more serious foot-and-mouth disease. Cases of SVA have been identified as early as 1988 with outbreaks on the rise since July 2015.[1,2]

Originally identified in 2002 as a tissue culture contaminant at Genetic Therapy, Inc. (Gaithersburg, MD, USA).[3] The source of the contamination was believed to from animal derived components used in cell culture. While porcine derived products undergo exogenous agent testing, the testing does not cover several viruses including SVA, porcine parvovirus, porcine circovirus 2. Several methods are employed to limit any contaminants in animal derived products without a significant impact on performance, including filtration, heat inactivation and gamma-irradiation. In this study we investigate the presence of SVA in archival porcine serum and porcine derived commercially available trypsin.

Biosecurity addresses the conditions surrounding animal health, pest control, housing, feed and human traffic throughout production facilities. Measures have drastically increased over the years to limit accessibility, improve disinfection of vehicles and equipment, monitor employees movement throughout facilities and even feed quality. The balance comes in managing the cost of steps and practices to implement greater biosecurity measures against the threat of disease introduction. Prevention is significantly less than control and eradication.

One area of risk management that needs additional attention is that of animal derived reagents. They comprise a large component of our animal health management by means of vaccine programs. In 2014, the European Medicines Agency, Committee for Medicinal Products for Human Use (CHMP) issued a draft document outlining the concerns of viral particle resistance to gamma-irradiation at a level of 25-30 kGy.[4] In light of this, raw material testing has greater relevance given the detection of such viruses as Porcine Parvovirus and Porcine Circovirus may not be inactivated. SVA is included in viruses that may not be inactivated by 25-30 kGy.

This study examined the presence of SVA in porcine derived reagents using an in-house developed semi-quantitative PCR. The current levels of screening do include extraneous agent testing, however, several viruses may elude this testing. Additional steps must be taken to ensure the purity of swine derived biologicals.

Materials and methods

Samples

Trypsin samples were purchased from Atlanta Biologicals (Flowery Branch, GA) (Cat#B81710) and Sigma Aldrich (St. Louis, MO) (Cat# T5266). According to the manufacturer's product insert, samples received 25-40 kGy of gamma-irradiation and general extraneous agent testing. From the 10X stock concentrations, 100 X, 10X, 5X and 1X dilutions were made for testing. Eight porcine sera dating back to 2001 were screened by PCR.

Briefly, samples were incubated to mimic production conditions and isolated using magnetic bead kits according to the manufacturer's recommended protocols.

Real-time PCR primers and probe design

Senecavirus A primers were designed using the reference sequence NC_011349.1 with the online NCBI primer blast tool. Using the Primer 3 algorithm, 10 possible PCR target locations were chosen. Using DNAStar alignment functions (MegAlign, Madison, WI), the NC_011349.1 sequence, a second SVA sequence (USDA KC667560.1) and the in-house sequence were aligned (Fig. 1) and a conserved region was determined. Out of the pool of primers developed there was a focus on one in the 5' UTR with the idea being it would be more conserved. The primers and the probe amplify the polyprotein gene and fall in-between the known SNPs that allow for a broad spectrum of detection. Primers and probes were purchased from IDT (Coralville, IA).[5] Sensitivity was determined by standard curve from virus isolated from chicken embryo fibroblasts titered to ~ $TCID_{50}$. A standard curve was generated using a starting concentration of 1000 viral particles of RNA/ 2.5 µl and serial diluted 10^{-7} representing ~0.0001 viral particles of RNA/ 2.5 µl using the virus titer.

This PCR was then tested against a panel of common swine viruses and bacteria to determine specificity for SVA. There was no cross reactions detected(data not shown).

Whole genome sequencing

Briefly, barcoded sequence libraries for both trypsin lots were prepared using the Ion Total RNA-Seq Kit v2 (Life Technologies, Carlsbad, CA and were loaded on an Ion 318 Chip v2 (Life Technologies). Samples were run on the Ion Torrent Personal Genome Machine (Life Technologies).

DNA sequence analysis

Proprietary algrothims evaluate the signal processing and base calling algorithms of the raw sequence reads (Life Technologies). Quality assessed raw sequences were analyzed using reference guided assembly (NGen software, DNASTAR, Madison, WI), to SVV-001 (GenBank accession # NC_011349.1) and NADC SVV 11-55910-3 (GenBank accession # KC667560.1) as scaffolds. INDEL and SNP calls were made using SeqMan Pro (DNASTAR). Phylogenetic trees were analyzed using MegaAlign (DNASTAR) and the ClustalW statistic including various regional sequences from GenBank.

Results and discussion

PCR confirmed the presence of SVA in both commercial lots of trypsin were PCR. Of the eight serum samples, six were PCR positive for SVA. Whole genome sequencing gave full read sequence including the poly-A tail. Although, not part of the discussion here, samples were then placed on primary cell culture and and cytopathic effect was observed, indicating the SVA was active in the trypsin and serum samples.

A second round of 25-40 kGy gamma-irradiation proved to inactivate SVA in the trypsin samples without a significant decrease in activity. The samples were also negative for PPV and PCV2 by qPCR.

Of note is where SVA-715 falls on the phylogenetic tree (Figure 1). SVA-715 has a common ancestry with isolates sequenced that originated in Brazil. The current outbreak isolates all share a common ancestry with SVA-715. In light of the lower costs and easier workflow of WGS, should we be screening our raw materials more closely for current and emerging infectious diseases? Our biosecurity measures must extend to all aspects of animal health and include evaluating the risk raw materials pose in our production facilities.

Table 1: PCR Ct values for original trypsin and serum.

Sample date	Lot #	qRT-PCR*
2001	102070	ND
2002	201040	+
2010	015280	+
2011	115140	+
2012	213270	+
2013	313220	+
2014	418190	+
2015	515110	ND

* ND = Not detected, + < 40 Ct

Figure 1: Phylogenetic tree.

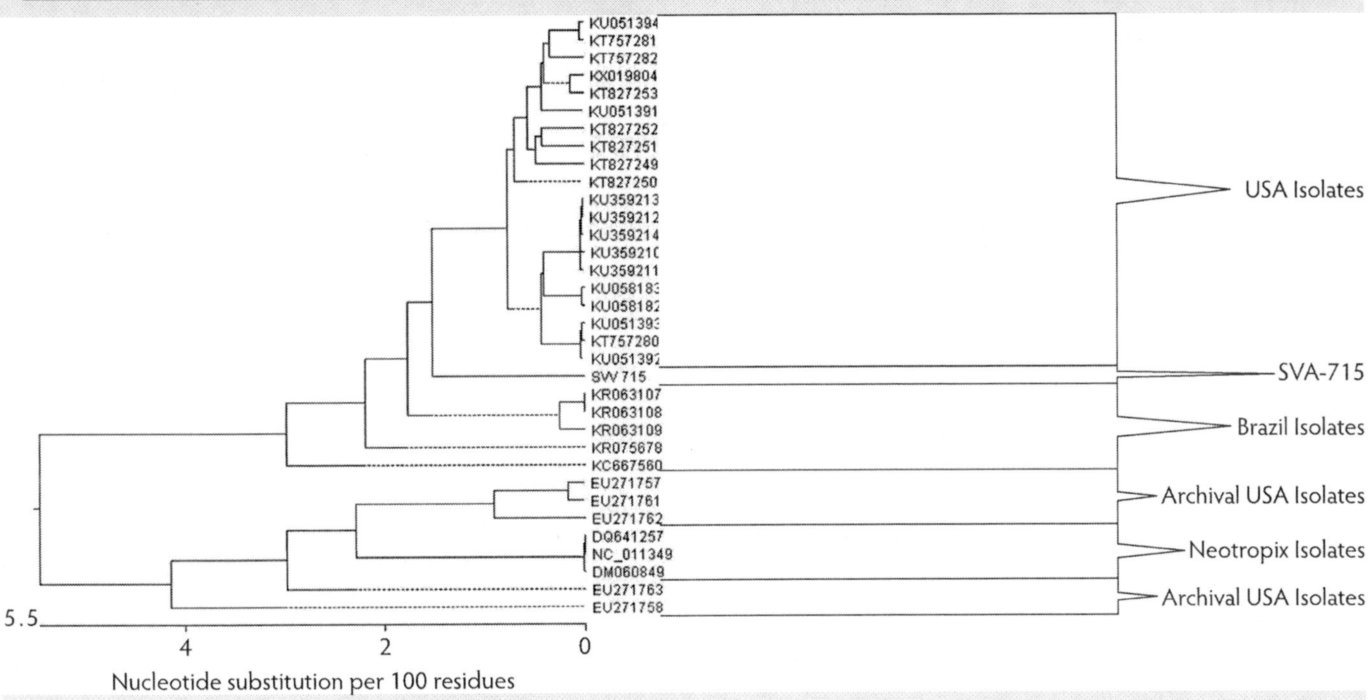

References

1. Hales LM, Knowles NJ, Reddy PS, Xu L, Hay C, Hallenbeck PL. Complete genome sequence analysis of Seneca Valley virus-001, a novel oncolytic picornavirus. Journal of General Virology. 2008;89(5):1265-1275.

2. Ameican Association of Swine Veterinarians.2016. http://nationalhogfarmer.com/health/seneca-valley-virus-may-be-down-it-s-not-out. Accessed November 10, 2016.

3. Knowles, N.J. (2005). A pan-picornavirus RT-PCR: identification of novel picornavirus species. EUROPIC 2005: XIIIth Meetingof the European Study Group on the Molecular Biology of Picornaviruses, Lunteren, The Netherlands, 23-29th May 2005.Abstract A06.

4. Overview of comments received on 'Guideline on the use of porcine trypsin used in the manufacture of human biological medicinal products'. EMA/CHMP/BWP/574033/2013. 20 February 2014.

5. Clampitt, et. al. Detection of Senecavirus A (SVA-715)in porcine-derived trypsin. Submitted to Vaccine. 2016.

Specialized VaxLiant® adjuvant for DNA vaccine development using influenza A virus-swine hemagglutination (HA) gene

M. Inman, PhD; M. Pfannenstiel, PhD; D. Petrik, PhD; L. Trygstad; T. Miller, PhD
AgriLabs Corporation

Introduction

Vaccine platform technologies have evolved quickly both in the elements used to construct the vaccine platforms but also the strategies for delivery. VaxLiant® has developed a group of adjuvant and delivery formulations that have shown to be safe, stable, and efficacious for several types of conventional and novel vaccine platforms, including DNA vaccines. One of these adjuvant formulations has consistently provided the best response in both chickens and pigs using avian and swine influenza virus as a model. As an example, Influenza A virus is an important infectious respiratory disease agent in swine in North America and throughout the world. In addition, the virus has a zoonotic potential which can impact public health. The HA antigen is a common focus of vaccine development as this antigen usually induces the strongest antibody response after infection of swine with wild-type SIV. Antibodies closely-matched to the infecting virus HA protein can be sufficient to prevent infection and can also contribute to clearing virus from the lungs. Influenza A viruses are known to undergo both antigenic shift and antigenic drift resulting in viruses for which a pre-existing immune response in a given pig or herd is no longer effective. This propensity of the virus to "change" its antigenic make-up can cause sudden disease outbreaks in previously immune herds.

DNA vaccines serve as new weapons against SIV drift because SIV HA-based DNA vaccines, unlike more conventional SIV vaccines, are readily modifiable to adapt to drift of HA protein composition of a newly problematic field virus. A newly emergent distinct HA gene can rapidly be sequenced, synthesized, and substituted into the DNA production platform, thereby specifically addressing disease outbreaks caused by "new" strains expressing slightly different protein sequences in the HA surface protein. To this end, a DNA vaccine was developed in AgriLabs using a well-characterized DNA backbone plasmid and E. coli host cell together as a DNA production platform. The plasmid backbone DNA was used to carry a synthetic gene encoding the sequence of an H3 antigen identified from a recent swine influenza H3N2 Cluster IV strain (A/swine/South Dakota/A01280095/2013). The H3-HA specific plasmid backbone was mixed with a VaxLiant® adjuvant and inoculated into weaning pigs. The HA specific antibody was evaluated for cross recognition to several H3 SIV serotypes by HAI. The results demonstrate the use of DNA vaccines with an adjuvant delivery system can be used as a means to quickly provide new vaccines against emerging disease variants in swine.

Materials and methods

The plasmid backbone 8685-eRNA41H-KP307984 was used to produce a DNA vaccine encoding a synthetic gene for the sequence of an H3 antigen identified from a recent swine influenza H3N2 Cluster IV strain A/swine/South Dakota/A01280095/2013. The final plasmid construct is 5334 nucleotides long and for this study contained the HA gene from A/swine/South Dakota/A01280095/2013, H3N2. The plasmid DNA was purified by affinity chromatography to 99% purity. The plasmid expressing the H3-HA specific DNA was inoculated into 28 weaned piglets (SIV antibody negative) from four sows sourced from a high health herd, and were randomly assigned to one of seven treatment groups. Six of the groups were vaccinated intramuscularly (IM) with 100ug of the plasmid DNA combined with one of the following adjuvants, 1- ENABL S3, 2X ENABL S3, 1X ENABL LS5, 2X ENABL LS5, 1X ENABL LS6, or 2X ENABL LS6. The remaining group was vaccinated with PBS using the same procedures. Vaccination occurred on study days 0, 14, and 35. Blood for serology was collected on study days 0, 14, 28, 42, and 49. One milliliter of each test article was administered intramuscularly (IM) in the lower neck region behind the ear. The first vaccination (day 0) was administered on the right side; the second (day 14) was administered on the left side and the third vaccination (day 35) on the right side. Blood was collected 14 days after the initial vaccination (day 14), 14 days after the second vaccination (day 28), and 10 days after the third vaccination (day 49). Serum samples were clarified by centrifugation and the supernatants were stored at 18 ± 5°C.

Hemagglutination inhibition (HAI) assays were used to compare swine influenza viruses that vary in pair-wise sequence homology and structural homology of the HA gene (Table 1). HAI was performed at Benchmark Biolabs using standard procedures. For samples tested by NVSL, test serum samples from study days 28 (post 2 doses) and 49 (post 3 doses) were tested in an HAI test using the challenge virus A/swine/Iowa/A01480656/2014 (H3N2) (IA/14). For samples tested by the Veterinary Diagnostic Laboratory at ISU, test serum samples from study days 14 (post 1 dose), 28 (post 2 doses) and 49 (post 3 doses) were tested by HAI using challenge viruses IA/14 and A/swine/South Dakota/A01280095/2013 (H3N2) (SD/13). The SD/13 isolate is the homologous virus to the DNA vaccine insert.

For immunofluorescent assays, cells were seeded into 96-well plates at 2.0×10^5 cells/mL for ~24 hours 75% confluent. A

Table 1: Comparison of H3N2 variants by hemagglutination inhibition antibody titer using serum generated from pigs vaccinated with a DNA vaccine containing the H3-specific HA gene of A/Swine/SD/2013 swine influenza. Amino acid pair-wise homology was taken from GenBank accession numbers or ISU SIV surveillance database. Structural homology was taken from Lewis, et. al. 2014. J. Virol. 88: 4752-4763.

Strain ID	Result (HAI GMT)	Pair-wise homology	Structural homology (Lewis, et al. J. Virol)
A/Swine/MN/2013[1]	Poor (HAI <4)[2]	99	100
A/Swine/MO/2005	Poor (HAI <10)[2]	88	33
A/Swine/NC/98	Mid (HAI 40)[2]	89	17
A/Swine/TX/98	Mid (HAI 20)[2]	89	33
A/Swine/IA/2014	Good (HAI 80)[2]	93	33
A/Swine/SD/2013	Best (HAI 160)[3]	100	100

[1] Virus grown to > 7 log TCID50 has low HA titer
[2] Pooled sera from samples with positive HAI titer from pigs vaccinated with A/Swine/SD/2013 tested by diagnostic lab
[3] GMT from serum derived from pigs vaccinated with A/Swine/SD/2013

transfection master mix was prepared with 0.05ug/well of plasmid DNA (SIV H3N2; or plasmid expressing green fluorescent protein - EGFP), 5ul/well of jetPRIME buffer, and 0.1ul/well of transfection reagent. After 10 minutes incubation, 5ul of master mix was added to appropriate wells of the 96 well plates. Transfected cells were incubated overnight (16-24 hours) at 37 ± 5°C, 5 ± 2% CO_2 until > 50% EGFP-transfected cells were fluorescent. Cells expressing HA protein were detected by washing the monolayer with PBS, fixing with 80% acetone, and then blocking with Starting Block. Serum (test samples, positive control, and negative control) was diluted two-fold in Starting Block, added to wells starting at 1:4, and incubated for 30 minutes at room temperature. Cells were washed with PBS, goat anti-swine FITC conjugate was added at a 1:100 dilution, and incubated for 30 minutes at room temperature. Cells were washed with PBS once, washed with water twice, and then plates were read for fluorescence. For hemadsorption, 0.4% cRBC were added to transfected cells when 50% of the cell monolayer was positive for EGFP fluorescence. After incubation for 30 minutes at 37C, the cells were rinsed twice with DPBS and observed by light microscopy at 100X.

Results and discussion

VaxLiant® adjuvants are a copolymer and lipid matrix in an aqueous buffered background. The biopolymer forms a matrix shell in aqueous environment (Figure 1a) that becomes more opaque as lipid is added to the mixture (Figure 1b) which then forms a spherical or ellipsoidal particle (Figure 1c) that is fully loaded with lipid on a biopolymer scaffold. Graphically, this is depicted in Figure 1d where the lipid component can also be loaded with lymphocyte agonists, antigens, or other immune modulating agents (Figure 1e) which then can be displayed bound to both the internal and outer particle surface (Figure 1f). The adjuvant as well as adjuvant with protein antigen have been shown to be stable at both room temperature and at refrigerated temperatures (Figure 2) for over two years with three-year data pending. The HA (H3) protein was evaluated for expression by transfecting the plasmid DNA into 293T human kidney cells. The HA protein expression was detected by immunofluorescent assay using antibody specific to H3 (Figure 3a, 3b) or by hemadsorption (Figure 3c, 3d). The DNA vaccine is easily assembled in that the aqueous form of the adjuvant is mixed with the appropriate dose of the DNA by gentle agitation or swirling motion. The When mixed in aqueous form the final assembled vaccine is very stable as shown in Figure 4. After over two years in storage at 2-7°C, the DNA retained its covalently closed supercoiled CCSC format without any appreciable degradation.

To evaluate the adjuvant delivering the best serological response, weaning pigs were inoculated with a DNA vaccine expressing the H3 specific HA gene of A/swine/South Dakota/A01280095/2013 mixed with various formulations of the ENABL adjuvant. Previous data in chickens had indicated that the 05 formulation series of ENABL adjuvant provided the best immunological response by HAI serology (Figure 5a) when 14-day old SPF chickens were inoculated IM with a DNA vaccine expressing the H5-specific HA gene of A/gryfalcon/WA/41088-6/2014. For comparison, weaning pigs were inoculated 1, 2 or 3 times IM, with 100ug of DNA vaccine expressing the H3 specific HA gene A/swine/South Dakota/A01280095/2013. The resulting data indicates that the 05 formulation series of ENABL also provided for the best serological response for pigs as measured by HAI (Figure 5b). The serum derived from the A/swine/South Dakota/A01280095/2013 DNA vaccination was evaluated on several H3-specific A-type swine influenza viruses which are shown in Table 1. The highest homology and highest serological response by HAI assay was to the homologous challenge virus A/swine/South Dakota/A01280095/2013. The next highest

Figure 1: Scanning electron micrograph (SEM) and computer graphic of ENABL adjuvant. (a) biopolymer in aqueous solution; (b) partially formed biopolymer and lipid particle; and, (c) fully formed particle with standard ratio of polymer and lipid. Computer depiction of the polymer (d) being loaded with lipid containing biomolecules (e), which can then assemble on the surface and intra-particle to form an adjuvant particle than can present and/or deliver the biomolecule (f).

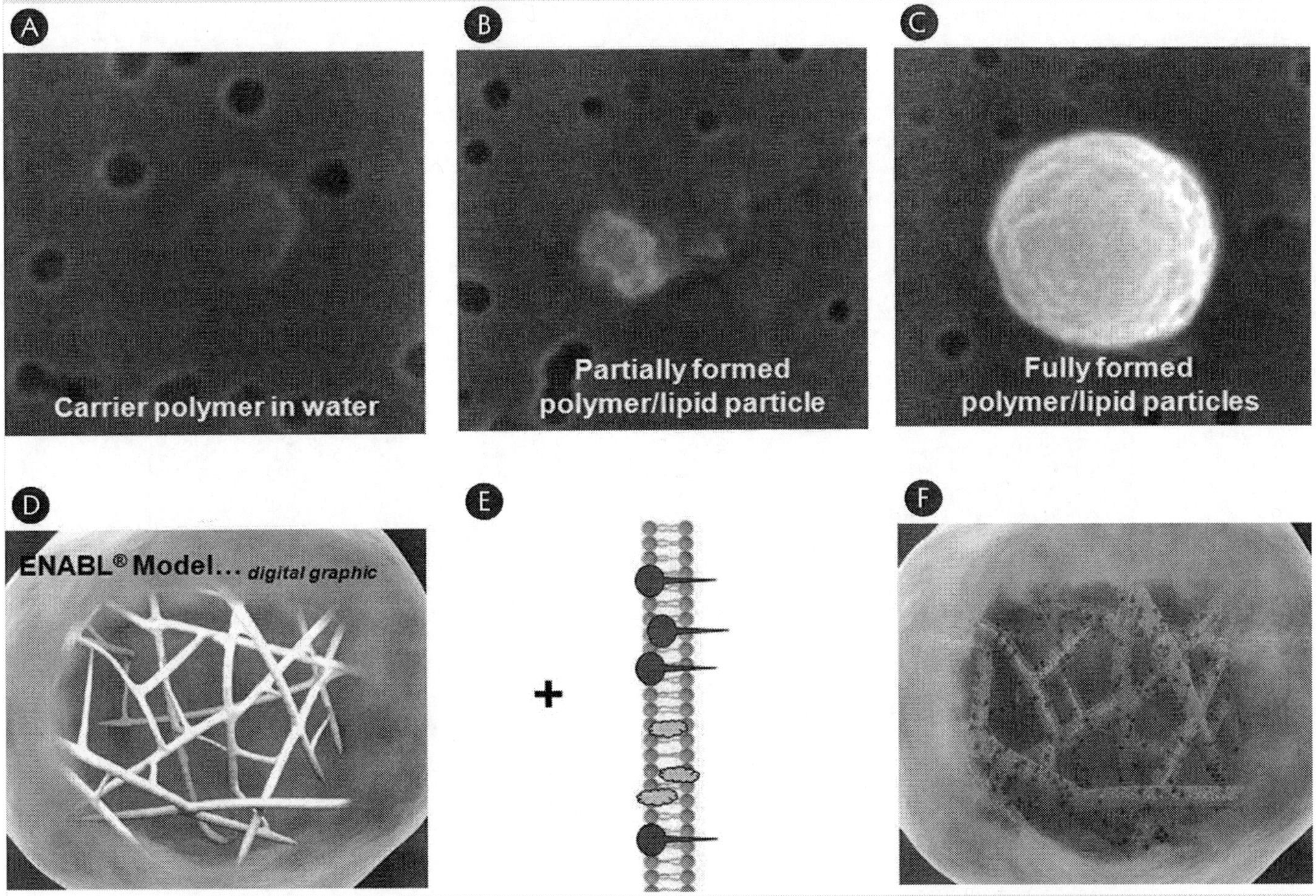

serological response by HAI assay was observed with A/Swine/IA/2014, which was only 93% homologous by pair-wise match and only 33% homologous by structural homology as described by Lewis et al, 2013. In contrast, the A/Swine/MN/2014, which is 99% homologous by pair-wise match and 100% by structural homology match, was the lowest in cross recognition by HAI assay results, although the A/Swine/MN/2014 isolate does not have a good hemaglutination titer compared to the A/Swine/SD/2013 or the A/Swine/IA/2014.

These results indicate that pigs can be immunized with plasmid DNA expressing the H3-specific HA gene with a priming and boosting dose with a low dose of plasmid DNA (100ug). The 05 series adjuvant of ENABL has unique features of stable binding, delivery, mixing, and inducing an immunological with DNA vaccine response in pigs. It only takes a matter of days for a DNA sequence to be incorporated into a DNA plasmid and produced to high copy number in a highly purified CCSC format. This makes the DNA vaccine platform ideal for correlating different molecular details of amino acid structural and pair-wise variants of SIV, or other emerging zoonotic diseases of swine.

Figure 2: Two year Stability of ENABL adjuvant at 2-7 C and room temperature. Adjuvant (06 series) was incubated by itself or in presence of antigen, bovine serum albumin (BSA). At various times after incubation the assembled mixture (antigen plus adjuvant) was inoculated into mice and immunological response to BSA was measured using a BSA specific capture ELISA. The data show that after two years there was no loss of antigenicity of the BSA, also adjuvant incubated alone for two years and mixed with fresh BSA had the same immunological response as BSA assembled in the adjuvant and incubated at two temperatures over a two-year period. Fresh = adjuvant stored at indicated temperature, BSA formulated fresh on day of inoculation.

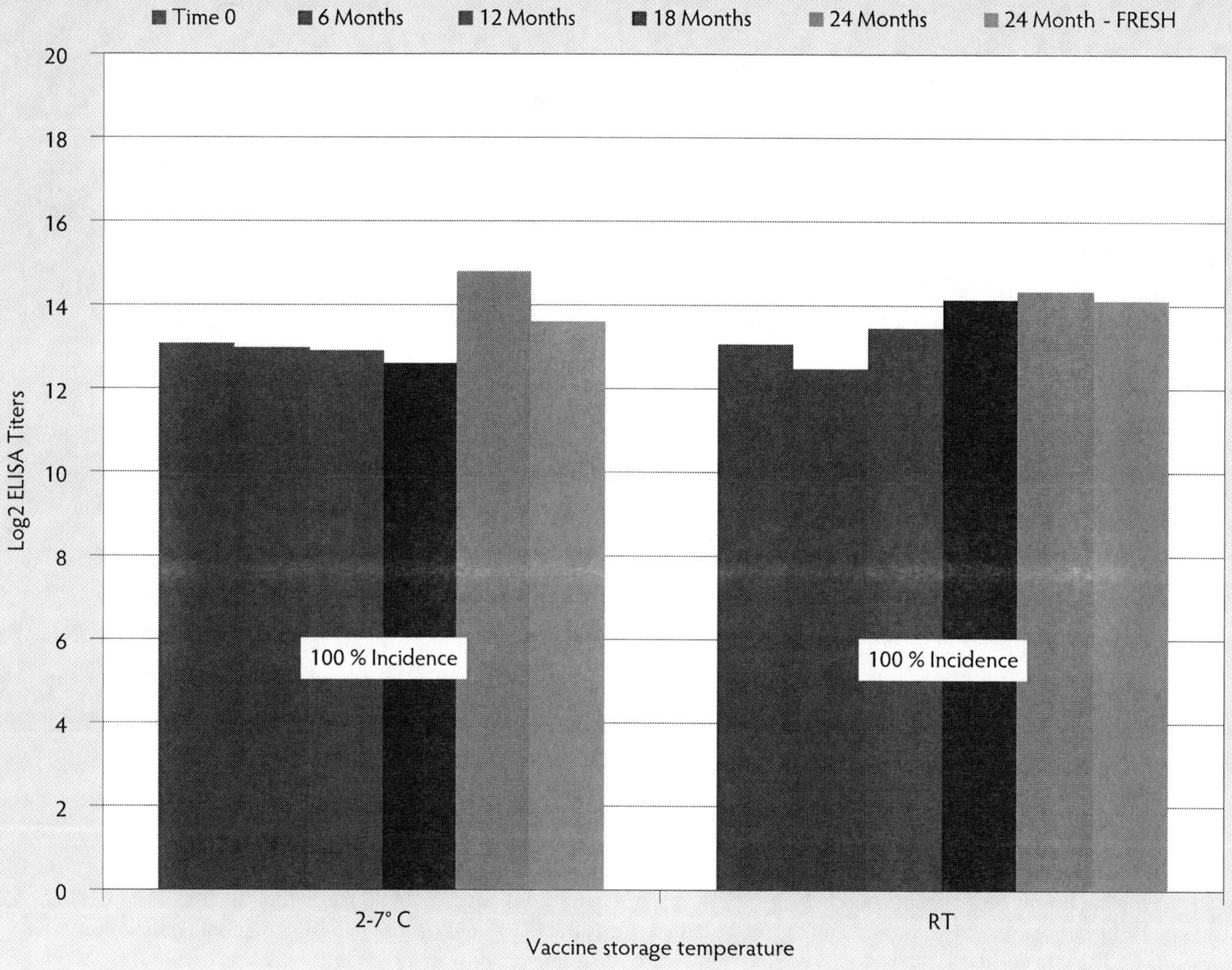

Figure 3: Potency of DNA vaccines by immunofluorescence and hemadsorption from transfection of 293T cells. (a) 293T cells non-transfected stained for AIV H5-specific HA by immunofluorescence. (b) 293T cells transfected with vaccine DNA containing AIV H5-specific HA gene A/gryfalcon/WA/41088-6/2014 and stained with H5 specific antibody. (c) 293T cells non-transfected and treated with 0.4% chick red blood cells. (d) 293T cells transfected with vaccine DNA containing H5-specific HA from A/gryfalcon/WA/41088-6/2014 and treated with 0.4% chick red blood cells.

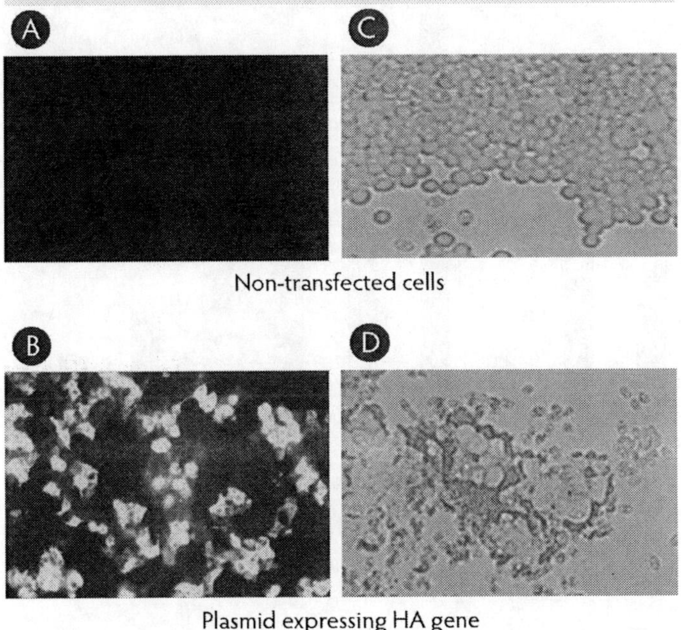

Figure 4: Two year stability of DNA vaccine stored 2-7 C in presence of ENABL adjuvant analyzed by agarose gel electrophoresis. Lane 1. One Kb DNA markers. Lane 2. Super coiled DNA ladder markers. Lane 3-5. Plasmid DNA stock at 400, 200, and 100 nanograms per lane, respectively. Lane 6-8, plasmid DNA stock mixed with ENABL adjuvant for 2 years at 400, 200, and 100 nanograms per lane. Lane 9-10, plasmid DNA mixed with buffer at 400, 200, and 100 nanograms per lane.

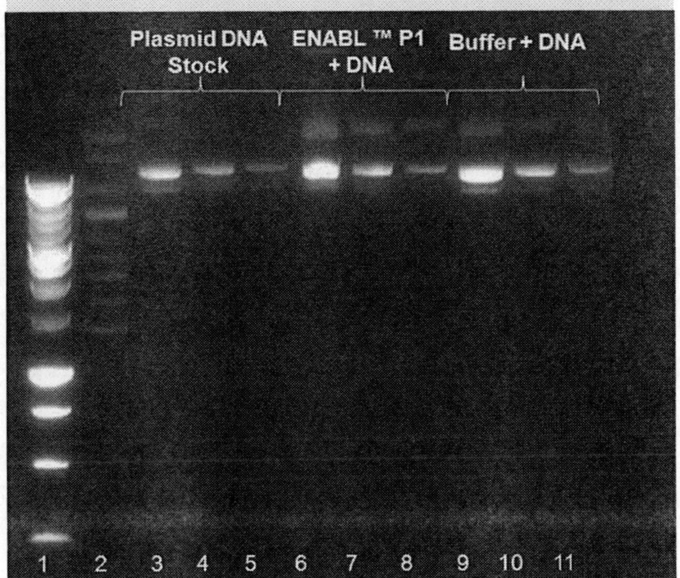

Figure 5: Optimization of ENABL adjuvant formulations by comparison of hemagglutination inhibition (HAI) antibody titers to HA protein expressed from plasmid DNA vaccination in chickens and pigs. (A) HAI antibody titer of negative control compared to plasmid DNA vaccine expressing H5-specific HA gene of A/Tk/WI/68. DNA vaccine was inoculated into Chickens IM two times 14 days apart and serum was collected 14 days after second inoculation. Adjuvant included ENABL 01, 02, 03, 05, and 06 series labeled BBL 01 derivatives a, b, and c; BBL 02 derivatives a, b, and c.; BBL 03 derivatives a, b, and c; BBL 05 derivatives a, b, and c; and, BBL 06 derivatives a, b, and c. (B) HAI antibody titer of negative control compared to plasmid DNA vaccine expressing H3-specific HA gene of A/Sw/SD/13. DNA vaccine was inoculated into 2-3 week old pigs IM two times 14 days apart and serum was collected 14 days after second inoculation. Adjuvants included ENABL 03, 05, and 06 series labeled 03a, 05a, and 06a.

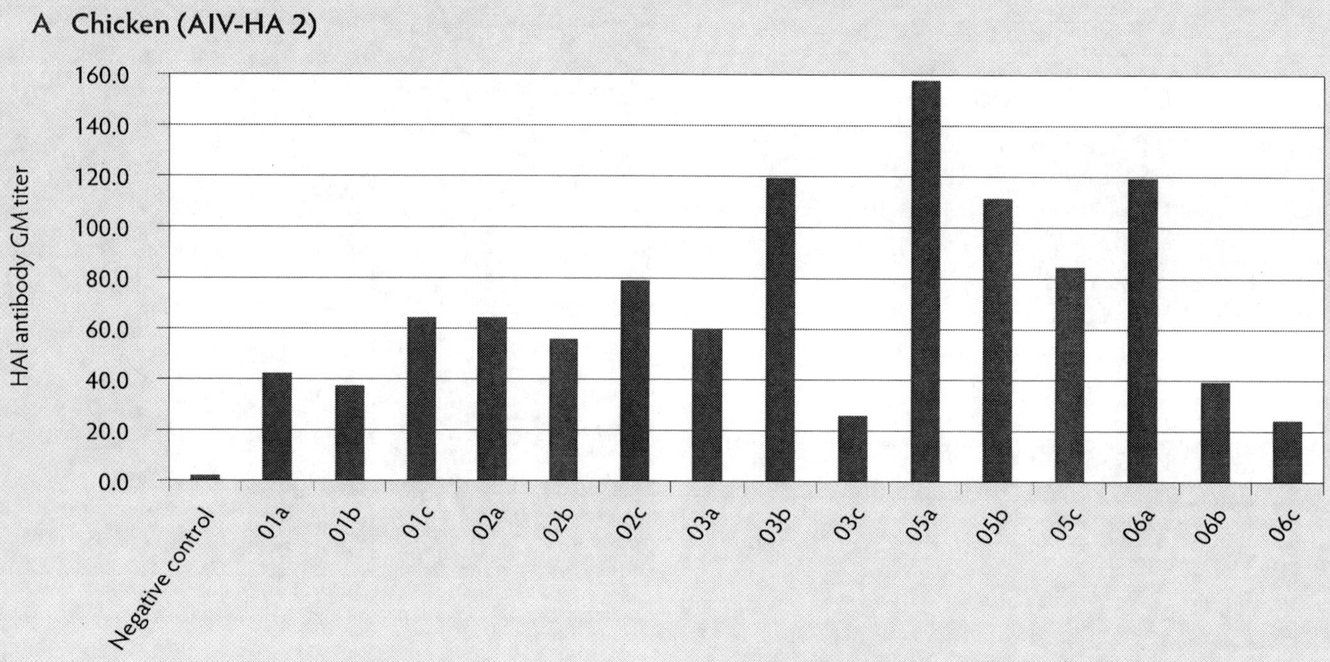

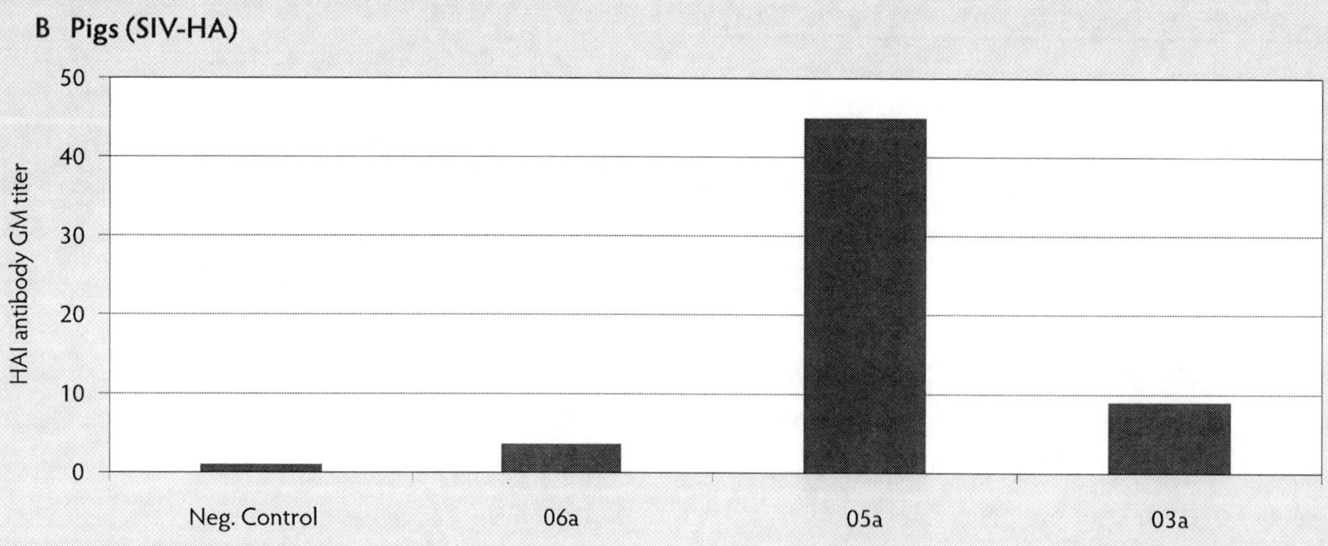

Feed hygiene: A layered approach

J.A. Jendza, PhD
BASF Corporation, Florham Park, New Jersey

Introduction

Feed is a primary route for introduction of pathogens onto swine farms. Feed ingredients are often secured from a wide geographical area and many are less-valuable by-products of other industries with biosecurity coming as a lower priority to manufacturing throughput. Unlike modern wean to finish facilities, feed mills are not operated on all-in all-out principles, which means that once pathogens are established within a mill, decontamination can be difficult. No system wide biosecurity plan can therefore be complete without consideration and mitigation of risks posed by the feed mill.

Feed processing

A popular intervention at feed mills is to pellet the feed, particularly at higher temperatures. It is well documented that most bacterial pathogens are temperatures sensitive.[1] In addition, pelleting has been shown to increase feed intake, and feed efficiency under optimal conditions. However, pelleting trials done under experimental conditions as opposed to under the tight time constraints of a high throughput industry mill can mislead as to the practical value of higher pelleting temperatures with regards to biosecurity.

Davies and Wray[2] took over 3,000 samples from various points in 9 industry feed mills and reported that *Salmonella* prevalence was highest in the feed intake pits, followed closely by the pellet cooler at 25 and 20%, respectively. This is despite *Salmonella* prevalence being only 11% in the mixer and < 8% in the pellet mill. The pellet cooler is a pretty optimal environment for the growth of bacteria with warm temperatures, high moisture (steam from pelleting process), a plethora of recently released nutrients (starch gelatinization), and a recent reduction in competition as a direct result of the heat stress of pelleting. Paradoxically, incomplete bacterial kill as a result of pelleting can lead to an increase in the pathogen load in feed leaving the feed mill.

Another limitation of pelleting is that it doesn't offer on-going protection. As demonstrated by Davies and Wray,[2] the reduction in *Salmonella* prevalence between the mixer and the pellet die did not translate into lower prevalence in the pellet cooler. Even under ideal conditions, where the pelleting process results in perfectly clean pellets moving into a perfectly clean pellet cooler, the sterility of these hypothetical pellets offers no protection against re-contamination further down the supply chain from the feed mill, to the transportation system, or on farm feed delivery system.

Chemical treatments

Chemical treatment of feed in the mill can be complementary to feed processing techniques as these rely on a different mode of action from simple heat. Whereas pelleting is a single point-in-time intervention, chemical treatments like organic acid or formaldehyde can detoxify feed already carrying undesirable levels of bacteria at the time of treatment, but can also protect against re-contamination so long as the product is still present and free to interact. For example, formaldehyde is used for the control of *Salmonella* for up to 21 days.[3] The 21-day limit stems from the volatility and evaporation of formaldehyde over time after treatment. Feed acidifiers tend to require more time to achieve the same level of microbial control, but persist essentially indefinitely in feed.

Acidifiers

Generally organic acids are preferred for feed acidification over inorganic acids, despite inorganic acids like HCl being stronger acids as measured by pKa.[4] In part, this is due to organic acids having the ability to act on both sides of the cell membrane. Whereas inorganic acids completely dissociate into their component ions in an aqueous environment, organic acids will only partially dissociate depending on the pH of the solution. As with the completely dissociated ions of the inorganic acids, the proportion of the organic acid in the dissociated state can only act on the extra-cellular environment because the highly polar charges of the two ions prevent them from entering into the cell. The extra cellular effect consists largely of altering proton pump efficiency, and damaging pH sensitive extracellular structures. However, the undissociated portion of organic acids in solution are non-polar and fat soluble, thus enabling them to diffuse across the cell membrane. Once inside the cell, the undissociated acids are exposed to a higher pH than extracellularly, and spontaneously dissociated. This reduces the intracellular pH, and can effect energy metabolism of the cell as it tries to restore the optimal intracellular pH. *In vitro* studies[5] have demonstrated that formic acid, lactic acid, and propionic acid all required less than 1 hr to kill all of the *Campylobacter* at a pH of 4.0, but that HCl required 4 hrs at the same pH. They also reported that at a pH of 4.5 the time to kill was less than 2 hr for all 3 of the organic acids, but it was 8 hours for HCl.

Formic acid

Formic acid is the simplest organic acid consisting of a single carboxyl group with a hydrogen ion attached at the R-side chain ($HCOO^- + H^+ \leftrightarrow HCOOH$). It is also one of the more potent organic acids in terms of both pKa and density (Table 1). The minimum inhibitory concentration (MIC) for formic acid is approximately half that of propionic acid, and one third that of lactic acid against a range of potential pathogens including *Salmonella, E. coli, Clostridia,* and *Campylobacter* (Figure 1).[6] As a

Table 1: Strength (pKa) and density of common acids.

Acid	pKa	g/Mol	Mol/kg
Hydrochloric[a]	-6.3	36.5	27.4
Formic	3.75	46	21.7
Acetic	4.75	60	16.7
Nitric[a]	-1.4	63	15.9
Propionic	4.8	74	13.5
Butyric	4.82	88	11.4
Lactic	3.86	90	11.1
Sulfuric[a]	-3.0	98	10.2
Benzoic	4.19	122	8.2
Ascorbic	4.10	176	5.7
Citric	3.08	192	5.2

[a] Strong inorganic acids are defined as those with a pK < 0.

Figure 1: Minimum inhibitory concentration of formic, propionic, and lactic acid for 8 potential feed borne pathogens (adapted from Strauss and Hayler, 2001).

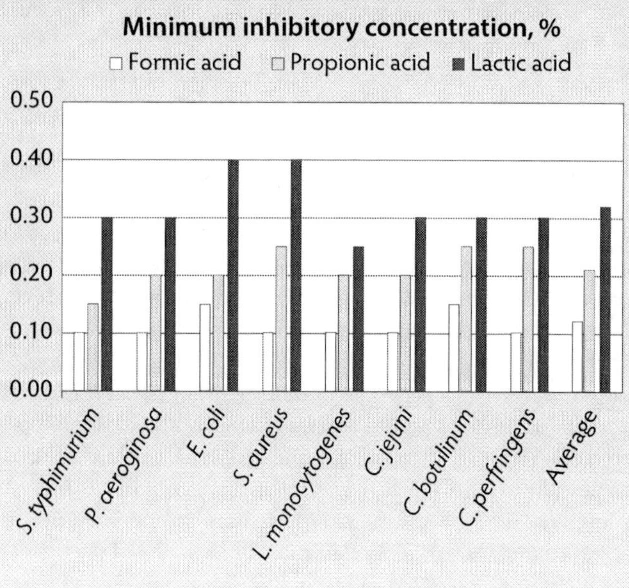

result, the use formic acid for feed acidification, whether by itself or as part of a blended product, has become a standard tool for swine production in the EU where swine are fed more complex diets, *Salmonella* surveillance programs have real teeth, and growth promoting antibiotics are not available to swine producers. In the US there are three forms of formic acid approved for use in swine feed, formic acid (< 85%), ammonium formate, and (recently approved on Sept. 30, 2016) feed grade sodium formate. The efficacy of the ammonium and sodium formate feed additives are driven by their respective formate contents, but are buffered to reduce the corrosivity for practical and worker safety reasons.

Multiple reports have demonstrated that formic acid based acidifiers are capable of reducing the levels of bacterial contaminants in animal feed. Up to 1.5% formic acid in feed has been shown to reduce *S. gallinarium* abundance by 1 log10 on the day of challenge and 2.5 log reductions within 7 days.[7] This was confirmed in a recent (Figures 2 and 3; unpublished) trial at Kansas State University, where pelleted feed containing 1% feed grade sodium formate was spiked with 7 log CFU/g of *Salmonella* (simulated post-pelleting recontamination) showed a 4 log reduction in *Salmonella* recovery after 7 days than the untreated control pellets.

Unlike formaldehyde, which is not reported to directly impact animal performance, feed acidification has been shown to enhance swine performance. Particularly in early weaned piglets. Acidification of feed with formic acid reduces the pH of the feed and thus the acid binding capacity of the feed, this in turn reduces the incidence and persistence of diarrhea, and ultimately feed efficiency of piglets.[8] The connection between these 4 criteria stem from the relative immaturity of the piglet stomach at weaning. They are relatively limited in their capacity to produce HCl, and thus ensure a sufficiently acidic environment to kill bacteria present in their feed. Lower gastric kill rates in the stomach due to high acid binding capacity feeds, results in higher bacterial loads in the small intestine where the bacteria undergo logarithmic growth leading to increased competition for nutrients, and ultimately scours. By supporting the barrier function of the gut, acidification reduces bacterial load along the GIT, which in turn reduces diarrhea prevalence and increases feed efficiency. A meta-analysis of a number published reports on the acidification of swine feed has pegged this feed efficiency response at 4%-points of gain-to-feed per 1% formic acid in the feed (Figure 4, unpublished).

Additionally, protein digestion and nutrient absorption are influenced directly by pH changes in the gut, with a lower pH generally favoring digestion and nutrient absorption.[9] For example, a National Pork Board[9] review of feed acidifiers found that formic acid improved CP digestibility by 1.64%, but that fumaric acid only improved it by 0.57% ($P < 0.05$). Improved nutrient availability has been shown to improve performance during lactation.[10]

Conclusions

Feed hygiene has always been important, but with the changes headed our way, it will only become more important with time. Pelleting feed is a good point-in-time mitigation strategy if it is utilized effectively. However, it is tricky to get it right all of the time. Furthermore, the hygiene protections afforded by pelleting end at the face of the pellet die. Chemical treatment of feed can both support pelleting by decreasing the likelihood that a pathogen will survive the pelleting process in the first place, as well as working to prevent re-contamination of the resulting pellets as they move from the mill to the feeder. Additionally, some treatments like acidification, can also improve performance, nutrient availability and efficiency of feed utilization independent of their effects on biological feed contaminants.

Figure 2: *Salmonella* abundance in mash feed containing graded levels of feed grade sodium formate after being contaminated with 4 log10 cfu/g of *Salmonella*. Simulating pre-pelleting contamination.

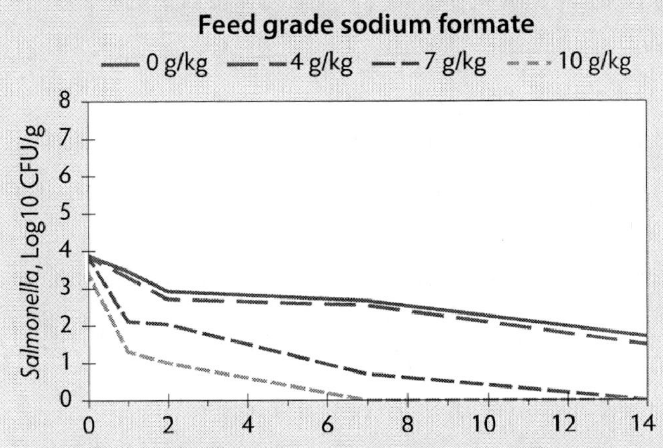

Figure 3: *Salmonella* abundance in pelleted feed containing graded levels of feed grade sodium formate after being contaminated with 7 log10 cfu/g of *Salmonella*. Simulated post-pelleting recontamination.

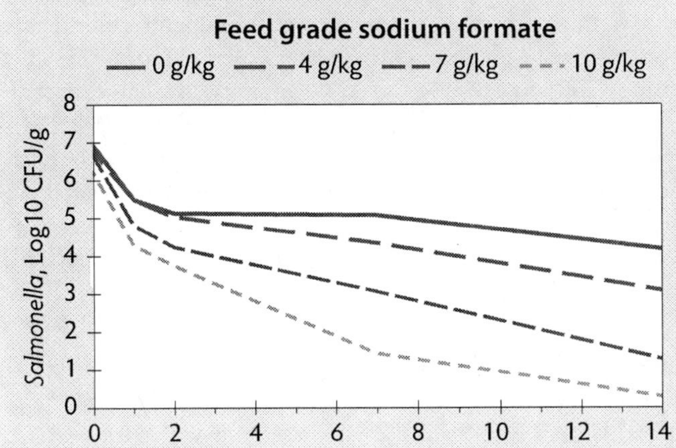

Figure 4: Predicted change in feed efficiency of pigs fed formic acid containing diets. Based on a meta-analysis of previously published literature. (y = -0.12 + 3.99x; r2=0.699).

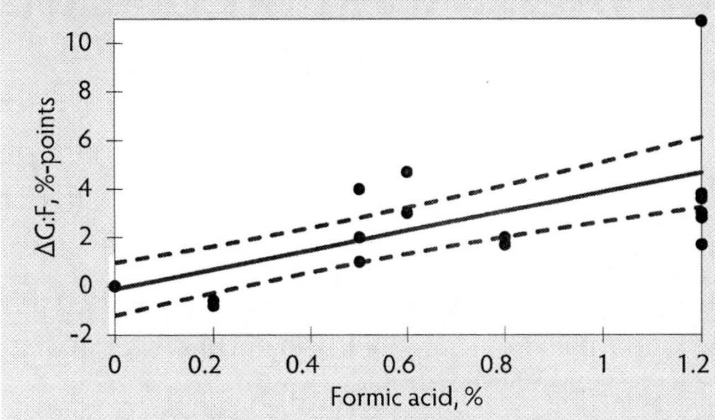

References

1. AFIA (2010). Salmonella Control Guidelines. URL: http://www.afia.org/rc_files/789/SALMONELLA%20CONTROL%20GUIDELINES%202010.pdf Access date: Nov. 1, 2016

2. Davies, R. H. and C. Wray (1997). "Distribution of salmonella contamination in ten animal feedmills." Veterinary microbiology 51: 159-169.

3. US Code of Federal Regulations section 573.460. Feed Grade Sodium Formate. Updated September 30, 2016.

4. Suiryanrayna, M. V. and J. V. Ramana (2015). "A review of the effects of dietary organic acids fed to swine." Journal of Animal Science and Biotechnology 6: 45.

5. Caveerach, P., et al. (2002). "In Vitro Study on the Effect of Organic Acids on Campylobacter jejuni/coli Populations in Mixtures of Water and Feed." Poultry Science 81: 621-628.

6. Strauss, G. and R. Hayler (2001). Effects of organic acids on microorganisms. Feed Magazine. 4: 147-152.

7. Al-Natour, M. Q. and K. M. Alshawabkeh (2005). "Using Varying Levels of Formic Acid to Limit Growth of Salmonella gallinarum in Contaminated Broiler Feed." Asian-Australasian Journal of Animal Science 18(3): 390-395.

8. Eckel, V. B., et al. (1992). „Zum Einfluß von Ameisensaure auf tagliche Zunahmen, futteraufnahme, futterverwertung und Verdaulichkeit." Journal of Animal Physiology and Animal Nutrition 67: 93-100.

9. Tung, C. M. and J. E. Pettigrew (2006) Critical Review of Acidifiers. URL: http://www.pork.org/ResearchDetail/541/CriticalRevieHwofAcid.aspx Access date: Sept. 18, 2014.

10. Jendza, J. A., et al. (2016). "Sodium buffered formic acid improves sow and piglet performance when fed during lactation and the nursery periods." Journal of Animal Science 94(Supl. 2):117.

Effect of combined supplementation of 25-hydroxycholecalciferol and cholecalciferol on sow and piglet Vitamin D status and progeny muscle cell parameters

J. Hahn, PhD
DSM Nutritional Products, Ames, Iowa

Introduction

Although the use of 25-hydroxycholecalciferol (25OHD3) has only been available for investigative purposes in commercial swine production in the United States for a limited time, its use as an alternative to cholecalciferol (VITD3) as a source of Vitamin D activity in swine has been investigated elsewhere.[1-4] This paper is a review of recently published original work by Zhou et al,[4] which evaluated the supplementation of additional Vitamin D activity from 25OHD3 to gilts already supplemented with a level of VITD3 in excess of National Research Council (NRC) recommendations. The research objective was to determine if there was any additional benefit on Vitamin D status and skeletal muscle development to feeding the combination of 25OHD3 and VITD3 versus feeding the higher level of VITD3 alone. The control treatment utilized 2000 IU vitamin D/kg feed from VITD3, a level 2.5 times the current NRC recommendation of 800 IU/kg,[6] and the experimental treatment was the combination of the 2000 IU/kg VITD3 and 50 µg/kg of 25OHD3 (HyD, DSM). The two treatment regimens were fed to gilts during both the gestation and lactation periods, with treatments being initiated post-mating and ending with the weaning event. Serum 25OHD3 levels were measured for gilts and piglets as an indicator of Vitamin D status, and multiple parameters such as muscle cross sectional area, fiber density, and fiber number for the *psoas major* and *longissimus dorsi* muscles were measured in piglets at birth and weaning.

Materials and methods

Twenty Large White gilts with initially similar body weight (140.31 ± 4.08 kg) were used, and each was randomly assigned to one of two dietary vitamin D groups (10 gilts/group) immediately after mating. The control diets were formulated to contain 50 µg/kg VITD3 and the experimental contained 50 µg/kg VITD3 + 50 µg/kg 25OHD3 purchased from DSM Nutritional Products Ltd (Shanghai, China). All matings were performed twice by artificial insemination with semen from a single boar. The day of last insemination and day of farrowing were defined as d1 of gestation and d1 of lactation, respectively. Dietary treatments were fed throughout gestation and lactation, and weaning was performed on d28 of lactation. Diets were designed to meet or exceeded the nutrient requirements as recommended by NRC.[6] Gilts were provided the same amount of diets during gestation and allowed *ad libitum* access to diets during lactation. Water was provided *ad libitum*. No creep feed was provided to piglets and no cross-fostering took place during lactation.

Determination of body weight and feed intake

Gilts were weighed in the morning before feeding at mating, within 24h after farrowing, and at weaning. Feed intake during lactation was determined by measuring feed disappearance between farrowing and weaning. Piglets were individually weighed at birth and weaning.

Collection of samples

Gilts' blood was collected by jugular puncture in the morning before feeding at mating, farrowing, and weaning. Umbilical cord bloods were collected immediately after farrowing from as many umbilical cords as possible and an attempt was made to collect at least 10ml of umbilical cord bloods from each gilt. At birth and weaning, one male piglet whose body weight was closest to the average body weight of the litter was selected and slaughtered after blood collection. *Psoas major* (PM) and *longissimus dorsi* (LM) were excised from the both sides of the carcass. The weight of each whole muscle was recorded and 2cm × 1cm × 1cm clips of PM and LM were excised at middle level and the last rib level, respectively. The cross-section of left muscles from same respective locations were outlined on cellophane paper, the outlines were scanned, and muscles cross sectional area (MCSA) were obtained by using Image J 1.49 (NIH shareware). Blood samples were centrifuged (2,500 × g for 10 min at 4°C) and serum was harvested and frozen at -80°C.

Determination of serum 25OHD concentrations

Serum 25OHD concentrations were determined by the researchers using an EIA kit (IDS Immunodiagnostic Systems Ltd., Tyne and Wear, UK). The minimum detectable level was 2ng/ml.

Histological examination of muscle samples

The fixed PM and LM samples were embedded in paraffin blocks, sectioned to 5µm thickness, and stained with haematoxylin and eosin. Each section was photographed with a Zeiss Axiocam mounted on a Axiovert 200 inverted microscope (Carl Zeiss,

NY, USA) at different microscopic fields after inter fiber network extraction. Magnification was adjusted to make each photograph contain 100-300 fibers. Number of fibers in each photograph was counted. Muscle fiber density was defined as number of muscle fibers per mm^2 of MCSA. The total fiber number of each muscle was estimated by extrapolation from the fiber numbers counted over the selected photograph (field) and the MCSA.

Results and discussion

No differences in feed intake or body weight gain of gilts or piglets ($P > .20$) were observed between treatment groups. Gilt and piglet serum data for the experiment are shown in Table 1. Serum levels of 25OHD3 were not different at mating ($P = .41$). Gilt serum 25OHD3 concentrations at farrowing and weaning were increased ($P < .01$) for the gilts receiving both VITD3 and 25OHD3 versus the serum 25OHD3 levels of control gilts receiving only VITD3. Serum levels of 25OHD3 were also increased ($P < .01$) by the combination of the two vitamin D sources in the umbilical cord and in the piglet serum at birth. Piglet serum 25OHD3 levels were not different ($P = .61$) at weaning.

Piglet muscle data results are shown in Table 2. In newborn piglets, PM muscle cross sectional area (MCSA) was not different between treatment groups ($P = .59$). Fiber density and fiber number were both increased ($P < .09$) in the newborn piglets from gilts receiving the additional 25OHD3 versus piglets from gilts receiving the control diets. In the LM muscle MCSA, fiber density, and fiber number were all increased ($P < .05$) in piglets from gilts receiving the additional 25OHD3 versus piglets from gilts receiving the control diets.

In weaned piglets, PM muscle fiber density was not different between treatment groups ($P = .22$). MCSA and fiber number were both increased ($P < .03$) in the newborn piglets from gilts receiving the additional 25OHD3 versus piglets from gilts receiving the control diets. Longissimus dorsi MCSA and fiber number were both increased ($P < .04$) in piglets from gilts receiving the additional 25OHD3 versus piglets from gilts receiving the control diet. LM muscle fiber density was not different between weaned piglet treatment groups ($P = .21$).

References

1. Flohr, J. R., Bergstrom, J. R., Woodworth, J. C., Tokach, M. D., DeRouchey, J. M., Dritz, S. S., & Goodband, R. D. Evaluating the impact of maternal dietary vitamin D supplementation on sow performance, serum 25OHD3, and subsequent pig performance. Journal of Animal Science. 2016, 94:125-125.

2. Hines, E. A., J. D. Coffey, C. W. Starkey, T. K. Chung, and J. D. Starkey. Improvement of maternal vitamin D status with 25-hydroxycholecalciferol positively impacts porcine fetal skeletal muscle development and myoblast activity. J. Anim. Sci. 2013, 91:4116–4122.

3. Lauridsen, C., U. Halekoh, T. Larsen, and S. K. Jensen. Reproductive performance and bone status markers of gilts and lactating sows supplemented with two different forms of vitamin D. J. Anim. Sci. 2010, 88:202–213.

4. Lauridsen, C., and P. K. Theil. Influence of dose and forms of vitamin D on early reproduction and gene expression in uterus of gilts. Proc. 62nd Annual Meeting, European Federation of Animal Science, Stavanger, Norway. 2011, (Abstr.) p. 354.

5. Zhou H, Chen Y, Lv G, Zhuo Y, Lin Y, Feng B, Fang Z, Che L, Li J, Xu S, Wu D. Improving maternal vitamin D status promotes prenatal and postnatal skeletal muscle development of pig offspring. Nutrition. 2016, 32:10, Pages 1144–1152.

6. NRC. 2012. Nutritional requirements of swine. 11th rev. ed. Natl. Acad. Press, Washington, DC.

Table 1: Effects of maternal vitamin D supplementation on serum 25OHD concentrations of gilts, umbilical cords and piglets*

Item	Control	25OHD3	*P*-value
Gilts (ng/ml)			
Mating	37.55 ± 1.94	36.85 ± 1.74	0.41
Farrowing	34.61 ± 2.12	65.36 ± 2.85	< 0.01
Weaning	36.95 ± 3.29	62.48 ± 3.70	< 0.01
Umbilical cord	13.10 ± 0.92	37.93 ± 0.76	< 0.01
Piglet			
Birth	15.59 ± 1.93	32.62 ± 2.18	< .01
Weaning	13.60 ± 1.39	13.87 ± 0.98	.61

Control = gilts fed diets with 50 µg/Kg vitamin D3
25OHD3 = gilts fed diets with 50 µg/Kg vitamin D3+50 µg/Kg 25-hydroxyvitamin D3
25OHD = 25-hydroxyvitamin D
* Data are presented as mean ±SD

Table 2: Effects of maternal vitamin D supplementation on muscle histological characteristics of newborn and weaning piglets*

	Newborn piglets			Weaned piglets		
Item	**Control**	**25OHD3**	**P-value**	**Control**	**25OHD3**	***P*-value**
Body weight, slaughter (kg)	1.29 ± 0.22	1.27 ± 0.09	0.80	5.83 ± 0.73	6.06 ± 0.66	0.47
PM						
Weight (g)	6.2 ± 1.1	6.8 ± 1.3	0.33	15.3 ± 3.0	19.37 ± 5.0	0.05
MCSA (mm^2)	99.2 ± 19.2	103.8 ± 16.7	0.59	257.0 ± 30.6	296.0 ± 37.6	0.03
Fiber Density ($10^3 \times mm^{-2}$)	4.30 ± 0.79	5.38 ± 0.76	0.01	1.77 ± 0.31	1.93 ± 0.22	0.22
Fiber Number (10^4)	43.2 ± 15.1	56.5 ± 15.3	0.09	45.2 ± 5.0	57.3 ± 11.5	0.02
LM						
Weight (g)	22.5 ± 5.3	26.3 ± 2.93	0.05	62.2 ± 7.2	77.1 ± 13.4	0.02
MCSA (mm^2)	85.5 ± 14.6	97.9 ± 10.5	0.04	218.2 ± 41.3	263.2 ± 31.6	0.03
Fiber Density ($10^3 \times mm^{-2}$)	9.74 ± 0.66	10.51 ± 0.80	0.05	3.92 ± 0.35	4.13 ± 32.9	0.21
Fiber Number (10^4)	83.2 ± 14.5	102.8 ± 12.9	< 0.01	85.5 ± 16.9	109.0 ± 20.3	0.02

Control = gilts fed normal vitamin D diets with 50 µg/Kg vitamin D3
25OHD3 = gilts fed high vitamin D diets with 50 µg/Kg vitamin D3+50 µg/Kg 25-hydroxy vitamin D3
PM, psoas major; MCSA, muscle cross-sectional area; LM, *longissimus dorsi*

- Data are presented as mean ±SD

Effect of Tonisity Px™ administration on pre-weaning mortality and weight gain

Ava M. Firth[1], BS, DVM, MVS, MANZCVS, DACVECC, DECVECC; René Mozo Martín[2], DVM MSc PhD; Guillermo López Cano[2], DVM MSc; Alberto Morillo Alujas[2], BS DVM MS PhD MBA
[1]Tonisity International, Dublin, Ireland; [2]Tests and Trials S.A., Monzón, Huesca, Spain

Introduction

Improving intestinal health in piglets is important to achieving optimal productivity in the farrowing house. Px is an oral isotonic protein drink designed to provide microenteral nutrition to suckling pigs. The aim of this study was to assess the impact of Px on suckling pigs.

Materials and methods

212 sows (Landrace × Large White or Danbred) and their litters (1496 piglets) from 2 farms were used in the study. Farm A was a commercially managed farrow-to-nursery operation of 2500 sows with an average live-born of 13.1 pigs and a historic pre-weaning mortality (PWM) of 13.7%. The farm had a history of clinically significant diarrhea during farrowing with *Clostridium difficile, Clostridium perfringens, Escherichia coli* and type A rotavirus. Farm B was a small family-owned farrow-to-finish farm with 400 Landrace × Large White sows with an average live-born of 13 pigs, a historic PWM of 10-12% and batch farrowing. Farm B had a historically low incidence of diarrhea during farrowing, but documented pathogens included *E. coli* and *Clostridium difficile*, though not rotavirus A.

Sows and their litters were randomised to one of two groups. On the day of farrowing, piglets were individually ear-tagged and weighed. Starting on day 2 of age, litters in the Px group received 500 mL of Px in an open pan, once daily until day 8 of age. Litters in the control group received no extra supplementation. All litters were allowed to suckle normally and had access to an automatic drinker. The amount of Px consumed was recorded daily by weighing the administered volume and any residual liquid in the pan.

Piglets were again weighed at day 8 (D8) and at weaning (D19 Farm A, D17 Farm B) of age. Weaning day was considered the end of study (ES). Creep feed was started in all litters at day 10.

Data collection

The volume of Px consumed was recorded daily. Mortality and apparent cause of death were recorded daily. The presence and severity of diarrhea in each litter was recorded three times per week during the study. Diarrhea was scored from 0 to 3 using the following scale: 0: normal, healthy; 1: a few pigs affected, pasty faeces; 2: most pigs affected, liquid faeces; 3: all pigs affected, liquid faeces, piglets gaunt.

Statistical analysis

The experimental unit for mortality, scour incidence, BW and ADG was the litter. Fisher's exact test was used to assess the relationship between mortality vs treatment and scour vs treatment for each time period (D8, ES).

Scour incidence and severity was analyzed by a generalized linear model using treatment group, sow body condition and parity as fixed effects. When Farm B results were analyzed, the farrowing week (batch) was also included as a fixed effect.

The body weight was analyzed by a linear mixed model including treatment group, gender and sow's parity as fixed effects, and litter as random effect. BW on D1 was included as a covariate (for body weight on D8 and at ES). The number of days from D1 to ES was included as covariate for the analysis of body weight at ES. When Farm B results were analyzed, the farrowing week (batch) was also included as fixed effect.

The average daily gain was analyzed by a linear mixed model; including treatment group, gender and sow's parity as fixed effects and litter as random effect. BW on D1 was included as covariate. The number of days from D1 to ES was included as covariate for the analysis of the ADG until ES. When Farm B results were analyzed the farrowing week (batch) was also included as fixed effect.

For mortality, scour incidence, BW and ADG, the farm and the interaction between the farm and the treatment group were also included when Farm A and Farm B combined results were analyzed. Tests were two-tailed and carried out with a risk $\alpha = 5$ %. P-values of ≤ 0.05 were considered statistically significant, while $0.05 < P \leq 0.10$ was considered a near-significant trend. All statistical analyses were performed with R software.†

Results

Px consumption

On Farm A, consumption of Px increased quickly during the first week of life. By day 3 of age, the median consumption of Px was

† R Core Team (2015). R: A language and environment for statistical computing. R Foundation for Statistical Computing, Vienna, Austria. ISBN 3-900051-07-0, URL `http://www.R-project.org`.

Figure 1: Px Consumption by litters

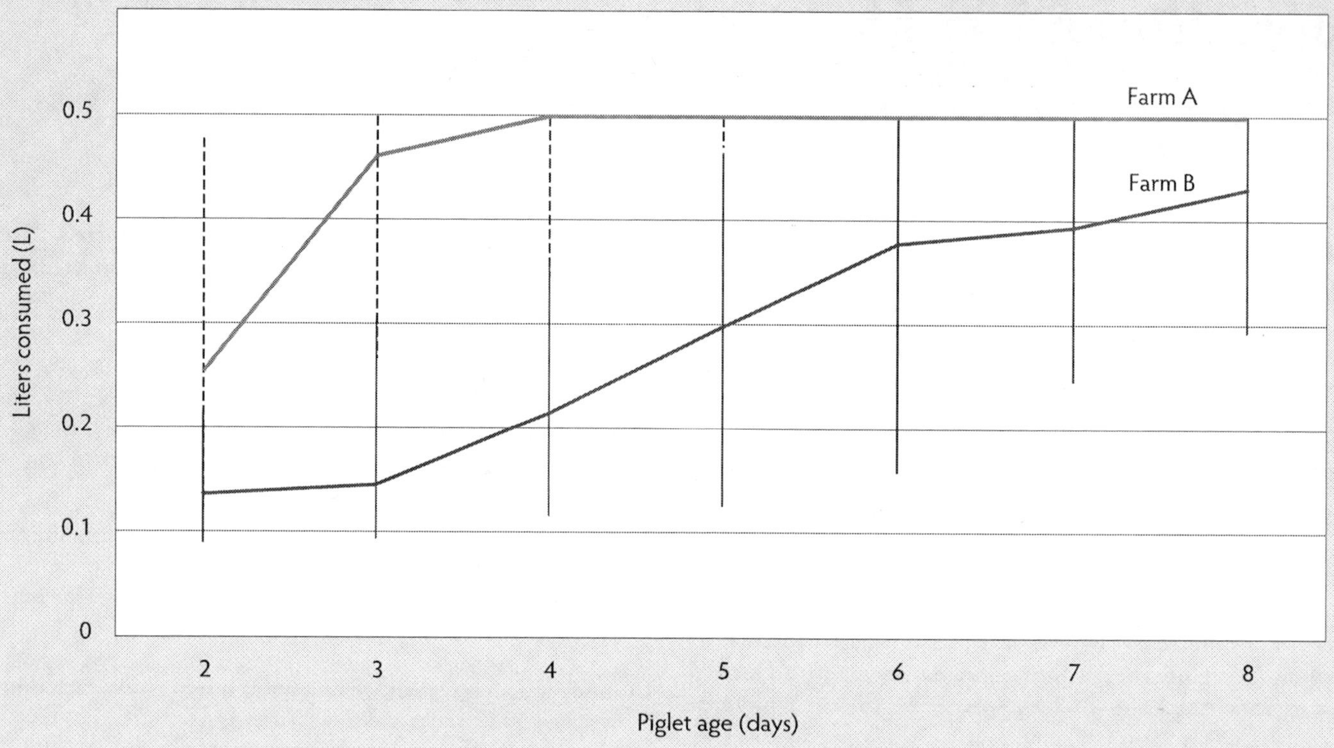

500 mL/litter/day. Litters contained an average of 14 piglets, with a median consumption of 36 mL/pig on day 3 of treatment. See Figure 1.

On Farm B, the consumption of Px also increased steadily during the first week of life, reaching a median consumption of 430 mL/litter on day 8. Litters contained an average of 12 piglets, with a median consumption of 18 mL/pig on day 3 of treatment.

Mortality

Using the litter as the experimental unit, pre-weaning mortality was calculated at ES for both farms combined (Table 1). There was a significant difference in PWM between the two treatment groups, in which Px litters had a lower mortality by 3 percentage points (9.99% v 6.91%, $P = 0.04$). This equates to a 31% reduction in mortality.

Weight gain

The body weight on day 1 was significantly different between farms. Therefore, body weight and ADG were analysed separately for each farm using the litter as the experimental unit.

On Farm A, piglets in the Px group tended to weigh more than the control group at day 8 ($P < 0.1$) and weighed significantly more at day 19 (240 grams, 0.53 lb, $P < 0.05$). Pigs in the Px group had a higher average daily gain both in the first week ($P < 0.1$) and until day 19 ($P < 0.05$). See Table 2. Farm B results were the combination of 2 farrowing batches and showed no significant differences in body weight or ADG.

Incidence and severity of scour in litters

There was no significant difference in the incidence or severity of scour (scour score ≥ 2) between the control and Px litters on either farm (Table 3). However, on farm B, an abnormally high incidence of scour (24-25%) occurred in all litters during the time of the study, which was December – January.

Discussion

Px consumption

This study has shown that piglets less than one week old will drink significant volumes of liquid Px from an open pan. Various factors may have contributed to this, including palatability and texture of the liquid.

Delivering the product in an open pan facilitates the pigs natural curiosity and eliminates the learning and tactile barriers to drinking that are posed by mechanical nipple drinkers. Open pans do allow the possibility of faecal contamination and do require daily cleaning.

Pre-weaning mortality

This study demonstrated that Px had a significant impact on the pre-weaning mortality when given during the first week of life. The impact seen on pre-weaning mortality is interesting and

Table 1: Pre-weaning mortality rates

	Mean	95% CI	*P* value
Control	9.99%	7.6 – 12.3%	**0.0431**
Px	6.91%	4.6 – 9.2%	

Table 2: Body weight and ADG

	Total n	Px*	Control*	*P* value
Body weight d1 (kg ± SE)	1462	1.41 ± 0.04	1.37 ± 0.04	NS
Body weight d8 (kg ± SE)	1376	2.29 ± 0.05	2.23 ± 0.05	*P* < 0.1
Body weight d19 (kg ± SE)	**1330**	**4.25 ± 0.11**	**4.01 ± 0.11**	**P < 0.05**
ADG d1 – d8 (kg/d)		0.123 ± 0.007	0.114 ± 0.007	*P* < 0.1
ADG d1 – d19 (kg/d)		**0.158 ± 0.006**	**0.145 ± 0.006**	**P < 0.05**

* Values are least-squares means ± standard error. NS = not significantly different.

Table 3: Litter incidence of scour (scour score ≥ 2)

Site	Treatment group	Number of litters without scour (%)	Number of litters with scour (%)	*P* value
Farm A	Control	45 (83.3%)	9 (16.7%)	*P* = 0.801
	Px	47 (81.1%)	11 (19.0%)	
Farm B	Control	39 (75.0%	13 (25.0%)	*P* = 1.0
	Px	38 (76.0%)	12 (24.0%)	

could be explained by several different mechanisms. The 3% Px solution does not contain a significant number of calories, so extra caloric intake is not a likely explanation. The quantities consumed per piglet equate to 3-5% of a 1 kg piglets' bodyweight, so it is unlikely that correction of dehydration is the primary mode of action. It is possible however that some piglets may be subclinically dehydrated, in which case Px could be exerting an effect. The effect could also be due to the ingredient profile of Px, which contains key amino acids to support the metabolic requirements of intestinal enterocytes.

Weight gain

Significant differences in the weights on both day 8 and at weaning were seen on Farm A. On Farm B, it is likely that the seasonal increase in scour affected the results of the study. Farm B litters also had a lower consumption of Px which may have reduced the beneficial impact of the product.

Scour

The incidence of scour in Farm B was much higher than usual for that farm but was the same in both the control and treatment groups. The increase was attributed to winter conditions. The optimal use of Px in supporting pigs with scour was not investigated in this study. A second concern with any protein source, especially before weaning, is that it may contribute to the development of scour. The protein levels in Px are very modest but it was appropriate to examine this possibility. However, there was no significant difference in scour scores between treatment groups on either farm, thus reducing this cause for concern.

Conclusion

Px was well accepted by neonatal pigs. Px administration was associated with a significant reduction in pre-weaning mortality, as well as improved weaning weights.

Effect of gruel and Tonisity Px™ on feed intake and weight gain at weaning

Ava M. Firth[1], BS, DVM, MVS, MANZCVS, DACVECC, DECVECC; Guillermo Lopez Cano[2], DVM MSc; Alberto Morillo Alujas[2], BS DVM MS PhD MBA
[1]Tonisity International, Dublin, Ireland; [2]Tests and Trials S.A., Monzón, Huesca, Spain

Introduction

The transition to weaning and dry food intake is a well-documented problem in pig management. Weight gain in the first week post-weaning has a large impact upon subsequent performance and economics of the pig.[1] Decreased feed intake after weaning results in a post-weaning growth check that causes significant production losses and has negative impacts on gut health.[2] The immediate post-weaning period may also see a significant proportion of pigs that have subnormal water intake for up to 48 hours.[3]

Increased post-weaning diarrhea also results in increased antibiotic usage requirements. As producers aim to better manage their antibiotic usage, it is appropriate to investigate whether or not the use of semi-liquid feed supplementation in the period immediately before and after weaning is beneficial.

A novel isotonic protein-electrolyte solution (Px) has recently been developed and has the characteristics of being highly palatable as well as providing key amino acids for support of intestinal function. It was hypothesised that using Px solution to make a gruel with creep feed could result in improved feed intake post-weaning and improved health.

Materials and methods

The study was performed on a 400 sow (Landrace × Large White/Pietrain) farrow-to-finish operation with batch farrowing and regular weaning at 3 weeks of age. The farm was positive for PRRS and APP. The farm had a low incidence of scours in farrowing, but the most recent surveillance sample had documented *E. coli* (strains F41 and *gad*) as well as *Clostridium difficile*. Surveillance samples were negative for rotavirus A and PEDV on PCR.

52 sows and their litters (608 piglets) were enrolled in the study and randomized to one of two groups, control (group A) or supplementation with 3% Px solution (group B). Groups were balanced by sow's parity, number of piglets and piglets' weight. Cross-fostering was permitted only within the treatment group and within the first 24 hrs. All piglets were weighed at day 2 of age. Starting approximately 24 hours after birth (Day 2 of age), group B litters were given 500 mL of 3% Px solution daily in an open pan, in their farrowing crate. Group B litters continued to receive 500 mL of Px daily until day 8 of age. Control litters (group A) were given no extra supplementation. All litters had access to fresh water through a drinker nipple and were allowed to suckle the sow normally.

The pre- and post-weaning feeding plan is summarized in Table 1. Four days before weaning (approximately 15 days of age, study day -4), all pigs were weighed and litters were allocated to one of 3 new treatment groups - dry creep feed (D), water-gruel (WG) or Px-gruel (PG). Group A and B pigs were allocated equally across the 3 new treatment groups. These new groups were again balanced by sow's parity, number of piglets and piglets' weight on SD -4. On SD -4, D litters received dry creep feed, WG litters received 500 mL of water and PG litters received 500 mL of 3% Px solution. All feeds and liquids were served in an open pan.

On SD -3, SD -2, and SD -1, D litters continued to receive dry creep feed. WG litters received a gruel of creep feed mixed with water and PG litters received a gruel of creep feed mixed with Px solution. All gruel was made using the ratio of 1 kg of dry feed to 1.5 L of liquid (water or 3% Px solution), i.e., a 1:1.5 w/v ratio.

On SD 0 (19 days of age), pigs were weaned, weighed and sorted into 53 pens of 10-11 pigs each. Pigs were also sorted into pens by bodyweight (heavy (H), medium (M), light (L)) while still remaining within their feed groups. Heavy was defined as > 5.8 kg, medium 4.1 – 5.8 kg and light < 4.1 kg.

Both WG and PG pigs continued to receive their gruel ad libitum for at least 1 day after weaning and were then tapered off gruel over another 3-4 days according to body weight. Light and medium pigs were tapered more slowly.

The quantity of dry creep feed and gruel to be delivered was calculated to achieve an ad libitum consumption. This calculation was based on the average bodyweight (BW) of the piglets in the litter/pen calculated from individual BW measured on SD-4 and SD 0, the expected dry matter intake (DMI) based on the average BW, the number of piglets in the litter/pen, and the % dry matter (DM) of the creep feed used. The DM of the creep feed was based on proximate analysis (moisture, crude protein, crude fat and ash) as measured by near infrared spectroscopy. A margin of +20% was added in order to ensure that there was enough gruel available. During the days that gruel/feed was administered ad libitum, if the leftover gruel/feed of a pen was 0, the quantity of gruel/feed delivered was increased by another 20% on the next day in that pen. The total quantity of gruel or dry creep feed in open pans was divided into two doses per day.

Additionally, creep feed was offered ad lib in separate feeders to all groups from SD1 to at least SD7. Pre-starter feed was offered to M and H pens from SD7 and to L pens from SD10.

Table 1: Feeding plan

Age (days)	Study day	Individual weights	In LITTERS		
2		x	**Group A** Normal suckling control group		**Group B** Suckling + 500 mL Px solution/ litter/ day
8		x	↓	↓	
			In LITTERS		
15	SD -4	x	Dry creep feed	500 mL water	500 mL Px solution
16	-3		Dry creep feed ad lib	Water-Gruel ad lib	Px-Gruel ad lib
17	-2		Dry creep feed ad lib	Water-Gruel ad lib	Px-Gruel ad lib
18	-1		Dry creep feed ad lib	Water-Gruel ad lib	Px-Gruel ad lib
			Into PEN BLOCKS*		
19	Weaning SD 0	x	**Light**	**Medium**	**Heavy**
			Gruel ad lib	Gruel ad lib	Gruel ad lib
20	1		Gruel ad lib	Gruel ad lib	Gruel ad lib
21	2		Gruel ad lib	70% of the gruel intake on SD1	60% of the gruel intake on SD1
22	3		70% of the gruel intake on SD2	50% of the gruel intake on SD1	40% of the gruel intake on SD1
23	4		50% of the gruel intake on SD2	30% of the gruel intake on SD1	20% of the gruel intake on SD1
24	5		30% of the gruel intake on SD2	10% of the gruel intake on SD1	Dry creep feed only
25	6		10% of the gruel intake on SD2	Dry creep feed only	↓
26	7	x	Dry creep feed only	Prestarter diet	Prestarter diet
29	10		Prestarter diet	↓	↓
33	14	x	↓	↓	↓
40	21		Starter diet	Starter diet	Starter diet
55	36	x	↓	↓	↓

* All pens had dry creep feed available ad lib in feeders, alongside the gruel feeders.

Data collection

Pigs were individually weighed on day 2 and day 8 of age, then SD -4, SD 0, SD 7, SD 14 and SD 36. Feed intake was measured for each group and normalised to dry matter intake (DMI), which equalised differences between the volume and weight of the gruel and dry feed. Feed intake was calculated for the pre-weaning period, 7 days post-weaning and 14 days post-weaning. Average daily gain was calculated for the same periods. The number of pigs with positive ADG in the first week and second post-weaning was calculated.

The number of pigs requiring antibiotic treatment during the study period was also recorded. Antibiotic treatment for diarrhea was at the discretion of farm staff and consisted of Marbocyl 2% by injection.

The number of piglets per pen with diarrhea and their individual scour score was recorded daily from SD1 to SD14. The scale used was 0 = no scour, 1 = pasty faeces and 2 = liquid scour. The total scores per pen were then summed and divided by the number of pigs in the pen to create a pen faecal score.

Statistical analysis

The experimental unit was the litter until weaning, and the pen after weaning. Tests were two-tailed and carried out with a risk $\alpha = 5$ %. P-values of ≤ 0.05 were considered statistically significant, while $0.05 < P \leq 0.10$ was considered a near-significant trend. Least squares means (LSM) ± standard error (SE) were used to calculate 95% confidence intervals. Logistic regression was used to calculate the differences between treatment groups for positive and negative ADG. All statistical analyses were performed with R software†, using packages GLM and lme4 as appropriate.

Results

Day 2-8 consumption of Px

The mean consumption of Px solution was 38 mL/piglet on day 2 of age, which increased to a mean of 45 mL/pig at day 8, i.e., all 500 mL was consumed by the litter. The mean consumption per kg BW was 188 mL for the week.

Dry matter intake (DMI)

The PG litters had an average DMI of 75 ± 5 g/pig/day in the pre-weaning period (SD -3 to SD -1, which was significantly higher than the WG litters (58 ± 4.9 g/pig/day) or the D litters (54 ± 5.7 g/pig/day) ($P < 0.05$).

After weaning the DMI was tracked both by type of feed and weight class (Heavy, Medium, Light). Table 2 shows total DMI of gruel and creep feed consumed in the first week after weaning. The most marked differences were seen in the Light and Medium pigs in the first week after weaning. Both gruel groups in these weight classes had significantly higher DMI compared to the dry creep feed group. Light-weight Px-gruel pigs had an average DMI of 164 g/pig/day compared to 150 g/pig/day for water-gruel and 126 g/pig/day for dry creep in the first week post-weaning.

Average daily gain

When all weight groups of pigs were compared (Table 3), the PG group tended to have a higher percentage of pigs with positive ADG in the first week post-weaning compared to the D group (OR 1.66, 95CI 0.95 – 2.92, $P = 0.076$. The PG group also had a significantly higher percentage of pigs with positive ADG in the first week post-weaning compared to the WG group (OR 1.79, 95CI 1.05 – 3.04, $P = 0.031$). There was no significant difference between the WG and D group in this parameter (OR 0.93, 95CI 0.55 – 1.58, $P = 0.788$).

When analyzed by weight class (Table 4), the Medium and Light pigs in the PG group also out-gained their counterparts in the other feed groups. 88% of the Medium pigs and 92% of the Light pigs in the PG group had positive ADG in the first week post-weaning ($P < 0.05$). This difference was not seen in the Heavy pigs.

Odds ratios were also calculated for each group to assess the likelihood of an effect. Light pigs in the PG group were more likely to have positive ADG than the WG group (odds ratio 3.25, 95% CI 1.00 – 10.56, $P = 0.050$). Medium pigs in the PG group were more likely to have positive ADG than the D group (odds ratio 2.44, 95% CI = 1.08 – 5.52, $P = 0.032$).

Antibiotics and diarrhea

Though no significant differences were seen in scour scores between treatment groups or weight classes in either the first or second week post-weaning, PG pigs tended to be less likely to be treated with antibiotics in the post-weaning period compared to D pigs (odds ratio 0.59, 95% CI 0.33 – 1.04, $P = 0.068$). Similarly, WG pigs were significantly less likely to be treated with antibiotics in the post-weaning period compared to D pigs (odds ratio 0.51, 95% CI 0.28-0.93, $P = 0.028$). There was no significant difference in the incidence of antibiotic treatment between WG and PG pigs (odds ratio 1.16, 95% CI 2.13, $P = 0.645$).

Discussion

This study again showed that newborn piglets less than one week old will consume substantial volumes of 3% Px solution. It was also clear that piglets found Px-gruel more palatable than water-gruel or dry creep feed, especially in the immediate pre-weaning period. Feeding Px-gruel and water-gruel in the peri-weaning period had a significant impact on the DMI and ADG of medium-weight and light-weight pigs, which comprised 75% of the population. Light-weight Px-gruel pigs consumed 30 g/kg BW/day more food than the same water-gruel pigs and 74 g/kg BW/day more food than the same dry creep pigs in the first week post-weaning. This was reflected in the overall trend towards increased ADG in the Px-gruel pigs. Further studies are planned to investigate how best to optimise the effect of Px on heavy-weight pigs.

The reduction in antibiotic use suggests that gruel may be a useful strategy for decreasing antibiotics in the post-weaning period and should be investigated further. Feeding semi-liquid food may also ameliorate the possible dehydration associated with weaning and moving into pens.

Future studies are required to investigate whether or not gruel feeding can increase the percentage of pigs that are 'eaters' of creep feed, encouraging more suckling pigs to eat that will be better adapted to weaning.[4]

References

1. Pluske, J.R., D.J. Hampson, I.H. Williams. 1997. Factors influencing the structure and function of the small intestine in the weaned pig: a review. Livest. Prod. Sci. 51:215-236.

2. Lallès J-P, G. Boudry, C. Favier, N. Le Floch, I. Luron, et al. 2004. Gut function and dysfunction in young pigs: physiology. Anim. Res. 53:301-316. doi:10.1051/animres:2004018.

3. Brooks P, S. Russell, J. Carpenter. 1984. Water intake of weaned piglets from three to seven weeks old. Vet. Rec. 115:513-515.

4. Sulabo R. 2009. Influence of creep feeding on individual consumption characteristics and growth performance of neonatal and weanling pigs. PhD thesis, Kansas State University.

† R Core Team (2015). R: A language and environment for statistical computing. R Foundation for Statistical Computing, Vienna, Austria. ISBN 3-900051-07-0, URL `http://www.R-project.org`.

Table 2: Average daily DMI of gruel and/or creep feed per piglet (g/pig/day), by feed group and size

		Dry creep (D)		Water-gruel (WG)		Px-gruel (PG)	
Size	Days post-weaning	LSM (SE)	CI 95%	LSM (SE)	CI 95%	LSM (SE)	CI 95%
H	0 to 6	142 (9.3)	124 - 161	149 (9.6)	130 - 168	148 (8.7)	131 - 165
M	0 to 6	**120^{b} (5.3)**	**109 - 130**	**150^{a} (4.6)**	**141 - 159**	**141^{a} (4.4)**	**133 - 150**
L	0 to 6	**126^{b} (9.8)**	**107 - 145**	150^{ab} (9.3)	132 - 168	**164^{a} (8.2)**	**148 - 180**

H: heavy (pens with average BW > 5.8); M: medium (pens with average BW of 4.1 to 5.8 kg); L: light (pens with average BW < 4.1 kg). T1: dry feed; T2: gruel with water; T3: gruel with 3% PX solution. a, b different superscript in the same row indicates statistical differences ($P \leq 0.05$). t in the same row indicates statistical tendency ($P \leq 0.10$).

Table 3: Number and percentage of piglets with positive ADG in the first week post-weaning, by treatment group

	Dry creep (D)	Water-gruel (WG)	Px-gruel (PG)
ADG ≤ 0	31 (22%)	41 (23%)	28 (14%)
ADG > 0	113$^{ab\,t}$ (78%)	139^{b} (77%)	**170$^{a\,t}$ (86%)**

^{a,b} different superscript in the same row indicates statistical differences (P≤0.05). t in the same row indicates statistical tendency ($P \leq 0.1$).

Table 4: Number and percentage of piglets with positive ADG in the first week post-weaning, by treatment group and size (H, M and L).

Size		D	WG	PG	Comparison	Odds ratio	CI 95%	*P*-value
H	ADG ≤ 0	9 (24%)	15 (31%)	11 (30%)	D vs. WG	0.70	0.27 – 1.84	0.474
	ADG > 0	29 (76%)	34 (69%)	26 (70%)	D vs. PG	0.73	0.26 – 2.05	0.555
					WG vs. PG	1.04	0.41 – 2.64	0.930
M	ADG ≤ 0	17 (25%)	17 (19%)	12 (12%)	D vs. WG	1.45	0.68 – 3.11	0.338
	ADG > 0	**51 (75%)^{b}**	74 (81%)^{ab}	**88 (88%)^{a}**	**D vs. PG**	**2.44**	**1.08 – 5.52**	**0.032**
					WG vs. PG	1.68	0.76 – 3.75	0.202
L	ADG ≤ 0	5 (13%)	9 (23%)	5 (8%)	D vs. WG	0.52	0.16 – 1.73	0.287
	ADG > 0	33 (87%)^{ab}	**31 (78%)^{b}**	**56 (92%)^{a}**	D vs. PG	1.70	0.46 – 6.30	0.430
					WG vs. PG	**3.25**	**1.00 – 10.56**	**0.050**

^{a,b} different superscript in the same row indicates statistical differences ($P \leq 0.05$).

Relationship between LP5™ and litter uniformity and birth weight

T. Rathje, PhD; C. Abell, PhD
DNA Genetics, LLC, Columbus, Nebraska

Introduction

Selection programs have historically, and in many current instances, focused on improving total number of pigs born in a litter. Phenotypic and genetic relationships between large litter size and pig birth weight and survival are undesirable meaning that as litter size increases in a population, the mean pig birth weight and survival rate decreases resulting in a diminishing return to improving litter size. Altering a selection program to improve litter size and pig quality at the same time can offset these negative effects. Changing the selection emphasis to "Live pigs at day 5"™ (LP5™) and measuring the component trait of pig birth weight have proven to be effective methods to capture the full value of improved litter size.

Background

DNA Genetics initiated selection on LP5™ in 2004 following a 3 pig improvement that resulted from more than 10 years of selection for total born (Su et al., 2007). LP5™ is measured as number born alive less pig death loss that occurs from birth to the 5th day of life. By definition, this trait is complex in that it encompasses both litter size and survival with selection acting upon both characteristics (Nielsen et al., 2013). A myriad of component traits can be conceived that contribute to LP5™, of which the most promising is pig birth weight. Pig birth weight recordings began in September 2013. The results presented derive from nucleus data on these traits from both the DNA Line 200 Yorkshire and DNA Line 400 Landrace populations. The genetic model for pig birth weight was a maternal effect model with appropriate fixed effects and covariates. Models describing the relationship between birth weight and litter size were regression models controlled for the fixed effect of parity. The variation of birth weight within litters was demonstrated as the standard deviation of birth weight for pigs born within various litter size categories.

Results and discussion

Pig birth weight is largely a maternal trait (trait of the dam). The estimated heritability for the dam additive effect was 0.22 in Yorkshire and 0.16 in Landrace. The heritability of individual pig birth weight was 0.06 and 0.08 in the two breeds, respectively. The low heritability of individual birth weight and moderate heritability for the genetic effect of the sow on birth weight indicate that selection for the genetic effect of the dam will improve the mean birth weight of pigs within litters more efficiently than selecting for individual birth weight.

The relationship between total born, LP5™ and birth weight are shown in Table 1. For each additional total born, the mean birth weight of pigs in the litter declined by 28 grams in Yorkshire and 27 grams in Landrace. This relationship changed dramatically when measured against LP5™ with mean birth weight changing very little as LP5™ increased (-7 grams and -10 grams, respectively). Figures 1 and 2 show graphically the relationship between LP5™ and birth weight in the Yorkshire and Landrace breeds. Abell et al. (2015) demonstrated the strong positive relationship between pig size at birth and survival to weaning.

Table 2 displays the standard deviation of birth weights within litters across a range to total born within the two breeds. As litter size increases, it is a common statement in the industry that variability of pig size increases. In these two populations selected for the past 12 years for LP5™ variation of pig size within litters is largely invariant with litter size. The proportion of small versus large pigs within litters is relatively constant over the range of total born between 10 and 20 pigs.

LP5™ has been a proven strategy that acts upon both litter size and pig survival, resulting in improvement in both traits and without an increase in the variability of pig size within litters. LP5™ acts directly upon birth weight as a component trait as demonstrated by the results showing sows with high values of LP5™ must produce litters of higher average birth weight due to birth weight's impact on pig survival. Sows with large litters of small pigs are eliminated by selecting for LP5™ whereas they would be retained if selection was for total born. A selection scheme that combines LP5™ with birth weight as a trait of the sow will result in more direct selection pressure for the component trait of birth weight and produce faster improvement in pig quality while maintaining improvements in litter size.

Table 1: Relationship between total born and Live Pigs at Day 5™

	Breed	
Trait	**Yorkshire**	**Landrace**
Total number born	-.028 (± .001)	-.027 (± .001)
Live pigs at day 5™	-.007 (±.001)	-.01 (± .001)

™ Registered trademark of DNA Genetics, LLC, Columbus, NE, USA

Figure 1: Relationship between birth weight and Live Pigs at Day 5™ for Yorkshire.

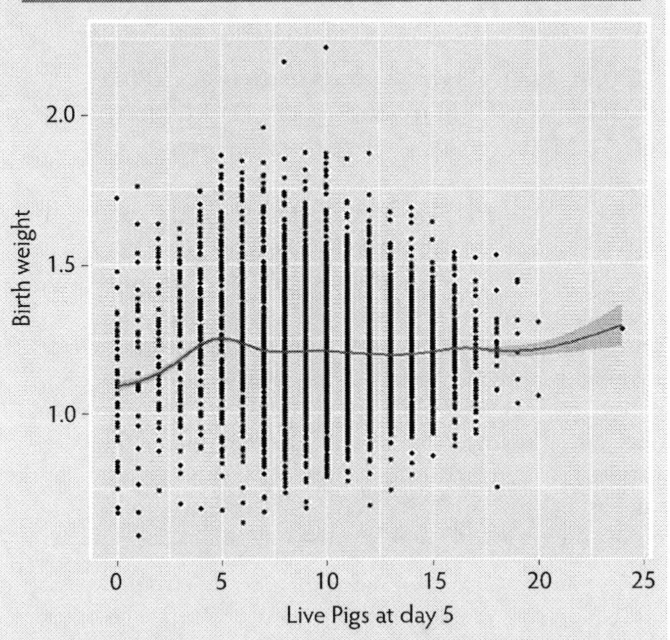

Figure 2: Relationship between birth weight and Live Pigs at Day 5™ for Landrace.

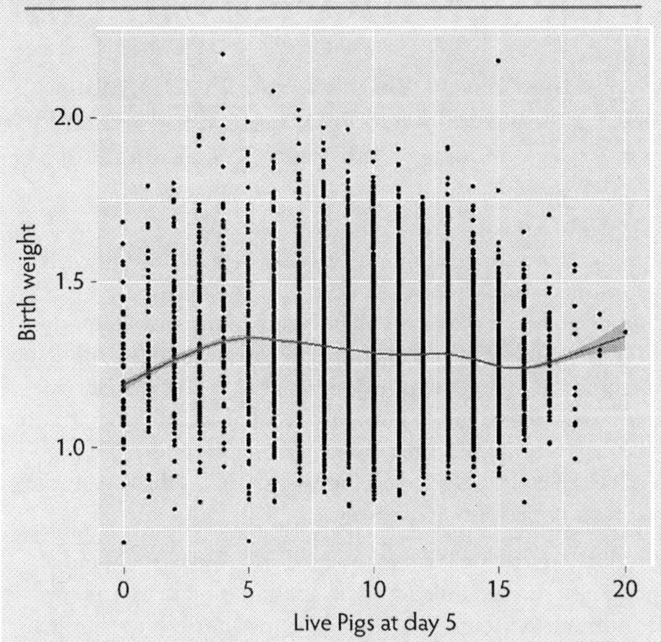

Table 2: Variation of pig birth weight within litter size

Number born alive	Standard deviation of birth weight, yorkshire	Standard deviation of birth weight, landrace
10	0.32	0.31
11	0.31	0.31
12	0.30	0.30
13	0.29	0.30
14	0.29	0.30
15	0.30	0.30
16	0.30	0.30
17	0.30	0.29
18	0.31	0.35
19	0.31	0.30
20	0.31	0.28

References

1. Su et al. Selection for litter size at day five to improve litter size at weaning and pig survival rate. *J. of Anim. Sci.* 2007;85:6.

2. Nielsen et al. Selection for increased number of piglets at day five after farrowing has increased litter size and reduced pig mortality. *J. of Anim. Sci.* 2013;91:2575.

3. Abell et al. Relationship between pig birth weight and survival in a swine nucleus herd. AASV 2016.

Reducing antibiotic use: Educational materials for swine veterinarians and producers

Mike Tokach[1], PhD; Jose Soto[1], MS; H. Morgan Scott[2], DVM, PhD; Raghavendra Amachawadi[1], PhD; Bob Goodband[1], PhD; Jason Woodworth[1], PhD; Joel DeRouchey[1], PhD; Steve Dritz[1], DVM, PhD
[1]Kansas State University, Manhattan, Kansas; [2]Texas A&M University, College Station, Texas

Introduction

Antibiotics are important for maintaining the health and wellbeing of animals and people. The development of resistance to antibiotics is a major concern, and has brought increased attention to antibiotic use. The guiding principle for antibiotic use in swine production is still to 'use as little antibiotics as possible, but as often as necessary'. Future introduction of new antibiotics for use in animals is projected to be limited. Therefore, efforts are needed to ensure that antibiotics available today remain effective for the treatment or prevention of diseases.

The USDA National Institute of Food and Agriculture funded a project to further understand the use of antibiotic alternatives on antibiotic resistance in pigs and cattle. One of the prime goal of this project is to develop educational materials that can be used by producers and veterinarians to reduce antibiotic use and to promote education about antibiotic resistance in livestock production. This presentation will highlight some of those educational materials and promote their use by veterinarians and their clients.

Factsheets

A series of fact sheets have been developed to provide easy to read resources on antibiotic resistance. The factsheets include:

1. Why livestock producers should care about antibiotic resistance
2. How bacteria become resistant to antibiotics
3. Swine management practices that can reduce the need for antibiotics
4. Beef management practices that can reduce the need for antibiotics
5. Alternatives to antibiotics: overview of options

Additional fact sheets on antibiotic alternatives continue to be added to the list.

Website: `www.KSUantibiotics.org`

The website: `www.KSUantibiotics.org` was created as a launching point to find information about antibiotics. There is a major section on antibiotic resistance where the KSU fact sheets can be found. There are also web links to sites that provide an overview of antibiotic resistance, mechanisms, the knowledge about resistance in livestock production, the USDA and FDA action plans concerning resistance, and news feeds where you can find the latest information on the topic. Each of the subpages contains the major agency (ex. WHO, FAO, CDC), producer group (National Pork Board), and Journal publications on the topics.

Besides information on resistance, there are links to the major rules concerning Veterinary Feed Directives (VFD) and other antibiotic regulations. Another section leads the user to links to reviews on the main alternatives to antibiotics that have been tested. The last section provides links to information on management practices that can reduce the need for antibiotics.

We encourage you to visit this site, use the factsheets to educate producers, and refer producers to this site for their use in learning about antibiotic resistance and alternatives to antibiotics.

This work is/was supported by the USDA National Institute of Food and Agriculture, AFRI Food Safety Challenge Grant project 2013-68003-21257. The contents are solely the responsibility of the authors and do not necessarily represent the official views of the USDA or NIFA.

Use of an autogenous PRRSV vaccine in conjunction with MLV elicits broadly neutralizing antibodies

Paulraj Lawrence, PhD; Eric Bumgardner, MS
Biological Development and Research Department, Newport Laboratories Inc., Worthington, Minnesota

Introduction

Porcine reproductive and respiratory syndrome virus (PRRSV) is among the most important viruses impacting the swine industry on a worldwide basis.[1] Since its widespread emergence in the late 1980s the virus has evolved rapidly resulting in tremendous genetic diversity.[2] Most recently, a new PRRSV strain, identified by RFLP pattern "1-7-4" has emerged as a significant threat to the health of both the growing-pig and breeding herds in the United States.[3] Constant antigenic drift within circulating strains affects vaccine efficacy and hinders the development of new vaccines in a timely manner. Commercial modified live virus (MLV) PRRSV vaccines can offer homologous protection. However, they often exhibit reduced efficacy when used to combat disease caused by heterologous strains circulating in the field.[4] Inactivated autogenous vaccines can be manufactured to match the outbreak or circulating field strain, however, their efficacy has been debated.[4] The objective of this study was to broaden the antibody repertoire by using modified live vaccines in combination with a PCV2 vaccine and a bivalent autogenous PRRSV vaccine in a prime-boost strategy. The results demonstrate that the inactivated vaccine was able to elicit neutralizing antibodies with broader specificity than those made in response to the MLV vaccine alone.

Materials and methods

Four-week old weaned pigs were divided into five treatment- and a base-line control group as follows- A) Commercial MLV vaccine; B) Commercial MLV vaccine rehydrated in Circovac® (Merial Inc.); Autogenous PRRSV vaccine delivered at a separate site; C) Experimental MLV Vaccine; D) Experimental MLV mixed with Circovac®; Autogenous PRRSV vaccine delivered at a separate site; E) Mock-vaccinated controls; F) Non-treated/non-challenged controls. Three weeks after their initial vaccination, groups B and D received a booster vaccination consisting of Circovac® and the autogenous PRRSV vaccine delivered at separate injection sites. Six weeks after the initial vaccination animals were challenged with a virulent 1-7-4 RFLP PRRSV strain homologous to one of the strains included in the autogenous vaccine. Over the course of challenge, nasal swabs were collected to monitor shedding of the challenge virus using semi-quantative real time polymerase chain reaction (qRT-PCR, Figure 1). Fifteen days after the challenge, serum was drawn from all pigs before euthanasia. Lung tissue was excised and sent for examination to Iowa State University for PRRSV specific immunohistochemistry (IHC) (Figure 2). Serum was sent to South Dakota State University for fluorescent focus neutralization (FFN) assay (Figure 3). FFN results were analyzed using a Student's t-test to determine whether significant differences were present between the treatment groups. The qRT-PCR results were evaluated using a *Ct* cut-off of 37 to determine whether a sample was positive or negative for PRRSV.

Results and discussion

Analysis of the FFN data revealed a significant increase in neutralizing antibody titer against the challenge virus in both groups that received Circovac® and autogenous PRRSV vaccine. The results of the IHC analysis demonstrated that fewer animals were IHC positive in groups that received the autogenous vaccine in comparison to the corresponding groups that received only MLV. At eight days post challenge, nasal shedding was reduced in pigs that received the autogenous vaccine alongside the commercial MLV vaccine relative to those that received only the commercial MLV vaccine. The commercial MLV vaccine was better at reducing nasal shedding than the experimental MLV. However, some of the pigs that received the experimental MLV vaccine were still shedding virus at the time of challenge, possibly due to replication and shedding of the MLV virus. Irrespective of this, when the experimental MLV was used in combination with Cirocvac® and the autogenous PRRSV product, its performance was equivalent to the commercial vaccine used with the same vaccination strategy at the end of the challenge period. In conclusion, the autogenous PRRSV vaccine, when used in a prime-boost combination strategy, improved the protective capacity of both the MLV vaccines. The broad cross-reactivity and high titer of the neutralizing antibodies elicited by treatment groups B and D suggests that farms experiencing sub-clinical disease caused by circulating quasi- or dominant PRRSV species could be effectively managed with this vaccination strategy.[5]

® Circovac is a registered trademark of Merial, Inc.

Figure 1: Results of qRT-PCR testing in nasal swabs. Positive vs negative status for each pig was determined using a *Ct* cut-off threshold of 37.

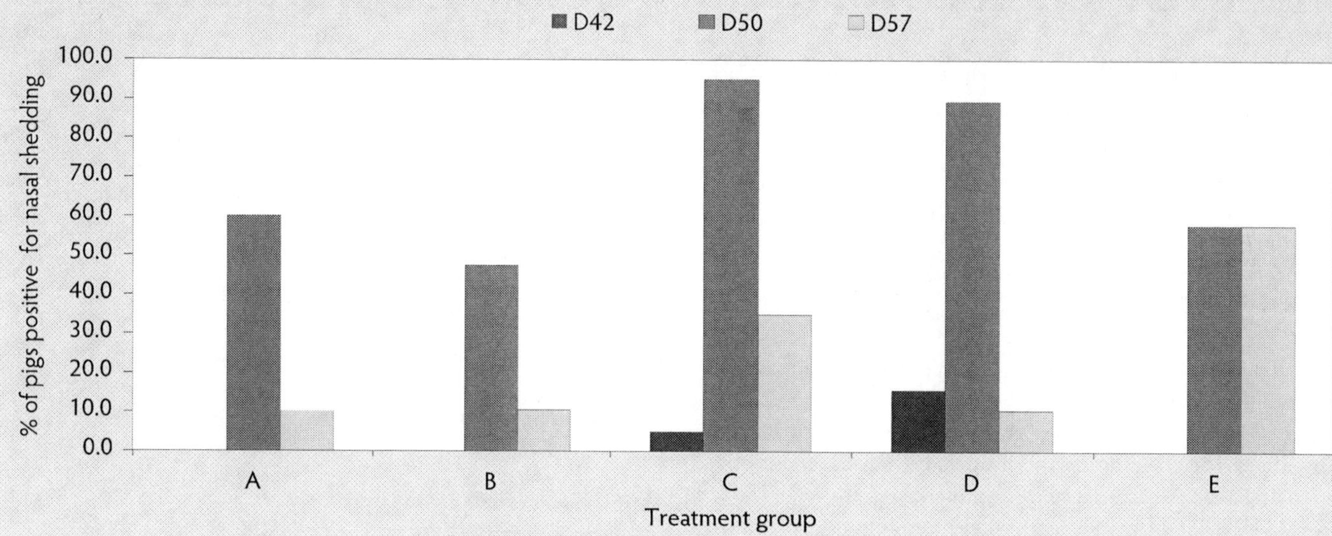

Figure 2: Results of PRRSV specific immunohistochemistry (IHC) testing.

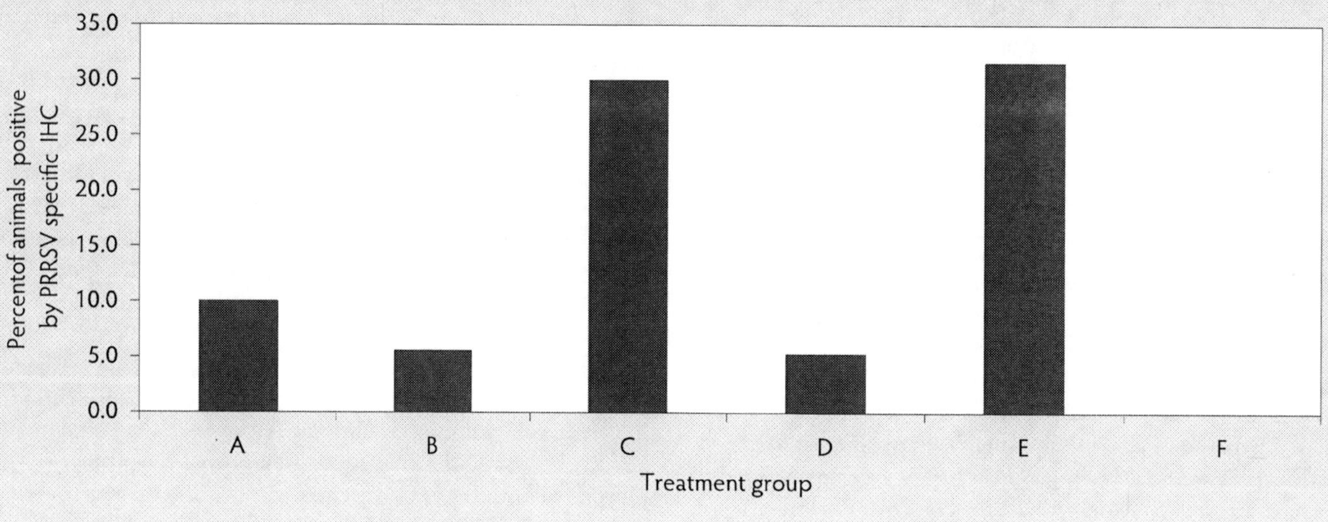

Figure 3: Results of FFN testing versus antigenically diverse strains of PRRSV using serum drawn two weeks post-challenge. (*) denote the results of a Student's t-test comparing groups that received MLV only versus corresponding groups that received MLV with inactivated virus and Circovac® ($\alpha = 0.05$). The same results were obtained comparing groups to the non-vaccinated controls. Error bars indicate standard error. FFN titers were not determined for non-vaccinated/non-challenged controls (Group F).

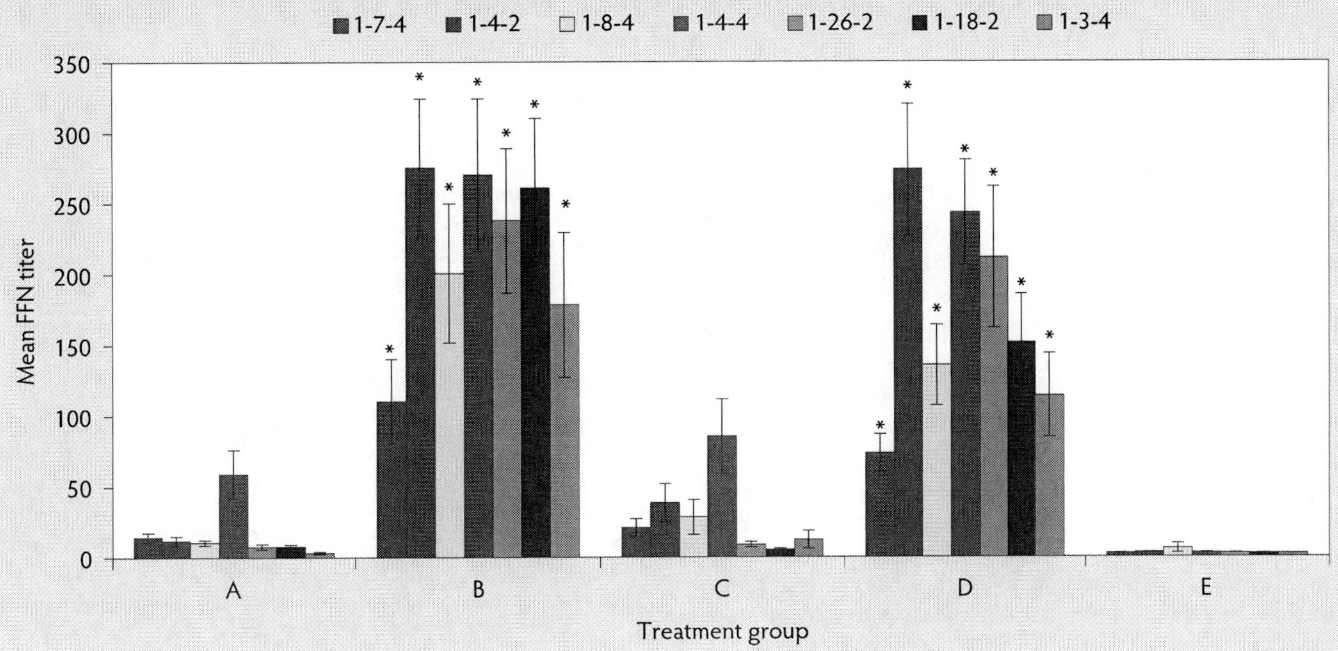

References

1. Holtkamp DJ, Kliebenstein JB, Neumann EJ, Zimmerman JJ, Rotto HF, Yoder TK, Wang C, Yeske PE, Mowrer CL, Haley CA: Assessment of the economic impact of porcine reproductive and respiratory syndrome virus on United States pork producers. *J Amer Vet Med Assoc*, 2013, 21:72-84.

2. Murtaugh MP, Stadejek T, Abrahante JE, Lam TTY, Leung FCC: The ever-expanding diversity of porcine reproductive and respiratory syndrome virus. *Virus Res.*, 2010, 154:18-30.

3. Alkhamis MA, Perez AM, Murtaugh MP, Wang X, Morrison RB: Applications of bayesian phylodynamic methods in a recent U.S. porcine reproductive and respiratory syndrome virus outbreak. *Front. Microbiol.*, 2016, 7:67.

4. Murtaugh MP, Genzow M: Immunological solutions for treatment and prevention of porcine reproductive and respiratory syndrome (PRRS). *Vaccine*, 2011, 29:8192-8204.

5. Lopez OJ, Oliveira MF, Alvarez Garcia E, Kwon BJ, Doster A, Osorio FA: Protection against porcine reproductive and respiratory syndrome virus (PRRSV) infection through passive transfer of PRRSV-neutralizing antibodies is dose dependent. *Clin Vacc Immunol.*, 2007, 14:269-275.

Antibiotic-free production in the face of an acute health challenge

Jon A. De Jong[1], MS, PhD; Casey R. Neill[1], MS; Barry R. Kerkaert[1], DVM; Joe P. Ward[1]; Dan B. Hanson[1]; Roger A. Arentson[2], MS, PhD; Keith D. Haydon[2], MS, PhD
[1]Pipestone Grow Finish, Pipestone, Minnesota; [2]Elanco, Greenfield, Indiana

Introduction

Historically, antibiotics have been used throughout the swine and livestock industries as either a growth promotant[1] at sub-therapeutic levels or as a means to prevent and treat disease when given at therapeutic levels.[2,3] Today there is pressure by both the consumer and the regulatory agencies, as seen by directives 209 and 213,[4] for our industry to reduce the risk for antibiotic resistance through the judicious use of antibiotics. Various programs have been implemented across production systems ranging from "never ever" antibiotic usage to "judicious use" plans.

Materials and methods

In the current trial at the Pipestone Grow Finish research barn, two treatment programs were implemented: 1) "Never Ever" antibiotic free (ABF), and 2) "Judicious Use" antibiotic use (AB). Treatments for the trial started at the sow farm where pigs from the ABF treatment were not allowed to be treated with antibiotics beginning at parturition. Pigs from the AB treatment were treated with antibiotic injectables in the farrowing house. Pigs were moved from the sow farm and placed into a 2400 hd wean-to-finish barn at ~21 d of age. Pigs were placed in pens with a 4-hole dry feeder and dual cup waterer with 27 pigs/pen. Animals were fed common diets for the duration of the trial with the only difference being the use of chlorotetracycline and tiamulin in the feed for the AB treatment in the first three diet phases after weaning. Pigs from the AB treatment were also treated with oxytetracyline and tiamulin via the water and injectable medications as needed. The ABF treatment pigs were prescribed flunixin meglumine as an anti-inflammatory for a first treatment and if the condition worsened were treated with injectable ceftiofur or enrofloxacin. Pigs were fed to a common market weight of ~275 lbs, and then they were harvested and carcass data was collected at a commercial abattoir. Due to the large number of removals and slower growth rate from the pigs subjected to the ABF treatment, pigs were not marketed similarly across treatments in order to reach as similar a market weight as possible.

Health challenge

During the trial, significant health challenges affected the experimental pigs. During the pre-wean period, the source farm was diagnosed with influenza (H1N2), and diarrhea (both Rotavirus Group C and *Clostridium perfringen* was present in the farrowing rooms. During the pre-wean period, a small group of the ABF pigs inadvertently received injectable antibiotics over a weekend period which led us to mass inject all of the pigs from the trial to maintain homogeneity across treatments. At placement into the finisher, pigs were again challenged with diarrhea and a subsequent respiratory challenge which lasted through the duration of the trial. Both *Actinobacillus* and *Streptococcus suis* were identified, as well as *Mycoplasma hyopneumonia* and a positive PCR test for PRRSV. The PRRSV detected was 99.3% homologous to the PRRS MLV used for vaccination upon arrival to the wean-to-finish barn, but was still present 18 weeks after vaccination. This would suggest that virus was circulating through the pigs the entire trial, and this was possibly caused by the large number of injections being given in the ABF treatment pigs.

Results and discussion

Pigs from the AB treatment had more ($P < 0.001$) full value pigs and fewer ($P < 0.05$) culls and removals (Table 1). Pigs from the AB treatment were significantly ($P < 0.001$) heavier at placement into the wean-to-finish barn compared to pigs from the ABF treatment. This was most likely due to the significant health challenge present at the sow farm. Start weight was not used as a covariate in the analysis based on the assumption that initial differences were treatment driven. Overall, pigs from the AB treatment had improved ($P < 0.05$) ADG and FG. They also had higher ($P < 0.05$) market weights, carcass ADG, and improved ($P < 0.05$) days on feed (DOF) and carcass FG. Pigs from the AB treatment had significantly greater ($P < 0.001$) yield, and had less ($P < 0.02$) back fat and greater ($P < 0.007$) percent lean which agrees with other immune activation research[5,6] showing immune challenged pigs deposited more fat than lean tissue. Lastly, pigs from the AB treatment had numerically higher feed cost, but significantly greater ($P < 0.001$) MOFC. In conclusion, in the face of an acute health challenge, pigs had significantly improved growth performance, had more full value pigs, and greater MOFC when a judicious use antibiotic program was utilized.

References

1. Holt JP, van Heugten E, Graves AK, See MT, Morrow WEM (2011) Growth performance and antibiotic tolerance patterns of nursery and finishing pigs fed growth-promoting levels of antibiotics. Livest Sci. doi: 10.1016/j.livsci.2010.09.007

2. Apley, M.D. et al. (2012) Use estimates of in-feed antimicrobials in swine production in the United States. Foodborne Pathog. Dis. 9, 272–279.

3. Cromwell, G. L., Why and how antibiotics are used in swine production, Anim. Biotechnol. 13 (2002) 7–27.

4. Wagstrom, L. 2012. Multiple challenges – legislative, legal, regulatory and marketplace – to antibiotic use. Allen D. Leman Swine Conference Proc. Vol. 39.

5. Williams, N. H., T. S. Stahly, and D. R. Zimmerman. 1997. Effect of level of chronic immune system activation on the growth and dietary lysine needs of pigs fed from 6 to 112 kg. J. Anim. Sci. 75:2481–2496.

6. Williams, N. H., T. S. Stahly, and D. R. Zimmerman. 1997. Effect of chronic immune system activation on body nitrogen retention, partial efficiency of lysine utilization, and lysine needs of pigs. J. Anim. Sci. 75:2472–2480

Table 1: Growth performance of pigs raised with or without the use of antibiotics

Item	AB[1]	ABF[2]	Probability, $P <$
Growth performance			
ADG, lb	1.68	1.61	0.001
ADFI, lb	4.38	4.29	0.11
FG	2.60	2.67	0.01
Beginning wt., lb	15.2	12.90	0.001
Final wt., lb	273.4	263.9	0.001
DOF[3]	164.1	166.5	0.01
Carcass characteristics			
Carcass ADG, lb	1.30	1.22	0.001
Carcass FG	3.37	3.51	0.007
Yield, %	75.09	74.42	0.001
Back Fat, mm	16.89	17.38	0.02
Loin depth, mm	64.60	63.40	0.051
Percent lean, %	56.90	55.90	0.007
Marketing performance			
Full value pigs, %	87.8	74.3	0.001
Cull, %	1.6	3.5	0.01
Removals, %	10.5	21.8	0.001
Economics			
MOFC, $	102.62	82.83	0.001

[1] AB = Pigs raised with antibiotics given via feed, water, and injectables.
[2] ABF = Pigs raised without antibiotics.
[3] DOF = The average days on feed of each treatment.

Managing enterotoxigenic *Escherichia coli* post-weaning diarrhea with Avilamycin

Robert Evelsizer, DVM; J. Mark Hammer, DVM; Marcos Rostagno, DVM, PhD
Elanco Animal Health, US, Inc. Greenfield, Indiana

Introduction

Enterotoxigenic *E. coli* (ETEC) is an important cause of illness and death in recently weaned pigs. Avilamycin (Kavault®) is a recently approved in-feed antimicrobial labeled for the reduction of incidence and overall severity of diarrhea in the presence of pathogenic *E. coli* in groups of weaned pigs. The presentation will review factors involved in successfully using this novel compound in helping to manage the occurrence of ETEC in weaned pigs.

E. coli, clinical picture and diagnosis

Strains of ETEC that cause diarrhea in pigs possess two types of virulence factors, adhesins and enterotoxins, both of which are essential for disease to manifest.[1] The *E. coli* most commonly associated with post-weaning diarrhea are the enterotoxigenic *E. coli* with F4 (K88) or F18 fimbriae. F4 (K88) *E. coli* are most likely to occur early following weaning, whereas F18 *E.* coli infections are more commonly expressed 10-14 days post-weaning. Mixed infections of F4 (K88) and F18 can occur within the same group of pigs.[2]

The fimbrial adhesins target specific receptors on porcine intestinal brush border enterocytes enabling the bacteria to colonize the cell surface and there excrete toxins manifesting in cellular disruption and pathology. Expression and exposure of these receptors is restricted by both age and lineage.[1] Age related expression becomes important in timing of clinical disease and management strategies.

Making an accurate diagnosis in cases of post-weaning diarrhea is important in selecting appropriate interventions. Diagnosis begins with obtaining an accurate and complete history. Then, because enteric and viral infections can be transient, it is important to sample untreated animals within the first 24 hours of diarrhea. Table 1 describes the recommended sampling procedure for a complete enteric evaluation. A presumptive diagnosis of most post-weaning *E. coli* infections can be made by bacterial culture of intestine or with rectal swabs (multiplex PCR). A definitive diagnosis is best made with accompanying histopathology.[2,3]

Avilamycin (Kavault®)

Avilamycin (Kavault®) is an oligosaccharide in the orthosomycin class having primarily gram + activity. It decreases fimbriae development of *E. coli,* and thus minimizes *E. coli* attachment to the enterocytes and prevents disease in post-weaned pigs.[4,5]

Table 1: Suggested submission for post-weaning diarrhea[3]

	Preservation	Samples	Tests
From euthanized untreated animals: 2 acute and 1 chronic	Fixed in 10% NBF within 15 minutes of death	2 1" segments of colon	Histopathology and immunohistochemistry
		6 1" segments small intestine	
		Lung	
		Brain (if nervous clinical signs are observed)	
	Fresh Quickly chilled, ship quickly – avoid shipping on weekends	2/3 colon	Bacterial culture
		2 6" segments of small intestine	Bacterial culture PCR for *E. coli* genotyping
		Cecal contents or feces	agELISA for Rotavirus PCR for TGEV
		Lung	PCR for PRRSV
		Brain (if nervous clinical signs are observed)	Bacterial culture
Pooled samples: groups of 3-4 acute animals	Fresh quickly chilled	Feces	agELISA for Rotavirus PCR for TGEV
Fecal swabs from several untreated acute animals	Keep moist and quickly chilled	Fecal swab	Bacterial culture PCR for *E. coli* genotyping

Studies conducted with weaned pigs naturally challenged with (ETEC) *E. coli* demonstrated a significant reduction in severity and incidence of diarrhea when pigs were fed diets containing 73 g/ton of avilamycin (Kavault®) compared to those fed control diets.5

Avilamycin is animal use only and is poorly absorbed by the gut (Table 2, Figures 1,2).

Managing ETEC

Environmental factors

Maintaining a warm, draft free environment is always a challenge in buildings housing weaned pigs. Providing supplemental heat is important year round. Chilling of weaned pigs decreases gut motility and can be a major contributor to development of *E. coli* disease. Make sure that adequate supplemental heat is provided to the micro-environment of pigs. The pigs will seek the warmth and you should manage according to the pigs lying patterns.

Sanitation

Minimizing environmental exposure to *E. coli* is important, and pigs should be provided with clean disinfected pens. Using compatible detergent and disinfectant to sanitize and disinfect pens will accomplish this objective. Soaking the pen to remove gross manure and using a detergent prior to pressure washing will enhance cleaning. As little as 5% of the manure remaining can inactivate or overwhelm disinfectants. Note: 5% manure is not visible in a pen. Once pens are pressure washed, a compatible disinfectant can be applied.

Feeder management

Providing fresh feed daily to weaned pigs will encourage all pigs to start consuming dry feed. Weaned pigs are called by the sow to suckle, and this auditory eating stimulus is not available upon weaning. Piglets need to be stimulated to recognize hunger and find fresh feed in the feeder with multiple feedings until all pigs are consuming adequate feed. This is a few days for most pigs, but some pigs will need more attention to start consuming adequate feed. Feeder management will change through the nursery period. Provide access to feed until all pigs are adequately consuming feed and then managing the feeder to minimize feed wastage is important.

Table 2: Least square means for the effects of avilamycin inclusion level on incidence and overall severity of diarrhea caused by pathogenic Escherichia coli in groups of post-weaned pigs.

	Avilamycin premix inclusion level, g/ton			
Item	**0**	**73**	**SEM**	***P*-value**
No. of Pens	40	400	-	-
No. of pigs starting study, day 0	400	400	-	-
No. pig removals, non-E. coli reasons[1]	18	19	-	-
No. pigs in efficacy analysis	382	381	-	-
No. pig removals, E. coli diagnosis[2]	72	56	-	-
Bodyweight, lb				
Start of 21 day treatment period	12.04	12.05	0.52	-
End of 21 day treatment period	17.88[c]	19.67[d]	2.35	0.03
Diarrhea incidence, %	90.78[a]	62.44[b]	4.00 and 10.73[3]	< 0.01
Mean diarrhea severity score	0.87[a]	0.43[b]	0.01	< 0.01
Mean depression score	0.18[c]	0.12[d]	0.01	0.04
Mean gauntness score	0.25	0.16	0.01	0.10
ADG, lb/d	0.29[c]	0.38[d]	0.08	0.05
ADFI, lb/d	0.51	0.55	0.10	0.09
G:F, lb:lb	0.51	0.57	0.08	0.47

a,b Least square means are significantly different ($P < 0.01$).
c,d Least square means are significantly different ($P < 0.05$).
1 Non-*E. coli* No. pig removals and deviations were not included in the efficacy analysis.
2 *E. coli* No. pig removals, based on *E. coli* diagnosis, were included in the efficacy analysis.
3 Diarrhea Incidence %, SEM of Avilamycin 0 and 73 g/ton respectively.

Figure 1: Diarrhea incidence, %[1,2,3].

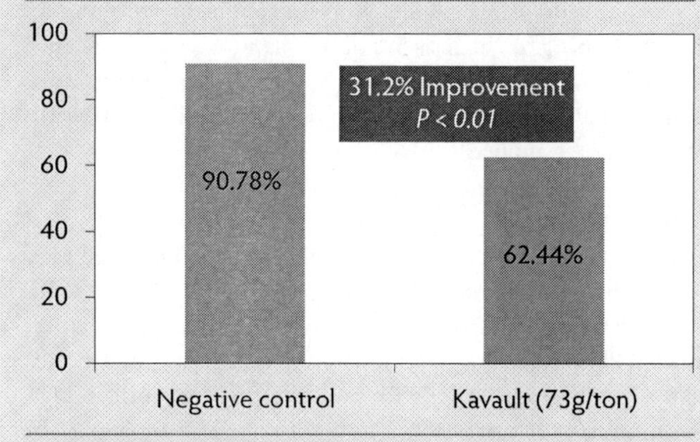

1 Diarrhea incidence rate = (# of animals with diarrhea score > 2)/(total animals at risk) × 100.
2 Freedom of information summary. Original New Animal Drug Application. NADA 141-438. Kavault. Avilamycin Type A Medicated Article, May 8, 2015.
3 Treatment LS means different: $P < 0.01$; SEM of Avilamycin 0 and 73 g/ton respectively 4.00 and 10.73.

Figure 2: Mean diarrhea severity score[1,2,3].

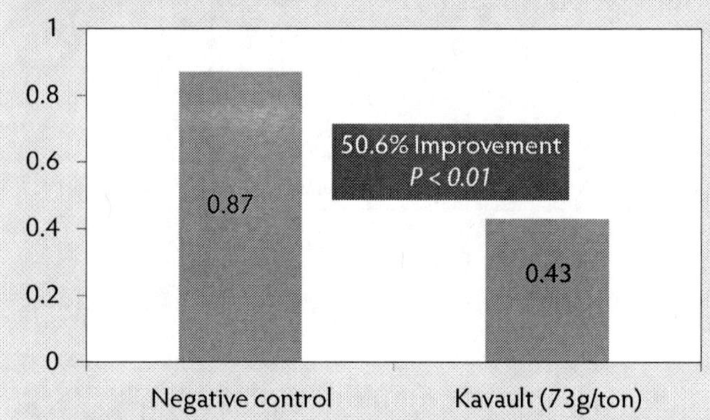

1 Mean diarrhea severity score = (sum of diarrhea severity scores)/(total head days).
2 Freedom of information summary. Original New Animal Drug Application. NADA 141-438. Kavault. Avilamycin Type A Medicated Article, May 8, 2015.
3 Treatment LS means different: $P < 0.01$; SEM 0.01.

Using Avilamycin (Kavault®) successfully as part of an ETEC strategy

When encountering post-weaning diarrhea, it is important to first get a definitive diagnosis. When enterotoxigenic *E. coli* are found to be causative, then appropriate treatment, prevention, and control programs can be properly prescribed and implemented. Prevention and control of ETEC usually involves multiple strategies.[6,7] As outlined above, sanitation, biosecurity, and control of environment are still basic to any control program as well as general good pig management.

When implementing avilamycin (Kavault®) for the first time, it is generally recommended to start before clinical signs exist, and then to continue beyond when clinical disease is expected. Our rule of thumb is to first determine the efficacy of an intervention, and then determine the most economical way to apply it.

A clinical efficacy study showed a trend of increasingly lesser severity and incidence of diarrhea in avilamycin (Kavault®) fed pigs the second and third weeks of treatment when compared to controls (Figures 3-5).

Actual availability of medicated feed in the feeder of at risk pigs can vary. A '1 lb./pig feed budget' of pre-starter diet may actually be fed much longer to the smaller, most at risk pigs at weaning. It is recommended when experiencing early ETEC to have avilamycin (Kavault®) in the pre-starter and starter diets.

Be aware of non-medicated diets within a feed budget when designing and implementing avilamycin (Kavault®). From experience, 'it doesn't work if it is not in the feed'.

When determining duration, remember that expression and exposure of enterocyte receptors is age related and may be affected by diet changes. Avoid withdrawing avilamycin (Kavault®) from the diet at the same time as a major nutritional change. In cases where ETEC occurs late, an avilamycin (Kavault®) medicated and non-medicated version of the last diet may need to be considered to manage the disease.

Summary

Approaches to managing ETEC disease and its impact include controlling associated environmental, nutritional, and stress factors occurring around weaning, as well as the use of many different specific interventions. Nevertheless, post-weaning diarrhea continues to challenge the pork industry. Avilamycin (Kavault®) is a recently approved in-feed antimicrobial labeled for the reduction of incidence and overall severity of diarrhea in the presence of pathogenic *E. coli* in weaned pigs. Diagnostics in confirming ETEC (and any other concurrent pathogen) is the first step in determining the appropriate control strategies. Timing and duration of Avilamycin (Kavault®) in feed as well as other factors need to be considered when application of this technology is deemed appropriate.

References

1. Francis DH. Enterotoxigenic *Escherichia coli* infection in pigs and its diagnosis. *J Swine Health and Production* 2002;10(4):171-175.

2. Rossow K. Post-weaning enteric infections. AASV. 2010;523-525.

3. Stevenson GW. Trends and diagnostic tips for enteric diseases. ISU Swine Disease Conference for Swine Practitioners. 2009;46-51.

4. Rostagno M et al. Exposure to the antibiotic avilamycin inhibits *E. Coli* fimbrae and attachment. IPVS. 2014;238

5. Marsteller T et al. Effectiveness of avilamycin (Kavault®) for reducing the incidence and severity of colibacillosis in post-weaned pigs. AASV. 2016;155-157.

6. Kurt J. Post-weaning enteric infections: interventions with impact. AASV. 2010;531-534.

7. Rowles C. Management of hemolytic *E. coli* in recently weaned pigs. AASV. 2014;563-564.

Figure 3: Number of animals per diarrhea score by treatment, days 1-7.

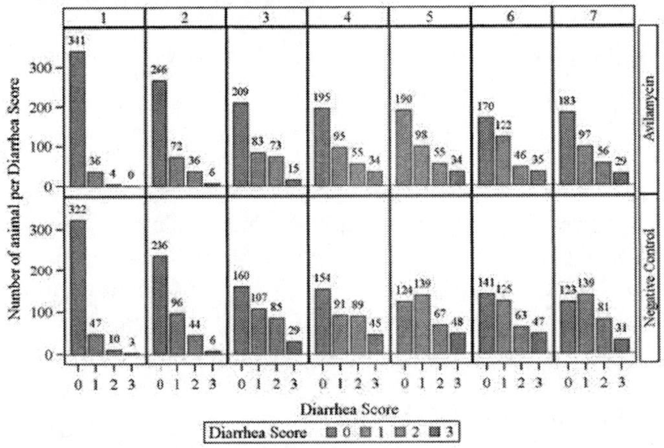

Figure 4: Number of animals per diarrhea score by treatment, days 8-14.

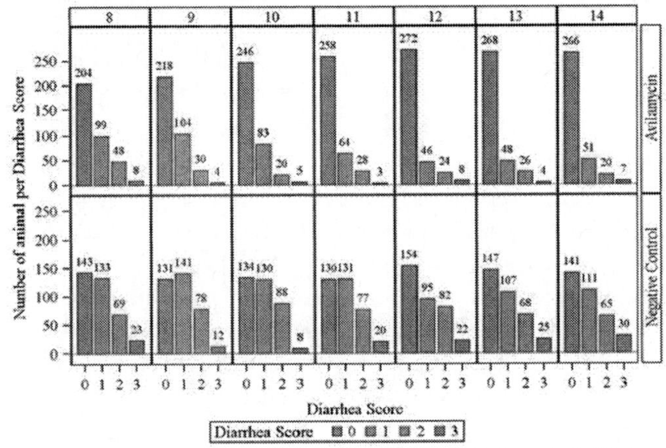

Figure 5: Number of animals per diarrhea score by treatment, days 15-21.

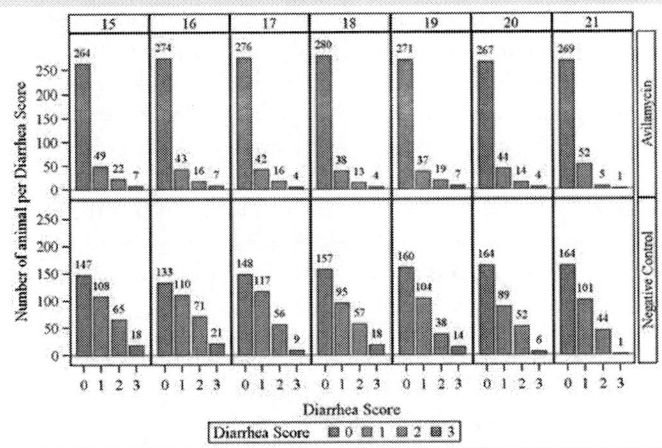

1 Elanco. 2011. "Efficacy of Avilamycin Administered In Feed For Reduction Of Incidence And Severity Of Nursery Pig Colibacillosis." Study Numbers. T4EUS100011, T4EUS100012, T4EUS110002, T4EUS110003. Data on file.

Pursuing American Board of Veterinary Practitioners (ABVP) certification in Swine Health Management

Jeffrey A. Husa, DVM, Diplomate ABVP
Swine Health Management representative to the Council of Regents

Introduction

Are you interested in board certification? How might ABVP credentials benefit you and your clients? An informal 2016 survey of current Swine Health Management (SHM) Diplomates netted the following responses: "motivated me to sharpen critical skills", "recognized for excellence", "challenged me to broaden expertise for my producers", and "opened doors for career advancement." A more extensive survey of current and past Diplomates is presently underway and will be reported during this presentation at AASV 2017. No matter what may have sparked your interest in this topic, I personally encourage you to take the ABVP plunge. I have found the entrance, credentialing, and maintenance processes to be challenging and rewarding. Becoming a Diplomate required that I push my professional limits harder than ever, and has been far more valuable to me, my clients, and the animals I care for than I ever expected.

Applicant requirements*

To be eligible for certification, a veterinarian must:

A. Be a graduate of an AVMA accredited / approved college of veterinary medicine, or hold a certificate from an Educational Commission for Foreign Veterinary Graduates (ECFVG).
B. Be legally qualified to practice in the US or other country.
C. Demonstrate unquestionable moral character and ethical professional behavior.
D. Have a background in swine medicine and production in one of the following ways:
   a. At least 5 years of clinical practice experience prior to first credential submission, and at least 6 years in practice prior to sitting the credentialing examination. The first year need not be in swine practice, however the applicant must have worked in a predominantly swine-oriented practice during the 5 years prior to examination.
   OR:
   b. Complete an ABVP-approved Residency or MS degree program, plus at least one year of swine practice.

The path to credentials begins with an entrance exam

So how do you become credentialed? Board certification in Swine Health Management begins by successful completion of an entrance examination. This multiple choice test is offered each Spring at the AASV Annual Meeting. We begin with this exam to assess your basic knowledge of key topics such as reproduction, disease, pathology, economics, epidemiology and more. This test also helps to set the stage for your success throughout the remainder of the credentialing process. To sit the exam, you must submit an application and $110 fee, due January 15th of the year in which you intend to test. Additional details about the examination, schedule, topics covered, and more can be found at **www.abvp.com**. The Applicant Handbook is available at this address, and is an excellent guide throughout the entrance and credentialing processes. Applicants who fail may retake the exam in succeeding years through the same process (application plus fee).

The certification process

After you pass the entrance exam, you will select an ABVP advisor, subject to the approval of the SHM Regent or their delegate. You will then complete a plan of study developed in collaboration with your advisor. During the next 34 months, you will complete the requirements for your credentials application. You will write three publications: two case reports written in the ABVP format (Applicant Handbook pg 16), and one peer-reviewed article accepted for publication. You will also begin studying for the qualifying examination during this time. Many Diplomates have found these studies to go hand-in-hand with daily clinical experiences. The credentialing process will push, stretch, and energize you!

Then in September of your third year, 30 months after the entrance exam, you will submit an on-line application. This alerts ABVP of your impending submission. Your advisor and other Swine Health Management Diplomates will be available to you throughout the journey. In particular, we recommend review of your materials prior to submission to better ensure compliance with ABVP guidelines. You will submit your credentials not later than 11:59PM on January 15th of the third year (34 months) after your entrance exam. Then, it's time to study for the next 14 months. The qualifying exam is administered during our AASV Annual Meeting. It includes written and oral evaluations with short answer and problem solving components. A graphic representation of the credentialing timeline is shown in Table 1.

Maintenance of certification*

Once you have successfully obtained credentials, they will remain valid for 10 years with submission of annual dues. Between the 8th and 10th year after you initially credential, you will recertify

* Note: For current application, credentialing, and maintenance requirements, see the ABVP website[1] for handbooks and other guidance.

Table 1: Timeline for successful board certification in Swine Health Management through the American Board of Veterinary Practitioners

	Mar	Apr	May	Jun	Jul	Aug	Sep	Oct	Nov	Dec	Jan	Feb
Year 1	Pass entry exam – AASV Annual Meeting											
Year 2												
Year 3							Submit application by 11:59PM, Sep 1				Submit credentials by 11:59PM, Jan 15	
Year 4												
Year 5	Sit for qualifying exam – AASV Annual Meeting											

through the accumulation of points or by reexamination. Points are earned through a variety of options including CE attendance, refereed publication, presenting lectures, writing case reports, and more. All topics attended, presented, or published must be relevant to swine practice and meet ABVP guidelines. Complete details about either pathway can be found in the Maintenance of Certification Handbook, also available at the ABVP website under Forms and Documents on the Members-Only page. Credentials may be maintained for the life of the Diplomate through this rolling 10 year process. The Council of Regents urges all Diplomates to submit in their 8th year to best ensure continued success.

Conclusion

So why should you consider certification in Swine Health Management through the American Board of Veterinary Practitioners? It will stretch you! You will be recognized among your peers as an expert in this field. The journey will help you to hone new skills and enhance old ones. You will be asked to demonstrate not only your knowledge, but also your ability to solve critical problems...problems which may make all the difference for you, your producers, their pigs, and even consumers world-wide. Come join us!

References

1. Website: www.abvp.com.

2. Applicant handbook: http://www.abvp.com/uploads/file/ABVP%20Applicant%20Handbook%20(Summer%202016).pdf.

3. Maintenance of Certification (MOC) Handbook: http://www.abvp.com/uploads/file/ABVP%20Applicant%20Handbook%20(Summer%202016).pdf.

Impact of piglet birthweight on post-weaning performance and system profitability

D.W. Newcom[1]; C. Witte[2]; M. Brubaker[2]; S.E. Lawrence[2]
[1]National Swine Registry, West Lafayette, Indiana; [2]PureTek Genetics, Albion, Indiana

Introduction

Profitability of commercial swine producers is dependent upon several key factors: reproduction, post-weaning performance, carcass quality, health, and cost-management. Reproduction can be broken down into several factors as well: farrowing rate, litter size at birth and weaning, pre-weaning mortality/morbidity, litter weight at birth and weaning, lactation feed intake, wean-to-service interval, farrowing interval, lactation length, and parity distribution are examples. Some of these traits also have a marked impact on both subsequent reproductive performance as well as post-weaning performance of pigs in the litter in addition to lifetime reproductive performance of gilts in the litter.[1]

Litter size at birth has been improved through genetic selection, but piglet size at birth has become smaller and generally more variable.[2] It is well documented lighter piglets at birth have a lower probability of survival to weaning and greater probability of becoming a less than full-value market pig.[3] Greater mortality and morbidity, reduced growth rate and/or more days on feed are a drag on the productivity and profitability of a system. Systems designed for producing a certain number of pigs, marketed in specific weight windows, in a specified period of time would have the added expense of additional sows and/or additional finishing spaces or in some cases purchase extra pigs to fill holes in production.

The goal of this analysis is to show the relationship between litter size and birth weight in a commercial test herd, the impact of birth weight on subsequent pig performance, and the variation that exists in performance across progeny of known sires in a commercial setting.

Materials and methods

Data from a Commercial Test Herd (CTH) was utilized in this study. Commercial pedigreed Yorkshire x Landrace females were bred to known pedigreed purebred Duroc boars via single-sire inseminations. Piglets were identified at birth and tracked through the system to market. Birth and weaning weights were collected by litter (not individuals). Cross-fostering was performed to balance litters, but was not tracked for individual pigs – therefore data for number weaned and weaning weight will not be shown and no inferences will be made. Removals were recorded and noted in on-farm herd management software. Pigs were reared in commercial facilities where weight, 10th rib off-midline backfat and loin muscle depth was collected and lean percentage was calculated. Groups were off-tested on a single day to minimize extra labor.

A total of 27,350 piglet birth records, sired by 126 Duroc sires, from 1,772 dams and 2,349 litters were utilized in the study. Litter birth (LBWT) and weaning weights were averaged by number born alive and number weaned, respectively. Age at removal (AR) was calculated as days from birth to removal. Days to market (DAYS), backfat (BF), loin muscle depth (LD), and lean percentage (LP) were adjusted to 127 kg. Pigs born, pigs marketed, and percent marketed were calculated for each sire. Sire-group averages were also calculated for Total Number Born (TNB) and Number Born Alive (NBA) per litter and Average Piglet Birth Weight (ABWT). Numerical differences were also calculated for TNB, NBA, Litter Birth Weight (LBWT), and ABWT, based on their litter of birth, between pigs marketed and pigs removed. Removals were also divided based on the age at removal into groups corresponding to stage of production: day 1 – 21 (PWM), day 22 – 56 (NW), and day 57+ (FM).

Results and discussion

Descriptive statistics are shown in Table 1. Utilizing raw means, traits have similar variability as reported by Fix.[3] However, when reporting sire-group averages the variability calculated is reduced (Table 2). When graphing these results, the impact of birthweight on growth and carcass traits indicate the relationship is weak or non-existent – we believe this is an artifact of the data structure and not a true result. Results in Table 2 also indicate that on average pigs are being removed prior to weaning, where the pig is the least valuable (ie hasn't consumed expensive feed yet). Figure 1 shows the impact of litter size on average piglet birth weight. The impact is similar to the result reported by Patterson[1], however, using sire-group averages instead of individual piglet weight, shows the impact is reduced. We would propose this is simply a data collection and data structure issue and not a true reduction in the impact of litter size on piglet birth weight.

Table 2 illustrates the differences in performance across sire groups. In terms of percent marketed (born alive to market weight), pigs from the best sires survive to market almost 90% of the time, while pigs from the worst are more than 30% lower. These results also indicate some sires lose most of their pigs prior to weaning and others are substantially older when removed. Our goal is to reduce mortality and removals, and the most expensive removals are closer to market weight when pigs have all of the fixed costs in addition to almost all of the variable costs associated with each market pig. This can be seen more clearly in Figure 2, where sire-groups were ranked by percent marketed (as a deviation from the mean) and plotted against age at removal. The desirable sires are those at the far right

Table 1: Descriptive statistics for litter reproductive measures (1,772 dams) and post-weaning performance.

	Number	Average	St. deviation
Total born	2,349	13.43	2.97
Born alive	2,349	12.59	2.83
Litter birth weight, kg	2,349	19.94	4.38
Average birth weight, kg	2,349	1.61	0.29
Removal age, d	5,765	19.94	28.02
Days to 127 kg, d	21,585	191.3	20.1
Backfat @ 127 kg, mm	21,374	22.47	5.01
Loin Depth @ 127 kg, cm	21,374	6.25	0.49
Lean percentage @ 127 kg, %	21,374	51.58	0.03

Table 2: Descriptive statistics for sire-group averages from 126 purebred sires of commercial offspring.

	Average	St. deviation	Minimum	Maximum
Pigs born alive	217.1	60.3	102	469
Pigs marketed	171.3	49.9	73	356
Percent marketed, %	78.8	5.8	56.3	87.4
Removal age, d	19.7	9.3	5.89	59.42
Total born	13.34	0.85	11.42	15.72
Born alive	12.50	0.80	10.42	14.87
Birth weight, kg	1.62	0.11	1.33	1.89
Days to 127 kg, d	191.5	7.7	175.8	209.6
Backfat @ 127 kg, mm	22.10	2.47	10.11	27.39
Loin depth @ 127 kg, cm	6.19	0.37	3.40	6.68
Lean percentage @ 127 kg, %	51.7	1.25	48.5	54.5

of the graph, but even those can be sorted by age at removal – those in the right quarter with the lowest age at removal would prove more valuable to the system. There also exists differences from best to worst sires in terms of growth (30 days to market) and carcass lean (6%) (Table 2).

Table 3 shows results for average performance differences between pigs marketed and pigs removed. We only investigated litter of birth records, because the pigs removed don't have performance testing information. Pigs surviving to market tended to be from slightly smaller litters for both TNB and NBA, but were slightly heavier at birth (0.03 kg).

Correlation coefficients between litter traits, removal and post-weaning traits are illustrated in Table 4. As expected, TNB was highly correlated with both NBA and LBWT, but negatively correlated with ABWT. The same holds true for the relationship between NBA, LBWT, and ABWT. Larger litters tend to be heavier at birth (as a whole), but average piglet birth weight is reduced. TNB and NBA are also positively correlated with DAYS, indicating pigs from larger litters tend to grow more slowly. The relationship between ABWT and DAYS (-0.11) shows there is a negative (beneficial) correlation between the two; heavier pigs at birth grow faster, taking fewer days to reach 127 kg. Results also indicate heavier pigs at birth may indeed be leaner at the 10th rib and have slightly greater lean percentage.

Table 5 demonstrates differences in litter traits from pigs removed based on age at removal. Piglets removed prior to weaning are from larger litters (both TNB and NBA), from slightly lighter litters, and are smaller, on average, at birth. Pigs removed during the nursery and finishing phase appear to be from litters of similar size and weight. As expected, a majority of the removals occur prior to weaning, accounting for over 65% of total removals.

Figure 1: Sire-group average mean piglet birth weight vs. total number born per litter.

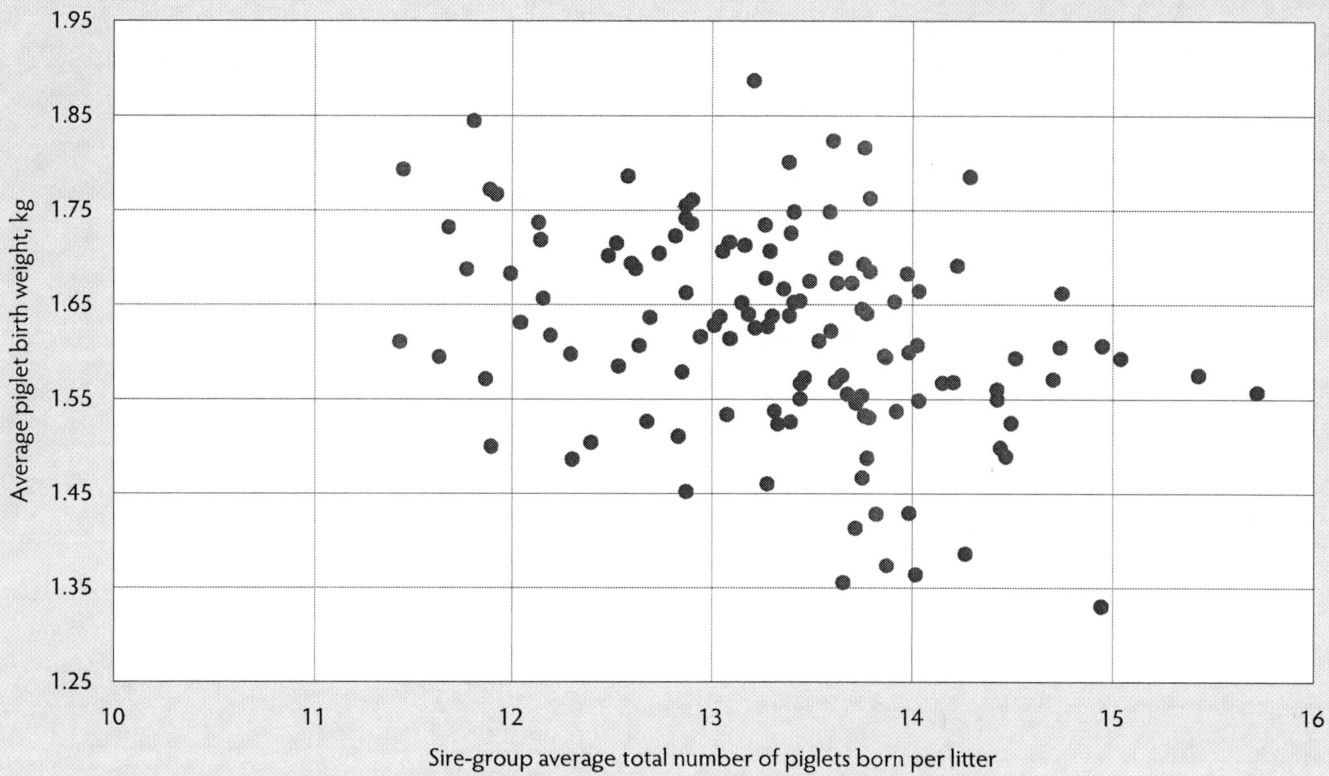

Figure 2: Sire-group average percent marketed as deviated from the mean and age at removal.

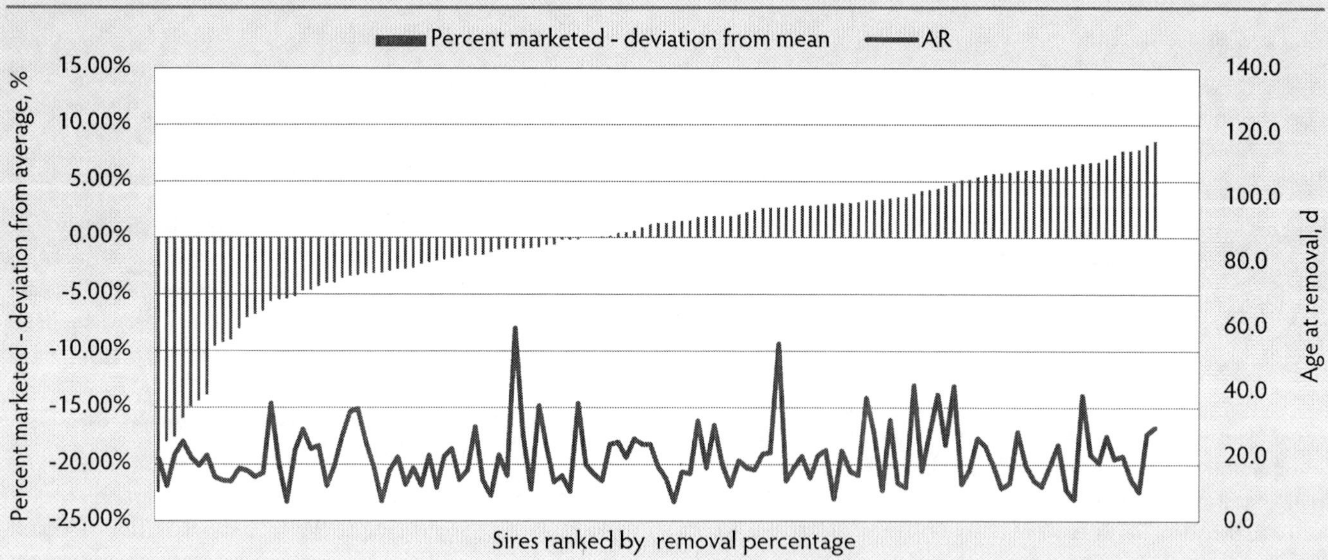

In conclusion, results indicate our study's data structure was not conducive to determining the true impact of birthweight on subsequent pig performance; within litter variation in piglet birthweight cannot be estimated. This may also be due in part to larger than average piglet birth weight compared to industry average and extremely good post-weaning performance. Small effects of average piglet birthweight on survival and post-weaning performance were detected but individual piglet birthweight data is needed to define the true impact on post-weaning performance and survival, and detect differences of economic importance between sires or sire lines. However, results do show large amounts of variability across sires for traits of economic importance – survival possibly being the most valuable (percent marketed and age at removal). Utilizing a commercial testing program can exploit this variability to add survival to an already robust breeding objective to further enhance genetic progress. Testing offspring from known sires in a commercial environment may be key to making rapid genetic progress that consistently reaches the commercial level of production as quickly as possible.

Table 3: Differences in reproductive performance of their birth litter between pigs marketed and pigs removed.

	Difference[1]
Total born	-0.41
Born alive	-0.33
Litter birth weight, kg	-0.09
Average birth weight, kg	0.03
Number weaned	0.43

Number marketed = 21.585; Number removed = 5,765
[1] Difference = Marketed – Removed

References

1. Patterson, J. and G. Foxcroft. 2016. Large litter size vs. low average birth weight – it's a production trade-off. Nat'l Hog Farmer. October: 20-24.

2. Quiniou N, Dagorna J, Gaudre D. 2002. Variation of piglets birth weight and consequences on subsequent performance. Livest Prod Sci. 78:63–70.

3. Fix, J.S. 2010. Relationship of Piglet Birth Weight with Growth, Efficiency, Composition, and Mortality. PhD Thesis, North Carolina State University. Raleigh, NC.

Table 4: Pearson correlation coefficients between birth, removal, and post-weaning traits.

	NBA	LBWT	ABWT	AR	DAYS	BF	LD	LP
TNB	0.92	0.61	-0.46	-0.06	0.07	0.03	0.01	-0.03
NBA		0.69	-0.45	-0.05	0.06	0.04	0.03	-0.03
LBWT			0.31	0.03	-0.02	-0.07	-0.02	0.06
ABWT				0.10	-0.11	-0.13	-0.06	0.12

TNB = Total number born; NBA = Number born alive; LBWT = whole litter birth weight, kg; ABWT = Average piglet litter birth weight, kg; AR = Age at removal, d; DAYS = Days to 127 kg; BF = 10th rib off-midline backfat depth adjusted to 127 kg, mm; LD = 10th rib off-midline loin depth adjusted to 127 kg, cm; LP = Lean percentage adjusted to 127 kg.

Table 5: Average litter of birth performance for pigs removed during different phases of production

	TNB	NBA	LBWT	ABWT	AR	%RM
PWM	13.87	12.97	19.9	1.56	7.81	65.6
NM	13.53	12.61	20.2	1.63	26.75	27.1
FM	13.44	12.65	20.4	1.65	103.72	7.3

PWM = Removal age < 21 d; NM = Removal age from 22 to 56 days; FM = Removal age > 56 d; TNB = Total number born; NBA = Number born alive; LBWT = whole litter birth weight, kg; ABWT = Average piglet litter birth weight, kg; AR = Age at removal; %RM = Percentage of removals in each stage of production.

Just how soluble are the water solubles?

D. Nolan[1], DVM; J. McCoy[2], CWS; R. Campbell[3]
[1]Huvepharma, Inc., Peachtree City, Georgia; [2]MWI Animal Health, Boise, Idaho; [3]Dosatron USA, Clearwater, Florida

Introduction

The solubility of water soluble antibiotic powders has been a concern since they were first introduced to the market more than 6 decades ago. Veterinarians and producers are less concerned with how a given product will dissolve in distilled water, or if solubility is effected by various water quality specifications, but with the practical. "How will it dissolve in the water on 'my farm'?" The concern around solubility is in reality a concern over making sure the medication reaches the pig in the proper dose. In other words; solubility impacts effective dose and is therefore an integral part of the judicious use of antibiotics. It is common to hear rumors of solubility issues; however, details that allow for a thorough follow up in the field are generally elusive.

Objective

To determine if common water soluble antibiotics will exhibit signs of insolubility when they are added to water solutions that replicate common water impurities encountered under field conditions.

Materials and methods

Fiji water was selected as a control because it was readily available, contained enough dissolved solids to represent field conditions (TDS 222) and had a guaranteed analysis (Table 1). Using the control water, solutions were created to represent water that is high iron, high chlorine, high calcium (measured as CaCO3), low pH, high pH. 30ppm iron solution was created by adding 300mg/l 10% iron chelate. 20ppm chlorinated water solution was created by adding .4ml/l 8.25% sodium hypochlorite (bleach). 500ppm Ca solution was created by adding 500 mg/l CaCO3 and 500mg/l citric acid. The high pH solution was created by adding 1,000mg/l high test lye (98.5% NaOH) to result in a 12.3 pH. The low pH solution was created by adding 1000mg/l food grade citric acid to result in a 3.5 pH. The label dose on each of the antibiotics chosen was used to calculate the amount of product needed in 100ml of a stock solution administered at a rate of 1:128 through a proportioner. The antibiotic powder was weighed into the beaker and 100ml of each of the treatment waterers were added. Immediately upon adding the treatment water to the antibiotic powder in the beaker the solution was stirred by hand for 30 seconds using a wooden coffee stir stick. A visual scoring method was created (Table 2). The score was then assigned to the mixture based on the amount of product estimated to have not gone into solution. The visual score was repeated at 15 minutes with no additional mixing. All solutions were at room temperature of approximately 72°F.

Table 1: Water solutions tested

Treatment	Ppm	Preparation
Control	Bicarb 152.0 pH 7.7 Mg 15.0 Ca 18.0 Si 93.0 TDS 220	Fiji water
Fe	30.0	Control +
Cl	20.0	Control +
$CaCO_3$	500.0	Control +
Hi pH	12.3	Control +
Lo pH	3.5	Control +

Table 2: Visual solubility score (subjective)

0	100% soluble
1	> 98% soluble
2	> 95% soluble
3	> 90% soluble
4	> 50% soluble
5	0% soluble

Results and discussion

Neomycin sulfate

Neomycin has a swine label that adjusts for expected water intake differences at various weights. The instructions call for 1, 1.5 or 2 100g packets (71.5g neomycin sulfate) per gallon of stock solution (1:128) for pigs weighing 25-50 lbs, 50-100 lbs, or over 100 lbs respectively. The label allows for further adjustments in concentration if needed for variation in water intake to achieve 10mg/lb. For this study we chose the more concentrated solution of 2 packets per gallon of stock solution which when administered at 1:128 in the drinking water will result in the desired dose of 10mg/lb of neomycin sulfate if a 111.7lb pig were to consume 1 gallon of medicated water. After 30 seconds of stirring the solution there was very little (< 1%) of the neomycin sulfate that did not go into solution and after 15 minutes the solution was complete with no settling. No significant differences between water types or brand of product (Table 3).

Tylosin tartrate

Tylosin has a swine label for 250mg/gal of drinking water (1-100g packet per 3.1 gallon of stock or 1-256g packet per 8 gallons of stock solution at 1:128), this concentration was used. Significant differences existed in how the three products went into solution (Table 4). Product A appeared to immediately form a suspension and then quickly moved to solution. Product B was slower to go into solution and some of the product adhered to the stir stick or beaker until it was dissolved and Product C formed a 'ball' of material and stuck to anything it touched or floated on top until it slowly went into solution. All products were fully in solution by 15 minutes. Care must be taken when mixing Product B or C as they could pose an issue if the product stuck to the bucket or stir stick outside of the stock solution level.

Lincomycin hydrochloride

Lincomycin has a swine label for 250mg/gal drinking water equivalent to 3.8mg/lb lincomycin (1-160g packet of per 2 gallon of stock solution at 1:128). The label allows for further adjustments in concentration if needed for variation in water intake to achieve 3.8mg/lb. The label assumes a water intake of 1g/150lb. At the label dose all products almost immediately went into solution. There appeared to be very minute crystals in most samples that settled to the bottom and then went into solution within 7-8 minutes with no further mixing (Table 5).

Penicillin G potassium

Potassium penicillin does not have a swine label. The turkey label dose is 1,500,000 IU/gal (1-0.5 billion IU packet per 2.6 gallon of stock solution at 1:128). The penicillin was weighed to convert IU to g and the turkey label dose was used to test solubility. Product A was noticeably more soluble during the initial stirring (Table 6). Both products went completely into solution in all water types by 7-8 minutes. Care must be taken to assure all product remains in contact with the water and not adhered to the side of the mixing container or stir stick.

Sodium sulfamethazine

Sulfamethazine sodium has a swine label of 108mg/lb for day one of treatment and 54mg/lb for days 2-4 of treatment and does not contain any label instructions for use in a water proportioner. The label instructs for the 1lb packet to first be mixed in one gallon of water to create a 12% solution and then to administer the solution at 3oz per 100lb of body weight for the first day of treatment and then 1.5oz per 100lb for days 2-4. To achieve the lower of the 2 doses (54mg/lb) a proportioner could be used that injects the 12% (1lb/gal stock) solution at 1.5% (1:85) or attempt to mix 1.5-1lb packets per gallon of stock solution, which we tried (Table 7). After 30 seconds of mixing almost 50% of the original volume of the antibiotic powder still remained in the beaker, mostly adhered to the bottom. After 15 minutes of setting in solution there was still approximately 25% of the original product remaining undissolved. To further test solubility a magnetic stirrer was used on

Table 3: Neomycin sulfate equivalent to 143g per gallon of stock solution

	30s stirring		15m setting	
	Product		Product	
Trt	A	B	A	B
Cntrl	.5	0	0	0
Fe	.5	.5	0	0
Cl-	.5	.5	0	0
$CaCO_3$	.5	.5	0	0
Hi pH	.5	.5	0	0
Lo pH	0	1	0	0

Table 4: Tylosin tartrate equivalent to 32.25g per gallon of stock solution

	30s stirring			15m setting		
	Product			Product		
Trt	A	B	C	A	B	C
Cntrl	0	2	4	0	0	0
Fe	0	2	4	0	0	0
Cl-	0	1	4	0	0	1
$CaCO_3$	0	0	4	0	0	0
Hi pH	0	0	4	0	0	0
Lo pH	0	0	4	0	0	0

Table 5: Lincomycin HCL equivalent to 80g per gallon of stock solution

	30s stirring		15m setting	
	Product		Product	
Trt	A	B	A	B
Cntrl	0	*	0	0
Fe	*	*	0	0
Cl-	*	*	0	0
$CaCO_3$	*	*	0	0
Hi pH	*	*	0	0
Lo pH	*	*	0	0

* Minute crystals formed and settled to bottom. All crystals dissolved in 7-8 minutes

Table 6: Penicillin G potassium equivalent to 192 million IU per gallon of stock solution

	30s stirring		15m setting	
	Product		Product	
Trt	A	B	A	B
Cntrl	0	1	0	0
Fe	0	1	0	0
Cl-	.5	.5	0	0
CaCO3	0	1	0	0
Hi pH	.5	1	0	0
Lo pH	0	.5	0	0

Table 7: Sulfamethazine sodium equivalent to 680g per gallon of stock solution

	30s stirring		15m setting	
	Product		Product	
Trt	A	B	A	B
Cntrl	4	4	3.5	3.5
Fe	4	4	3.5	3.5
Cl-	4	4	3.5	3.5
CaCO3	4	4	3.5	3.5
Hi pH	4	4	3.5	3.5
Lo pH	4	4	3.5	3.5

sulfamethazine in the control water and it went completely into solution after approximately 30 minutes of constant mixing. No differences in solubility were noted with water type; however, the chlorinated water did turn a pale yellow color after the addition of the sulfamethazine. The 3lb/gal solution was not attempted. In order to properly use a water proportioner you will likely need at a minimum powered agitation and time to prepare the stock solution at least 30 minutes ahead of time. Test the day-one dose to see if it is even possible in your system. The best solution may be a water proportioner set at an injection rate of at least 2 or 3%.

Chlortetracycline hydrochloride

Chlortetracycline has a swine label for 10 mg/lb of body weight, has water proportioner instructions if used at 1:128 and allows for adjustment based on water intake. According to the instructions, at the assumed 1gal of water per 100lb of body weight, 128 g/gal of stock solution, (5-25.6 oz packets mixed in 4 gallons of stock solution treats 512 gallons of water (1000 mg/gal of treated water). Three products were tested. Product A and C appeared to very quickly go into suspension during the initial stirring with product B going into suspension as well, except a small portion that stuck to the bottom of the beaker. All products had sediment (Table 8) after the stirring had ended, and the majority of the sediment that initially settled remained at the end of the 15 minutes of observation. The procedure was repeated at a common extra-label rate of 2-25.6 oz packets per gallon of stock solution (204.8 g/gal stock) or (1600 mg/gal of treated water) and the same relative portion of antibiotic settled after stirring ended (Table 9). All 3 products at the 204.8 g/gal rate were further subjected to mixing with a magnetic stirrer. Product A went into solution with no sediment at ~15 minutes, Product B went into solution with no sediment at ~30 minutes, and Product C had the sediment reduced but still had a solubility score of 1 after 3 hours of continuous mixing. It was noted that the process of dissolving was endothermic therefore the temperature of the water could play a significant role in solubility but was not investigated.

Oxytetracycline hydrochloride

Oxytetracycline has a swine label for 10 mg/lb of body weight and suggests using a water proportioner set at 1:128; however, the packet contains the following statement pertaining to mixing the stock solution: "Varies with age & water consumption". The decision was made to test stock solutions (128 g/gallon and 204.8 g/gallon) that when injected at 1:128 would yield 1000 mg/gal and 1600 mg/gal oxytetracycline respectively in the treated water. At the rate of 128g/gal the majority of the two products tested almost immediately went into solution with only trace floating particles remaining on the surface of the solution (Table 10) at both 30s and 15m post mixing. There was little or no effect of water treatment on the solubility. When mixing the stock solution at the equivalent rate of 204.8 g/gallon there was a much greater amount of product that did not dissolve and remained in solid form; however, it was clear that the majority of the extra oxytetracycline had gone into solution compared to the 128 g/gal mixture (Table 11). One of the investigators noted for the 204.8 g/gal stock solution, "There just doesn't appear to be enough water". If the higher dose levels did indeed limit solubility due to the amount of solution, an easy fix is to inject at a higher rate and reduce the concentration in the stock solution. An injection rate of 1:50 would require the stock solution mixed at 80 g/gal to deliver 1600mg/gal, well below the level of 128 g/gal which was already demonstrated to be soluble. Product A appeared to mix better in the control but was similar in the other water treatments. Further testing with the magnetic stirrers was not performed on the oxytetracycline products.

Increased dilution

To test if increased water for dilution could increase solubility, Chlortetracycline was mixed at a rate of 128 g/3gal of stock to take advantage of newer medicators that can apply solutions up to 3:128 ratios. According to the instructions, at the assumed 1gal of water per 100 lb of body weight, 128 g/3gal of stock solution, (5-25.6 oz packets mixed in 12 gallons of stock solution treats 512 gallons of water if applied at 3:128). Three products were tested. All products

Table 8: Chlortetracycline HCL equivalent to 128g per gallon of stock solution

	30s stirring			15m setting		
	Product			Product		
Trt	A	B	C	A	B	C
Cntrl	*	2*	*	2	2	2
Fe	*	2*	*	2	2	2
Cl-	*	2*	*	2	2	2
$CaCO_3$	*	2*	*	2	2	2
Hi pH	*	2*	*	2	2	2
Lo pH	*	2*	*	2	2	2

* Suspension

Table 9: Chlortetracycline HCL equivalent to 204.8g per gallon of stock solution

	30s stirring			15m setting		
	Product			Product		
Trt	A	B	C	A	B	C
Cntrl	*	2*	*	2	2	2
Fe	*	2*	*	2	2	2
Cl-	*	2*	*	2	2	2
$CaCO_3$	*	2*	*	2	2	2
Hi pH	*	2*	*	2	2	2
Lo pH	*	2*	*	2	2	2

* Suspension

Table 10: Oxytetracycline HCL equivalent to 128g per gallon of stock solution

	30s stirring		15m setting	
	Product		Product	
Trt	A	B	A	B
Cntrl	0	.5	0	0
Fe	.25	.5	.25	.25
Cl-	.5	.5	.5	.5
$CaCO_3$	.5	.5	.5	.5
Hi pH	.5	.5	.5	.5
Lo pH	.5	1	.5	.5

had small amounts of sediment (Table 12) after the stirring had ended, albeit lower than the original testing and all of the sediment had gone into solution before the 15 minute observation. This shows the alternative method of increasing the stock solution size, to correlate with an increased injection rate could potentially improve solubility of products that do not readily go into solution for 1:128 application.

Conclusion

The water treatment (often referred to as 'water quality') had little overall effect on the solubility of the water soluble antibiotic powders tested in this bench top exercise. We did not combine water quality concerns (ie. high pH, hard water with iron and chlorine) therefore the overall effect of water quality is not fully understood and will require additional work. Additionally, adjusting the rate of dilution and rate of application showed changing the way the industry applies water soluble therapies may

Table 11: Oxytetracycline HCL equivalent to 204.8g per gallon of stock solution

	30s stirring		15m setting	
	Product		Product	
Trt	A	B	A	B
Cntrl	1	2	1	2
Fe	2	2	2	2
Cl-	2	2	2	2
$CaCO_3$	2.5	2	2.5	2
Hi pH	2.5	2	2.5	2
Lo pH	2	2	2	2

Table 12: Chlortetracycline HCL equivalent to 128g per 3 gallons of stock solution

	30s stirring			15m setting		
	Product			Product		
Trt	A	B	C	A	B	C
Cntrl	*	*	*	0	0	0
Fe	*	*	*	0	0	0
Cl-	*	*	*	0	0	0
$CaCO_3$	*	*	*	0	0	0
Hi pH	*	*	*	0	0	0
Lo pH	*	*	*	0	0	0

* Suspension

increase solubility of some products if the industry changed the application rates. The amount of sediment noted by the authors on several of the products and dose levels is of some concern in being able to obtain the desired level of medication delivered to the pig and could theoretically become problematic with some water delivery systems (drinker lines, nipples, screens, regulators, etc.). Product differences in solubility were noted on several of the antibiotic soluble powders tested and should be taken into account when designing an effective treatment protocol to also include water quality, the use of an agitator, and selection of the proper proportioner and rate of application.

Masked trichothecene mycotoxins in US feed and corn samples from 2014 to 2016

E.G. Hendel[1], VMD, PhD; T. Jenkins[2], PhD; S.M. Mendoza[1], PhD; G. R. Murugesan[1], DVM, PhD
[1]BIOMIN America, Inc., San Antonio, Texas; [2]BIOMIN Holding GmbH, Getzersdorf, Austria

Introduction

Mycotoxins are harmful fungal metabolites commonly found in commercial crops. Type B trichothecenes, mycotoxins produced by *Fusarium* fungi, are detrimental to gut health, decreasing mucus production, compromising intestinal epithelial tight junctions,[1] increasing intestinal inflammation,[1] and suppressing immune function.[2]

The Type B trichothecenes (B-Trich) include deoxynivalenol (DON), nivalenol (NIV), and fusarenon-X. Additionally, fungi produce acetylated forms of DON (3-ADON, and 15-ADON). Mycotoxins can also masked by plants with added glucose or sulfate molecules, rendering them undetectable to conventional analysis.[3] Masked B-Trich (DON-3-glucoside; DON 3-G, and NIV-glucoside; NIV-G), are broken down to DON and NIV, respectively, within the gastrointestinal tract and are thus equivalent in toxicity to unmodified toxins.[4] These masked mycotoxins present a challenge for mycotoxin risk management; these metabolites are not tested by most US diagnostic laboratories.

The University of IFA-Tulln, Austria, has been conducting multi-toxin analysis on commercial feed materials and compound feeds since 2014. The purpose of this study was to evaluate these results for the prevalence of masked typed B trichothecene mycotoxins in samples from the United States.

Materials and methods

94 samples from the United States were submitted from 2014-2016 for multi-mycotoxin testing at the University of IFA-Tulln, Austria (Spectrum 380®) (Figure 1). These samples consisted of corn, corn silage, and complete feed for multiple livestock industries (cattle, swine, and poultry) (Figure 2). Samples were analyzed for more than 350 bacterial and fungal metabolites utilizing Liquid Chromatography Tandem Mass Spectrometry (LC-MS/MS). Sample results for DON, 3-ADON, 15-ADON, DON-3-G, and NIV-G were selected for additional analysis. Equivalent DON -3-G and NIV-G were calculated utilizing molecular mass ratios, allowing for summation of total type B tricothecene contamination.

Results and discussion

Previous studies utilizing the multi-mycotoxin method at the University of Tulln revealed a prevelance of 75% (83 samples) of DON-3-G.[5] These 83 samples were of global distribution, and thus it was of interest to determine the presence of masked mycotoxins in US corn and feed samples.

Figure 1: Sample number per year

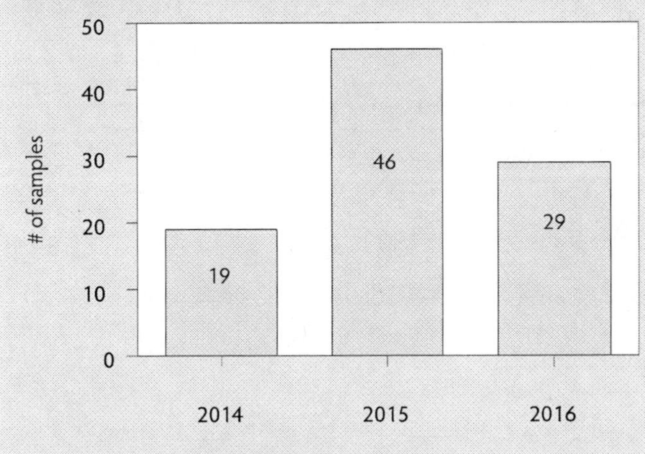

Samples submitted for analysis for multi-mycotoxin analysis in 2014, 2015, and 2016.

From 2014 to 2016, a total of 94 samples were submitted. The majority of samples (46) were from 2015 (Figure 1). These samples were of multiple material types; 52% corn and corn silage, and 37 complete feed samples (Figure 2). 41% of the submitted samples were for ruminants, 22% for poultry, 10% for swine, and 22% were labeled for multi-species use (Figure 3).

Of the 94 samples 63% had detectable DON, 5% 3-ADON, 12% 15-ADON and 49% had detectable DON-3-G. The median positive level of DON-3-G was 37 ppb with a maximum of 326 ppb. NIV was less commonly detected than DON (21% of samples) and only 2% of samples tested positive for NIV-G (Figure 4, Table 1).

Although there are limitations in the small sample size of this study, it is important to consider the potential risk of masked mycotoxins in US corn and subsequently finished feeds. 5% of submitted samples were positive for DON-3-G without detectable DON, and the summed levels of Type B trichothecenes were on average 20% higher compared to DON alone. Type B tricothecene mycotoxins are the most commonly occurring mycotoxin in US corn samples identified by the annual Biomin US Corn Survey for 2016 thus far (Figure 5). Collectively these results suggest an increased risk to animal gut health beyond what is suggested by current detection methods in the US.

Table 1: Contamination levels of type B trichothecenes in US corn and feed samples from 2014 to 2016.

MTX	%POS	MED	MAX
DON	65%	305	2703
DON3G	50%	37	326
DON3G corrected		20	210
NIV	21%	10	1072
NIV-G	2%	17	34
NIV-G corrected		11	22

Figure 2: Material type distribution of samples.

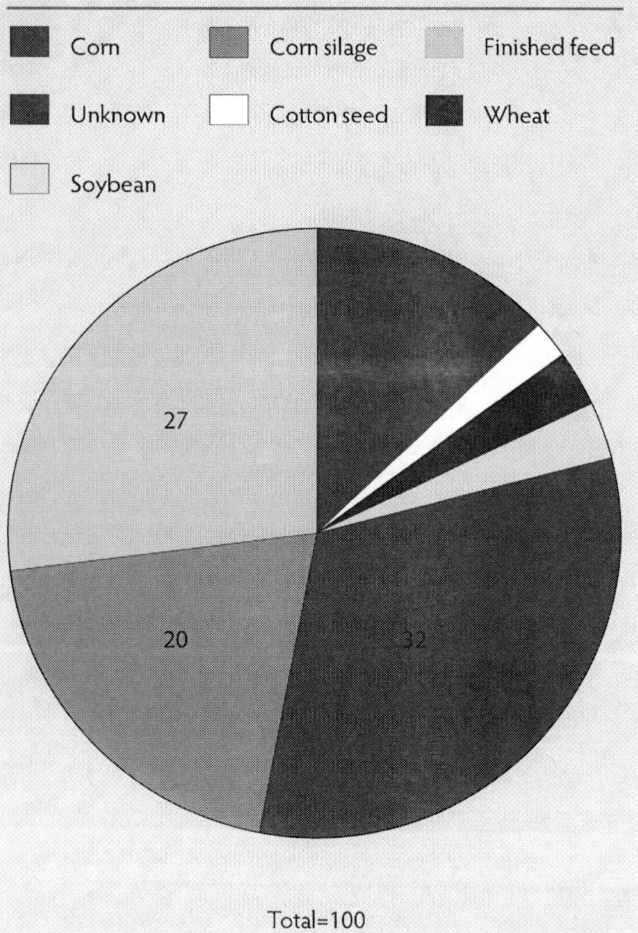

Percent of total samples divided into sample description. Most samples consisted of corn, corn silage, and finished feed.

Figure 3: Species distribution of sample

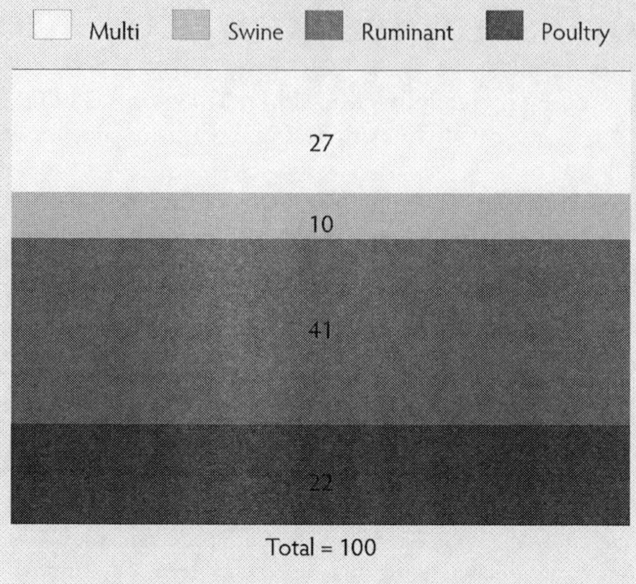

Percent of total samples divided into species. Ruminant, swine, and poultry were indicated, as well as samples marked for multi-species use.

Figure 4: Detected levels of type B trichothecenes

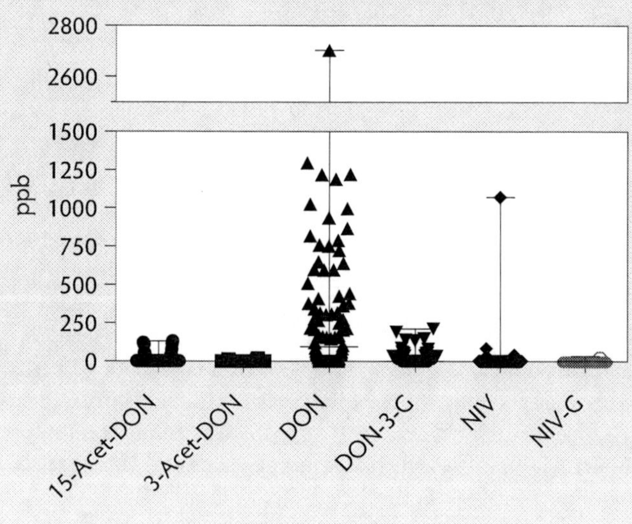

Scatter plot of all detected levels of mycotoxins (ppb) in US samples submitted for multi-mycotoxin analysis.

Figure 5: Sample distribution of the major mycotoxin groups in 2016 North American corn. Total number of samples analyzed for major mycotoxins with percent positive and negative samples.

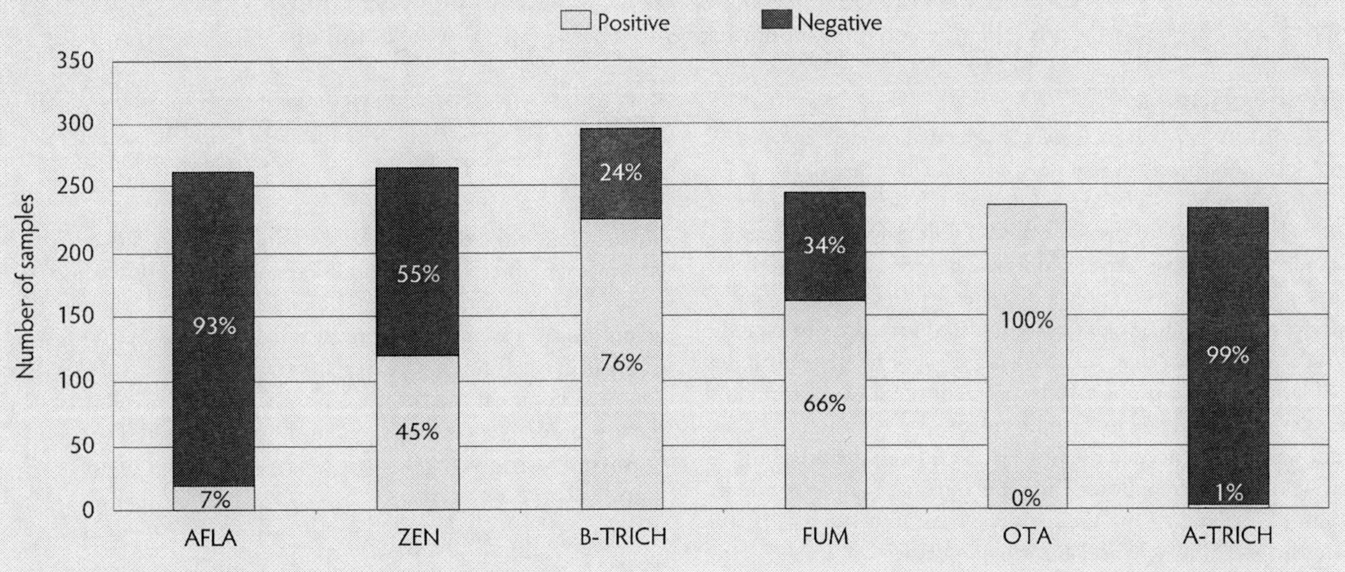

References

1. Pinton et al. Toxicity of Deoxynivalinol and its acetylated derivatives on the intestine: differential effects on morphology, barrier function, tight junction proteins, and mitogen-activated protein kinases. *Toxilogical Sciences*. 2012; 130(1): 180-190

2. Pierron et al. Impact of mycotoxin on immune response and consequences for pig health. *Animal Nutrition*. 2016; 2: 63-68.

3. Berthiller et al. Masked mycotoxins: A review. *Mol. Nutr. Food Res.* 2013; 57: 165-168.

4. Papyros et al. Toxicology of deoxynivalenol and its acetylated and modified forms. *Arch. Toxicol.*2016; 90: 2931-2957.

5. Streit et al. Multi-mycotoxin screening reveals the occurrence of 139 different secondary metabolites in feed and feed ingredients. *Toxins* 2013.

Senecavirus A, a genetic supplier perspective

J. Geiger[1], DVM, MS; P. Graham[2], DVM, MS; JP. Cano[1], DVM, PhD
[1]PIC, Hendersonville, Tennessee; [2]Ghrist Vet Clinic, Pittsfield, Illinois

Introduction

Senecavirus A (SVA) has caused considerable concern in North American pork production since a dramatic increase in cases in 2015, though it has been identified in US swine herds since 1988. The inability to distinguish SVA lesions from reportable vesicular diseases raises the significance of SVA. For a genetic supplier, with a focus on preventing the introduction of new pathogens, the primary goal is to establish biosecurity to exclude SVA from naïve populations. When SVA appears in a multiplier population, steps must be taken to prevent the spread within and between flows and downstream to customers. A structured elimination process is needed to eliminate SVA from the breeding herd and downstream flow. This report describes an SVA case in a multiplication flow and the steps taken to eliminate the virus from the breeding herd and growing population.

Materials and methods

A 4000 sow breed-to-wean multiplier herd became infected with SVA in September 2016. A mild diarrhea and increased pre-weaning mortality were the first indications. The herd was known to negative for porcine reproductive and respiratory syndrome virus (PRRSV), *Mycoplasma hyopneumoniae* (Mhp) and swine enteric coronaviruses (SECs). Investigating the diarrhea, the veterinarian identified vesicular lesions on snouts and feet and contacted the USDA. Breeding stock sales from the multiplier herd and its two off-site breeding stock gilt finishers were immediately halted. A quick response and appropriate diagnostics confirmed the absence of Foot and Mouth Disease and the presence of SVA within eight hours.

With a solid SVA diagnosis, aggressive efforts began immediately to expose the resident multiplier herd and its on-site quarantine facility to the virus. Fecal material from clinically ill sows and intestinal tracts from scouring piglets were used in feedback and sows with vesicles were allowed to roam the alleys of the gestation stall barn. The quarantine was emptied and filled with replacement gilts which were then also exposed to SVA by the same means.

Forty miles away from the sow herd, the two wean-finish sites which receive breeding gilt candidates operate as continuous flow. The two sites are located 0.75 mile apart, with separate caretakers but the caretakers live in the same home. When SVA was diagnosed, weanings were diverted away from these sites. Vesicles appeared in pigs at the youngest site (North) two weeks after the sow herd. Vesicles appeared in the older pigs in the second finisher (South) two weeks after the North site, approximately a month after the sow herd.

Building on the sampling/testing model accepted by AASV for PRRSV elimination,[1] a strategy for SVA elimination began. The program establishes a negative status over time based on the absence of clinical lesions (vesicles) in the adult population, negative PCR testing of weaning-aged piglets, the introduction of naïve sentinels with no clinical evidence (vesicles) and negative sentinel PCR testing 4-6 weeks after exposure. Conditionally negative piglets were weaned into depopulated, washed, disinfected and dry finisher sites. Close observation for SVA lesions and SVA testing continued in the finisher populations.

Results and discussion

No downstream herds were exposed to SVA through delivery of breeding stock. Further details of the elimination process and diagnostic results will be shared during the oral presentation.

References

1. Holtkamp DJ et al. Terminology for classifying swine herds by porcine reproductive and respiratory syndrome virus status. *J Swine Health Prod.* 2011;19(1):44–56.

The impact of trained and untrained personnel on product implementation and subsequent pig health

Sarah Probst Miller[1], DVM; Ron Ketchem[2]
[1]AgCreate Solutions, Inc; [2]Swine Management Services

Introduction

It takes people to make pork™ and impact healthy swine in our industry. As a veterinarian, there were times when my strategy for influencing human behavior fell short. The desire to successfully impact change in human behavior has influenced AgCreate Solutions, Inc's (AgC) subsequent educational development approach. It has been the mission of AgC to create innovative learning experiences that respect the science of how adults learn. As a result, over 80 lessons covering all aspects of pork production have been created in English and Spanish in AgCreate's sister company, AgriSchool Party, Lmt.

We are starting to see the results of applied educational science in our industry. This is the first study in a series looking at the impact of trained and untrained personnel on product implementation and subsequent pig health. In this observational study, farm owners were expanding and wanted to implement products associated with Day 1 Pig Care on a start-up farm. This case study looks at the training strategy as relates to those Day 1 Pig Care products and tasks as well as subsequent results as benchmarked by Swine Management Services (SMS). SMS's benchmarking database contains 986 farms with 1,751,400 mated females.

Materials and methods

Production parameter goals were set and success of training was measured against these benchmarked goals. The goal set was to be in the top 10% of farms regarding piglet survival (100%-SB%-PWM) the first year and the top 10% of farms in pigs per sows per year. Training impact on health of pigs was measured by piglet survival as compares back to the SMS benchmark.

This case study allowed new personnel to be trained in Farm A prior to movement to Farm B which was scheduled to begin farrowing in a newly built facility in November of 2015. Personnel went through lessons relating to their anticipated role at Farm B. Personnel were mentored on these tasks and verified as physically capable of these tasks on Farm A prior to application on Farm B. SMS and on-farm leadership monitored lesson assignment and record of completion.

Lessons administered related to this subject and products used in Day 1 Pig Care were as follows:

- Day 1 Pig Care: Intro to Midwifery
- Day 1 Pig Care: Monitoring sows as they farrow
- Day 1 Pig Care: Colostrum Management
- Day 1 Pig Care: Fostering
- Take 5: Rupture Repair
- Treatment Management: Identification and Treatment of Sick Piglets
- Treatment Management: Identification and Treatment of Sick Sows
- NOTE: Other lessons were administered to breeding and gestation and farrowing staff related to breeding and gestation management and gilt management that impacted the PSY numbers. Full lesson list available here: `http://porkavenuetraining.com/site/wp-content/uploads/2016/11/Pork-Avenue-lesson-list-2016.pdf`

The Day 1 Pig Care program implemented included monitoring sows as they farrowed, warming baby pigs, split suckling qualifying litters, strategic minimal fostering, as well as identification and treatment of sick piglets throughout farrowing. Products used in this process included iron, drying agent, warming boxes, split suckle boxes and treatments for sows and piglets as prescribed by the farm veterinarian.

Results

The SMS database contains 1,751,400 mated females on 986 farms across the global industry. Farm B achieved a 96% ranking in the SMS database achieving their goal to be in the top 10% of farms in the database. Only 40 farms ranked above Farm B in this category. The average piglet survival of the SMS top 10% is 82.5% survivability, and the entire system survivability averages 78.80%. Farm B achieved 89.20% survivability (Table 1).

In regards to the goal of achieving the top 10% of pigs per sow per year, records over the past 13 weeks of production as well as the past 52 weeks of production were charted against the benchmark. The past 13 weeks achieved this goal with the result

Table 1: Piglet survival as compared with benchmarking results

Percentile	Piglet survivability
Farm B benchmarking percentile	96.00%
Farm B	89.20%
# of farms above	40
System top 10% result	82.50%
System all	80.00%
SMS all	78.80%

of 30.16 pigs weaned per mated female per year giving Farm B a 94% ranking in the SMS Benchmarking System (Table 2).

Discussion

An excellent training program is one of many factors that influence a strong start up. In addition to the Pork Avenue Training Program, Farm A and B were blessed with engaged experienced managers, hands-on owners, veterinary consult from Dr. Jack Coleman, as well as production consulting from Swine Management Services. Additionally, a training program is only successful with aforementioned veterinary and consultant implementation leadership as well as top-down buy in of on-farm leadership. This farm had an extremely competent and committed veterinarian, production consultants, and leadership team members to guide training and subsequent product implementation on farm.

References

1. Pork Avenue Training Portal. AgriSchool Party Lmt. 2016 Available at `http://www.porkavenuetraining.com/`.

2. Swine Management Services Benchmarking Tools. Swine Management Services. November 5, 2016. Available at `http://www.swinems.com/`.

Table 2: Farm B's past 13 weeks' performance and past 52 weeks' performance compared to benchmark

Week ending November 5, 2016	Past 13 week performance	Benchmark	Benchmark %	Past 52 weeks performance	Benchmark	Benchmark %
Pigs weaned/mated female/year	30.16	25.41	94%	26.94	25.28	68%
Pre-weaning mortality	9.4	14.1	87%	8	13.8	94%
Stillborns	3.2	6.5	87%	2.8	6.2	98%
Piglet Survival	87.3	79.4	94%	89.2	80	98%

Effect of direct fed microbial, Calsporin® (*Bacillus subtilis* C-3102), on enteric health in nursery pigs after challenge with porcine epidemic diarrhea virus (PEDV)

P. Canning[1]; C. Ruston[1]; D. Madson[2]; J Bates[1]; K. Skoland[1]; J. Davenport[1]; C. Wang[3]; Q. Chen[2]; J. Zhang[2]; L. Karriker[1]
[1]Swine Medicine Education Center, Iowa State University College of Veterinary Medicine, Ames, Iowa; [2]Veterinary Diagnostic Laboratory, Iowa State University, Ames, Iowa; [3]Veterinary Diagnostic and Production Animal Medicine, Iowa State University College of Veterinary Medicine, Ames, Iowa

Introduction

This study examined the effects of feeding *Bacillus subtilis* C-3102 (Calsporin® Calpis Co. Ltd., Japan) at the target inclusion rates of 0 CFU/g, 500,000 CFU/g and 1,000,000 CFU/g on intestinal health in weaned pigs after challenge with porcine epidemic diarrhea virus (PEDV).

Materials and methods

A two by three factorial design composed of three diets containing 0 CFU/g or 500,000 CFU/g or 1,000,000 CFU/g of Calsporin® and PEDV or sham challenge was conducted. Pigs were 14 days old at the start of the study and PEDV naïve. Ten pigs were randomly allocated to each treatment group and were housed in groups of five. Pigs were fed the treatment diets for 23 days. On day 19, pigs were challenged with PEDV positive or PEDV negative cell culture by oral gavage and necropsied at four days post inoculation (dpi). Five small intestine segments were collected per pig using a standardized technique. Histopathology slides were prepared for measurement of villi to crypt height ratios (VCR), and PEDV immunohistochemistry (IHC) and atrophic enteritis (AE) scoring for each section. For IHC, PEDV antigens were detected and semi-quantitatively scored based on the percentage of enterocytes showing a positive signal. Atrophic enteritis scores were based on the presence and severity of enteritis. All scoring was completed by a single blinded veterinary pathologist. Average daily gain (ADG) and mortality was compared across treatment groups at four dpi. Responses were analyzed using linear mixed models. Comparisons among groups were assessed using F-tests followed by Tukey's t-tests for multiple comparisons. Differences were considered significant at the level of $P < 0.05$. All analysis was performed on SAS® 9.4.

Results

There were significant differences in IHC, AE and VCR between PEDV positive and PEDV negative pigs fed 0 CFU/g Calsporin®. Within PEDV positive groups, there were significant reductions in IHC and AE between pigs fed 0 CFU/g Calsporin® and treatment groups that did receive Calsporin®. VCR was greater in PEDV positive pigs fed Calsporin compared to pigs on the 0 CFU/g diet. There were no significant differences in average daily gain and mortality between any groups.

Conclusions

The findings from this study support an association between administration of Calsporin® and improved IHC scores, histopathology and villi to crypt ratios in nursery pigs challenged with PEDV compared to cohorts that did not receive Calsporin®. The impact of these parameters on morbidity, mortality, ADG and feed efficiency during the entire fattening period is unknown and should be assessed with an additional study of longer duration and larger sample size.

Impact of spray-dried animal plasma species of origin on post-weaning pig performance

Chad Pilcher, PhD; Sabrina Williams, MS; Sara Ebarb, MS; Yanbin Shen, PhD; Doug Cook, PhD
Provimi, Brookville, Ohio

Introduction

Spray-dried animal plasma is a protein rich feed ingredient obtained from the fractionation of blood collected from harvested animals. Numerous studies have demonstrated improvement in post-weaning pig performance when spray-dried animal plasma is included in early nursery diets.[1] The recent emergence of porcine epidemic diarrhea virus in North American raised concerns about the risk of disease transmission with the use of slaughter-derived feed ingredients of porcine origin, including spray-dried porcine plasma. Spray-dried bovine plasma has been used as an alternative to porcine plasma in nursery pig diets, with mixed results when compared directly to plasma of porcine origin.[2] Spray-dried ovine plasma has recently become commercially available, mainly for pet food applications; however, to our knowledge, there are no published data on the impact of spray-dried ovine plasma on pig performance. The objective of this trial was to determine the effects of bovine, ovine, and porcine plasma sources on post-weaning pig performance.

Materials and methods

The experiment was conducted in a two barns at a commercial research nursery located in New Paris OH. A total of 1,760 pigs (initial BW = 12.7 lbs; wean age = 19 days) were utilized for this experiment. Pigs were housed in groups of 22 pigs/pen (2.72 ft^2/pig) and pens were allotted to one of 5 dietary treatments (16 pens/treatment): control (no spray-dried animal plasma) or 5% spray-dried animal plasma from bovine source A (SDBP A), bovine source B (SDBP B), ovine (SDOP), or porcine (SDPP) origin. Dietary treatments were fed for the first 10 days post-weaning (Phase 1) followed by common diets fed across treatments for Phase 2 (10.5 days) and Phase 3 (21.5 days). Body weight and feed disappearance were recorded by pen at the beginning and end of each diet phase in order to calculate ADG, ADFI, and gain:feed ratio. Pigs which were removed from test pens for poor health or general unthriftiness were recorded along with any mortality.

The control diet was formulated using fishmeal to replace amino acids supplied by plasma in the other experimental treatments. For each plasma source, equal amounts of 5 production lots were combined prior to addition to treatment diets. All diets were formulated with 3030 ppm Zn and 235 ppm Cu for Phases 1 and 2 and 144 ppm Zn and 235 ppm Cu for Phase 3. Diets were medicated with 50 g/ton carbodox for Phase 1 and 35 g/ton and 400 g/ton chloratetricycline for Phase 2. Phase 3 was non-medicated. Pens were equipped with one 5-hole feeder and two nipple waterers. Pigs were sourced from a commercial production system and were of a PIC genotype.

The trial data were analyzed using the mixed models procedure of SAS. The trial design was a replicated split-plot. Sex of pig (fixed effect) formed the whole plot structure. Whole-plots were blocked into groups/gender based on location in each barn. Within each block, one pen of pigs was allocated to each dietary treatment (sub-plot). Pen was considered the experimental unit. Orthogonal contrasts were used to test for the effects of plasma sources. Results were considered significant if the P-value was < 0.05 and trends if the P-value was ≥ 0.05 and < 0.10.

Results

All plasma sources improved ADG and ADFI over the control during the 10-day plasma feeding period ($P < 0.05$; Table 1), with SDPP increasing ADG by 28% over the control compared to 9% for SDBP A, SDBP B, and SDOP sources. There were no statistically significant differences in performance between plasma treatments and the control for Phases 2 and 3 ($P > 0.10$), in which common diets were fed among all treatments. For the overall nursery period, pigs fed SDPP during Phase 1 had significantly higher ADG, and ADFI than the control ($P < 0.05$), resulting in 1.2 lbs higher body weight at 42 days post-weaning ($P < 0.05$). The SDBP and SDOP treatments did not differ from the control treatment for the overall nursery period ($P > 0.10$).

Acknowledgements

The authors would like to thank nursery research technicians Craig Unger, Todd Kiracofe and Chad Ward for technical support and animal care.

References

1. van Dijk AJ et al. Growth performance of weanling pigs fed spray-dried animal plasma: a review. *Livest Prod Sci.* 2001;68:263–74.

2. Torrallardona D. Spray dried animal plasma as an alternative to antibiotics. *Asian-Aust. J Anim Sci.*23(1):131-148.

Table 1: Effects of plasma source on post-weaning growth performance

Phase 1 (10 days)	Control	Bovine A	Bovine B	Ovine	Porcine	SEM	Control vs Bovine	Bovine A vs. Bovine B	Control vs. Ovine	Control vs. Porcine
Initial wt, lbs	12.8	12.7	12.7	12.7	12.7	0.09	ns	ns	ns	ns
Final wt, lbs	15.9	16.2	16.2	16.1	16.8	0.09	**	ns	ns	***
ADG, lbs/d	0.32	0.35	0.35	0.35	0.41	0.009	***	ns	**	***
ADFI, lbs/d	0.34	0.38	0.38	0.37	0.44	0.011	***	ns	**	***
Gain to feed	0.94	0.92	0.92	0.95	0.92	0.020	ns	ns	ns	ns
Mortality & removals, %	0.0	0.3	0.5	0.3	0.3	0.3	ns	ns	ns	ns
Phase 2 (10.5 days)										
Initial wt, lbs	15.9	16.2	16.2	16.1	16.8	0.09	**	ns	ns	***
Final wt, lbs	26.5	26.7	26.8	26.9	27.3	0.68	ns	ns	ns	***
ADG, lbs/d	1.01	1.00	1.01	1.02	1.00	0.015	ns	ns	ns	ns
ADFI, lbs/d	1.17	1.16	1.16	1.18	1.16	0.011	ns	ns	ns	ns
Gain to feed	0.86	0.86	0.88	0.87	0.87	0.015	ns	ns	ns	ns
Mortality & removals, %	1.7	0.8	3.1	1.9	1.4	0.6	ns	**	ns	ns
Phase 3 (21.5 days)										
Initial wt, lbs	26.5	26.7	26.8	26.9	27.3	0.68	ns	ns	ns	***
Final wt, lbs	53.7	54.1	54.5	54.1	54.9	1.98	ns	ns	ns	**
ADG, lbs/d	1.27	1.27	1.29	1.27	1.28	0.033	ns	ns	ns	ns
ADFI, lbs/d	1.74	1.74	1.75	1.74	1.77	0.022	ns	ns	ns	ns
Gain to feed	0.73	0.73	0.74	0.73	0.73	0.013	ns	ns	ns	ns
Mortality & removals, %	2.8	2.5	2.6	2.6	1.1	0.8	ns	ns	ns	ns
Overall (42 days)										
Initial wt, lbs	12.8	12.7	12.7	12.7	12.7	0.09	ns	ns	ns	ns
Final wt, lbs	53.7	54.1	54.5	54.1	54.9	1.98	ns	ns	ns	**
ADG, lbs/d	0.97	0.99	0.99	0.98	1.00	0.022	ns	ns	ns	**
ADFI, lbs/d	1.26	1.26	1.26	1.26	1.30	0.011	ns	ns	ns	**
Gain to feed	0.78	0.78	0.79	0.78	0.77	0.014	ns	ns	ns	ns
Mortality & removals, %	4.4	3.6	6.1	4.7	2.2	1.1	ns	ns	ns	ns

Symbol key: $P > 0.1$ = ns; $0.1 > P > 0.05$ = *; $0.05 > P > 0.01$ = **; $P < 0.01$ = ***

Validation of an ileitis mucosal homogenate seeder pig model to study the natural transmission of *Lawsonia intracellularis* in grow-finish pigs

Nathan L. Winkelman, DVM; Adam D. Mueller, DVM; Robin L. Winscher, CVT
Swine Services Unlimited, Inc., Rice, Minnesota

Introduction

Ileitis or PPE (Porcine Proliferative Enteropathy) or *Lawsonia intracellularis* is considered the most common grow-finish enteric disease of North American swine. There are three forms of the disease: (1) acute (PHE-Porcine Hemorrhagic Enteropathy), (2) chronic (PIA-Porcine Intestinal Adenomatosis), and (3) subclinical. The subclinical form is by far the most prevalent. Ileitis mucosal homogenate challenge models have been used to test antibiotic efficacy since the mid-90's.[1] The classic FDA approved ileitis challenge model is given to 4-6 week old pigs at a 10^9 dose per pig to consistently cause 3-10% ileitis mortality and significant clinical signs of gauntness and/or diarrhea in about 80% of the control pigs.[1] Pigs challenged at older ages of 10-20 weeks do not reproduce similar clinical signs, with the majority of pigs being subclinical. However, performance loss and ileitis positive diagnostics are still very evident in these subclinical carrier pigs. Since the majority of ileitis occurs in the grow-finish pig and transmits naturally through a population, a mucosal homogenate seeder pig model given to 10 week old pigs was evaluated in this study.[2]

The objectives of this study are to:

- Validate a seeder pig model by administering a high dose *Lawsonia intracellularis* mucosal homogenate disease challenge to 10 week old seeder pigs. Natural transmission to contact pigs would be measured over a 70 day time period.
- Compare *Lawsonia* fecal PCR's quantitatively in seeder pigs vs contact pigs. The fecal PCR test is the gold-standard ante-mortem diagnostic to identify both clinical and subclinical ileitis. Current laboratory technology allows for quantification of *Lawsonia* bacteria per gram of feces directly via qPCR or indirectly via ct (cycle time) values.
- Compare IHC (immunohistochemistry) analysis of intestinal PPE pathology in pigs euthanized at different time points over the 10 week study. IHC is the gold-standard confirmatory post-mortem ileitis diagnostic test with a reported sensitivity of 90% and specificity of 100%.[3]
- Compare IPMA (immunoperoxidase monolayer assay) *Lawsonia* antibody titers to their endpoints in both seeder and contact pigs.
- Compare the prevalence of clinical vs subclinical infected seeder vs contact pigs.

Materials and method

Ninety-six PRRS, *M. hyopneumonia*, and *Lawsonia intracellularis* negative, 10 week old feeder pigs weighing 55 to 74 lbs were randomized into four pens of twenty-four pigs per pen (Table 1).

Two pigs per pen were arbitrarily chosen to be *Lawsonia intracellularis* challenged "seeder pigs" on Day 0. The remaining 22 "contact pigs" per pen were randomly selected to be euthanized for intestinal PPE lesion scoring on Days 27, 43, 56, and 70 at which time the study was terminated.

Lawsonia challenge

A pre-quantified mucosal homogenate and SPG (sucrose potassium glutamate) concentrate previously stored at -70°C was diluted with SPG to provide a 50ml direct oral gavage containing 3.0×10^7 *Lawsonia intracellularis* per ml (19.52 ct/40 U of MN VDL). The total *Lawsonia* dose of 1.5×10^9 was administered to two seeder pigs per pen on Day 0.

Clinical observations and ante-mortem diagnostics

Clinical observations were taken once each week for pig Demeanor Score (0-3), Abdominal Appearance Score-AAS (0-3), Fecal Consistency Score-FS (0-3), and Fecal Color Score-FC (0-3) to determine the presence of clinical or subclinical ileitis during the ten week study.

Fecal samples for *Lawsonia* indirect quantitative PCR (ct values) were obtained via direct rectal digital palpation to obtain ≥ 2 grams of feces at eleven time points for the seeder pigs, nine time points for the Day 70 (D70) contact pigs, and on the day of euthanasia for the Day 27, 43, and 56 contact pigs.

IPMA antibody titers were assayed on Day 0, 21, 27, 34, 43, 56 and 70 for the seeder and D70 contact pigs and on the day of euthanasia for the others. IPMA titers were carried out to their final endpoints to determine the maximum titer level achieved at the U of MN VDL. Titers can range from a low of 1:30 to a high of 1:122,880.

Post-mortem diagnostics

Approximately 22 contact pigs were euthanized, weighed, and intestines scored for gross PPE Lesions (0-3) in the ileum, jejunum, cecum, and colon. PPE lesion length (cm) was measured and a final PPE Lesion Severity Score was obtained by multiplying the lesion score times the lesion length in each section of intestine and added together.

A 2 cm piece of terminal ileum just proximal to the ileo-cecal-colonic junction was obtained and fixed in 10% buffered formalin for IHC PPE Lesion Scoring (0-4) as follows - 0: no Lawsonia infected cells; 1: 1-25% heavily infected *Lawsonia* cells; 2: 26-50% heavily infected enterocytes, 3: 51-75% infected cells; and 4: 76-100% infected cells. Comments were also made by the pathologist if most infected cells were in the crypt enterocytes or in the lamia propria of the intestine. The latter is indicative of a more chronic infection.

Results and discussion

Ileitis mortality

One seeder pig died from peracute PHE on Day 18 post-challenge with melena (Fecal Score-3, Fecal Color-3) and approximately 152 cm of Score 1 (mild) PIA lesion in the ileum and jejunum. The loss of this seeder pig did not negatively reduce *Lawsonia* transmission within this pen because there was another seeder pig present and significant bacterial fecal shedding had already occurred prior to its death.

Fecal PCR: Seeder pigs

In general the seeder pigs, which were directly challenged, showed more clinical signs of ileitis, slightly higher levels of fecal bacteria shed and much higher IPMA titers compared to the D70 contact pigs. Figure 1 compares the % fecal PCR positive seeders vs contact pigs and the mean ct value of seeders vs contacts on each collection day. Already by Day 11 PI (the first collection day for seeder pigs), seeders were shedding high levels of 18.32 to 24.15 ct 40. This PCR ct value corresponds to 6.14×10^7 to 1.75×10^6 *Lawsonia* per gram of feces (see Table 2 *Lawsonia* Quant Reference Chart) It is possible that seeder pigs were shedding already by Day 2 or 3 post-challenge as shown in other studies.[4]

All of the seeders were fecal PCR positive out to Day 43 PI. Thereafter they gradually decreased shedding so by Day 70 PI only one pig was still shedding. Peak shedding (low ct values) occurred in 6 of 7 seeders already between Day 11 and 21 PI. However, Pig 1-23 was a persistent "super shedder" shedding very high levels of *Lawsonia* out to Day 63 (ct 17.18 to 23.22). It was the only pig still fecal PCR positive (ct 34.75), IHC 2^+ positive, and with an IPMA titer of 61,440 on Day 70. Clinically, this

Table 1:

Pen #	# Seeders/pen	# Contact pigs/pen	Pig age	Challenge administered	Li dose
1	2	22			
2	2	22			
3	2	22	10 wks	Oral Gavage to 2 Seeders	≥ 10^9
4	2	22			
Total pigs	**8**	**88**			

Figure 1: % Fecal PCR positive and *Lawsonia* ct value in seeders vs. D70 contact pigs

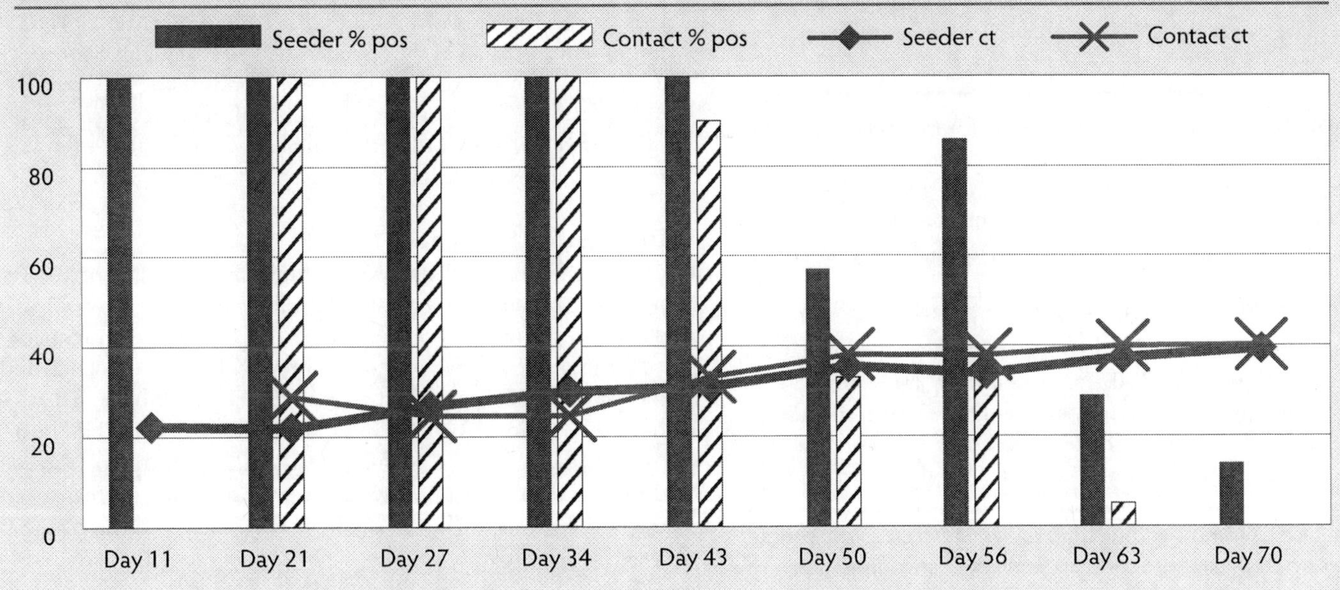

Table 2: *Lawsonia* Quant Reference Chart

Ct from real time	*Lawsonia*/ml
15	4.44E+08
16	2.42E+08
17	1.32E+08
18	7.19E+07
19	3.92E+07
20	2.14E+07
21	1.16E+07
22	6.35E+06
23	3.46E+06
24	1.88E+06
25	1.03E+06
26	5.60E+05
27	3.05E+05
28	1.66E+05
29	9.06E+04
30	4.94E+04
31	2.69E+04
32	1.47E+04
33	7993.0
34	4356.0
35	2374.0
36	1293.8
37	705.1
38	384.3
39	209.4
40	114.4

* U of M VDL Revision 9-17-14

pig had diarrhea and abdominal AAS for five weeks. By Day 63 it was clinically normal, however, its body weight was only 102 lbs-83 lbs below the mean B.W. of the other seeder pigs. It is an example of a severe PPE chronically infected pig with extensive PIA lesions that eventually became a normal, but stunted pig.

Fecal PCR – D70 contacts

Individual fecal and blood samples were taken weekly starting at Day 21 PI from the D70 Contact pigs. Figure 1 shows all contact pigs positive by Day 21 and through Day 43, then shedding decreased with only 1 pig shedding on Day 63. These contact pigs shed high levels of bacteria comparable to the seeders out to Day 34 with a mean ≤ 25 ct/40 level (1×10^6 per gram). Assuming the seeders were shedding high *Lawsonia* levels to the contact pigs by Day 7 and the contacts were likely shedding high levels by Day 14, then seeder pigs in this study shed *Lawsonia* for 5 to 9 weeks and contact pigs shed for 4 to 8 weeks.

IPMA titers in seeders and D70 contacts pigs

Seroconversion generally follows high dose *Lawsonia* exposure by 2-3 weeks. Routine laboratory IPMA serology usually is limited to positive/negative reporting, or if titers are requested they are plated out from 1:30 to 1:960. However, in this trial titers were taken to their final endpoints to see the extent of the IgG response. The antibody decay half-life is about 21 days, so in severely infected PPE pigs, titers could theoretically last up to 36 weeks post-exposure. It should be noted that the antibody titers are not correlated with fecal PCR shedding or presence of IHC lesions. Higher titers are not correlated with fecal shedding or lesions. However, pigs exposed to natural *Lawsonia intracellularis* infection are presumed to be protected for life. Therefore, the presence of a titer indicates prior exposure and protection from a re-break from subsequent ileitis exposure.

Table 3 compares the mean IPMA titers and % sero-positive seeder vs contact pigs from trial Day 21 to Day 70. Seeder pig titers were already high by Day 21 and stayed high throughout the study. They ranged from 480 to 61,440 on Day 70 post-challenge. Overall the contact pigs' titers were not as high as the seeder pigs. Seroconversion was just beginning on Day 21 with 36% negative and 64% (14 of 22) positive at a 1:30 titer. All but one pig seroconverted by Day 34 and titers ranged from 60 to 7,680 on Day 70.

Clinical/subclinical seeders and contact pigs

All but one of the seeder pigs had a clinical positive fecal diarrhea score and/or an abnormal abdominal appearance score on at least one day during the study. Therefore, there was only one subclinical seeder pig in the study. A subclinical pig is (strictly defined by the author) a pig with any clinical observation of an abnormal FS, AAS, or FC Score throughout the entire study. Another more general definition of subclinical pig is a pig without clinical diarrhea but having a positive fecal PCR, serologically titer, or IHC lesion score. Also, 3 of the 8 seeders had a Fecal Color Score of 2 or 3, indicative of varying degrees of bloody feces. Conversely, 40% of the D70 contact pigs were subclinical carrier pigs showing no positive FS, AAS, or FC clinical observation. Table 4 shows the Clinical Observations by trial day on the D70 contact pigs.

PPE lesion IHC score

Typically in pigs challenged with a high *Lawsonia* dose of 10^8-10^9 per pig, gross and histological PPE lesions will peak about 21-28 days post-challenge. Thereafter, lesions dissipate as the pigs natural immunity takes over. Figure 2 shows % IHC positive pigs and their mean IHC Score (0-4) on each respective euthanasia day. On Day 27, ninety-six percent (22 of 23) of pigs were IHC positive-all with 2^+-4^+ IHC Scores (mean = 2.57). The graph shows that both the number positive and the lesion scores decrease over the next 4 weeks to Day 56 and by Day 70 there were no pigs that were IHC positive.

Table 3: Mean IPMA titer and % sero-positive seeder vs. D70 contact pigs*.

	Day 0	Day 21	Day 27	Day 34	Day 43	Day 56	Day 70
Seeders (n=17)	Neg	1170 (100)	5706 (100)	468 (100)	1295 (100)	1170 (100)	1427 (100)
Range	--	480 - 3840	1920 - 7680	1920 - 7680	480 - 3840	480 - 3840	480 - 61,440
Contact (n=22/21)	Neg	23 (64)	37 (71)	512 (95)	530 (95)	646 (95)	870 (100)
Range	--	30-30	30 - 960	30 - 7680	60 - 7680	60 - 7680	60 - 7680

* % Positive in ( )

Table 4: Clinical observations by day on day 70 contact pigs.

	Fecal score*	AAS*	Fecal color
Day 21	2	0	0
Day 25	4	0	2
Day 39	2	5	0
Day 41	7	0	1
Day 48	1	0	0
Day 63	0	1	0
Day 70	0	0	0

* AAS=Abdominal appearance score
Fecal Score = Diarrhea (0-3)

Table 5 summarizes all parameters measured on all individual Day 27 pigs. The high prevalence and severity of fecal PCR levels, IHC scores, and gross PPE Severity Scores in the Day 27 contact pigs indicates that this model was very effective in natural transmission of *Lawsonia intracellularis* from seeder to contact pigs. All of the pigs have ct values between 18.19 to 31.86. Significant IHC lesion scores of 2^+ to 4^+ were present in all but one pig. The PPE Gross Severity Score had four pigs with very extensive severe PIA gross lesions for many feet of the ileum and jejunum with PPE Severity Scores from 300 to 930 in these pigs. However, note that 61% of the pigs on Day 27 are subclinical ileitis with no evidence of diarrhea (FS), gauntness (AAS), or abnormal fecal color, yet all of them are shedding high numbers of *Lawsonia* and all but one with confirmed IHC positive lesions. This is another illustration that subclinical ileitis is highly prevalent and greatly contributes to the economic cost of this disease.

Summary

This Seeder Pig Model was shown to effectively transmit *Lawsonia intracellularis* from seeder to contact pigs as measured by the prevalence and severity of Fecal PCR shedding, gross PPE lesions, IHC lesion scores, and IPMA serologic titers. The ratio of two seeder pigs to twenty-two contact pigs (9%) was used with a high *Lawsonia* mucosal homogenate dose of 1.5×10^9 given per seeder pig via direct oral gavage on Day 0. The clinical disease course, fecal PCR transmission, and intestinal PPE pathology was followed out to the disease endpoints until all pigs were negative for these parameters (except 1 seeder pig) at Day 70.

Implications and future research

1. Expanding the use of a validated Seeder Pig Models could enhance the study of the transmission, epidemiology, and pathogenesis of this little understood and understudied disease.
2. Utilization of quantitative fecal PCR's and IPMA serology with titers could help swine practitioners with ante-mortem ileitis diagnosis and interpretation.
3. Utilization of this Seeder Pig Model could allow for natural transmission of *Lawsonia intracellularis* in larger populations of grow-finish pigs at a lower cost to study the epidemiology of ileitis in market pigs.

Acknowledgements

The author would like to thank Merck Animal Health and specifically Drs. Robyn Fleck and Brad Thacker for their help with this project.

Figure 2: % IHC positive and mean IHC score.

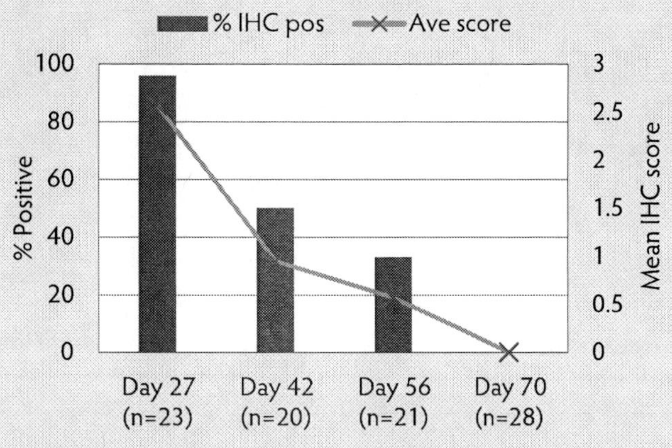

References

1. Winkelman NL, Hawkins PA, Evaluation of carbodox and neomycin-oxytetracycline for control of proliferative enteropathy (Ileitis) in swine. Proceedings of the 14th IPVS Congress Bologna Italy 1996 p278

2. Winkelman unpublished

3. Guedes, R. Update on epidemiology and diagnosis of porcine proliferative enteropathy. Journal of Swine Health and Reproduction. 2004; Vol. 12, Number 3:134-138

4. Beckler DC, Hammer JM, etal Spread of *Lawsonia intracellularis* in a pure culture seeder pig challenge study AASV Proc. 2011, p 51-55

Table 5: Day 27 euthanized contact pigs.

ID		BW day 0	BW Day 27	ADG	IPMA result		Fecal PCR result		IHC	PPE gross severity score	Clin Obs* Day 27
1	4	61.4	111.7	0.72	POS	480	POS	18.57	2	870	0-2-2
1	24	74.1	130.0	0.80	POS	30	POS	31.86	2	51	2-2-0
1	38	70.5	114.9	0.63	POS	240	POS	20.41	4	550	0-2-0
1	45	67.4	83.8	0.23	POS	30	POS	26.58	2	0	0-0-0
1	54	70.7	128.7	0.83	POS	240	POS	19.04	2	11	0-0-0
1	57	59.3	130.7	1.02	POS	240	POS	21.52	4	63	0-0-0
2	61	72.3	110.7	0.55	POS	60	POS	21.11	2	22	0-0-0
2	71	63.6	130.3	0.95	POS	60	POS	26.56	2	0	0-0-0
2	77	57.2	116.9	0.85	POS	120	POS	20.12	3	24	0-2-0
2	84	60.6	112.2	0.74	POS	30	POS	20.78	3	7	0-1-0
2	95	62.6	120.7	0.83	POS	30	POS	22.70	3	18	0-0-0
3	6	59.0	124.2	0.93	POS	30	POS	26.38	2	6	0-1-0
3	10	67.4	125.1	0.82	POS	30	POS	25.22	3	15	0-3-0
3	14	65.8	117.6	0.74	NEG	15	POS	26.45	2	0	0-0-0
3	21	64.6	109.3	0.64	POS	480	POS	23.38	2	0	0-0-0
3	47	61.0	123.9	0.90	POS	30	POS	22.81	4	13	0-0-0
3	49	61.8	111.8	0.71	POS	120	POS	26.65	2	10	0-0-0
4	7	62.0	116.5	0.78	POS	1920	POS	24.61	3	17	0-0-0
4	11	65.7	112.9	0.67	POS	240	POS	18.19	4	930	0-3-2
4	25	59.9	110.7	0.73	POS	120	POS	21.83	2	18	0-0-0
4	50	59.9	119.9	0.86	POS	60	POS	22.88	0	20	0-0-0
4	82	60.4	108.4	0.69	POS	120	POS	22.22	3	31	0-0-0
4	96	58.3	92.5	0.49	POS	7680	POS	23.54	3	300	1-2-0
Summary		**63.72**	**115.80**	**0.74**	**21/21**	**113.0**	**5/21**	**23.19**	**2.57**	**129.4**	**2/23-9/23-2/23**
									22/23	**19/23**	

* Clinical observations: AAS-FS-FC

A new PRRSV ELISA, highly correlated with IPMA

Eric van Esch, DVM, Technical Director
BioChek BV Reeuwijk, Netherlands

Introduction

There is a general belief that PRRSV ELISA's are showing a long response after infection or vaccination while IPA/IFA are showing a shorter response and therefore more preferable to show dynamics in herds. This paper describes the new BioChek PPRSV ELISA which is strong correlated with IPMA/IFA.

Materials and methods

Sera

A panel of 91well defined sera with known IFA and competitor ELISA results provided by the MSD Service Lab in Boxmeer, NL were used.

ELISA

The BioChek Porcine Respiratory and Reproductive Syndrome Virus Antibody Test Kit was used and performed according to the manufactures kit insert.

Software

To calculate correlation and R^2 Microsoft Excel 2010 was used. For correlation log2 titers of the ELISA and IPMA were used.

Results

Results are summarized in Table 1 and Figure1.

The correlation based on log 2 titers for both IFA and BioChek PRRSV ELISA was calculated as 0.982, R^2 was 0.58.

Table 1 and Figure 1 clearly show the high correlation of the BioChek ELISA with IFA and the low correlation of Brand X with IFA.

Discussion

ImmunoPeroxidaseMonolayerAssay (IPMA) or ImmunoFluorescenseAssay (IFA) are considered to be the Gold Standard for PRRSV serology and are still used to show herd dynamics. But IPMA or IFA are laborious and therefore not always available in laboratories. A PRRSV ELISA which can replace the IPMA/IFA will be in favor to show herd dynamics.

Lopez et.al. described two main different classes of antibodies, the total antibodies and the neutralizing antibodies. It is important to distinguish between the two classes, the neutralizing antibodies are considered to provide protective immunity. The best way to detect neutralizing antibodies is by IPMA/IFA or SN.

Neutralizing antibodies are shorter lasting than total antibodies. Under field conditions this is important to recognize, as neutralizing antibody response correlate with protection and as presence of neutralizing antibodies correlates with a recent infection.

Since the BioChek PRRSV ELISA is showing a strong correlation with IFA and a better discrimination of between total antibodies and neutralizing antibodies the BioChek PRRSV ELISA could be more in favor to overwatch herd dynamcis and protective immunity in herds.

Conclusion

The BioChek PRRSV ELISA correlates well with IPMA/IFA and therefore correlates well with neutralizing antibodies. The BioChek ELISA can be used to detect neutralizing antibodies.

Acknowledgements

BioChek likes to thanks MSD for kindly providing the sera.

References

1. O.J. Lopez et.al, (2004), Vet. Immunology and immunopathology, 102, 155-163.

Table 1: BioChek ELISA vs IPMA vs competitor ELISA (Brand X).

Result	BioChek PRRSV ELISA	IFA	Competitor ELISA (Brand X)
Positive	24	16	91
Negative	67	75	0

Figure 1: Correlation between BioChek PRRSV ELISA and Gold Standard test (IFA)

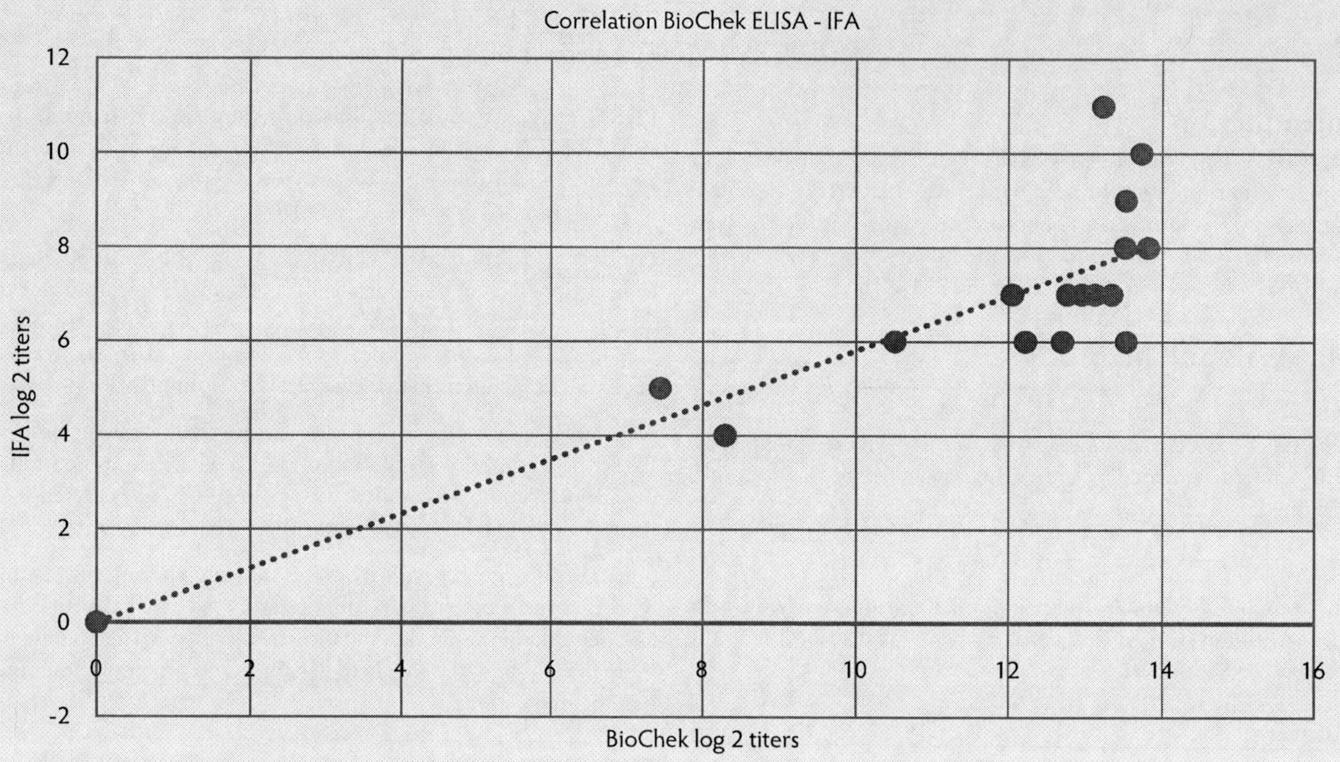

Study and field evaluation of an artificial intelligence system's ability to predict the good moment to breed sows: PigWatch® from Ro-Main

J. Labrecque[1], B. Eng.; J. Rivest[1], PhD; C. Klopfenstein[2], PhD, DVM
[1]Conception Ro-Main inc., St-Lambert-de-Lauzon, Quebec, Canada;
[2]Centre de développement du porc du Québec inc. (CDPQ), Quebec City, Quebec, Canada

Introduction

In order to have optimal reproductive performances, sow insemination has to happen within a relatively short time in the sow's estrus period. Studies have shown that the ideal time is approximately in the twenty-four hours before ovulation.[1-4] For the producer this ideal time is entirely theoretical because of the high level of difficulty to monitor the ovulation. What we do know is that ovulation occurs consistently around the two thirds through the estrus[5] and that the estrus period is characterized by distinctive symptoms such as responsiveness to boar exposure. The estrous usually lasts between 40 and 60 hours but it can be as short as 24 hours, or even extend up to 96 hours.[6] In short, it is nearly impossible for managers of farrowing operations to inseminate sows at the right moment in their estrus period.

To ensure fertilization of each sow, researchers often recommend doing two inseminations per estrus. Producers adapt these recommendations and develop various procedures that often require two and sometimes three or four inseminations per estrus period.[7] Generally, a first insemination is performed at the first behavioral symptoms associated with estrus, followed by another insemination every following 24-hour period up to estrus cessation. This method gives good results, provided that good estrus detection has taken place. For this reason, it requires the presence of skilled employees as well as the use of multiple doses of semen.

Ro-Main's PigWatch® system is an artificial intelligence innovation which goal is to predict the best time to inseminate weaned sows, based on real-time analysis of sow behavior in their stall. Product development has been achieved mainly through measurement of sow behavior before and during estrus, the development of electronic modules, and the design of artificial intelligence algorithms to analyze behavioral data and predict the best time for insemination. PigWatch® 2016 thus aims to inseminate each sow only once and at the best moment relative to ovulation.

The following sections present the results of a joint study made by the Centre de développement du porc du Québec inc. (CDPQ) and Conception Ro-Main inc. that aimed to assess the reliability of the PigWatch® (2016 Beta) system's insemination planning recommendations. Recent field results from commercial farms equipped with the PigWatch® 2016 system will also be presented and discussed.

Description of PigWatch® 2016

PigWatch® 2016 is a computerized AI management system designed to predict the best time to inseminate recently weaned sows. The PigWatch® 2016 system consists of motion sensors installed on top of every weaned sow's stall, plus a data analysis module and a computer software to interface with the producer. Figure 1 shows the architecture of the PigWatch® 2016 system and Figure 2 shows an actual PigWatch device in the farm.

The motion sensors allow constant and non-intrusive monitoring of sow behavior by assessing its real-time level of activity. The monitoring unit including the sensors is installed directly above each sow. A display system consists of LED's that indicate the sow's heat status, if it has already been inseminated, and whether to do the AI at that precise moment or not. The data analysis module is a computer that analyses the real-time behavioral data of each sow. It contains the artificial intelligence algorithm that is at the heart of the system and that performs complex calculations to determine the best time to do the insemination. The PigWatch® 2016 software provides a user interface that shows the information as graphs detailing the progression of each sow during estrus and as reports with the list of sows to be inseminated. Figure 3 shows an example of a PigWatch® 2016 behavioral graph in which the best moment to breed is indicated as a vertical line at 7:00 on November 11th.

The PigWatch® 2016 algorithm was optimized to predict the ideal time for insemination of sows, while taking into account the presence of farm workers in the barn. The aim of the system is to predict a single best time to inseminate each sow. Even so, very few operations employ farm workers around the clock during insemination periods. For this reason, at the end of each day, just before the farm staff leave the premises, PigWatch® 2016 tries to predict which sows should be inseminated in anticipation of having their optimal insemination time during the night. The next morning, the system evaluates the need to repeat the insemination if it judges that optimal insemination time did not occur overnight, as expected. Consequently, some sows will have two PigWatch® AI requests and be inseminated twice.

Figure 1: PigWatch® 2016 system architecture

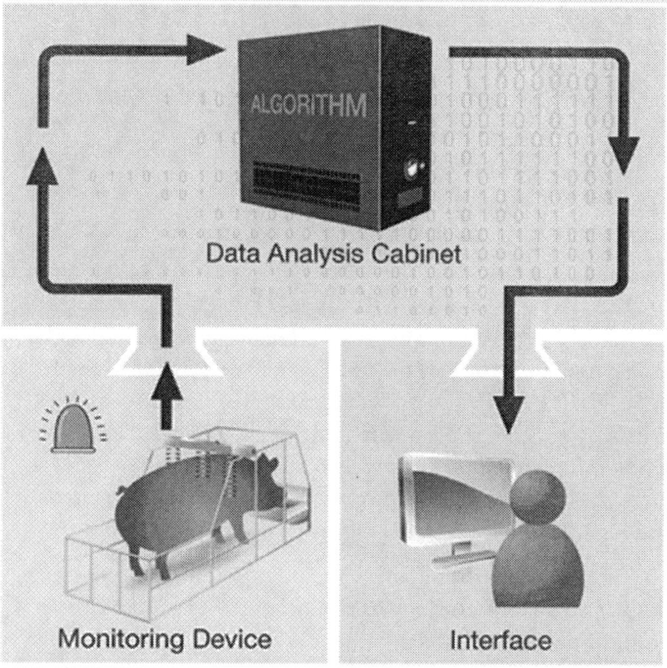

Figure 2: PigWatch® device above sow stall

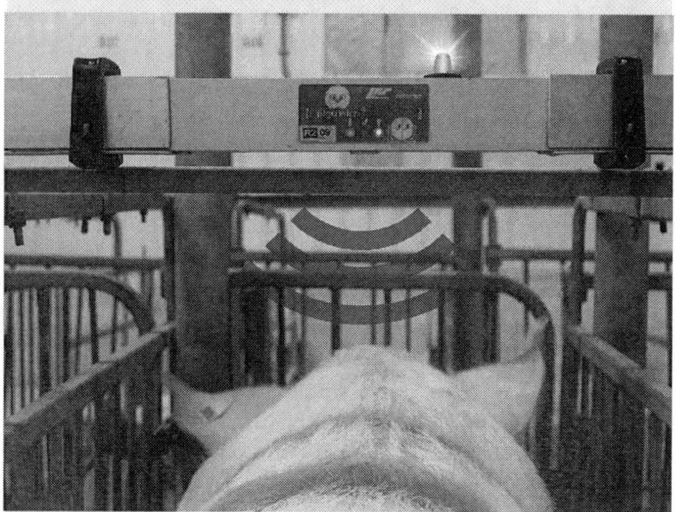

Study

Materials and methods

This study was conducted from September 2014 to March 2015. The project was carried out in a commercial breeding-gestation-farrowing facility, with a holding capacity of 1,600 sows (Genetiporc, Landrace × Large White), at Ferme A.G & R. Labrecque inc. in St-Bernard-de-Beauce, Quebec. This facility operates in four-week batches and weans piglets over three days (Wednesday, Thursday and Friday).

For each batch involved in the project, 39 of the 90 sows weaned on Wednesday morning were randomly selected and routed into the gestation room assigned to the project. Of the 39 selected sows in every batch, between 20 and 30 were chosen so that the necessary human interventions in the room were limited to one hour in the morning and one hour in the afternoon. Since the PigWatch® system monitors the natural behavior of the sow, it is advisable to avoid disturbing the sows as much as possible so that PigWatch® can analyze noise-free data.

Blood samples were taken from all eligible sows selected for the project (20 to 30 sows per batch) with the aim of establishing time of ovulation for each sow by analysis of the serum hormones, progesterone and estradiol.

The authors hypothesized a priori that the PigWatch® system issues a request for insemination before ovulation and that these requests are based on the analysis of sow behavior. Therefore, by sampling blood at the PigWatch® insemination request, it would be possible to confirm the decrease in serum estradiol and detect the increase in serum progesterone, the indicator retained to identify time of ovulation. To obtain sufficient data to estimate the time of ovulation from the serum levels of both estradiol and progesterone but to limit the disturbance to the sows, a minimum of five blood samples per sow were taken over a 48-hour period, at a rate of two blood samples per day.

Sows were fed twice daily on a fixed schedule (5:30 am and 2:00 pm) and farm workers minimized their daily activities in the room to reduce interference with the sows as prescribed by the PigWatch® protocol. Blood sample collection began immediately after feed distribution in the morning and in the afternoon.

All inseminations were post-cervical AI with 2 billion sperm cells per dose.

The total number of piglets born and sow fertility at 35 days served as indicators of reproductive performance. Sow fertility was assessed by transabdominal ultrasonography (diagnostic imaging) between 35 and 40 days post-insemination.

Results and discussion

122 sows were selected for analysis (20 sows in Batch 1, 22 in Batch 2, 27 in Batch 3, 23 in Batch 4, and 30 sows in Batch 5) and the time of ovulation was estimated from a statistical analysis of the temporal variation of the serum estradiol and progesterone levels. This time was chosen as the mid-point of the most likely period of ovulation as determined by the analysis.

The 122 sows were inseminated on average 1.16 times following requests from PigWatch®, and they had a 95.1% fertility rate and 15.29 piglets born per litter.

Figure 3: Example of a PigWatch® 2016 behavior graph

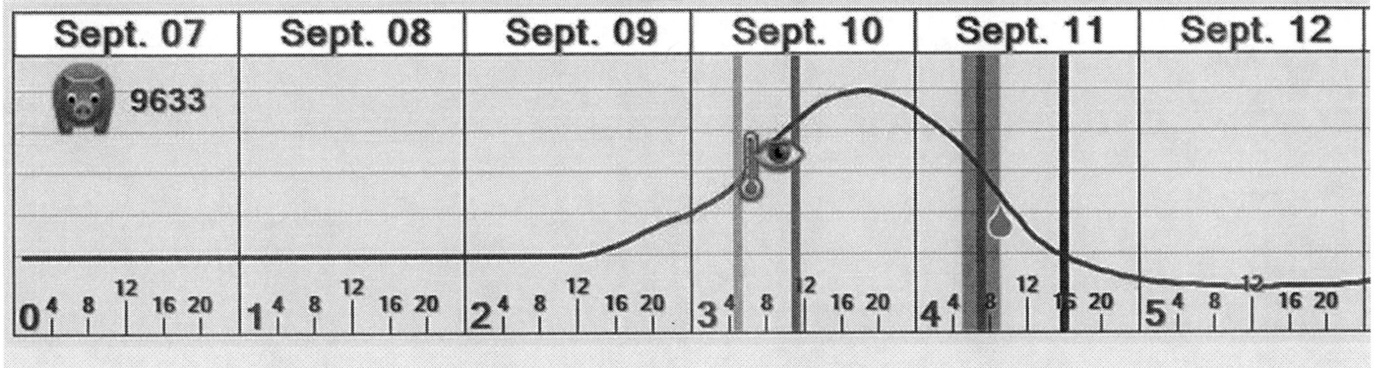

The sows were inseminated an average of 10.1 hours before ovulation; with a standard deviation of 10.8 hours (between 41 hours before and 11 hours after ovulation). Median is 10.65 hours before ovulation and 95% of insemination requests are between 31.5 hours before ovulation and 8 hours after. Figure 4 shows the distribution of the sows per the time interval between their PigWatch® insemination request and their ovulation time as estimated by the above-mentioned method. Time zero refers to the time of ovulation and negative time intervals mean that the PigWatch® insemination request was before the estimated ovulation time. Table 1 shows the results for mean total born piglets and gestation rates by group of eight hours.

Statistical analysis did not enable identification of any linear ($P = 0.7378$), quadratic ($P = 0.3379$), or cubic ($P = 0.1757$) relationship between gestation rate and the interval between the PigWatch® request and ovulation.

Statistical analysis identified a significant quadratic relationship ($P < 0.05$) between the number of total born and the time interval between the PigWatch® request and ovulation. This relationship indicates that the maximum number of total born is obtained when the PigWatch® request occurs 12.7 hours before ovulation.

Field evaluations

At the time of writing this paper, two 1600-sow units in Canada (Canada 1 and Canada 2) and one 200-sow unit in Belgium (Belgium) are equipped with PigWatch® 2016. The first 1600-sow unit in Canada is a 4-week batch farrowing operation installed with PigWatch® 2016 since July 2016. All the sows of this farm are not currently managed by PigWatch® 2016 because of the way the rooms are separated in the farm. The second 1600-sow unit in Canada is also a 4-week batch farrowing operation installed with PigWatch® 2016 since October 2016. The farm in Belgium is a 5-week batch farrowing operation using the complete PigWatch® 2016 method since September 2016. Table 2 shows the results of these farms with PigWatch® 2016 so far. The number of sows on PigWatch® is usually smaller at the beginning and increases over time as part of the PigWatch® integration and training process.

Canada 1 had a decrease in fertility between its first and second batch with PigWatch® 2016. Mycotoxins were identified as a potential source for many problems in the farm during that period including an abnormal number of abortions, low fertility in gilts, etc. The fertility results increased gradually after the problem occurred but the problem was still not completely controlled in November 2016.

The farm in Belgium had an average fertility rate of 86.9% in the five batches before the installation of PigWatch® 2016 (N = 259). The first fertility results with PigWatch® are encouragingly higher at 89.7%. These results were obtained with 1.6 inseminations per sow, which is higher than what is normally expected from PigWatch® 2016. This high number of inseminations can be explained by the system's difficulty to analyze the data that reflect disturbances to the sows. In fact, the PigWatch® 2016 integration technician identified a source of stress that affected several sows, thus affecting the quality of the data. That source of stress will be eliminated for the subsequent batches. This example shows the importance of strictly following the PigWatch® 2016 protocol to optimize the results.

A total of 1065 sows have been inseminated with PigWatch® 2016 in commercial farms so far with an average of 1.26 insemination per sow and the average fertility rate is 91.3%. Although no statistical analysis can be performed on these numbers because of the absence of controls, these results demonstrate that using PigWatch® in commercial setting can give good reproduction performance with fewer insemination doses per sow.

Conclusion

The results of the above-mentioned study show that the PigWatch® system makes insemination requests in a period around ovulation that is consistent with known information and is conducive to obtaining good reproduction performances. Although limited at time of writing this paper, the available commercial data tend to confirm that PigWatch® 2016 can

Figure 4: Distribution of the sows per the time interval between their PigWatch® insemination request and their ovulation time, where 0 is the estimated time of ovulation

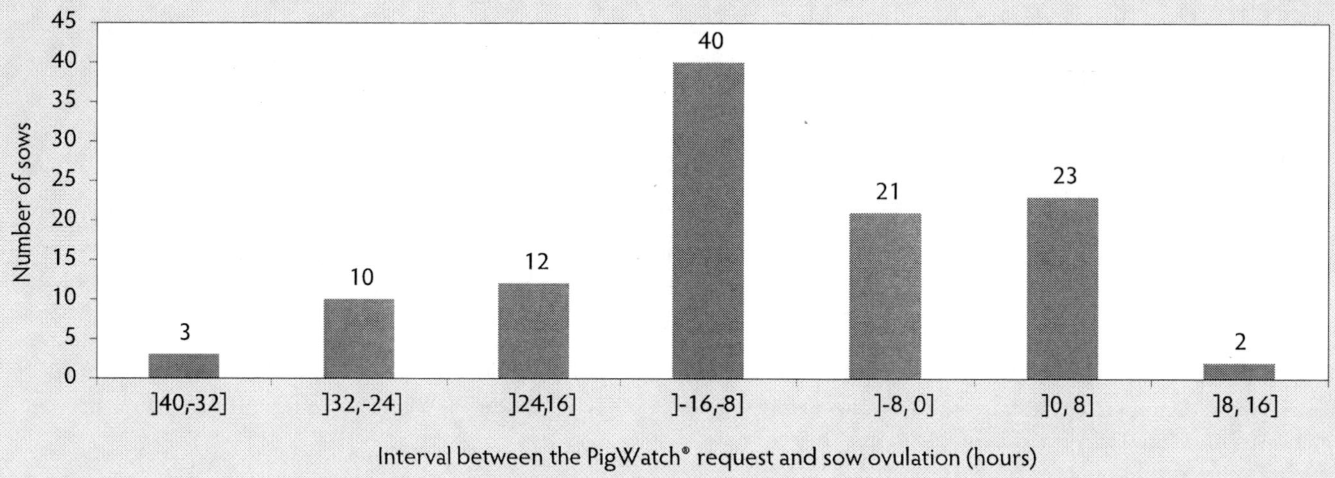

Table 1: Results for mean number of total born piglets and gestation rate by group of eight hours

Groups	Number of sows	Total born piglets	Fertility rate
]-40, -32]	2	10.50	100.0%
]-32, -24]	13	15.10	84.6%
]-24, -16]	12	15.18	100.0%
]-16, -8]	45	16.09	95.0%
]-8, 0]	22	16.24	100.0%
]0, 8]	24	13.74	95.8%
]8, 16]	3	15.50	66.7%

Table 2: Recent reproduction results of the farms equipped with PigWatch® 2016

Farm	Date	N	Dose/sow	Fertility rate	Total born piglets
Canada 1	07-2016	149	1.29	95.97%	15.37
Canada 1	08-2016	192	1.26	89.06%	N/A
Canada 1	09-2016	191	1.26	90.05%	N/A
Canada 1	10-2016	174	1.21	91.38%	N/A
Canada 1	11-2016	179	1,18	N/A	N/A
Canada 2	10-2016	141	1.26	N/A	N/A
Belgium	09-2016	39	1.62	89.7%	N/A

indeed yield good reproduction performances by targeting the good moment to breed sows. These findings indicate that it is possible to determine a good time to inseminate sows from the analysis of sow behavior with an artificial intelligence algorithm, as is done by the PigWatch®
2016 system.

Inseminating sows once at the right moment has the potential to improve the fertility rate and the number of total born piglets, while cutting nearly in half the number of insemination doses. Using less sperm cells per fertilization optimizes the use of the best boars and has the potential to increase the genetic improvement rate.

Acknowledgements

Portions of this work were funded by Agriculture et Agri-Food Canada through the Agri-Innovation Program and Conception Ro-Main inc. The authors would like to acknowledge Walter Signorini, Dr. Marie-Claude Poulin, and all CDPQ and Conception Ro-Main staff who contributed to these projects.

References

1. Soede, N.M., Wetzel, C.C., Zondag, W., de Koning, M.A. and B. Kemp. 1995. Effects of time of insemination relative to ovulation, as determined by ultrasonography, on fertilization rate and accessory sperm count in sows. Journal of Reproduction and Fertility, 104(1): 99-106.

2. Nissen, A.K., Soede, N.M., Hyttel, P., Schmidt, M. and L. D'Hoore. 1997. The influence of time of insemination relative to time of ovulation on farrowing frequency and litter size in sows, as investigated by ultrasonography. Theriogenology, 47: 1571-1582.

3. Almeida, F.R.C.L., Novak, S. and G.R. Foxcroft. 2000. The time of ovulation in relation to estrus duration in gilts. Theriogenology, 53(7): 1389–1396.

4. Hughes, P.E. and G. Pope. 2001. Estimating the optimum timing of insemination in pigs. Manipulating pig production VIII: 166-172.

5. Terqui, M., Guillouet, P., Maurel , M.C. and F. Martinat-Botté. 2000. Relationship between peri-estrus progesterone levels and time of ovulation by echography in pigs and influence of the interval between ovulation and artificial insemination (AI) on litter size. Reproduction, Nutrition, Development, 40: 393-404.

6. Soede, N.M. and B. Kemp. 1997. Expression of estrus and timing of ovulation in pigs. Journal of Reproduction and Fertility. Supplement, 52: 91-103.

7. Poilvet, D. 2003. Reproduction des porcs - Inséminez vos truies au bon moment. Available at: http://porc.reussir.fr/actualites/reproduction-des-porcs-inseminez-vos-truies-au-bon-moment:22836.html. Accessed 14 November 2016.

Fat-soluble vitamins: To esterify or not to esterify?

Robert L. Stuart, PhD

Stuart Products, Inc., Bedford, Texas

Introduction

Fat-soluble vitamins A, D, E and K are vital nutrients for all swine. They are routinely added to swine feeds at or above NRC recommended levels. Supplementation via drinking water and injection is also utilized. Vitamins A, D and E are classified as alcohols and vitamins K_1, K_2, and K_3 are classified as quinones. To improve stability of fat-soluble vitamins A and E, esters are produced. Esterification is the combination of an acid and alcohol. Vitamin A (retinol) and vitamin E (alpha-tocopherol) are esterified by the addition of a fatty-acid and vitamin D (cholecalciferol) is typically not esterified. Vitamin K feed supplements are usually salt complexes. Table 1 shows the forms and sources of fat-soluble vitamins presently used in swine feeds, water and injections.

Both naturally-occurring and supplemental fat-soluble vitamins go through various steps prior to absorption that is similar to fat metabolism. Prior to absorption, supplemental fat-soluble vitamins A and E need to be de-esterified by the action of intestinal esterases (pancreatic carboxyl ester hydrolase (CEH).[1] The esters are removed from the vitamin molecules prior to absorption. A second step for fat-soluble vitamins is micellization. Micelles are formed by the action of bile salts to create water-soluble droplets that are absorbed. Supplementing water with micellized, non-esterified vitamin E (d-alpha-tocopherol) assists in maintaining adequate vitamin E status of the post-weaned pig.[2] Stuart et al., 2015, reviewed surveys of fat-soluble vitamin status during various stages of production.[3]

The purpose of this paper is to review studies that showed lack of animal responses when vitamin E-acetate ester was added to feed and or injectable products.

After weaning vitamin E status in pigs diminishes dramatically

Several experiments have shown that after weaning, vitamin E status drops dramatically to a sub-optimum level below 1 µg/mL.[4-6] Regardless of level of supplementation of synthetic vitamin E-acetate, pigs were not able to maintain an adequate level of serum alpha-tocopherol. When vitamin E was supplemented in a micellized, non-ester form of vitamin E, pigs were able to maintain adequate serum alpha-tocopherol levels above 3 µg/mL.[2] These studies have shown that the best form of vitamin E to administer to newly-weaned pigs is the non-esterified source of vitamin E. van Klapen et al. reported that the bioavailability for oral *d,l-α*-tocopheryl acetate was 12.5%, whereas the area under the curve was only 5.4%. Vitamin E dosed as d-*α*-tocopherol at one-eighth the level of the oral source showed both a rapid and a very slow pool, whereas orally dosed vitamin E shows a single slow pool (Figure 1). The authors stated that the use of vitamin E-acetate brings into question the efficacy of typical use of vitamin E-acetate in swine diets for alleviating oxidative stress.[6]

In a recent experiment at Michigan State University, supplementing the water of newly-weaned pigs with micellized non-esters of *d*-alpha-tocopherol and cholecalciferol reduced oxidative stress in pigs. The enzymes superoxide dismutase (SOD) and glutathione peroxide (GPx) were determined in newly-weaned pigs supplemented with vitamin E and D in drinking water (EMCELLE E-D_3 Liquid). SOD was reduced ($P = 0.005$) and GPx activity was reduced due to the water supplementation of vitamins E and D. ($P = 0.0002$).[7]

Vitamin E-acetate ester appears to be the most critically under-utilized source of vitamin E by newly-weaned pigs. Lack of esterase enzymes for de-esterification and lack of bile salts for micellization

Table 1: Forms and sources of fat-soluble vitamins used in swine feeds, water and injections.

Vitamin	Feed	Water	Injection
Retinol (Vitamin A)	Retinyl-acetate Retinyl-propionate Retinyl-palmitate	Retinyl-acetate Retinyl-propionate Retinyl-palmitate	Retinyl-propionate Retinyl-palmitate
Cholecalciferol (Vitamin D)	Cholecalciferol(D_3) Ergocalciferol (D_2)	Cholecalciferol(D_3) Ergocalciferol (D_2)	Cholecalciferol(D_3)
Alpha-tocopherol (Vitamin E)	*d-α*-tocopheryl-acetate *d,l-α*-tocopheryl-acetate	*d-α*-tocopherol *d,l-α*-tocopheryl-acetate	*d-α*-tocopherol *d-α*-tocopheryl-acetate
Vitamin K	MSBC MPB	MSBC	Vitamin K_1

Figure 1: Modeled plasma concentrations of deuterated alpha-tocopherol after oral consumption of deuterated *d,l*-alpha-tocopheryl-acetate or intravenous administration of one-eighth the concentration of deuterated *d*-alpha-tocopherol (Van Kempen, et al. 2016)

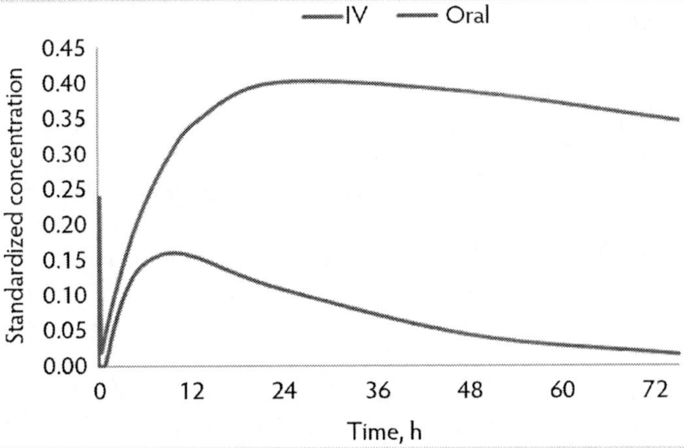

appear to be the reasons newly-weaned pigs are unable to utilize feed-grade vitamin E acetate. When d-α-tocopherol is administered either in drinking water or by injection, pigs' adequate vitamin E status is maintained. Injectable sources of fat-soluble vitamins are utilized approximately two times greater when compared to an equal oral dose.[8]

Vitamin D is typically supplemented in a non-esterified form

Due to lack of bioavailability of vitamin D esters, vitamin D is typically supplemented as the non-ester. Feed grade vitamin D is protected in a starch-gelatin matrix with added antioxidants. Liquid vitamin D products are also stabilized by the inclusion of antioxidants. EMCELLE TOCOPHEROL and EMCELLE E-D_3 Liquid both contain non –esters of vitamin E and vitamin D.

Some injectable vitamin A products contain the incorrect ester

Feed and liquid sources of vitamin A are esterified with either acetate, propionate or palmitate to improve stability. When fed, all are de-esterified in the small intestine and absorbed as retinol. After absorption, the retinol is re-esterified and transported to the liver for storage. The storage form of vitamin A in the liver is primarily retinyl palmitate. Some injectable vitamin A products contain retinyl propionate which is not a biologically active form of vitamin A to inject. Australian research has demonstrated that retinyl propionate is much less effective in improving reproduction in sows when compared to retinyl palmitate.[9] VITAL E-Repro and VITAL E-Newborn contain retinyl-palmitate, the biologically active form of vitamin A. Research has shown that sows injected with VITAL E-Repro containing d-α-tocopherol, retinyl palmitate, and cholecalciferol had reduced farrowing times.[10]

Conclusion

Of the three fat-soluble vitamins, vitamin E appears to be the most deficient in all post-weaning phases of swine production, while vitamin D supplementation may be most critical in the nursing pig due to low milk content. Reduced farrowing time due to pre-partum injection of fat-soluble vitamins and reduction in mulberry heart are areas affected by inadequate fat-soluble vitamin intake. Esterification of fat-soluble vitamins A and E is done to improve stability in complete feeds, but there are situations in swine production when it is best not to esterify fat-soluble vitamins, especially vitamin E. Stability of esters will be better, but at the expense of bioavailability.

References

1. Lauridsen, C., M.S. Hedemann, and S. K. Jensen. 2001. Hydrolysis of tocopheryl and retinyl esters by porcine carboxyl ester hydrolase is affected by their carboxylate moiety and bile acids. J. Nutritional Biochemistry. 12: 219-224.

2. Specht, T.A., D.C. Mahan, N.D. Fastinger and R.L. Stuart. 2003. Effects of d-alpha-tocopherol alcohol or acetate in water soluble or emulsified form to the drinking water of weanling pig. J. Anim. Sci. 81 (Suppl. 2): p 88 (Abstract No. 356).

3. Stuart, R. L., P. G. Brinkman, W. D. Safley, and J. J. Stuart. 2015. New findings for injectable and oral fat-soluble vitamins. Proc. AASV Annual Meeting.

4. Moreira, I. and D.C. Mahan. 2002. Effect of dietary levels of vitamin E (all-rac-tocopheryl acetate) with or without added fat on weanling pig performance and tissue alpha-tocopherol concentration. J. Anim. Sci. 80:663-669.

5. Lauridsen C., and S. K. Jensen. 2005. Influence of supplementation of all-*rac*-α-tocopheryl acetate pre-weaning and vitamin C post-weaning on α-tocopherol and immune responses of piglets.

6. van Kempen, T. A., M. H. Reijersen, C. de Bruijn, S. De Smet, J. Michiels, M. G. Traber and C. Lauridsen. 2016. Vitamin E plasma kinetics in swine show low bioavailability and short half-life of *all-rac* -α-tocopheryl acetate J. Animal Science. 94:4188-4195.

7. Copeland, K. R., D. B. Scales, G. M. Hill, J. E. Link, K. C. Turner, C. J. Rozeboom, and R. L. Stuart. 2017. Effects of vitamins E and D on performance and antioxidant enzymes in nursery pigs. Midwest ASAS Meeting, March 13, 2017.

8. Jang, Y. D., J. Y. Ma, J. S. Monegue, H. J. Monegue, R. L. Stuart, and M. D. Lindemann. 2015. Temporal plasma vitamin concentrations are altered by fat-soluble vitamin administration in suckling pigs. J. Anim. Sci. 93:5273-5282.

9. Robertson, J. A. 1997. Investigations of the action of vitamin A and beta-carotene on reproductive performance in pigs. Ph.D. Dissertation Victoria Univ. of Tech., Victoria, Australia.

10. Becerra, R., J., T. Gebhardt, J.E. Link, K. Turner, R. L. Stuart and G. M. Hill. 2015. Are porcine epidemic diarrhea virus (PEDv) exposed gilts and sows farrowing problems improved by vitamin E? J. Anim. Sci. Midwest Meeting. Mar. 15-18, 2015.

Safe-guard Aquasol® 20% suspension, a novel method of delivering fenbendazole through swine drinking water systems

R. Fleck, DVM; J. Slightom
Merck Animal Health, Madison, New Jersey

Introduction

Safe-guard Aquasol (SA) is a novel formulation of fenbendazole approved for both swine and poultry. SA is a 20% suspension concentrate (200 mgs fenbendazole per ml) and is designed to be delivered via water proportioners or bulk tanks. SA is labeled for the treatment and control of Lungworms: adult *Metastrongylus apri*, adult *Metastrongylus pudendotectus*; Gastrointestinal worms: Adult and larvae (L3, L4 stages, liver, lung, intestinal forms) large roundworms (*Ascaris suum*), nodular worms (*Oesophagostomum dentatum, O. quadrispinulatum*), small stomach worms (*Hyostrongylus rubidus*), Adult and larvae (L2, L3, L4 stages-intestinal mucosal forms) whipworms (*Trichuris suis*); and Kidney worms: adult and larvae *Stephanurus dentatus*. SA is formulated for a three day treatment regimen at 2.2 mg/kg per day and has a two day pre-slaughter withdrawal in swine. It is approved for all classes of swine except for suckling piglets.

Formulation

Fenbendazole is poorly soluble in water, a property which has up until now been a hindrance for the creation of a formulation that can be applied through drinking water systems. Merck Animal Health, Madison, NJ, USA has developed an innovative wet-milling technique which utilizes shear forces to create homogeneously distributed particles that are approximately 100 nanometers across. These tiny particles remain in suspension in concentrations high enough for use in water proportioners, thereby meeting the needs of modern swine operations. The small particles size also prevents plugging of nipple drinkers.

Licensing studies

Dosage characterization

A bioequivalence study was performed in which SA was dosed to pigs at a rate of 2.2 mg/kg and was found to have a similar rate of and extent of absorption to fenbendazole levels in pigs dosed with the 20% Type A medicated in feed at the labeled dose of 3.0 mg fenbendazole/kg body weight. Thus the dosing regimen chosen for fenbendazole oral suspension was 2.2 mg/kg body weight per day administered on 3 consecutive days.

Palatability

Seventy two pigs weighing between 90 and 117 kg were divided into 36 pens and divided into two groups (A and B). Using a cross over design, pigs were exposed to either unmedicated water or water containing 2.2 mg fenbendazole/kg BW per day offered over a five hour period. The average percent of fenbendazole medicated water consumed was 98.18% of the amount of non-medicated water consumed, demonstrating acceptable palatability.

Dose confirmation study

Two dose confirmation studies (A and B) were performed using a *Trichuris suis* challenge. Fifty six days prior to treatment administration, pigs were orally dosed with approximately 4,000 embryonated *T. suis* eggs. Fecal samples were obtained on days 46, 47 and 48 after inoculation. Animals with at least two fecal examinations positive to *T. suis* were enrolled in the study. Twenty four study eligible pigs per site were randomly assigned to two treatment groups (fenbendazole treated and non-medicated). Six pens of 2 pigs each were used per treatment group. Fenbendazole treated animals received 2.2 mg/kg body weight in the water over a 4-6 hour dosing period for three consecutive days. All pigs were necropsied 8-9 days after the last fenbendazole treatment and the large intestine removed for worm counts. Site A had 11 out of 12 control animals with *T. suis* worm counts of over 100 and site B had 9 out of 12 control animals with adequate worm counts, indicating successful challenge. Worm counts of medicated animals were compared to non-medicated animals and percent efficacy was calculated using Abbott's formula. On site A the percent efficacy was 98.5% ($P = 0.0006$) and on site B the percent efficacy was 98.6% ($P = 0.0003$). The results of these studies demonstrated that 2.2 mg/kg per day for three days was efficacious against the dose-limiting parasite, *T. suis*.

Tissue residue depletion study

For determination of the US pre-slaughter withdrawal, the target tissue is liver and the marker residue is parent fenbendazole. Based on changes on the acceptable daily intake for swine liver, the maximum tolerable concentration has been revised from 12 ppm in 1984 to 3.2 ppm in 2015. Forty four cross bred swine (22 castrated males and 22 females) weighing 102 to 125 kg were assigned to one of six treatment groups (Table 1). Animals in groups 2-5 received SA at a target dose of 2.5 mg/kg per day for three consecutive days. Swine were slaughtered according to the schedule listed in Table 1, and liver samples were collected. Fenbendazole concentrations were determined using a validated LC-MS/MS method.

Table 1: Residue depletion study

Study group	N pigs	Treatment	Withdrawal time (hours)	Fenbendazole in liver (ppm)
1	2	Unmedicated	12	0
2	12	Medicated	12	2.75 ± 0.63
3	6	Medicated	18	2.75 ± 1.1
4	6	Medicated	24	2.22 ± 0.25
5	6	Medicated	30	1.50 ± 0.54
6	6	Medicated	36	2.08*

* only one value over the limit of quantitation of 0.729 ppm.

Using an algorithm that determines the upper tolerance limit for the 99th percentile of the population with a 95% confidence, the data established in this study supported a two day withdrawal period for SA in swine when used according to label directions.

Product stability

Effect of chlorine

Fenbendazole is readily degraded by chlorine at levels of 2 ppm and greater. Many municipal water supplies have chlorine levels at 5-7 ppm. Thus if chlorinated water is used on the farm, a chlorine stabilizer must be added to the drinking water before the SA is added. A study was performed to determine the impact of commonly used chlorine binders on two final concentrations of fenbendazole. SA solutions of 200 ppm and 20 ppm were made up in deionized water (positive control) and in water containing 5 ppm hypochlorous acid (negative control). The chlorine binders Vac-Pac® and Vac-Pac Plus® (Animal Science Products, Inc. Nacogdoches TX) and Reload Pack (Boehringer Ingelheim Vetmedica, Inc. St. Joseph MO) were mixed according to label into water containing 5 ppm hypochlorous acid. After preparation of the chlorine binder solutions, SA was added to create the two concentrations of 200 ppm fenbendazole and a solution of 20 ppm fenbendazole. The solutions were tested for fenbendazole levels immediately after preparation and again at 24 hours. All solutions were prepared and tested in duplicate. The results of the study are depicted in Figures 1 and 2.

All three chlorine stabilizers prevented the degradation of fenbendazole when used per label. It can also be seen from the time zero 5 ppm HCLO that the effect of hypochlorite on fenbendazole is immediate. Thus the chlorine stabilizer must be added to the water before the SA is added.

Hard water

SA is stable in water with a hardness of 425 and 500 ppm $CaCO_3$ over 24 hrs.

Remains in suspension

A study was performed testing the SA stock solution made up of 2 ounces of SA per gallon stock solution (3128 ppm) using two separate on-farm medicators. Samples were taken from the stock solutions at 0 hours, 8 hours and 24 hours after preparation. The stock solution for each medicator was agitated continuously on three days and not agitated on three days. New solutions were made up daily. There was no effect of agitation on fenbendazole concentration in the stock solution at any of the time points (Figure 3). Medicated water samples were taken at the farthest nipple drinker in the barn at the same time points. There was greater variation in concentrations at the nipple drinker but no impact of agitation on nipple drinker concentration (Figure 4). SA remains in a homogeneous suspension when mixed at 2 ounces per gallon stock solution over the relevant time periods.

Shelf Life

The stability of unopened SA is two years. If the bottle is opened and stored at controlled room temperature (below 77°F), it is stable for six months. SA must not be allowed to freeze or the suspension will be destroyed and agglutination will occur.

Mixing Instructions

SA 20% suspension must be shaken well before use to ensure that it is fully re-suspended. Once shaken, the suspension will incorporate a significant volume of air in the form of bubbles and the air volume will overestimate the product volume. When SA is poured into an equal volume of water, the air dissipates from the water and an accurate measurement can be obtained. Thus, a 1:1 dilution should be prepared to ensure accurate measurement. If complete liters or gallons of SA are used, this step is not required since the total volume of SA is known. The 1:1 dilution of SA can then be added to stock solutions or bulk tanks. No additional stirring is required. Prepare a new batch of drinking water every 24 hours for three consecutive days. The 1:1 dilution with water is potentially stable for is 2-3 days; however, the fenbendazole will settle out of suspension in the 1:1 dilution after 8 hours and must be shaken before use.

Figure 1: Impact of chlorine binders on fenbendazole levels of 200 ppm.

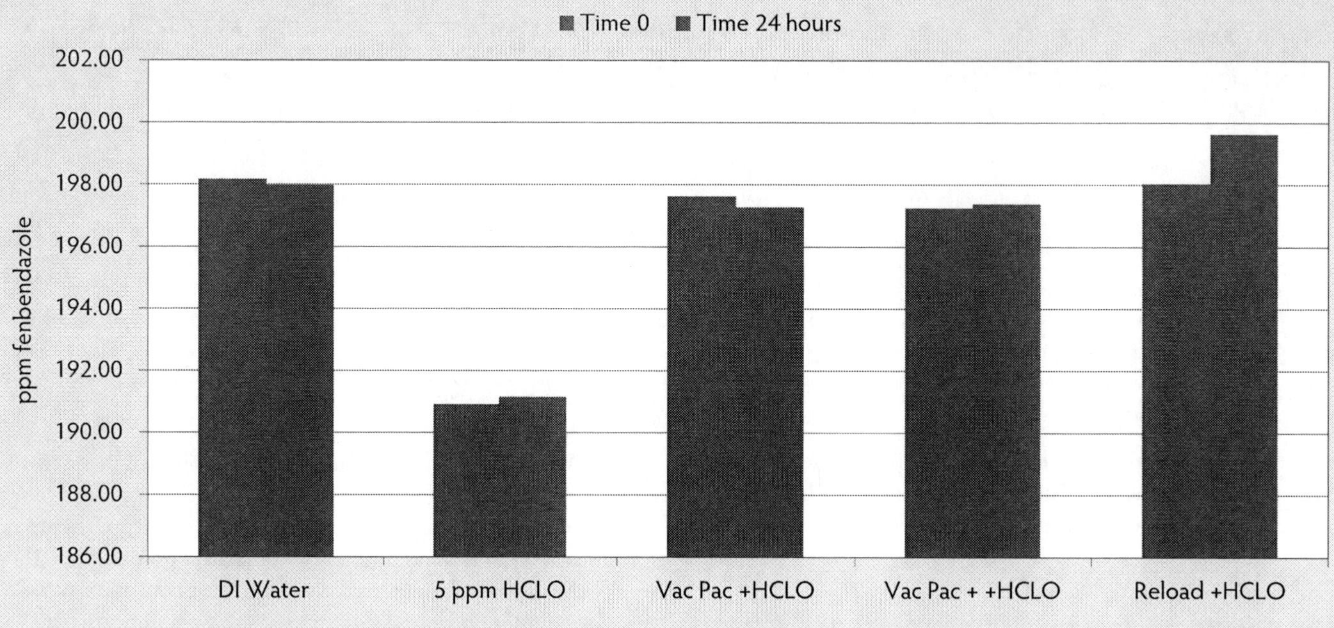

Figure 2: Impact of chlorine binders on fenbendazole levels of 20 ppm.

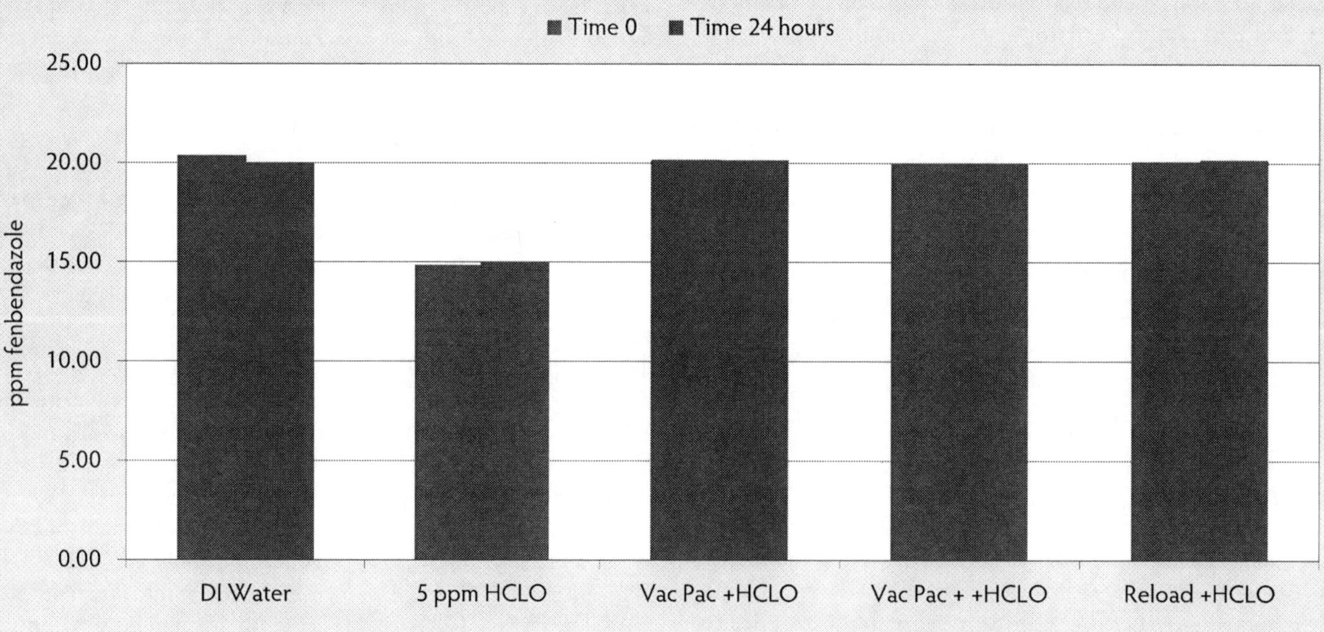

Figure 3: Impact of agitation on fenbendazole levels in stock solution.

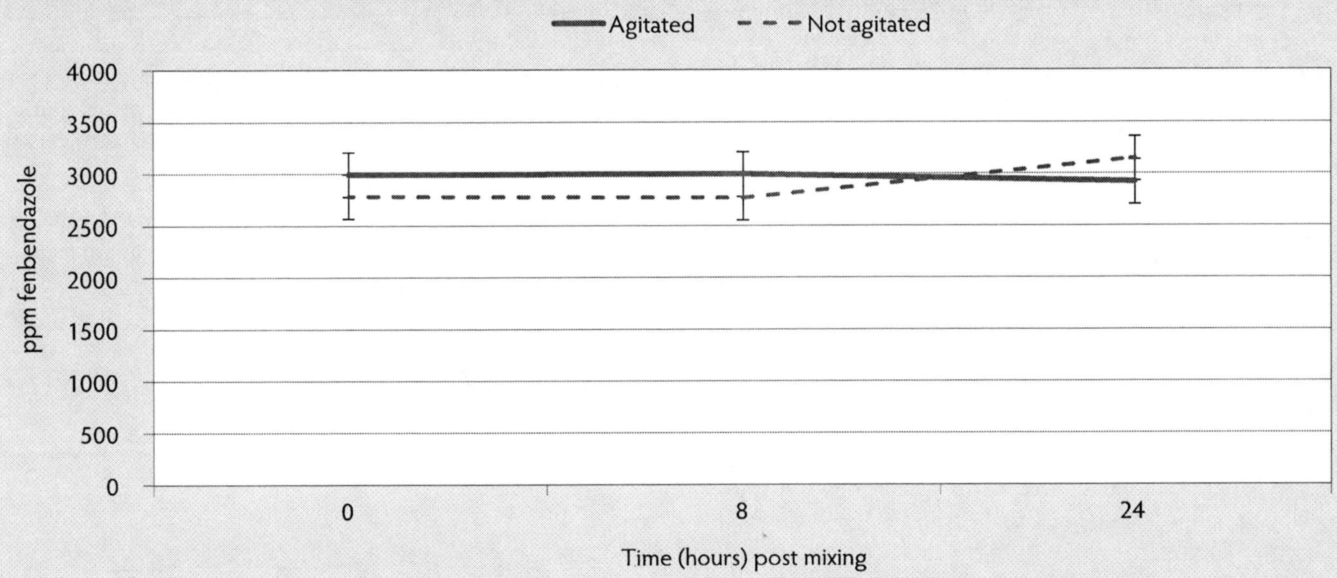

Figure 4: Impact of agitation on fenbendazole levels in final drinking water concentration.

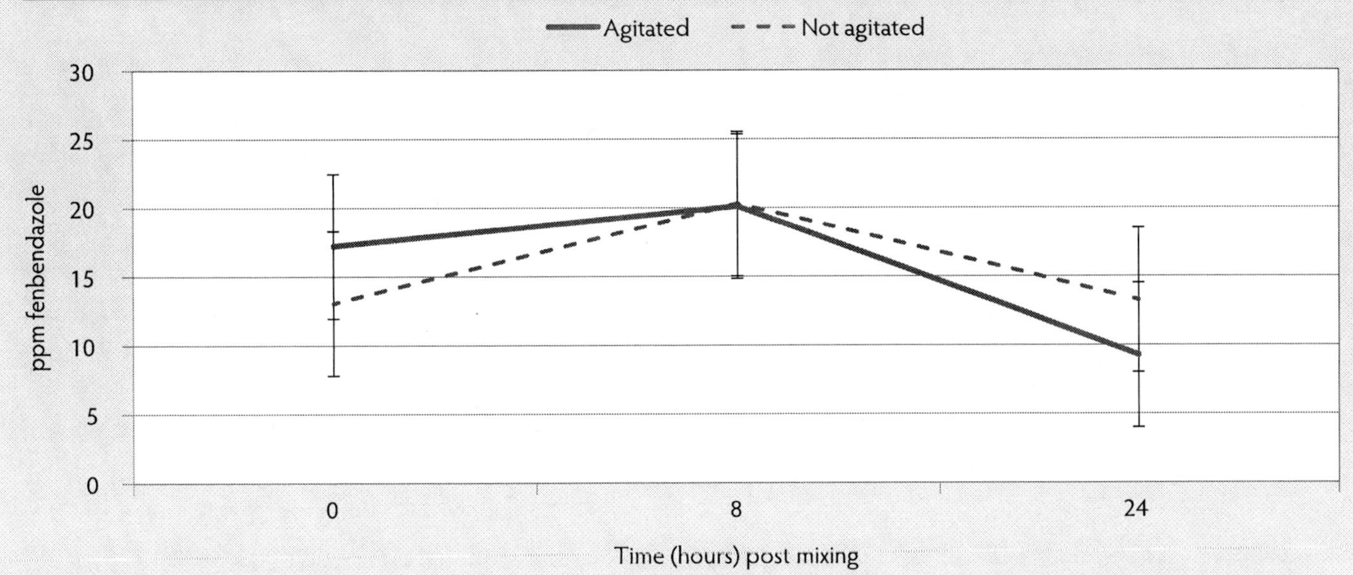

Water proportioners

For a 24 hour medication time frame, the target concentration of SA in drinking water is 93 mg/gallon (24.4 ppm) assuming that the pigs consume 8.5% body weight of medicated water. Shorter duration medication times can be achieved by increasing the concentration of SA in the stock suspension concentrate. On the day prior to treatment, determine by observation the total volume of stock solution that is consumed by the pigs in 24 hours or by dividing the total gallons of drinking water consumed in 24 hours by 128. Calculate the volume of SA required per day by multiplying the total gallons of stock solution consumed per day by 2 ounces of SA. Use 4 ounces of 1:1 dilution per gallon stock solution if using less than a liter of 20% SA at time.

Bulk tanks

Determine the total weight of animals to be treated. One mL of SA is required per 200 lbs. body weight per day for three days. Place the required amount of SA into the volume of water consumed by the animals over 24 hours. Mix up a new solution daily.

Conclusions

SA is a convenient, efficacious method to deworm pigs through either drinking water proportioners or bulk tanks. The tiny particle size allows the fenbendazole to remain in suspension in stock solutions prepared for consumption over 24 hours as well as in bulk tanks. Fenbendazole is the broadest spectrum anthelmintic available to swine through the water and has an excellent safety profile.

Safe-Gard Aquasol is a registered trademark of Merck Animal Health, Madison, NJ, USA. Vac-Pac and Vac-Pac plus are registered trademarks of Animal Science Products.

Experimental PCV2d challenge of pigs vaccinated with prototype vaccines containing ORF2 from PCV2a, PCV2b or PCV2d

E. Strait[1], DVM, PhD; A. Suddith[2], MS; M. Allen[1], MS; J.E. Bijlsma, PhD[3]; E.H.H. Sanders, BS[3]; Jessica Lathouwers-van Schijndel[3], PhD; D. Hesse[2], DVM, PhD
[1]Merck Animal Health, De Soto, Kansas; [2]Kansas State University, Manhattan, KS; [3]MSD, Boxmeer, The Netherlands

Introduction

This study sought to assess the efficacy of three experimental porcine circovirus type 2 vaccines in a PCV2d/PRRSV co-infection system. Commercially available PCV2 vaccines are all currently based on the PCV2a genotype. In recent years, PCV2d has emerged in the United States and is replacing PCV2b as the dominant genotype of PCV2.[1] Experimental studies have demonstrated efficacy against PCV2d by commercial vaccines as well as by an experimental replicon particle vaccine based on PCV2d.[2,3] To date, vaccines of different genotypes of PCV2, made by the same platform, have not been directly compared against a PCV2d challenge. In this study, subunit vaccines consisting of ORF2 full length capsid protein expressed in a baculovirus expression system were tested. The capsid protein of PCV2a, PCV2b, and PCV2d were used as the immunizing protein. The commercially produced product, Circumvent® PCV G2 was included as a positive control. Pigs were then challenged in a PCV2d/PRRSV co-infection challenge model and efficacy was evaluated.

Materials and methods

Seventy-five specific pathogen free pigs were transferred to a research facility at three weeks of age. The animals were randomly assigned to five treatment groups of fifteen pigs (Table 1). Each of the treatment groups were co-mingled in four pens. The pigs were bled and the serum was tested for the presence of PRRSV and PCV2 nucleic acids and antibodies. The experimental vaccines contained baculovirus-expressed capsid from the PCV2a, PCV2b, or PCV2d genotypes of PCV2. The capsid proteins were formulated with Microsol Diluvac Forte® (MDF) adjuvant, similar to the commercial product Circumvent PCV G2, which was included as a positive control. The fifth group of animals was similarly vaccinated with a sham vaccine containing MDF and PBS. Pigs were intramuscularly vaccinated in the neck with 2 mL at 3 weeks of age according to their assigned treatment group. All of the animals were challenged 48 days after vaccination. The challenge was administered intranasally, 3 mL/nare, and contained both PRRSV and PCV2d. Following challenge, the animals were bled weekly; nasal and fecal swabs were collected at the same time points. Weekly testing continued for five weeks, at which time the animals were euthanized and tissues were collected for immunohistochemistry, histopathology and PCR analysis. Antibody titers to PCV2b were determined by an immunofluorescence assay (IFA). Detection of nucleic acids of PRRSV and PCV2 in serum, tissues and fecal and nasal swabs was performed by qPCR. All sample testing was performed at the Kansas State Veterinary Diagnostic Laboratory.

Results and discussion

PCV2b IFA

At the beginning of the study, all of the animals, had IFA antibody titers to PCV2b of 40 or less except for one pig in group B with a titer of 1280. Group A (placebo) pigs were seronegative until 3 weeks post challenge when they had a group median titer of 5120 that was maintained until the end of the study. All of the vaccines generated IFA titers in the PCV2b-based IFA. Pigs in groups B-E had median titers of at least 1280 by 3 weeks post vaccination and all group medians were 5120 at the time of challenge. At each time point post challenge median titers in groups B-E were 5120, except at 1 week post challenge for group E and 5 weeks post challenge for group D which were 2560.

PRRSV qPCR

There was no detectable PRRSV by qPCR in any of the animals from the time of arrival until challenge. PRRSV was detected in all of the animals one week after challenge. There was no significant difference in the viremic load of PRRSV between any of the treatment groups.

PCV2 qPCR

From arrival until the time of challenge, none of the pigs had detectable amounts of PCV2 DNA in their serum. PCV2 DNA was detected in the serum of all of group A placebo animals, 2 weeks after receiving the challenge and continued through the remainder of the study in 6 of 15 of these pigs. No viremia was observed in vaccinated pigs, except at the two weeks post challenge time point in which two pigs in the positive control group E were weakly detected (Ct 37.2 and 38.0).

Animals in the placebo group began shedding PCV2d through both the nasal and fecal routes by 2 weeks after challenge. Nasal shedding was observed at least once in 9 of 15 pigs in group A. One animal in each of the experimentally-vaccinated groups (B-D) was found positive at a single time point, either two or four weeks

Table 1: Study design

Group	Vaccine	No. of pigs	Age of pigs at vaccination	Age of pigs at challenge	Necropsy	Serum collection	Fecal and nasal swabs
A	Placebo						
B	PCV2a						
C	PCV2b						
D	PCV2d	15	3 weeks	10 weeks	33 dpc	Pre-vacc, 3wpv, 5wpv, pre-chall, and 5, 12, 19, 26, 33dpc	Pre-vacc, 3wpv, 5wpv, pre-chall, and 5, 12, 19, 26, 33dpc
E	Circumvent® PCV G2 (PCV2a)						

post-challenge. Four pigs in the positive control group E had single, positive nasal swabs ($Ct \geq 37.3$). Nasal shedding was resolved by the fourth week after challenge in all groups. Twelve of 15 pigs in group A shed viral DNA through the fecal route for a duration of one to four weeks, while only two animals total from all of the vaccinated groups had detectable viral DNA in their feces at any time point (groups D and E).

The amount of PCV2d DNA in the tissues of group A was significantly higher ($P < 0.01$) than that found in any of the groups that were vaccinated with a vaccine containing PCV2 capsid protein. The average composite tissue qPCR value for group A was $Ct = 24.7$ versus the vaccinated groups that ranged from an average $Ct = 36.6$ to 38.7.

Immunohistochemistry and pathology

Five different tissues, liver, bronchial and mesenteric lymph nodes, spleen, and tonsil were collected at necropsy and evaluated for the presence of PCV2 via IHC. There was no PCV2 detected by IHC in any of the tissues collected from animals in any of the vaccinated groups. In contrast, twelve of fifteen pigs in the group A were positive by IHC in at least one tissue sample. In addition to IHC staining, the tissues were also evaluated for lymphoid depletion, macrophage replacement, and general pathology. Lymphoid depletion and macrophage replacement was only detected in tissues from animals in the placebo treatment group (10 of 15 pigs, each). The other pathological lesion noticed in the tissues was hepatitis in seven of the placebo animals and one animal in group B (PCV2a experimental). Aside from the one animal in group B with observed hepatitis, no other lesions were observed in any of the tissues from vaccinated pigs.

Summary

The PCV2d/PRRSV challenge resulted in the development of severe clinical disease in the placebo group, including one pig in group A which had to be euthanized 20 days after challenge. Animals in group A (placebo) had significantly more microscopic lesions than the vaccinated animals as evidenced by the presence of lymphoid depletion and macrophage replacement in tissues. None of the vaccinated animals developed any clinical signs and there was a complete absence of pathological lesions in any of their tissues with the exception of hepatitis observed in one animal which received the experimental PCV2a vaccine (Group B). Efficacy was further demonstrated by significantly reduced viremia, viral tissue load and viral shedding in vaccinated pigs versus the placebo group. Antibody titers generated through vaccination corresponded with protection and were similar in groups B-E. All of the PCV2 vaccines were effective in this challenge model and no difference in protection was observed between vaccinated groups.

References

1. Xiao CT et al. Global molecular genetic analysis of porcine circovirus type 2 sequences confirms the presence of four main PCV2 genotypes and reveals a rapid increase of PCV2d. *J Gen Virol.* 2015;96:1830.

2. Opriessnig T et al. A commercial vaccine based on PCV2a and an experimental vaccine based on a variant mPCV2b are both effective in protecting pigs against challenge with a 2013 U.S. variant mPCV2b strain. *Vaccine.*2014;32:230.

3. Opriessnig T et al. Commercial PCV2a-based vaccines are effective in protecting naturally PCV2b-infected finisher pigs against experimental challenge with a 2012 mutant PCV2. *Vaccine.* 2014;32:4342.

Field experiences with Porcilis™ Ileitis: The newest tool for controlling *Lawsonia intracellularis*

Brad Thacker, DVM, PhD; Greg Armbruster, DVM; Angela Baysinger, DVM, MS; Justin Cagle, DVM; Kimberly Crawford, DVM, MS; Jack Creel, DVM; Robyn Fleck, DVM; Kelly Greiner, DVM; Megan Inskeep, DVM; Doug King, DVM; Karen Lehe, DVM; Joe Roder,DVM, PhD; Channing Sebo, DVM
Merck Animal Health, Madison, New Jersey

Introduction

Porcilis™ Ileitis (Merck Animal Health, Madison, New Jersey, USA) was introduced to the US market in late 2015 as an aid in the control of ileitis caused by *Lawsonia intracellularis,* as an aid in the reduction of colonization by *Lawsonia* and as an aid in the reduction of duration of fecal shedding. Since then, the Merck Animal Health swine technical services team has conducted several field studies and diagnostic surveys to develop a thorough understanding of how to effectively utilize Porcilis Ileitis under field conditions. These activities to-date have focused on serological and molecular-based diagnostic testing strategies for understanding the pattern of infection in a herd, evaluating antibody responses following vaccination and exposure, and monitoring the level of fecal shedding to determine if a particular vaccination program is more effectively controlling *L. intracellularis* infection.

Licensing studies for Porcilis Ileitis demonstrated the induction of measurable antibody responses after administration of a single, 2 mL, intra-muscular dose. In addition, the level of fecal shedding was dramatically reduced in vaccinated pigs compared to controls.[1,2] Reduction in the amount of *L. intracellularis* shed in the feces was correlated to reduction in the level of tissue infection and lesion severity in the terminal ileum. Serological responses after vaccination and monitoring of fecal shedding should provide evidence that a Porcilis Ileitis vaccination program is working effectively in the field. Accordingly, the overall objective of our studies was to determine if the diagnostic testing observations made in the laboratory could be extrapolated to, and then used effectively, in the field to better manage Porcilis Ileitis vaccination programs.

Materials and methods

The data was obtained from three field studies conducted since Porcilis Ileitis became commercially available.

Study 1- This study was a cross-sectional survey of a finishing site that was endemically infected with *L. intracellularis* and contained multiple rooms with each room containing one week of production. This survey was performed to provide baseline data prior to implementing a Porcilis Ileitis vaccination program and to determine which test best described the dynamics of field exposure. At the time of the survey, the operation was using feed grade antibiotics up until 10 weeks of age to control ileitis. Fecal swabs, blood and oral fluids were collected on the same day from pigs at 4 weeks of age (1 week after weaning), and then every 4 weeks until 23 weeks of age (see Table 1). Within each age group, fecal swabs were collected from 5 pigs housed in 6 pens and tested by polymerase chain reaction (PCR) in pools of 5 samples by pen. Oral fluids (1 per pen) and sera (5 pigs per pen) were collected from the same pens and tested individually by PCR and ELISA, respectively. Results are presented as the number of positive samples per the number tested and the average PCR CT or ELISA percent inhibition for each age group.

Study 2- This study compared the two commercially available serological tests, IPMA (immunoperoxidase monolayer assay) and ELISA. The serum samples (n = 173) used for this comparison were obtained from pigs of various ages that were vaccinated with Porcilis Ileitis per label directions and then naturally exposed to *L. intracellularis*. The data is presented as positive and negative test results (see Table 2). In addition, for each IPMA reciprocal titer, the number of positive pigs per the number tested and the average percent inhibition as measured by ELISA was determined (see Table 3).

Study 3- For this study, fecal swabs and serum were obtained from a field vaccination – challenge study. Pigs were vaccinated with 2 mL of Porcilis Ileitis at 3 weeks of age (n = 31) or left as non-vaccinated controls (n = 15). At 10 weeks of age, pigs were challenged by oral gavage with approximately 6×10^8 organisms provided by and administered under the supervision of Dr. Dana Beckler, GutBugs, Inc., Fergus Falls, MN. Serum was collected at 3 (time of vaccination), 6, 10 (time of challenge), 14 and 18 weeks of age, which enabled evaluation of serological responses after vaccination and then following challenge. Serum was tested by ELISA and results are presented as the number of positive samples per the number tested and the group average percent inhibition at each sampling age (see Table 4). Fecal swabs were collected weekly from 1 to 8 weeks after challenge and tested by quantitative PCR (qPCR). Results are presented as the number of positive samples per the number tested and the group average qPCR and cycle times (CT). In addition, the percent reduction at each time point and overall was calculated as the difference in the qPCR values between the two groups (control group qPCR

™ Porcilis is a registered trademark of Merck Animal Health.

Table 1: Study 1 results - cross-sectional survey of a *Lawsonia intracellularis*, endemically infected finishing site containing multiple aged, non-vaccinated pigs

	PCR results				ELISA Results	
	Pooled feces (5 per pool)		Oral fluids			
Age in weeks	No. positive	Avg. CT	No. positive	Avg. CT	No. positive	Avg. % Inhibition
4	0/6	> 40.0	1/6	40.0	1/30	-1.9
7	0/6	> 40.0	3/6	37.9	0/30	-6.9
11	0/6	> 40.0	5/6	36.5	1/30	-1.6
15	5/6	36.5	6/6	26.8	10/30	19.4
19	2/6	38.7	6/6	29.0	30/30	59.4
23	1/6	39.9	6/6	27.0	30/30	53.2

Table 2: Study 2 – comparison of IPMA and ELISA positive/negative test results using serum from pigs collected at various time points following Porcilis Ileitis vaccination and field exposure

		IPMA results	
n = 173		Positive	Negative
ELISA results	Positive	44.5%	13.3%
	Negative	17.3%	24.9%

Table 3: Study 2 - comparison of IPMA titers to ELISA positive/negative results and percent inhibition using serum from pigs collected at various time points following Porcilis Ileitis vaccination and field exposure

		ELISA results	
No. samples tested	IPMA titer	% Positive	Avg. % Inhibition*
66	< 30 (Negative)	33.3%	22.5
20	30	90.0%	40.8
20	60	55.0%	37.0
32	120	59.4%	39.6
38	240	82.1%	50.1
7	> 480	85.7%	63.8

* Group means

value minus PI group qPCR value) divided by the control group qPCR value (see table 5).

Laboratory testing was done using standard methods. Serotesting by IPMA was performed at the University of Minnesota Veterinary Diagnostic Laboratory. The lowest titer tested was 1:30 and the highest titer was 1:480. Titers < 1:30 were considered negative and results are presented as the reciprocal titer. Serotesting by ELISA was performed at the Iowa State University Veterinary Diagnostic Laboratory (ISU-VDL) using a commercial assay (SVANOIR® L. intracellularis/Ileitis-Ab, Boehringer Ingelheim Svanova, Uppsalsa, Sweden). Suspect results were considered to be negative. *L. intracellularis* detection by PCR was performed by the ISU-VDL using either quantitative or semi-quantitative methods that provided results as copies per reaction or cycle times (CT), respectively. This assay was modified this past year such that the positive/negative CT cut-off was 40.0 for Study 1 and 35.0 for Study 3.

Table 4: Study 3 serology results - comparison of ELISA positive/negative results and percent inhibition between Porcilis Ileitis vaccinated and non-vaccinated pigs following vaccination at 3 weeks of age and subsequent challenge at 10 weeks of age

	ELISA results			
	Control		Porcilis Ileitis	
Age in weeks	No. positive	% Inhibition*	No. positive	% Inhibition*
3 - Vaccination	0/15	< 0.0	0/31	< 0.0
6	0/12	< 0.0	3/17	22.5
10 - Challenge	0/15	< 0.0	27/31	52.7
14	12/15	70.2	28/31	74.1
18	15/15	75.6	31/31	96.1

* Group means

Table 5: Study 3 fecal shedding results – *L. intracelularis* fecal shedding as measured by qPCR in pigs vaccinated with Porcilis Ileitis at 3 weeks of age compared to non-vaccinated pigs following challenge at 10 weeks of age

	Control			Porcilis Ileitis			
Weeks after challenge	No. positive	Avg. copies per reaction	Avg. CT	No. positive	Avg. copies per reaction	Avg. CT	Percent reduction*
1	13/14[a]	15,170,510[c]	27.3	27/31[a]	6,939,084[c]	29.3	54.3%
2	15/15[a]	946,407,372[c]	20.6	28/29[a]	38,576,516[d]	27.2	95.9%
3	15/15[a]	674,620,463[c]	20.9	28/31[a]	1,609,289[d]	29.9	99.8%
4	14/15[a]	158,907,409[c]	26.3	16/31[b]	41,650[d]	33.8	100.0%
5	15/15[a]	8,013,974[c]	28.2	16/31[b]	39,837[d]	33.7	99.5%
6	11/15[a]	1,580,560[c]	31.1	13/31[b]	21,856[d]	34.4	98.6%
7	8/15[a]	41,128[c]	33.6	3/30[b]	15,013[d]	34.8	63.5%
8	4/15[a]	9,827[c]	34.5	2/31[a]	1,219[d]	35.0	87.6%
Avg. 1-8	15/15[a]	227,362,169[c]	27.8	31/31[a]	5,639,879[d]	32.2	97.5%

* Based on qPCR data
[a,b] Values with different superscripts within a row were significantly different ($P < 0.05$) by Fisher's exact test.
[c,d] Values with different superscripts within a row were significantly different ($P < 0.05$) by Wilcoxon rank sum test.

Statistical analysis- *Statistix*™ 10 (Analytic Software, Tallahassee, FL) was used to perform statistical analysis on the fecal shedding data in Study 3. Positive/negative results were analyzed using Fisher's exact test. Copies per reaction were analyzed using the Wilcoxon rank sum test.

Results and conclusions

Study 1- The findings from the cross-sectional survey are presented in Table 1. *L. intracellularis* was detected in oral fluids in all age groups but the level of infection based increased dramatically between 11 and 15 weeks of age and similar levels were found at 19 and 23 weeks of age. Positive fecal swabs were detected starting at 15 weeks of age although the levels of *L. intracellularis* were relatively low and the number of positive fecal samples and the amount of *L. intracellularis* was even lower at 19 and 23 weeks of age. Seroconversion rates increased starting at 15 weeks of age and all pigs at 19 and 23 weeks of age were seropositive. In this survey, all three tests indicated that the timing of exposure occurred between 11 and 15 weeks of age although the positive oral fluids in younger age groups suggested that environmental contamination may have been present when the pigs were placed. Overall, serotesting by ELISA and PCR testing of oral fluids at the end of the finishing period can be used to characterize the *L. intracellularis* exposure status of non-vaccinated pigs. Fecal shedding at the end of finishing appeared to be a poor indicator of previous exposure.

Study 2- Although there appeared to be some correlation between the two tests, the relationship is not clear cut. The number and percentage of samples that were positive or negative by each test is presented in Table 2. The overall positive/negative agreement between the two tests was 69.4%. Table 3 presents the ELISA positive/negative results and the average percent inhibition at each IPMA titer. The level of agreement was greatest with higher titer samples and it appeared that the ELISA is less sensitive than IPMA with regard to detecting low levels of antibodies although 33.3% of the samples that were negative (< 1:30) by IPMA were seropositive by ELISA. The main advantage of the ELISA over the IPMA is test cost when the IPMA is used in a serial dilution format that provides a titer result. The test costs are similar when the IPMA is used as a positive/negative test where a single, 1:30 serum dilution is tested. In many situations, an antibody level or titer is preferred over a positive/negative result.

Study 3- The results of the field vaccination – challenge study are presented in Tables 4 (serology) and 5 (fecal shedding). Oral fluid testing (data not shown) indicated the pigs were not exposed to *L. intracellularis* prior to challenge. Accordingly, the antibody responses observed in the Porcilis Ileitis vaccinated pigs were solely due to vaccination. Control pigs remained seronegative until the time of challenge and then seroconverted at a high rate following challenge and the antibody levels expressed as percent inhibition were similar at 4 and 8 weeks after challenge. A low (17.6%) number of Porcilis Ileitis vaccinated pigs were seropositive at 3 weeks post-vaccination (6 weeks of age) but by 7 weeks post-vaccination (10 weeks of age), nearly all pigs (87.1%) were seropositive. At 4 weeks after challenge, antibody levels in vaccinated pigs increased although the number of seropositive pigs was similar to the time of challenge. By 8 weeks after challenge, all vaccinated pigs were seropositive and the antibody levels had further increased and were higher than control pigs. This data suggests that post-vaccination serological monitoring is feasible at 7 weeks post-vaccination but perhaps not at 3 weeks post-vaccination. Further evaluation of post-vaccination serological responses as measured by ELISA, on a weekly basis, is warranted.

The data generated in Study 3 supports the claim of aid in the reduction of duration of fecal shedding. Furthermore, the reduction in the cumulative amount of shedding in this study (97.4%) was similar to what has been observed in previous laboratory studies where the level of shedding was reduced by at least 95% using the same method for calculating the percent reduction.[3] The onset of protection following challenge provided by Porcilis Ileitis was relatively rapid in that the fecal shedding levels were reduced by over 50% at one week post challenge and then between 95-100% from weeks 2 to 6 after challenge. Thereafter, the levels in the control pigs declined at a more rapid rate than vaccinated pigs resulting in percent reductions of 63.5% and 87.2% at 7 and 8 weeks post challenge, respectively.

In summary, the data reported here provides background information for developing diagnostic strategies to enable the effective use of Porcilis Ileitis for controlling *L. intracellularis*. A high rate of seroconversion after vaccination with a single, 2 mL dose of Porcilis Ileitis at 3 week of age may take up to 7 weeks to fully develop in non-exposed pigs. Most importantly, Porcilis Ileitis vaccination dramatically reduced the level of fecal shedding in pigs directly challenged with *L. intracellularis*.

References

1. Roerink, F., Morgan, C., Knetter, S., Thacker, B., Strait, E. Porcilis® Ileitis: 20-week duration of immunity against *Lawsonia intracellularis* challenge. In: Proc. 46th Annual Meeting of the American Association of Swine Veterinarians, New Orleans, Louisiana, 2016, pp. 144-145.

2. Roerink, F., Morgan, C., Knetter, S., Segers, R.., Strait, E. Efficacy of Porcilis® Ileitis under field conditions, evaluated with an experimental challenge mode. In: Proc. 24th Congress International Pig Veterinary Society, Dublin, Ireland, 2016, p. 223.

3. Data on file.

Electronic sow feeding (ESF) vs. the gestation stall: Three head to head comparisons of performance

Jeff Schoening, Technical Sales Manager
Automated Production Systems

Introduction

Pork producers are being asked to move their gestating sows out of stalls and into loose housing. A variety of alternatives to the gestation stall are being tried and include small pen stanchion, small pen floor feeding, free access stalls and electronic sow feeding (ESF). Furthermore within each of these different approaches to loose housing there are differences in animal flow, pen design and equipment. However, very little objective data is available on how different sow housing systems compare. Between farm comparisons can be misleading if differences in factors such as nutrition, genetics or management are not taken into account. Here I present comparisons from 3 production systems that have both conventional stalls and at least one ESF farm. One advantage of making this type of comparison is that nutrition, genetics and management practices are similar within each of the three different production systems. I have gathered performance data from these systems and compared the overall system performance to the ESF farm within that system. The data examined included pigs per sow per year (PSY), feed usage per sow, as well as other standard metrics and was tabulated over a 1 year period ending in July of 2016.

System 1

System one has approximately 21,300 sows in production. The ESF farm is a 5600 sow farm that was populated in January 2014. Sows are moved into the pens after pregnancy diagnosis. The pens are rectangular in shape with the feeder placed on the aisle in the center of the long side (Figure 1). Sleeping bays are installed in the back portion of the pen. The bays are approximately seven foot deep and 10 feet wide. There are 43 AP/Schauer ESF feeders on the farm. Each pen has one feeder in it and sows are grouped with 80 sows per pen at 20 sq. ft. per sow. Gilts are raised and trained in an offsite gilt developing unit (GDU) with four ESF feeders. Gilts and small P1 sows are housed together. Older parity sows are then housed together. All pens are managed in a static flow. Production results from system one are summarized in Table 1.

System 2

System two has approximately 27,300 sows in production. The ESF farm in this system is a 2800 sow unit that was populated in April 2013. Sows are moved into the pens after pregnancy diagnosis. The pens are rectangular in shape with the feeder placed on the aisle in the center of the long side (Figure 2). Sleeping bays are installed in the back portion of the pen. The bays are approximately seven foot deep and 10 feet wide. There are 18 AP/Schauer ESF feeders on the farm. Each sow pen has one feeder in it and sows are grouped with 80 sows per pen at 20 sq. ft. per sow. Gilt pens have the divider gate removed which creates a pen with two feeders and approximately 160 gilts gestated in the pens. Replacement gilts are raised in an on-site GDU where there are three training feeders. Sows are housed in static groups. Gilts are flowed dynamically. Small P1 sows also are housed in the dynamic gilt pens. Production results from system one are summarized in Table 2. It is important to note that this particular farm started to implement an internal multiplication system. Due to some issue with semen sourcing this project was stopped. Had this not occurred we would expect the farrowing rate to be significantly higher. Feed data was not available on this farm.

Figure 1: Farm A layout

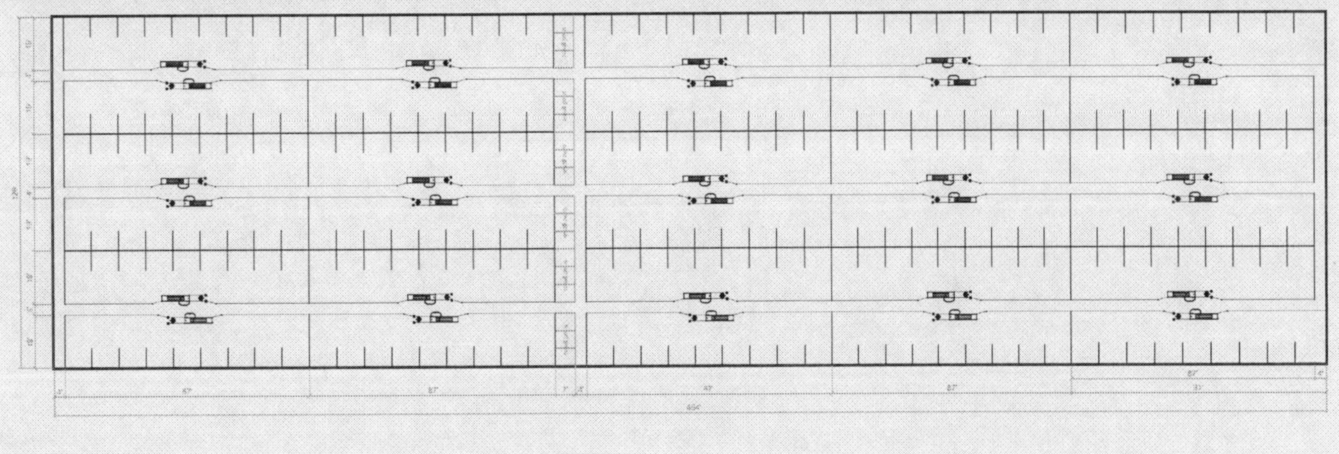

Table 1: Farm A Performance data

Criteria	Farm A ESF	Total system
Inventory	**6,045**	**21,302**
Total born	14.90	14.30
Live born	13.90	13.20
Farrowing rate	88.50	89.10
Weaned per litter	11.20	11.70
Litters per year	2.52	2.00
PSY	29.10	27.20
Feed per sow[gest]	5.31	5.43
Feed cost per pig	11.32	13.45

Figure 2: Farm B layout

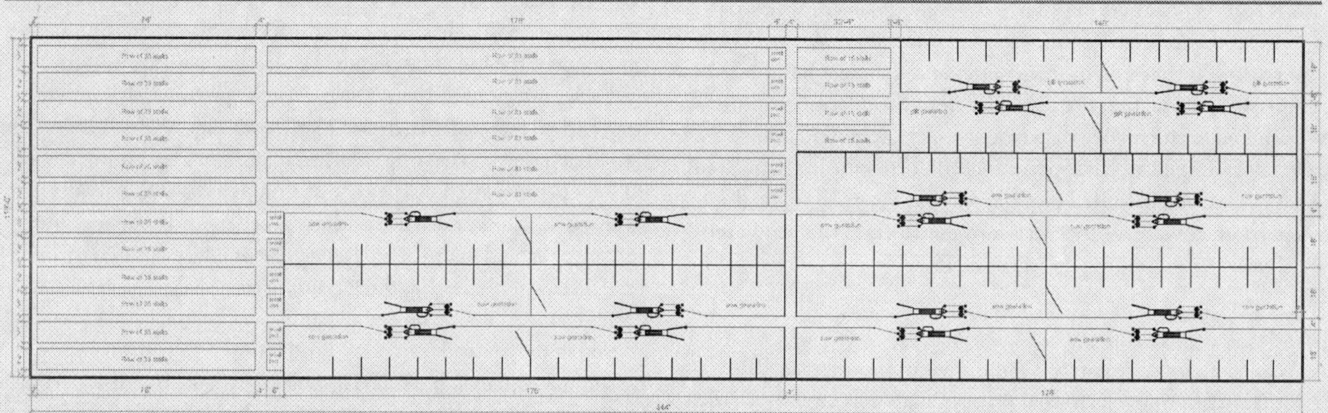

Table 2: Farm B Performance data

Criteria	Farm B ESF	Total system
Inventory	**2,707**	**27,323**
Total born	13.71	13.38
Live born	12.38	12.35
Farrowing rate	83.00	90.40
Weaned per litter	10.98	10.89
Litters per year	2.35	2.42
PSY	24.86	25.57

System 3

System three has approximately 12,600 sows in production. The ESF farm in this system is a 2800 sow farm that was populated in December 2012. Sows are bred in stalls and moved into pen as soon as estrus is not displayed. The pens are rectangular in shape with the feeder placed on the aisle in the center of the long side (Figure 3). Sleeping bays are installed in the back portion of the pen. The bays are approximately seven foot deep and 10 feet wide. There are 30 AP/Schauer ESF feeders on the farm. 16 of the feeders are in single feeder pens that house 80 sows at 20 sq. ft. per sow. There are four pens that have three feeders in each pen. Every week one pen of sows is housed in one of the single feeder pens. The remaining sows are house in one of the two pens with three feeders. Also the gilts are housed in one of the other two pens with three feeders. The single feeder pens are managed in a static flow. The three feeder pens are managed in a dynamic flow to better meet space requirements. There are two training feeders that are located in the gestation barn. Production results from system one are summarized in Table 3.

Conclusion

From this data collected over a one year period we see that ESF sow unit performance is very similar to the performance of sows housed in a conventional gestation stalls. The sow units that were examined use both static and dynamic flows. Sows are grouped within these three farms both pre-implantation and post-implantation. This demonstrates that various types of ESF farms can perform at high levels and reinforces the notion that management plays a key role in production.

Figure 3: Farm C layout

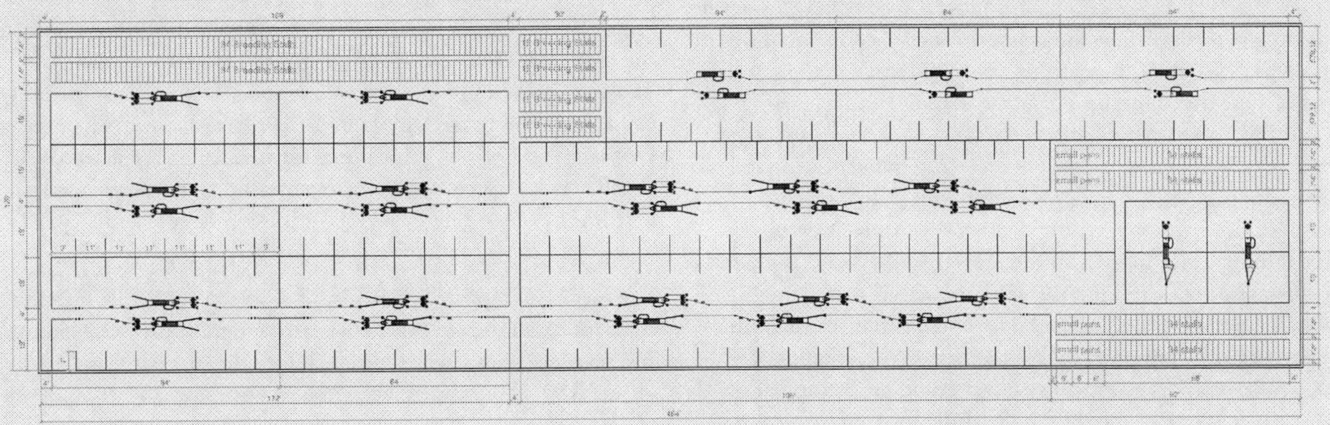

Table 3: Farm C Performance data

Criteria	Farm C ESF	Total system
Inventory	**2,693**	**12,591**
Total born	14.30	14.20
Live born	13.20	13.40
Farrowing rate	87.10	89.40
Weaned per litter	11.40	11.60
Litters per year	2.48	2.48
PSY	28.60	28.70
Feed per sow[gest]	5.06	5.49
Feed cost per pig	9.82	9.53

Comparative study of two vaccines against neonatal diarrhea on a Canadian commercial farm

W. Ariza[1], DVM; D. Robles[1], BSA; J. Badia[2], DVM; D. Torrents[3], PhD; I. Bernal[3], DVM
[1]Swine Health Professionals, Steinbach, Manitoba, Canada; [2]Hipra Animal Health Canada Inc., Ottawa, Ontario, Canada; [3]Laboratorios Hipra S.A, Amer, Girona, Spain

Introduction

Infection with pathogenic *Escherichia coli* is a common cause of diarrhea in suckling pigs worldwide and is becoming an important and devastating disease for swine producers. It is also responsible for a substantial economic impact on farms worldwide, due to high morbidity, mortality and reduced growth rates.[1] In some studies, conducted in Europe, diarrhea is estimated to account for 5-24% of overall pre-weaning mortality and to reduce average daily gain (ADG) by 8-14 g per day.[2,3] Based on these effects, the cost of neonatal diarrhea was recently estimated to be €134 per sow per year.[4] There are different products on the market that will help us to control neonatal diarrhea through vaccination. Suiseng®, which is one of the vaccines involved in this trial, is a subunits vaccine that contains purified adhesion factors (F4ab, F4ac, F5 and F6) and the heat-labile toxin (LT) of *Escherichia coli*, the β toxin of *Clostridium perfringens* type C, to prevent neonatal diarrhea in piglets through passive immunization, and the α toxin of *Clostridium novyi*, to protect sows suffering from sudden death by active immunization. Moreover Suiseng® includes a new, state-of-the-art adjuvant, Hipramune® G, based on ginsenosides, saponins from ginseng root, which enhances an animals' antigen presentation process.[5] An increase in the number of APCs (antigen-presenting cells) and the amount of antigen on their surface is one of the keys in the ability of adjuvants to increase the effectiveness of vaccines.[6] On the other hand, the other vaccine used in this comparative study, Vaccine B, which is a pure bacterin, contains *Escherichia coli* adhesion factors F4, F5, F6 and F41, and the β toxin of *Clostridium perfringens* type C, with an aluminum hydroxide adjuvant.

The aim of this study was to assess and compare the safety and efficacy performance of two commercial vaccines registered in Canada, under field conditions, on a commercial farm where *E. coli* was found to be the source of diarrhea.

Materials and methods

Experimental design and vaccination regimens

Sixty gilts were selected for this study from a farrow-to-wean farm with 1000 sows, located in Manitoba (Canada). Since the farm had a 4-week batch farrowing system, two consecutive batches of 30 gilts were selected and randomly divided into two groups. The first batch was divided into groups 1A, n = 15, Vaccine A (Suiseng®) and 1B, n = 14, Vaccine B, and the second batch was divided into groups 2A, n = 15, Vaccine A (Suiseng®) and 2B, n = 15, Vaccine B. The gilts enrolled in each group were identified by a spray marker, as were the piglets that those sows adopted from other sows not included in the study; they were ear notched so as not to include them in the final production parameters.

Among the selection criteria of the farm included in the protocol, the commercial farm under study had to be one affected by neonatal diarrhea caused by colibacillosis, i.e., more than 5% of the litters had to be affected by *E. coli* or *Clostridium perfringens*-induced neonatal diarrhea.

Fecal samples from piglets suffering from diarrhea were sent from the candidate farm to Hipra's Diagnos® Laboratory (Spain). Thus, samples from 9 piglets were sent on 3 FTA® (Fast technology for analysis of nucleic acids) ELUTE cards (Whatman Inc., Florham Park, NJ). A multiplex polymerase chain reaction (PCR) test, adapted from previous studies,[7,8] was performed to detect genes encoding F4, F5, F6 adhesion factors and LT of *E. coli*, and the β-toxin of *Clostridium perfringens* type C.

The vaccination protocol consisted of two doses by the intramuscular route 6 and 3 weeks before farrowing. Primary vaccination and revaccination were administered to the first batch on April 4th and 25th 2016, respectively; and May 2nd and 23rd 2016 in the second batch.

All the procedures described were performed following the guidelines from the Canadian National Farm Animal Care Council – Code of practice for the care and handling of pigs.

Safety and efficacy

Several parameters were assessed to evaluate and compare the safety and efficacy of both products. The first safety parameter evaluated was the body temperatures of all the animals involved in the study in relation to the first vaccination and revaccination. Temperatures were taken at the following times: the day before vaccination (-24h), time before vaccination (0h), 6 hours after vaccination (6h), one day after vaccination (24h) and two days after vaccination (48h). The basal temperature was obtained from the mean of temperatures recorded on -24h and 0h. These data were used to obtain the mean body temperatures of each of the groups and the temperature increases produced by vaccination.

The second parameter tested was that of local reactions at the injection site and two variables were used: Swelling or No swelling.

Lastly, the third parameter evaluated was that of general type reactions, using 7 different variables: Type 1 prostration, Type 2 prostration, Type 3 prostration, Anorexia, Death, Abortion or No systemic reaction.

To evaluate efficacy, piglets from gilts vaccinated with both products were monitored throughout the lactation period (mean of 19.49 ± 1.35 days). The following were evaluated as primary parameters: litters affected by diarrhea and piglets affected by diarrhea. The secondary parameters evaluated were: mortality of suckling piglets, the daily average weight gains of those piglets and, if they had diarrhea, to determine its causes.

Statistical methods

A statistical analysis of the results was undertaken at the University of Girona Statistics Department to compare the results of both the safety and efficacy of the vaccine. The statistical analysis consisted of a case-control study. SPSS software v.19 (SPSS Inc., Chicago, IL) was used to read, process and obtain results.

In the event that the quantitative variables were normal, a Student's t-test was used to compare the null and alternative hypotheses. In the case of non-normality, a non-parametric Mann-Whitney U test was used to compare the medians. The contingency table for two qualitative variables and chi-square test were used for qualitative variables to compare whether the two variables were independent or were related. Data are expressed as mean ± S.E.M. Statistics decisions were made using a significance level value of 0.05.

Results

Temperature measurements

Regarding the comparison of safety, the mean body temperature after vaccination always remained lower in the groups vaccinated with Suiseng® than in the groups vaccinated with Vaccine B. The mean body temperatures of all the groups before and after vaccination are summarized in Figure 1.

In the first batch vaccinated (baseline temperature was 37.98°C at the first vaccination and 37.95°C at revaccination), statistically significant differences were found at 6 hours after the first vaccination (Figure 1A): 38.41°C in 1A and 38.94°C in 1B (Mann-Whitney U test, P-value < 0.001) and 6 hours after revaccination (Figure 1B): 38.13°C in 1A and 38.40°C in 1B (Student's t-test, P-value < 0.05).

In the second batch vaccinated (baseline temperature was 37.94°C at the first vaccination and 37.93°C at the time of revaccination), statistically significant differences were found 6 hours after the first vaccination: 38.15°C in 2A and 38.51°C in 2B (Student's t-test P-value < 0.05) and 48 hours after the first vaccination (Figure 1C): 37.75°C in 2A and 38.10°C in 2B (Student's t-test, P-value < 0.05).

Regarding the body temperature increases that occurred after vaccination (data not shown), group 1A showed a mean increase of 0.98°C 6 hours after the first vaccination, whereas group 1B showed a mean increase of 0.86°C (Student's t-test, P-value < 0.05).

In relation to the other monitored parameters relating to safety, no local type reactions were reported in either group. With regard to general type reactions, results for all the groups are summarized in Table 1.

Remarkable general type reactions were only observed in one animal in group 2B that began with grade 2 prostration (response only to strong stimuli) and anorexia by 6 and 24 hours, respectively, after the primary vaccination. Subsequently, the animal exhibited vaginal discharge by 48 hours' post-vaccination and finally aborted its fetuses (16 fetuses were counted) (Figure 2).

Incidence of neonatal diarrhea

The 3 FTA cards evaluated were positive for the *E. coli* F4 adhesion factor; thus, in the end, it was decided to include this candidate farm in the study.

The results from litters and piglets with diarrhea were evaluated to compare vaccination efficacy. In the first batch, two litters from group 1A were affected (13.4% of litters) amounting to a total of 3 affected piglets (1.6% of the total of weaned piglets in the group). In group 1B, five litters were affected (35.7% of litters) amounting to a total of 39 animals (22.4% of the total number of piglets in the group).

In the second batch of gilts, a total of 7 piglets were affected by diarrhea (3.8% of all weaned piglets in the group) in two litters (13.4% of litters) in group 2A. In group 2B, 25 animals were affected by diarrhea (18.2% of weaned piglets in the group) in six litters (46.1% of litters).

The piglets with diarrhea did not require any specific antibiotic treatment and they were simply administered potato starch as an alternative approach to limit symptomatic incidence of scouring. In this study, the use of antibiotic was not at all needed, with no detrimental concerns on animal welfare. In addition, fecal samples were taken from 21 piglets that had diarrhea in both vaccinated groups and analysis showed that 86% of the samples were positive for F4, 57% were positive for F5 and 29% of the samples were positive for LT.

Production parameters

Production parameters are presented in Table 2. There were no substantial differences between groups regarding mortality and most animal deaths were due to them being non-viable or crushed. The reduced trend in born alive piglets, birthweight, weaned piglets and weaning weight in group B was due to the fact that there was initially one animal fewer in group 1B (random group split up) and two fewer animals in group 2B (one aborted 48 hours after the first vaccination and one animal euthanized at farrowing date due to dystocia).

Figure 1: The effect of vaccination on the body temperature of gilts vaccinated with Vaccine A (Suiseng®) or Vaccine B. Temperatures (°C) were monitored the day before vaccination (-24h), time before vaccination (0h), 6 hours after vaccination (6h), one day after vaccination (24h) and two days after vaccination (48h). Body temperatures of **(A)** batch 1 gilts after first vaccination, **(B)** batch 1 gilts after revaccination, **(C)** batch 2 gilts after first vaccination, and **(D)** batch 2 gilts after revaccination are shown. Values represent the mean ± SEM of 14-15 animals per group.

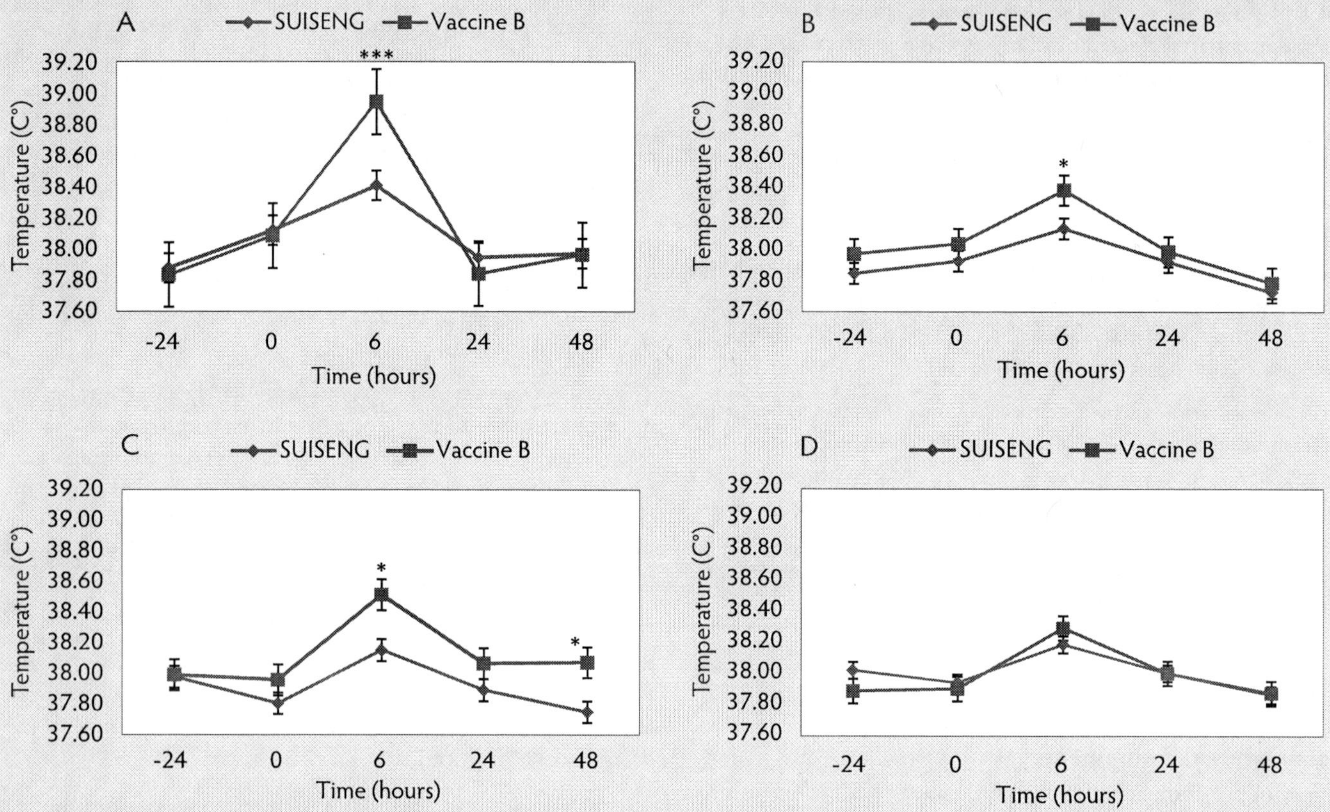

Statistically significant difference $^*P < 0.05$, $^{***}P < 0.001$.

Conclusions

Neonatal diarrhea currently continues to be one of the pathologies that has the highest cost for producers.[4] *E. coli* is still one of the primary triggers among the infectious agents that cause neonatal diarrhea in pigs.[9] Thus, current production conditions and the restrictions on the use of antibiotics to which we are subject, mean that, nowadays, vaccination of pregnant sows for controlling neonatal diarrhea is widespread.[9] This is largely due to the excellent results obtained when it is combined with good hygiene and management practices.[10]

The use of state-of-the-art vaccines that incorporate innovative adjuvants and new purification techniques for antigens are essential in cases where there is a high infection pressure or chronic colibacillosis.[10] It has previously been demonstrated that vaccination is an effective strategy to reduce mortality due to neonatal diarrhea on high infection pressure farms.[11] Along the same lines, the current study shows differing diarrhea-preventive effects of the two vaccines evaluated. Thus, Vaccine A (Suiseng®) exhibited a 6-fold lower incidence of diarrhea in suckling piglets from gilts vaccinated with it than in suckling piglets from gilts vaccinated with its competitor, Vaccine B. Moreover, Vaccine A (Suiseng®) showed greater safety than Vaccine B by causing a smaller increase in post-vaccination body temperature and by not showing any local or general type reactions, which may end in abortion and the resultant economic loss to the producer.

The current comparative evaluation indicates that immunization of sows with Suiseng® can effectively protect their offspring against neonatal *E. coli* diarrhea.

Acknowledgement

The authors would like to thank the Swine Health Professionals technical team for their help and cooperation in monitoring and carrying out tasks, the University of Girona Statistics Department for the statistical analysis performed, and the Diagnos® laboratory for the analytical support given in conducting this study.

Table 1: Summary of gilts' general adverse reactions recorded after first vaccination and revaccination with Vaccine A (Suiseng®) or Vaccine B.

	First vaccination				Revaccination			
	Gilt	6h	24h	48h	Gilt	6h	24h	48h
1A	6751	–	Pr2	Pr1	6751	Pr2	Pr1	–
1B	6764	Pr2	–	–		–		
	6759	Pr2	Pr1					
2A	6772	–	–	Pr1		–		
	6774	–	Pr1	–				
2B	6791	–	Pr1	–	6782	Pr2	–	–
	6784	Pr1	–	–		–		
	6795	Pr2	Pr2 + Ax	Pr2 + Ax + Ab				

– No systemic reactions; Ab, Abortion; Ax, Anorexia; D, Death; Pr, Prostration; Pr1, Inactive but responds to weak stimuli; Pr2, Does not respond to weak stimuli but does to strong ones; Pr3, Does not respond to strong stimuli.

Figure 2. Vaginal discharge image from one animal 48 hours post primary vaccination with Vaccine B. The gilt finally aborted 16 fetuses.

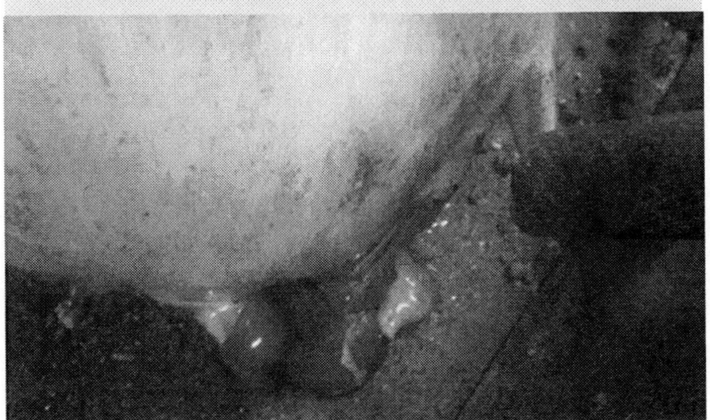

References

1. Shi, D. et al. Frequency of virulence factors in Escherichia coli isolated from suckling pigs with diarrhoea in China. *The veterinary journal*. 2014;199:286–289.

2. Westin, R. et al. Post-mortem findings and piglet mortality in relation to strategic use of straw at farrowing. *Preventive Veterinary Medicine*. 2015;119(3-4):141-152.

3. Kongsted, H. et al. The effect of New Neonatal Porcine Diarrhoea Syndrome (NNPDS) on average daily gain and mortality in 4 Danish pig herds. BMC *Veterinary Research*. 2014;10(1):90.

4. Sjölund, M. et al. Financial impact of disease on pig production. Part III. Gastrointestinal disorders. *Proceedings of 6th ESPHM*, Sorrento, Italy. 2014;189.

5. Rivera E. Ginseng extract in aluminium hydroxide adjuvanted vaccines improves the antibody response of pigs to porcine parvovirus and *Erysipelothrix rhusiopathiae*. Veterinary immunology and immunopathology. 2003;91(203):19-27.

6. Chase C. Swine immunology: What does it take to make the immune response take? AASV. 2012;411-416.

7. Zhang W, et al. Prevalence of virulence genes in *Escherichia coli* strains recently isolated from young pigs with diarrhea in the US. *Vet. Microbiol.* 2007;123(1-3):145-52.

8. Baums C, et al. Diagnostic multiplex PCR for toxin genotyping of *Clostridium perfringens* isolates. *Vet. Microbiol.* 2004;100(1-2):11-6.

9. Melkebeek V. et al. ETEC vaccination in pigs. *Veterinary Immunology and Immunopathology*. 2013; 152:37-42.

10. Guerra N. et al. Efficacy of Suiseng®: a novel friendly adjuvanted vaccine against neonatal colibacilosis and clostridiosis. *Proceedings of the 21st IPVS Congress*, Vancouver, Canada, 2010;778.

11. Bernal I. et al. Field study to assess the economic benefit of vaccination with Suiseng® on a Portuguese farm. *Proceedings of the 24th IPVS Congress*, 8th ESPHM, Dublin, Ireland, 2016;205.

Table 2: Summary of productive parameters obtained in sows vaccinated with Vaccine A or Vaccine B.

	Group 1A	Group 1B	Group 2A	Group 2B
Number of gilts	15	14	15	15
Total born alive	207	194	196	149
Total stillborn	7	9	11	6
Total mummified	5	5	5	5
Total birthweights (Kg)	301.4	280.8	302.8	229.2
Mortality	17	17	19	15
Litters with foster on	2	3	8	9
Litters with foster off	4	4	4	6
Total pigs weaned	190	174	180	137
Total weaning weight (Kg)	1613.5	1500.4	1013	790.2

1A, batch 1 animals vaccinated with Vaccine A (Suiseng®); 1B, batch 1 animals vaccinated with Vaccine B; 2A, batch 2 animals vaccinated with Vaccine A (Suiseng®); 2B, batch 2 animals vaccinated with Vaccine B.

The impacts of in-feed antibiotics, a unique organic acid blend, or an in-feed acidifier on nursery pig performance

K.M. Hogan[1], PhD; S. Gould[2], BS; Y. Han[3], MS, PhD; J. Patience[2], MS, PhD; C. Smits[3], MS, PhD; S. Webster[1], MS
[1]Trouw Nutrition North America, Highland, Illinois; [2]Iowa State University, Department of Animal Science, Ames, Iowa; [3]Trouw Nutrition Research and Development, Boxmeer, The Netherlands

Introduction

Due to socio-economic pressures and a changing regulatory landscape, the interest in effective feed additive alternatives to in-feed antibiotic supplementation in nursery pigs after weaning has increased. Elarom-F Plus (Selko Feed Additives; Trouw Nutrition) is a commercially available blend of short chain fatty acids (SCFAs), medium chain fatty acids (MCFAs) and slow release organic acids. Previous studies have demonstrated that the components present in Elarom-F Plus can help stabilize the microbiota of pigs and may improve gut barrier function.[1,2] Additionally, it is regularly recommended that starter diets for weaned pigs include an acidification component to aid in digestion.[3] Combiacid (Nutreco Canada) is a commercially available dry acidifier containing a blend of inorganic and organic acids. An integrated nutritional approach supports overall pig performance and could potentially be an effective strategy in antibiotic-free production systems. The purpose of this study was to assess the efficacy of a dry acidifier and/or an organic acid blend product with in-feed antibiotics.

Materials and methods

A total of 480 crossbred weaned piglets were selected for this experiment at Iowa State University Swine Nutrition Farm in Ames, IA, USA. Upon arrival, each pig was ear-tagged, individually weighed and vaccinated for PCV, PRRS, and *E. coli.* The average initial body weight was similar across all treatments (14.02 lb), with an equal number of barrows and gilts placed in each pen. The initial weight of the piglets was used to block pigs into experimental pens, which were randomly assigned to four treatments. Dietary treatments were fed in four phases. Phases 1, 2 and 3 were 7 days each, and phase 4 was 14 days. The dietary treatments consisted of a negative control, Elarom-F Plus, Combiacid or both Elarom and Combiacid. Each treatment was added at an inclusion rate of 4 lb/ton to the control diet at the expense of corn. The experiment was conducted under typical commercial conditions. Feed antibiotics were added during phases 2 through 4 (400 ppm chlortetracycline and 36 ppm tiamulin hydrogen fumarate during phase 2, 40 ppm carbadox during phase 3 and 4). All the pens had a four-space, dry self-feeder and two nipple drinkers to provide *ad libitum* access to feed and water with a floor space of 0.28 m^2/ pig (10 pigs per pen). The room temperature started at 85°F for the first 3 days. After day 3, the temperature gradually decreased by 0.5°F each day until the minimum temperature of 75°F was reached. The pigs were monitored for technical performance during the entire 5 week trial.

Results

The performance observed during this study was typical for nursery pigs of weaning age. On days 7 and 14, pigs receiving the diet with the addition of either Elarom-F Plus or Combiacid were observed to be heavier than the control ($P = 0.036$ and $P = 0.020$, respectively; data not shown). A 1.32 lb body weight increase was observed in the Elarom-F Plus treatment group at the end of the trial period, while a 1.46 lb body weight increase was observed in the animals fed a diet containing Combiacid (data not shown). The group receiving the combination treatment of Elarom-F Plus and Combiacid was observed to have the lowest F/G for the entire growing period (Table 1), with a decrease of 0.059 F/G when compared to the control diets. During the combined first two phases (days 0-14), it was noted that the group fed the Elarom-F Plus treatment had the highest ADG and lowest F/G (Table 2).

Significant treatment differences were obvious in phase 4 (days 22-35; Table 4), where an increased ADG of 30g was observed in the Elarom-F Plus treatment group when compared to the control group ($P = 0.017$). The observed differences in growth led to a decrease in F/G when fed a diet containing Elarom-F Plus. Additionally, when the diet contained both Elarom-F Plus and Combiacid, the F/G was decreased by 0.09 when compared to the control diets in the final growing period (Table 4).

Discussion

This experiment demonstrates the impact of diets containing Elarom-F Plus and/or Combiacid on performance, with significant differences observed in performance throughout each growing period, including the final diet phase. A similar formulation of the product has been assessed for many years overseas and in Canada, but the results have been difficult to replicate in the US. One interesting difference between this study and others recently performed in Europe and Canada with Elarom is the use of carbadox in the nursery diets. Carbadox is not approved in the EU and Canada, whereas it is approved and regularly used in the USA. Carbadox is a product that impacts enteric health and the microbiota, with both short term and lasting impacts on the

Table 1: Performance results for the full growing period (phases 1-4).

Day 0 - 35	Control	Combiacid	Elarom-F Plus	Elarom-F Plus + Combiacid	Std error	*P*-value
ADG (lb)	0.834	0.875	0.873	0.869	0.010	0.069
ADFI (lb)	1.268	1.308	1.316	1.276	0.033	0.243
F/G	1.526	1.497	1.503	1.467	0.017	0.097

Table 2: Performance results for days 0–14 (phases 1-2).

Day 0 -14	Control	Combiacid	Elarom-F Plus	Elarom-F Plus + Combiacid	Std error	*P*-value
ADG (lb)	0.489[a]	0.538b	0.542[b]	0.502[a]	0.019	0.026
ADFI (lb)	0.601	0.627	0.612	0.590	0.021	0.369
F/G	1.229[a]	1.166[b]	1.128[b]	1.188[ab]	0.016	< 0.001

[a,b] Means with differing lowercase letters were statistically different ($P \leq 0.05$).

swine gut microbiota.[3] An initial possible interaction of carbadox and Elarom-F Plus was observed in phase 3 (Table 3), but did not continue thereafter. Additionally, this potential interaction was observed in a trial at a different location. Further research may be conducted to determine if there is a microbiological or molecular impact from the use of the two products concurrently.

This trial result confirms the improved performance observed in previous global trials. Results from a meta-analysis of three trials performed in Canada resulted in an observed ADG increase and improved F/G when Elarom-Mn (Canadian formulation of Elarom-F Plus) was added to the diets (Table 5). Additionally, the results support the Trouw Nutrition R&D nursery pig global meta-analysis of several global trials, with significant increases in ADG and F/G (Figure 1).

References

1. Kien et al. Cecal infusion of butyrate increases intestinal cell proliferation in piglets. *J Nutrition.* 2007; 137:4.

2. Suiryanrayna and Ramana. A review of the effects of dietary organic acids fed to swine. *J Anim Sci Biotechnol.* 2015; 6:45.

3. Kansas State University. Starter Pig Recommendations. 2007. Available at http://www.asi.k-state.edu/research-and-extension/swine/research-and-extension/Starter%20Guidelines%20MF2300.pdf. Accessed November 7, 2016.

4. Looft et al. Carbadox has both temporary and lasting effects on the swine gut microbiota. *Frontiers in Microbiol.* 2014; 5:276.

Table 3: Performance results for days 15-21 (phase 3).

	Control	Combiacid	Elarom-F Plus	Elarom-F Plus + Combiacid	Std error	*P*-value
Day 15 - 21						
ADG (lb)	0.814[ab]	0.884[a]	0.752[b]	0.808[b]	0.029	0.007
ADFI (lb)	1.245	1.302	1.254	1.238	0.038	0.416
F/G	1.542[b]	1.484[b]	1.647[a]	1.535[b]	0.031	0.001

[a,b] Means with differing lowercase letters were statistically different ($P \leq 0.05$).

Table 4: Performance results for days 22–35 (phase 4).

	Control	Combiacid	Elarom-F Plus	Elarom-F Plus + Combiacid	Std Error	*P*-value
Day 22-35						
ADG (lb)	1.185[b]	1.210[ab]	1.267[a]	1.272[a]	0.030	0.017
ADFI (lb)	1.954[b]	1.997[b]	2.050[ab]	1.982[a]	0.050	0.194
F/G	1.646[a]	1.658[a]	1.624[ab]	1.559[b]	0.027	0.022

[a,b] Means with differing lowercase letters were statistically different ($P \leq 0.05$).

Table 5: Meta-analysis showing the impacts of Elarom-Mn* at 4 lb/ton in three separate nursery trials. Trials were performed at the Trouw Nutrition Agresearch farms in Quebec and Ontario, Canada.

Elarom-Mn* inclusion	Start weight	End weight	ADFI	ADG	F/G	Cost/lb	Mortality
	lb	lb	lb/d	lb/d	lb:lb	$/lb	%
0 lb/ton	12.68	49.82	1.31	0.946	1.384	1.76	1.7
4 lb/ton	12.59	50.71	1.32	0.974	1.357	1.75	0.6
Change		+ 0.89	+ 0.01	+ 0.028	- 0.027	- 0.01	- 67%
P-value		0.048	NS	0.014	0.003	NS	0.007

* Elarom-Mn is the Canadian formulation similar to Elarom-F Plus.

Figure 1: Meta-analysis of nine controlled research and development trials assessing Presan-FX** performed throughout Europe.

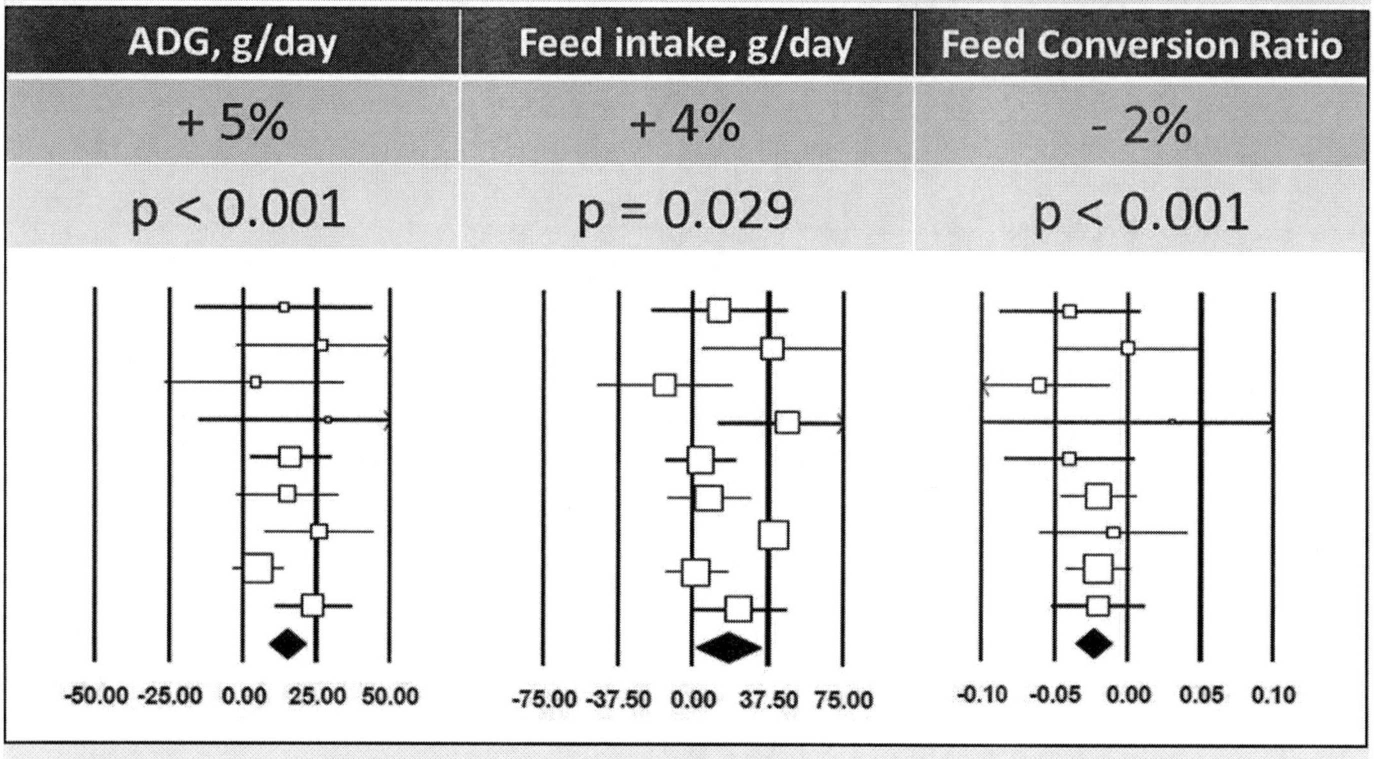

** Presan-FX is the European formulation similar to Elarom-F Plus.

Field evaluation of vaccination of piglets at processing using Fostera® PRRS

Keith Aljets[1], DVM; Jose Angulo[2], DVM
[1]Veterinary Medical Center and VMC Management, Williamsburg, Iowa; [2]Zoetis, Parsippany New Jersey

Introduction

Porcine reproductive and respiratory syndrome (PRRS) is still causing significant economic losses to the swine industry in North America. Controlling this disease is difficult, and there is a great variability on the efficacy of PRRS control strategies to reduce negative impact in production;[1] therefore, strategies for managing whole herd immunity along with bio-management practices to reduce resident Wild-Type PRRS virus (WT-PRRSv) circulation, and biosecurity protocols to prevent new WT-PRRSv introduction are critical for achieving long term success in controlling PRRS.

PRRS vaccination with a modified live vaccine is a common tool to homogenize immunity to achieve stability in the breeding herd,[1] and in growing pigs, to limit disease impact obtaining better weight gain during WT-PRRSv post-challenge period.[2] Fostera® PRRS has demonstrated efficacy in helping protect against reproductive disease[3] and safety and efficacy in piglets as early as 1 day of age.[4,5] In addition, Fostera® PRRS has shown cross-protection through reduction of lung lesions and viremia and increase of average daily gain (ADG) during the post-challenge period against different PRRS viruses including the most recent clinically important emerging virus PRRSv RFLP 1-7-4.[6,7]

The objective of this study was to evaluate the performance of pigs vaccinated at early age (3-5 days of age) with Fostera® PRRS.

Material and methods

This evaluation was conducted in a pig-flow from a 2,800-sow farm that sends weaned pigs to ten nursery sites in southeast Iowa. The breeding herd is PRRS positive and was infected with Wild-Type PRRS virus RFLP cut-pattern 1-26-2 in 2012, and started quarterly mass vaccinations with Ingelvac PRRS® MLV. Despite this vaccination protocol, the breeding herd was leaking this WT-PRRSv downstream in weaned pigs. The sow farm was not overtly clinical to PRRS virus; however, the presence of WT-PRRSv in weaned pigs was a concern, so a decision was made to use Fostera® PRRS under a whole herd protocol. Sows were vaccinated on February 4, 2015, with the goal of providing more stability and minimizing the leakage downstream of resident WT-PRRSV. The herd vaccination strategy involved alternating both modified-live PRRS vaccines (Ingelvac PRRS® MLV and Fostera® PRRS) every quarter. In addition to breeding herd vaccination, piglet vaccination with Fostera® PRRS (2 mL) at processing (3-5 days of age) was implemented immediately following the first sow vaccination with Fostera® PRRS.

Reproductive performance measured as conception rate was closely monitored before and after introducing Fostera® PRRS to evaluate any potential negative impact due to new vaccine strain introduction. Monitoring using Individual-Chart (I-Chart) Statistical Process Control (SPC) for Key Performance Indicators (KPIs) in the sow farm and nursery sites was put in practice on an on-going basis. Nursery mortality was analyzed for 13 months (November 2014 to February 2016) including eight nursery turns (turn = 10 individual close-outs from nursery sites). This monitoring included 2 turns with non-vaccinated pigs and 6 turns with vaccinated pigs and WT-PRRSv was detected at different levels during the monitoring process. Generalized Linear Mixed Models for binomials was used for statistical analysis in nursery mortality.

Results and discussion

There was no reported or measured negative impact on reproductive performance and early vaccinated suckling pigs after implementation of Fostera® PRRS (Figures 1and 2). During the period of nursery performance evaluation, the

sow farm became infected with a non-resident WT-PRRSv RFLP 1-7-4 in December 2015. The negative impact of this new WT-PRRSv introduction to the breeding herd was considered moderate based on abortions and pre-wean mortality percent (PWM%) (Figure 3). Time to Baseline Production (TTBP) defined as the time to recover to the number of pigs weaned prior to PRRS outbreak was achieved in 19 weeks (Figure 4) and Time to Stability (TTS) defined as the time in weeks it takes to wean PRRS PCR negative pigs at weaning during three consecutive sampling points was achieved in 26 weeks (Figure 5). A clear improvement in nursery mortality (reduction of average and variability) in vaccinated over non-vaccinated pigs is shown in figure 6. This improvement was regardless of different detection levels of WT-PRRSv at weaning (1-26-2 & 1-7-4) at different turns as indicated in figure 6. The reduction in nursery mortality was statistically significant (*P*-value 0.0001) when vaccinated turns vs non-vaccinated turns were compared having 6.08% vs 2.01% respectively.

Conclusions

This evaluation demonstrated no negative impact on reproductive performance in the breeding herd following the use of Fostera® PRRS and no detrimental impact in vaccinated young pigs (3-5 days of age) measured as pre-wean mortality and clinical observations confirming safety results from previous study.[4] Additionally, there was a significant improvement in nursery mortality in early vaccinated pigs with Fostera® PRRS under different scenarios of WT-PRRSv natural challenge. These results support the value

Figure 1: Run chart for conception rate before and after implementing Fostera® PRRS vaccine for first time in the breeding herd (MINITAB® 17.1.0).

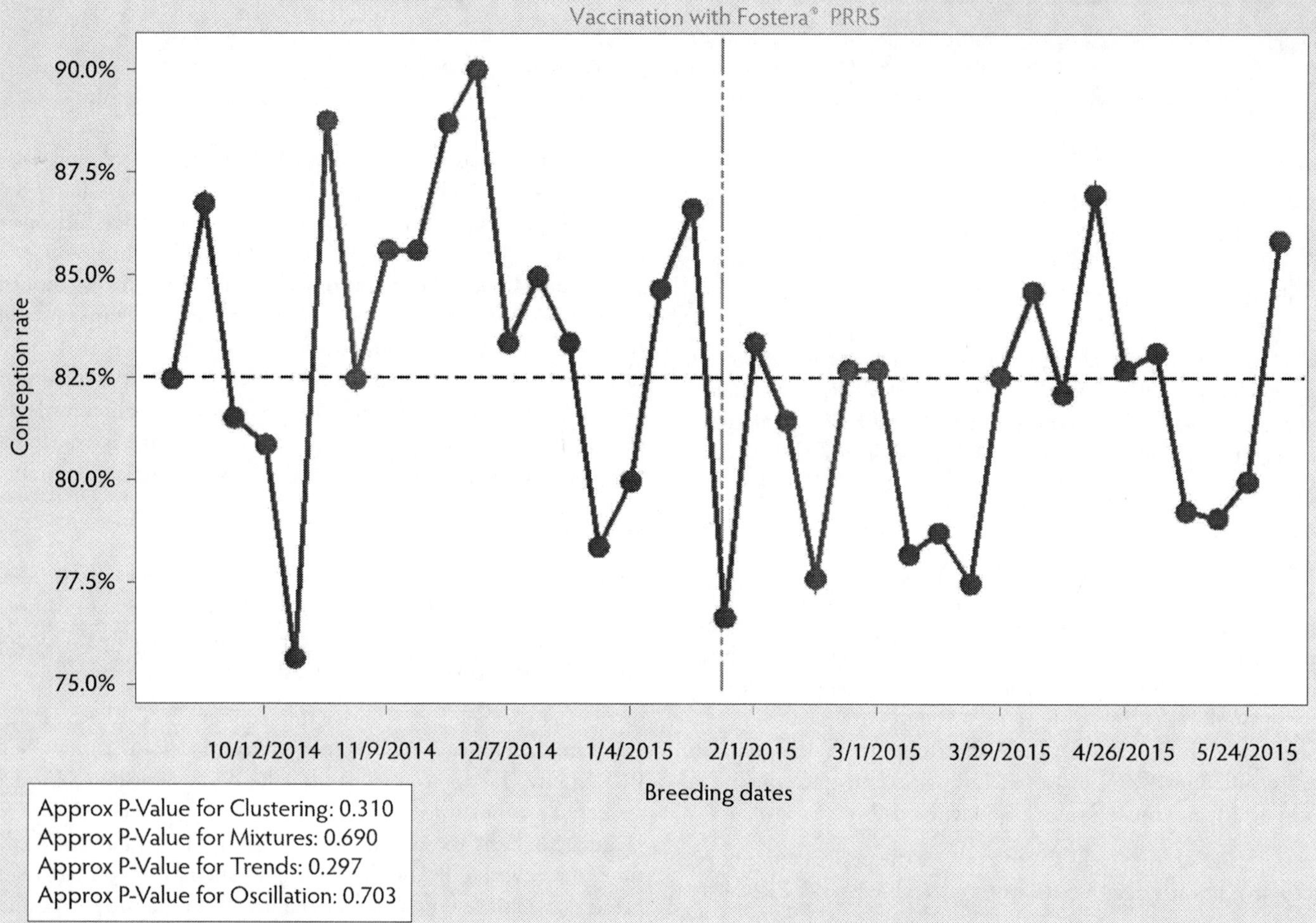

of PRRS vaccination with Fostera® PRRS at processing under commercial conditions and confirm the benefits of PRRS control strategies around whole herd immunity to maximize the performance of tools available under field conditions.

References

1. Linhares C.D, Cano J.P, Torremorrel M., Morrison R.B. Comparison of time to PRRSv-stability and produciton losses between two exposure programs to control PRRSv in sow herds. Preventive Veterinary Medicine 116, 111-119. 2014.

2. O'Brien B. Fostera® PRRS keeps pigs growing in the face of heterologous challenge. AASV Proceedings. 2015. p. 537-238.

3. O'Brien B. Fostera® PRRS for the aid in prevention of reproductive disease due to virulent PRRS in late gestation gilts. AASV proceedings. 2015. p. 367-368.

4. Ankenbauer RG, Keith ML, TL Marin. Field safety of a PRRS MLV vaccine administered once intramuscularly to pigs of 1 day of age. IPVS proceedings. 2014. Poster 392.

5. Ankenbauer RG, Keith ML, TL Marin. Efficacy of a PRRS MLV vaccine administered once intramuscularly to pigs of 1 day of age. IPVS proceedings. 2014. Poster 393.

6. Angulo J., Allison JRD., Calvert JG., O'Brien B., Ankenbauer RG. Efficacy of a PRRS MLV vaccine against a genetically diverse range of PRRSv isolates. Allen D Leman Swine Conference proceedings. 2015. p.16.

7. Angulo J., Jay J.G., Nemechek G., Senn MK., Keith ML., Taylor LP, Pearce DS., Brice C, Lenz MC. Evaluation of Fostera® PRRS cross-protection against a contemporary lineage 1 (RFLP 1-7-4) PRRS virus. AASV proceedings 2016. p. 170-174.

Figure 2: Individual SPC chart for pre-wean mortality before and after piglet PRRS vaccination at processing with Fostera PRRS (MINITAB® 17.1.0).

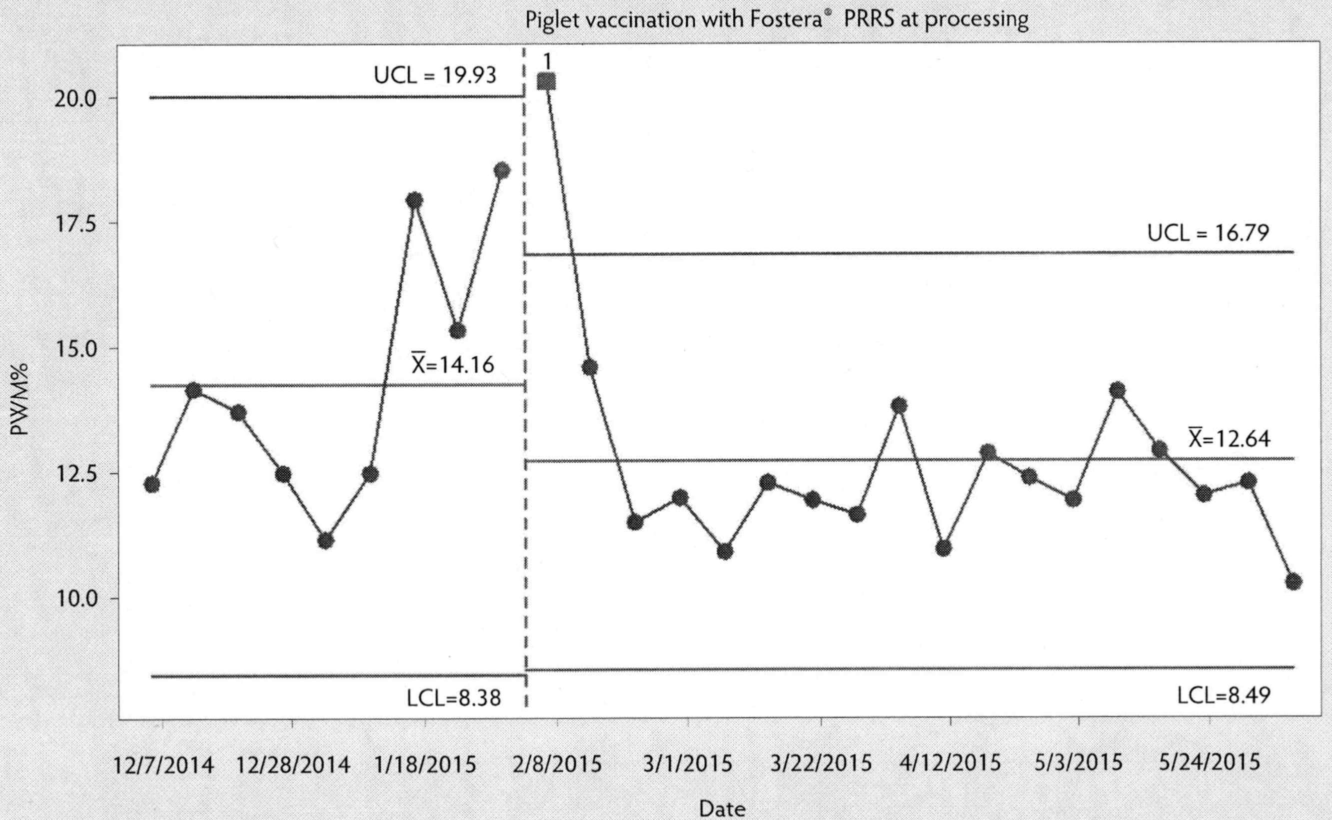

Figure 3: New WT-PRRSv introduction impact measured by Statistical Process Control (SPC) (MINITAB® 17.1.0). **A)** I Chart of abort rate/1000 sows, 12 months **B)** I Chart of PWM(%). Preweaning mortality, 12 months.

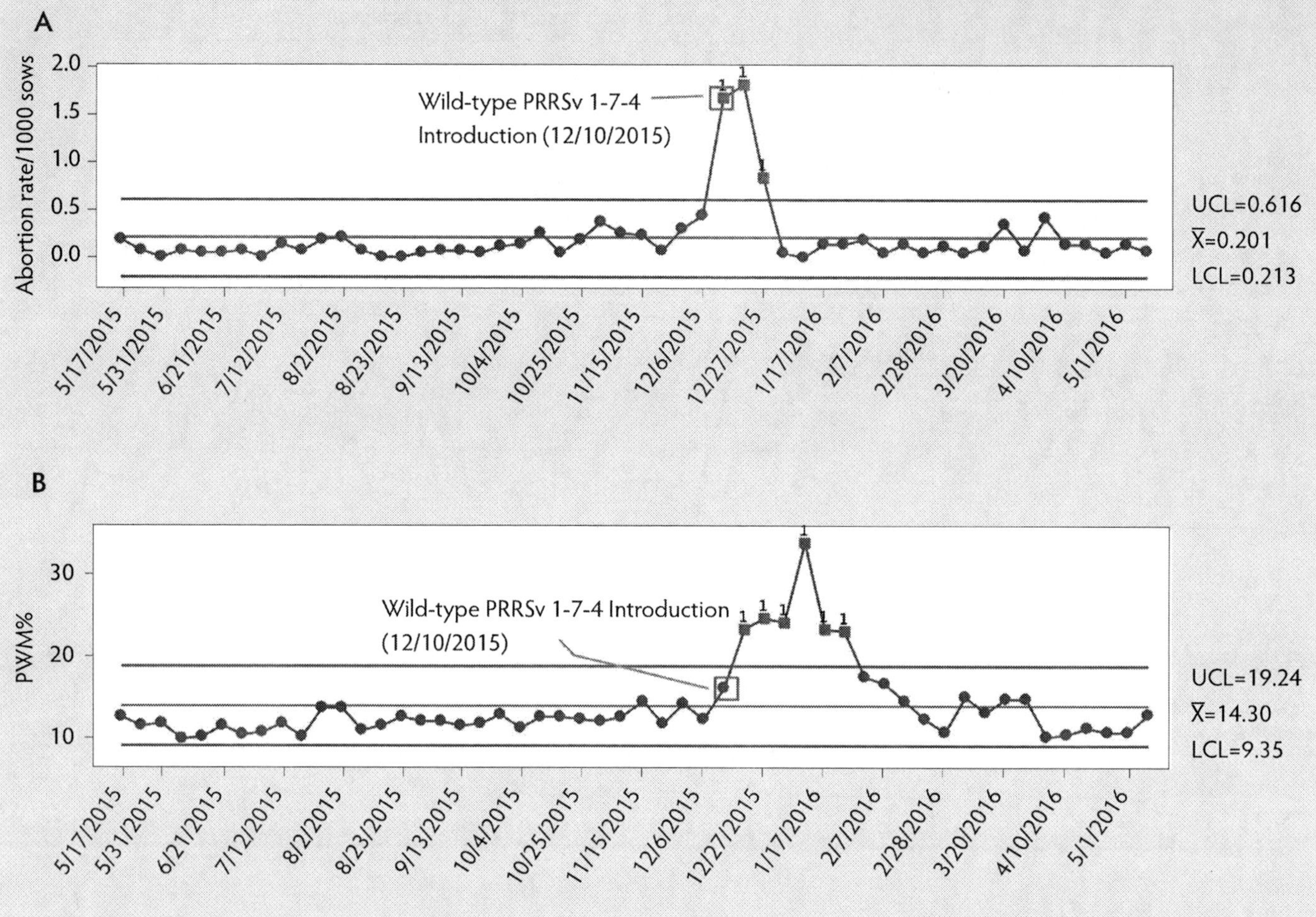

Figure 4: Time to Baseline Production (TTBP) after new WT-PRRSv introduction (MINITAB 17.1.0).

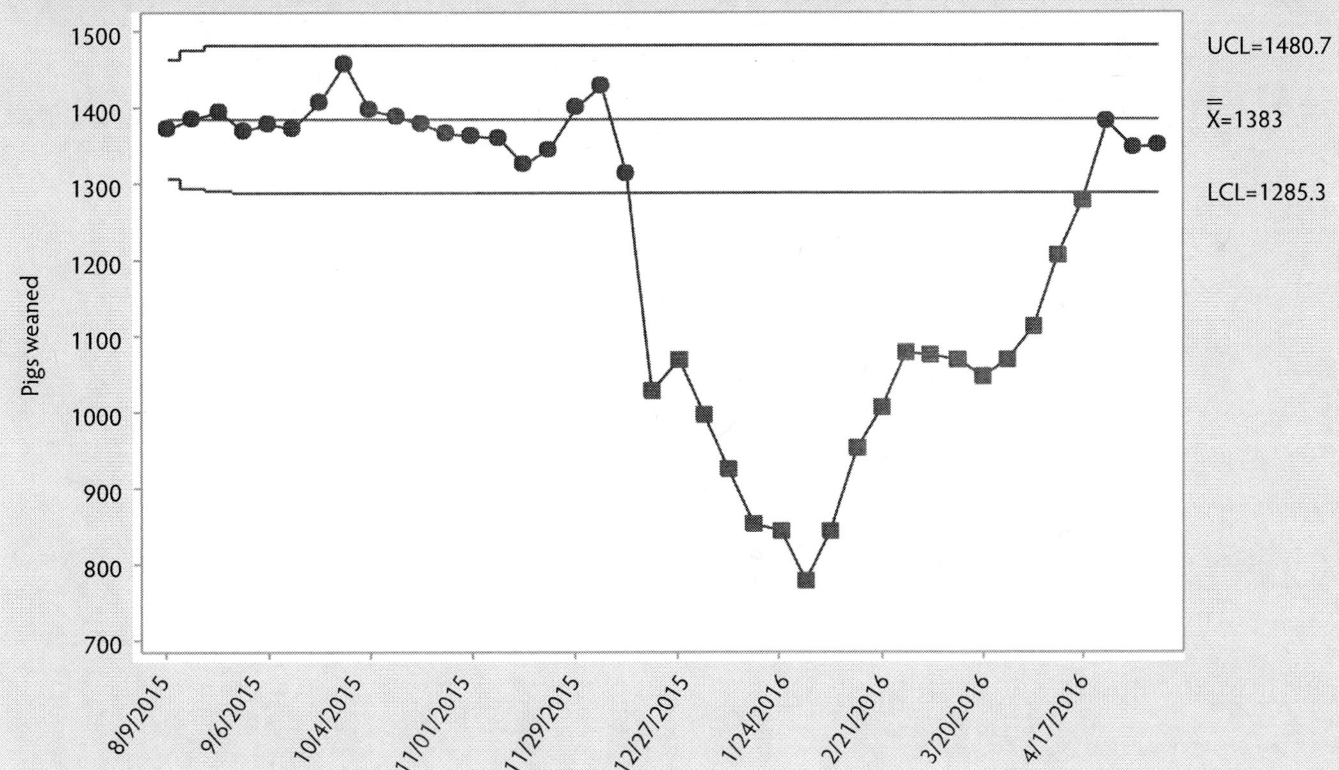

Figure 5: PRRS PCR monitoring in suckling pigs.

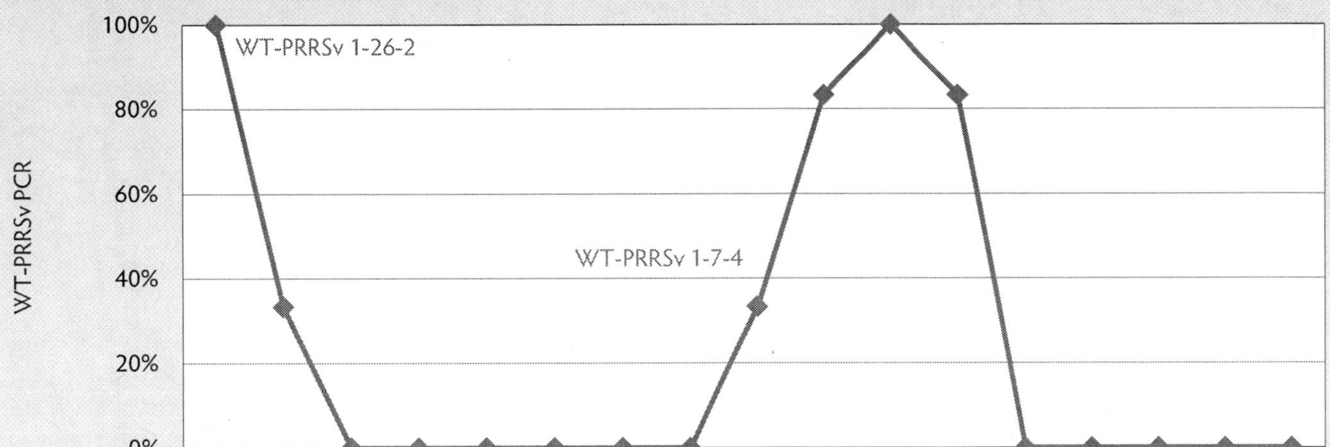

	5/15/2015	8/12/2015	9/3/2015	9/18/2015	10/1/2015	10/29/2015	11/12/2015	12/1/2015	12/10/2015	3/17/2016	3/30/2016	4/14/2016	4/27/2016	5/26/2016	6/21/2016	7/21/2016	8/18/2016
WT-PRRS	100%	33%	0%	0%	0%	0%	0%	0%	33%	83%	100%	83%	0%	0%	0%	0%	0%

Figure 6: Nursery mortality monitoring in early vaccinated and non-vaccinated pigs with different WT-PRRSv detection levels around weaning (MINITAB® 17.1.0).

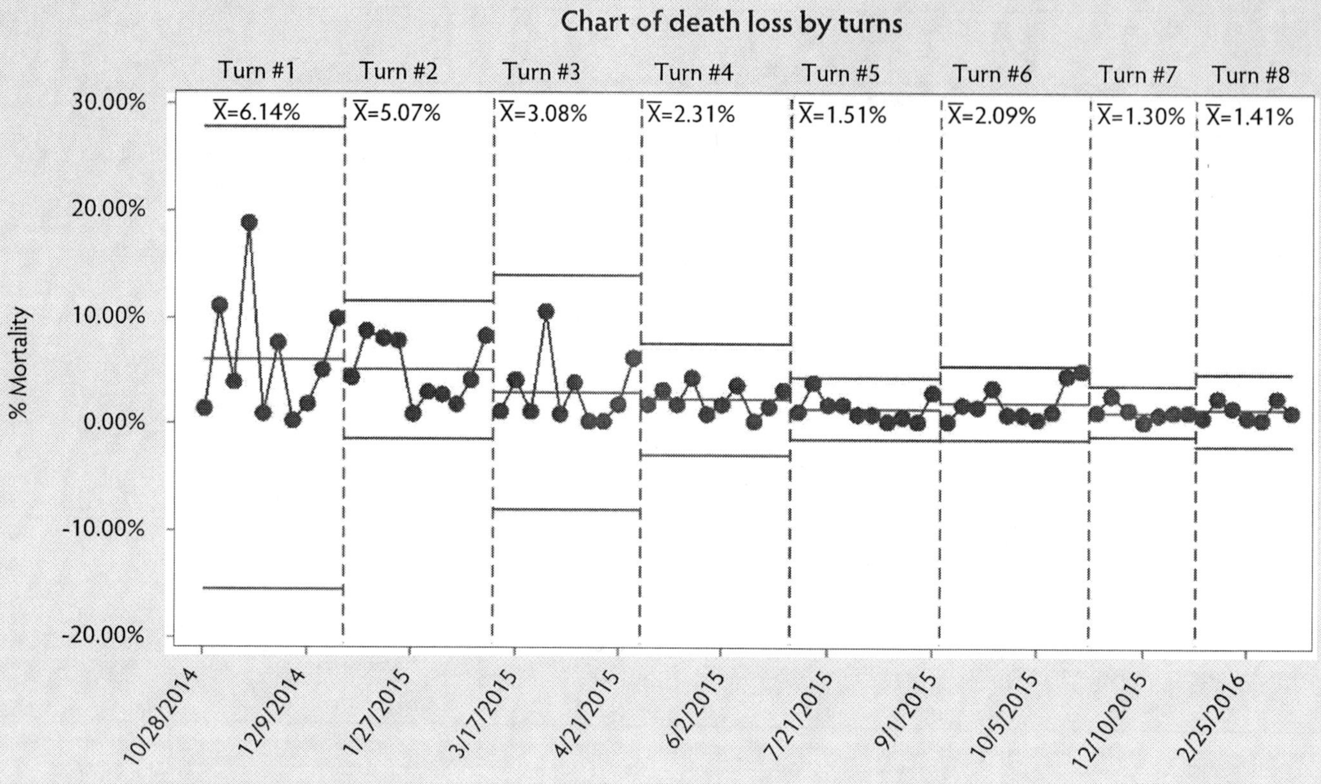

Turn #1:
No PRRS Vx
+/- WT-PRRSv
1-26-2 at weaning

Turn #2:
No PRRS Vx
+/- WT-PRRSv
1-26-2 at weaning

Turn #3 and #4:
Fostera PRRS
+/- WT-PRRSv
1-26-2 at weaning

Turn #5 and #6:
Fostera PRRS
Negative WT-PRRSv
at weaning

Turn #7 and #8:
Fostera PRRS
+ WT-PRRSv
1-7-4 at weaning

Determining *Mycoplasma hyopneumoniae* status in commercial breeding herds

E. Spronk[1], DVM; N. Garbes[2], DVM; L. Galina Pantoja[2], DVM, PhD
[1]Eichelberger Farms, Wayland, Iowa; [2]Zoetis, Parsippany, New Jersey

Introduction

For pig producers and veterinarians, the health and economic impact of *Mycoplasma hyopneumoniae* (MH) is clear. In the US, enzootic pneumonia caused by MH is considered one of the "big three" respiratory diseases in swine, following the porcine respiratory and reproductive syndrome (PRRS) and influenza viruses. Less clear, however, is how to manage this costly and persistent pathogen. The first step to manage the disease is to understand the current status within the herd. An MH herd status classification was introduced at the AASV 2016 conference.[1] By following well-defined diagnostic requirements, breeding herds can be categorized as positive stable, positive unstable, provisionally negative or negative. The objective of this study was to adhere to these diagnostic requirements in multiple breeding herds within a swine operation and to characterize their MH status in a commercial setting. To achieve this goal, we focused on the ability of detecting positive samples and testing diagnostic indicators of uniform immunity.

Material and methods

The diagnostic criteria used adhered to previous recommendations to determine the herd status classification of breeding herds.[1] Forty five laryngeal swabs were taken from wean age pigs and the same site was tested (also with wean age pigs) 30, 60 and 90 days later. A sample size of 45 swabs was chosen to detect a 10% prevalence of colonization in a population of 1000 pigs.[2]

Breeding herd sites

In this study, a sow herd was defined as a population of animals that included the breeding animals and their offspring. If a gilt development unit (GDU) was located on the premises, the GDU was considered part of the breeding herd. The study was conducted in a commercial production system which included 8 breeding herds: M, OB, LV, CL, H, HR, FC and B. These breeding herds were located in various counties in Iowa and housed from 1,200 to 5,600 sows. Some sow farms had attached gilt development units (GDU), one had an onsite detached GDU, and some herds used gilts from GDUs located at different geographic locations. All gilts entered the sow population at 28-30 weeks of age. To minimize PRRSV breaks during the winter months, the sow farms were closed at the end of October 2015 and opened up to new gilt introductions in the middle of March 2016.

Mycoplasma hyopneumoniae vaccination

Prior to moving to the breeding herd sites, gilts were vaccinated at 3 days of age with an MH vaccine and at 10 and 23 weeks of age with a PCV2, MH and PRRSV combo. Piglets born at the breeding herd received PCV2 and MH vaccinations at 3 days of age and 3 weeks of age and subsequent boosters with MH and PRRS MLV vaccines at 6 weeks of age. The MH status of the gilts prior to entry to the breeding herds was unknown and will be the subject of future investigation.

Sampling protocol

Laryngeal swab samples were collected in May, June, July and August of 2016. The 45 swabs were obtained via 15 swabs from each P1, P2 and ≥ P3 litters (parities 3-6). One pig per litter was targeted and fall back pigs were avoided. The oldest due-to-wean pigs on the farm were chosen and their age ranged from 12-28 days old. Laryngeal swabs were submitted for MH real time PCR[3] testing in pools of 5 to the diagnostic laboratory and when a pool was positive all samples were individually tested.

To evaluate other MH diagnostic indicators of immunity, in addition to evaluating colonization in suckling pigs, at farms B and FC, sera samples were also obtained from gilts that had recently entered the breeding herd (July sampling). At farm B, samples were collected from open gilts that were on farm between 6 and > 10 weeks. At farm FC, samples were collected between 7-14 weeks post entry. In the same farms serum samples were collected from suckling pigs close to wean age. These antibody tests were conducted to learn if the serological profiles were indicators of herd exposure/vaccination and colonization in young pigs. At these 2 sites, the biggest pigs in the litter were chosen as it was assumed that the pigs had adequate colostrum intake and therefore reflected the immune status of the breeding herd. Pigs selected for bleeding were 15-25 days of age. Serum samples were submitted for testing with an MH *IDEXX ELISA*. All tests were conducted at the Iowa State University Veterinary Diagnostic Laboratory.

Data analysis

The proportion MH positive samples were estimated at 0, 30, 60 and 90 days within a site and across sites. Based on the results farms were categorized as positive unstable (> 10% of wean-age pigs positive for antigen within respiratory tract), positive stable (< 10% of wean-age pigs positive for antigen within respiratory tract), provisionally negative or negative. The distribution of positive samples was estimated by parity to determine if certain parities were at a higher risk. The serological profiles of newly introduced gilts were characterized by estimating the percent positive and average S/P ratio.

Results and discussion

The herd-status classification focused on the breeding herd, which is defined as a population of animals that includes the breeding animals and their offspring. If a gilt development unit (GDU) was located on the premises, the GDU will be considered part of the breeding herd. In this study, samples were only collected from the breeding herd, but the MH health status reported included attached or detached GDUs on the same site.

A few deviations were made to the initial sampling protocol as the study progressed. To increase the chances of finding positive sampling, after the first round of sampling a decision was made to focus on P3-P4 litters (vs. P5-6) in the ≥ P3 category if they were available in the oldest room. On some of the smaller farms, multiple pigs out of one litter had to be collected primarily in P1 litters.

A total of 1,440 swabs from 8 sow farms was collected over 90 days. The use of wean pig laryngeal swabs detected positive samples in 6 out of 8 farms. Similar to previous reports the authors found laryngeal swabbing to be a satisfactory sampling technique to monitor colonization.[4] The proportion of positive samples by herd from all four collections is shown in Figure 1. The results indicated that farms CL and FC never showed a positive sample. These 2 farms can be considered provisionally negative farms and become negative if they meet further criteria previously reported (i.e., herds that have gone through a complete sow rollover for at least one following an MH elimination protocol and test negative by MH ELISA).[1] The remaining 6 farms were MH positive and farms B and HR showed the highest percentage of positive samples (8% and 6% respectively). All positive herds had <10% prevalence falling into the category of positive stable.

The number of positive samples at each of the four collection times is shown in Table 1. Our results showed variability in the proportion of positive samples in single sites when tested across time. The biggest variability was observed at farm HR where there were 0, 1, 13 and 0 positive pigs at 0, 30, 60 and 90 days. Our results are in agreement with previous reports of variability in the MH prevalence in groups of wean age pigs[5] and underscores the importance of conducting multiple group tests to determine the status of a single herd.

When combining data from all farms, more positive samples were collected in the month of July compared to May, June and August (Figure 2). A hypothesis could be that heat stress exacerbates *M. hyopneumoniae* dissemination; however, these observations should be replicated and the biological explanation for these findings remains to be determined. Another possible explanation could be that dams than farrowed in July did not have proper MH acclimation. It is more likely that the increase in the number of positive samples is related to the timing of the reopening to new introductions at the sow farms in of March 2015 and therefore the introduction of gilts with a different M. hyo status than the breeding herd. Another interesting finding was that the percentage of positive samples across all collection times was influenced by parity. Parity 1 litters were 18% positive, while other parities had lower number of positive pigs (P2: 5%, P3: 9%, P4: 7% and P5:1%). These observations need further validation but illustrate the importance of proper exposure and immunization of younger females prior to their first farrowing.

Figure 1: Percent *M. hyopneumoniae* PCR positive laryngeal swabs -average from 4 consecutive samplings at 0, 30, 60 and 90 days. The percentages were rounded up to the nearest number.

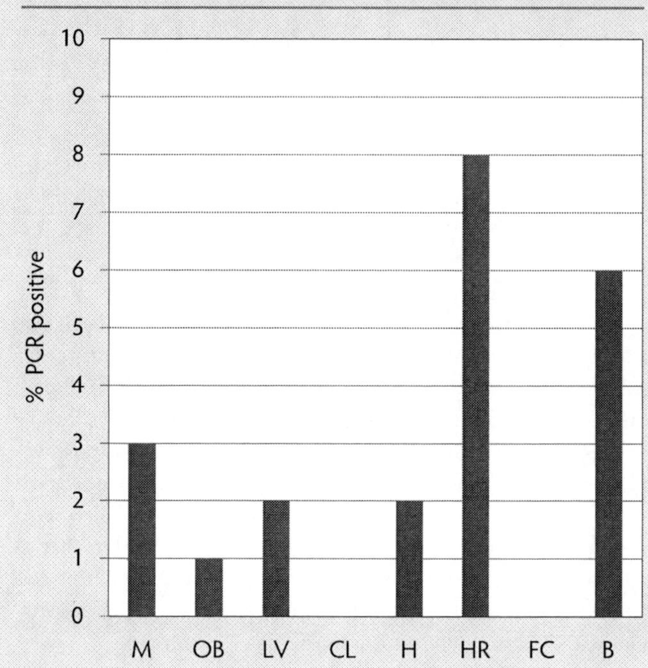

Serological information was obtained at 2 selected farms. Our results indicated that at farm B 60% of recently introduced gilts had MH antibodies (average S/P ratio 0.77) while in farm FC 100% of the gilts had detectable antibodies (average S/P 1.15). Serological profiles in gilts newly introduced to the sow farm could be an indicator of exposure and immunization. In this study 40% of gilts at farm B did not show evidence of MH antibodies and had the highest MH levels of colonization at weaning. Furthermore, while serological results from suckling piglets from farm B and farm FC were similar (73% positive with an average S/P ratio of 0.697 for farm B and 71% positives with an average S/P ratio of 0.657 in farm FC) parity appeared to have an effect. Notably, at farm FC piglets from younger parities had a higher percent of positive samples compared to farm B (P1: 95 vs.76%; P2: 64 vs. 55% and ≥P3: 53 vs. 87%. Since farm B had more pigs colonized at weaning, it could be speculated that seroconversion in newly introduced gilts and detection of antibodies in P1 and P2 litters can be an indicator of *Mycoplasma hyopneumoniae* natural or induced and passive immunization. Clearly this observation requires further investigation. Farm FC was provisionally negative (based on negative PCR results in wean aged pigs) so it

Table 1: Number of *M. hyopneumoniae* (MH) positive laryngeal swabs from wean-aged pigs. The results are presented by collection date and include the total of positive samples and percent of positive samples across 4 samplings.

Farm	Number and percent of MH positive laryngeal swabs in wean age pigs						
	Day 0 May (n = 45)	Day 30 Jun (n = 45)	Day 60 Jul (n=45)	Day 90 Aug, (n = 45)	Total number of positives (n = 180)	Percent positives across all samplings (n = 180)	MH herd status
M	0	0	3	2	5	2.7%	Stable
OB	0	0	2	0	2	1.1%	Stable
LV	1	1	2	0	4	2.6%	Stable
CL	0	0	0	0	0	0%	Provisionally negative
H	0	0	0	4	4	2.2%	Stable
HR	0	1	13	0	14	8.0%	Stable
FC	0	0	0	0	0	0%	Provisionally negative
B	0	5	5	0	10	5.6%	Stable
Total	**1**	**7**	**25**	**6**	**39**	**3.0%**	

Figure 2: Percent *M. hyopneumoniae* PCR positive laryngeal swabs collected across eight sow farms.

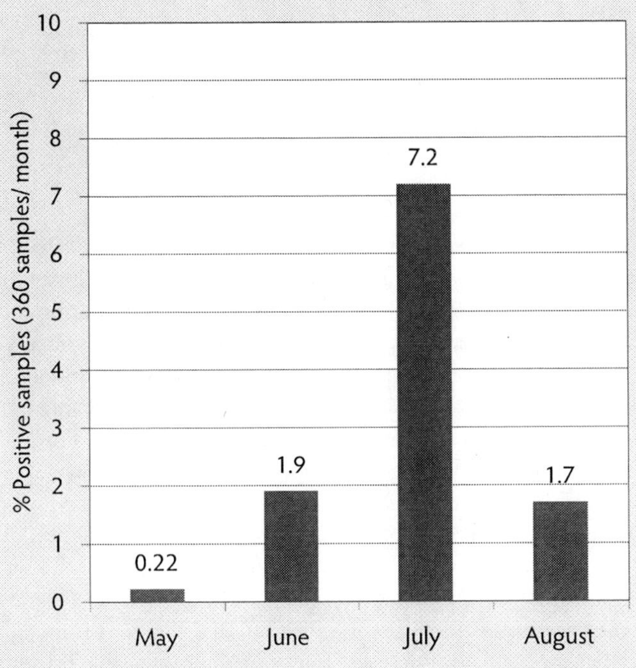

is possible that the exposure/vaccination program that gave gilts adequate immunity entering the herd may be an important factor to control colonization at weaning.

In conclusion, it was possible to detect positive MH samples in commercial settings following previously described classification criteria to separate status. Two out of the 8 sow farms were categorized as provisionally negative and six as positive stable. Serological testing showed evidence of incomplete immunization in the farm with the highest number of positive cases and could be another tool to evaluate the status of breeding herds. The next steps of this investigation will be reviewing gilt acclimation protocols, monitoring grow-finish sites and determining the role of clinical MH in the flows downstream.

Acknowledgements

The authors are thankful for the support provided by Eric Greiner, Jake Keppy, Jennie Green and Dr. Keith Aljets.

References

1. Galina L; Clavijo MJ. Establishing *Mycoplasma hyopneumoniae* herd status classification criteria for breeding herds. AASV. 2016;167-169.

2. EpiTools epidemiological calculators. Accessed November 18, 2016. http:epitools.ausvet.com/au/content.php?page=PrevalenceSS.

3. Strait E et al. Real-time PCR assays to address genetic diversity among strains of *Mycoplasma hyopneumoniae*. J Clin Microbiol. 2008;46(8):2491-2498.

4. Pieters M et al. Comparison of various samples types for detection of *Mycoplasma hyopneumoniae* in recently infected pigs. Leman Conference. 2013;75-76.

5. Fano E, et al. Effect of *Mycoplasma hyopneumoniae* colonization at weaning on disease severity in growing pigs. Can J Vet Res. 2007;71:195-200.

Efficacy of FluSure XP® swine influenza vaccine in pigs challenged with a cluster IV-A H3N2 swine influenza virus

M.C. Lenz[1], MS; V.J. Rapp-Gabrielson[1], DVM, PhD; A. Heinz[1], BS; LP Taylor[1], MS; J. Johnson[1], DVM; T. Hildebrand[1], MA; K. Segur[1]; A Kaliyati[1]; Y. Diamondidis1, DVM: T. Wolff[2], DVM, MPH
[1]Veterinary Medicine Research & Development, Zoetis, Kalamazoo, Michigan; [2]Zoetis, Parsippany, New Jersey

Introduction

Characterization of Cluster IV H3N2 influenza A viruses (IAV-S) from US swine has demonstrated the emergence of six phylogenic sub-clusters, IV-A through IV-F.[1-3] The average pair-wise nucleotide distance between sub-clusters is 5.4% to 7.6% and the most prevalent sub-cluster is IV-A, which represented 76.68% of the H3N2 viruses in the NVSL-USDA-APHIS-VS Influenza Surveillance database from January 2013 to March 2014.[3] The objective of this study was to evaluate the efficacy of newly updated FluSure XP® in pigs challenged with a heterologous H3N2 Cluster IV-A field virus.[4]

Materials and methods

Forty-five weaned, 3-week-old IAV-S negative pigs were randomly assigned to treatment using a generalized block design. Study animals were pigs of high health status, mixed breed and sex. The study enrolled only clinically healthy animals that were serologically negative to IAV-S, as defined by both IDEXX Influenza A multiscreen NP ELISA and hemagglutination inhibition (HI) negative to vaccine and challenge viruses. Pigs were vaccinated twice, on Days 0 and 14, with either an Amphigen® adjuvant placebo (T01, n = 21), FluSure XP® (T02, n = 21), or held as non-vaccinated, non-challenged controls (NTX, n = 3). NTX pigs were necropsied on Day 27 and not included in the analysis. The T01 and T02 pigs were challenged on Day 28 with A/swine/Indiana/A01271853/2012 (H3N2). This virus, received from National Veterinary Services Laboratory, Ames, Iowa, is a Cluster IV-A field isolate that contains the pH1N1 M gene. The HA genes of the vaccine and challenge viruses displayed 97.3% nucleotide similarity. Pigs were necropsied 5 days post-challenge. The primary variable analyzed was lung lesions at necropsy. All pigs were sampled for virus isolation from nasal swabs (daily) and bronchial alveolar lavage fluids (BALF) at necropsy. Daily clinical observations and rectal temperatures were observed. HI antibody titers to the challenge virus and all vaccine strains were also analyzed. The HI testing was completed at the Zoetis Veterinary Medicine Research & Development facility. The level of significance was set at $\alpha = 0.05$. This study was conducted in accordance with guidelines of the Zoetis Kalamazoo IACUC.

Results

This study was valid in that there was no serological, clinical or virological evidence of exposure to IAV-S prior to challenge. Vaccinated pigs had a significantly lower percentage of lung with lesions at necropsy, significantly higher HI antibody titers to the challenge virus, a significantly lower percentage of pigs shedding virus and a significantly lower percentage of pigs positive for virus isolation from BALF at necropsy (Table 1). Compared to 25% of vaccinates, 86% of placebo pigs had extensive lung involvement with lesion scores ≥ 10% (Figure 1). All pigs were negative for IAV-S isolation from nasal swabs collected pre-challenge on Day 27. The back-transformed LS mean $\log_{10}TCID_{50}$/mL virus titers from nasal swabs of vaccinates were significantly lower than controls on all five post-challenge days and by area under the curve analysis ($P \leq 0.0001$) (Figure 2). Baseline rectal temperatures (Days 26, 27, 28) in both groups of pigs were somewhat elevated, likely due to excitement or stress. Following challenge, the back-transformed LS mean rectal temperatures on post-challenge Days 29, 31 and 33 were significantly lower for vaccinates compared to controls ($P \leq 0.0002$) (Figure 3). The back-transformed mean percent of pigs that ever showed clinical signs of respiratory disease post-challenge was 79.3% for T01 and 39.2% for T02. The back transformed mean percent of days with clinical signs of respiratory disease was 21.77% for T01 and 5.59% for T02.

Discussion

This study is an example of the use of an influenza vaccine that contains a strain that resides in the same cluster or sub-cluster as the challenge strain and represents one actively circulating in swine populations. Proper strain selection is rigorous and an important aspect of vaccine development. Strain selection for updating FluSure XP is a long and arduous process involving ongoing surveillance, laboratory and clinical evaluation. Previous methods of evaluating protection focusing on nucleotide sequence similarity may not provide a complete picture and other methods such as serologic cross-neutralization and evaluation of key antigenic epitopes can provide additional insights. Ongoing work is needed to assure better understanding of critical aspects of protection and Zoetis is committed to this effort. Although vaccination against influenza does not prevent infection, it can help reduce clinical signs, lesions and viral shedding in pigs challenged with influenza. This reduction in shedding reduces the risk both to the population within a farm and to the greater population of swine in the neighborhood therefore aiding in regional control strategies. Finally, due to the zoonotic risk of influenza, it is our responsibility as swine veterinarians to mitigate the propagation of

Table 1: Summary of results of the Cluster IV-A H3N2 IAV-S challenge of FluSure XP® vaccinated and placebo pigs.

Group	Percent lung lesions*	HI titer**	Percent pigs ever postive: Nasal shedding	BALF
T01 - Placebo	16.15%	5.0	100%	100%
T02 - Vaccinates	4.67%	95.1	15%	0%
P value	< 0.0001	< 0.0001	0.0001	0.0001

* Back transformed LS Mean
** Back transformed geometric mean HI titer to challenge virus A/swine/Indiana/A01271853/2012 (H3N2) on Day 27

Figure 1: Summary of lung lesion percentage categories.

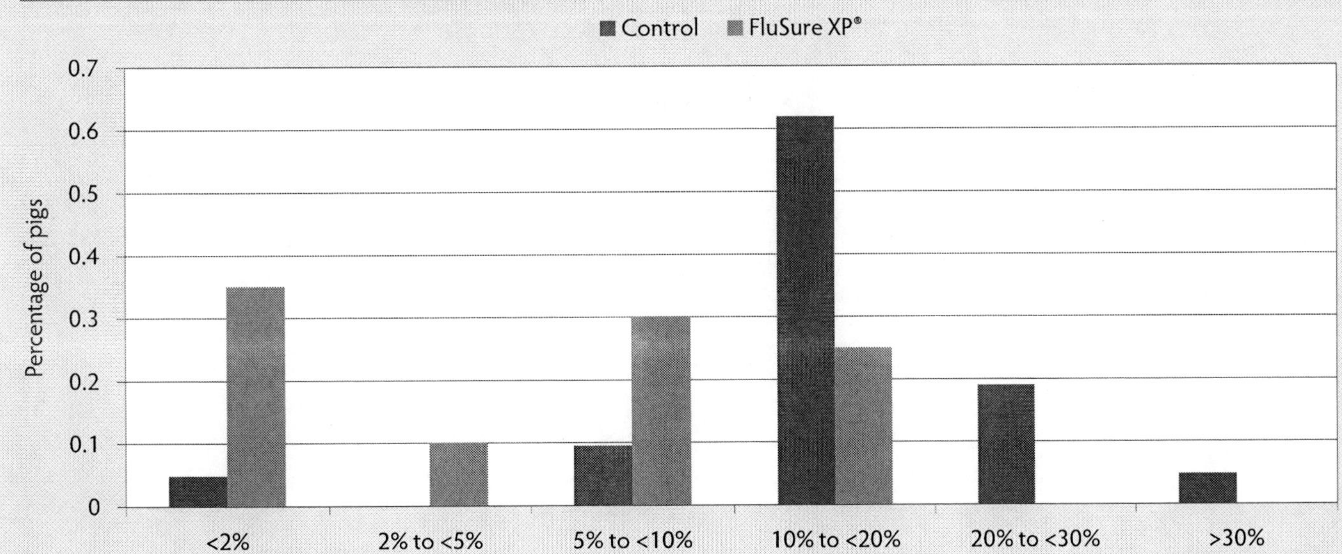

influenza in swine. FluSure XP® is an important tool in influenza control along with proper biosecurity, gilt management, pig flow, and people management.

Conclusions

The experimental Cluster IV-A H3N2 challenge was severe because it produced gross lung lesions in 100% of placebo controls with a back-transformed LS mean of 16.15%. FluSure XP vaccinated pigs had a significant reduction in lung lesions at necropsy ($P \leq 0.0001$); a significant reduction in the percent of pigs ever shedding virus; and in the mean virus titers in nasal swabs on all days post-challenge ($P \leq 0.0001$). Additionally, virus was not isolated in BALF collected from any vaccinate at necropsy compared to 100% of controls ($P \leq 0.0001$); and there was significantly lower rectal temperatures on 3 of 5 days post-challenge ($P \leq 0.0002$). Although not analyzed, ~ 50% fewer vaccinates ever showed clinical respiratory disease, and the number of days with respiratory disease was reduced ~ 75% in the vaccinates. Thus, under the conditions of this study, the newly updated FluSure XP® helped to protect pigs against challenge with a virus representing the most prevalent H3N2 sub-cluster currently circulating in US swine.

Acknowledgements

The authors acknowledge the excellent contributions of the ARS Staff at Zoetis for the execution of the in-life phase of this study, the PASS lab for processing all of the samples, and the SSG staff at Zoetis for the data management phase of this study.

References

1. Kitikoon P, et al. 2013. J. Gen. Virol. 94, 1236–1241, doi.org/10.1099/vir.0.51839-0.

2. Anderson TK, et al. 2013. Influenza Other Respir. Viruses 7 (Suppl. 4),42–51, doi.org/10.1111/irv.12193.

3 Anderson TK, et al. 2015. Virus Res. 201:24-31, doi.org/10.1016/j.virus-res.2015.02.009.

4. Data on file, Study Report No. B820R-US-14-436, Zoetis LLC.

Figure 2: Virus isolation from nasal swabs collected from pigs challenged with Cluster IV-A H3N2 IAV-S.

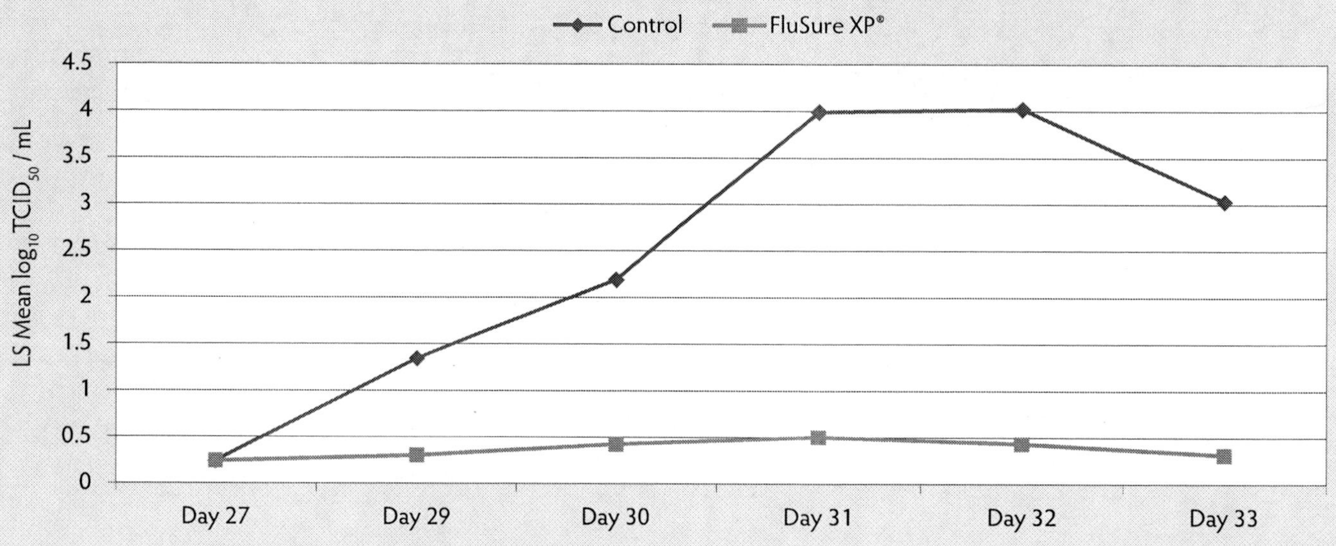

Figure 3: Rectal temperatures (°F) of pigs challenged with Cluster IV-A H3N2 IAV-S.

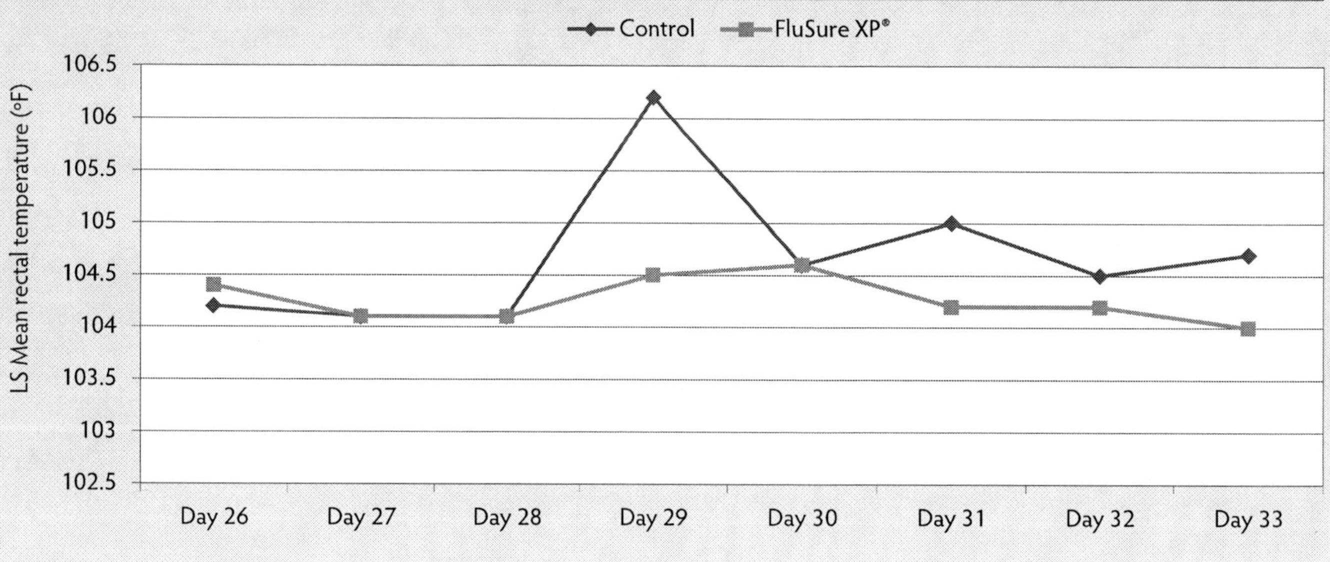

The impact of activated polyphenols on performance of weaned pigs

D. Little, MS, DVM
D.E. Little, DVM, Brookings, South Dakota

Polyphenols are ubiquitous in fruits and vegetables. Recent patents by LiveLeaf, Inc have resulted in the commercial availability of stable, activated polyphenols (LiveExtract®) that when ingested by animals and humans, are oxidized to prophenones that effectively stabilize the gut prior to and during times of digestive stress. The oxidation of LiveExtract® is triggered by the release of peroxidase enzymes from damaged mucosal cells. Internal studies indicate that once oxidized, these potent prophenones reduce gut permeability, inhibit swarming of coliform bacteria, decrease inflammation, and support the integrity of the mucosal barrier.

Supplementation to nursing piglets has been demonstrated to decrease diarrhea scores after a single, oral dose of Grazix Porcine F.[1] Supplementation of weaned piglets has been shown to increase average daily gain in piglets challenged with K88 *E. coli*.[2] Current studies are evaluating the effect of LiveExtract® (Porcine W) supplementation in the water to piglets during the immediate post-weaning phase. Results will be presented from two controlled studies evaluating the effect of Porcine W on average daily gain and mortality in piglets with known histories of K88 *E. coli* and rotavirus infection.

References

1. Internal studies. Grazix Animal Health. 2011.

2. Bontempo, V. et al. Administration of a novel plant extract product via drinking water to post-weaning piglets: effects on performance and gut health. Animal, 2014: 8(5), pp. 721–730.

® Registered trademark of Grazix Animal Health

The economic impact of OvuGel®

Dennis DiPietre[1], PhD; Dean Dau[2], DVM; Charlie Francisco[3], DVM, MS; Lance Mulberry[1], BA
[1]Economist, KnowledgeVentures, LLC; [2]Director of Sales, JBS United Animal Health;
[3]Director of Technical Services, JBS United Animal Health

Introduction

OvuGel® is a GnRH agonist administered to sows intravaginally four days post-weaning. OvuGel® stimulates the release of LH, inducing ovulation 24-56 hours post-treatment with the majority between 40-48 hours post-treatment. Since ovulation is synchronized in OvuGel® administered females, the group can be inseminated without reference to estrus thereby eliminating the requirement of heat detection as a signal of impending ovulation on most animals. The normal variance associated with the timing of post-weaning ovulation, estimated on average at 70% through the highly variable estrus duration among breeding group individuals, can be reduced with the administration of OvuGel®. This enables the farm to eliminate multiple matings of individual breeding females and employ single, fixed time insemination for most of the breeding group. This saves time and reduces the cost associated with multiple inseminations throughout the duration of estrus.[1] Reducing the variance in the wean-to-farrowing period produces economic benefits in finishing through harvest. These benefits are increased when combined with induced farrowing and other revised standard operating procedures to optimize the average total lactation length. This also reduces the variance of lactation days among the farrowing group and tends to maximize the exploitation of the limited farrowing house occupancy time allocated to the group. While inducing farrowing is not required to achieve significant economic benefits at harvest, it can add to these benefits. Synchronizing produces a more predictable estimate of true biological gestation length for each female thereby reducing the risks of inducing farrowing too early resulting in loss of the litter.

Methodology

A stochastic simulation model of a modern sow herd was developed to estimate the economic benefit from two basic strategies when deploying OvuGel®. We measure the increased profitability of single, fixed time insemination of the majority of the breeding group by optimizing the increased mean weight and reduced weight variation it produces in the wean group through harvest. The strategy focuses on optimizing the mean and measures of variance associated with several sequential biological interval distributions to achieve higher average lactation days, higher weaning weight and lower wean-weight variance, reduced cost of production, reduced incidence of culls at weaning and an increase in profitability of the animals at harvest. These biological intervals include the wean-to-service period, the insemination-to-ovulation interval, the gestation-to-farrowing period, and the average total lactation length. Alternative standard operating procedures were specified for the use of OvuGel® without altering the size or assets of the farm (e.g., number of gestation crates, farrowing crates, size of the gilt pool, etc.) or its production interval (weekly system). The productivity distributions of the diverse parity structure, the wean-to-service interval, observed estrus-to-ovulation interval, gestation length and lactation length distributions were all modelled.[2] The distributions of time associated with these intervals, the diverse biological performance and their correlations within parity structure and across wean-to-service intervals were obtained from qualified records among over one million sow records contained in a large, international swine record system.[3] Trial data from the use of OvuGel® was also used. The model was deployed to simulate thousands of replications of a complete sow-herd turn (weaning to subsequent weaning) for both the standard breeding protocols in wide use today in the United States (without the use of OvuGel®) and a modified set of protocols designed to optimize profitability using OvuGel®. The significant changes in the standard operating protocols included administering OvuGel® on the morning of the fourth day post weaning, instituting single, fixed-time insemination 24 hours later of as many of the target breeding group as possible and in the second case, inducing farrowing for all females which had not farrowed by day 115 of gestation. In the first case, we model the outcomes from first service through harvest without farrowing induction and in the second example, with the induction of farrowing at day 114 of gestation for the OvuGel® administered animals.

The economic model

We modelled a 3,300 head breeding herd with an historical annual average litters per breeding female of 2.35. For simplicity and to avoid the confusion associated with accommodating the wide variety of idiosyncratic farrowing house management strategies, we modelled a weekly batch system and estimated the weekly breeding group size at 149 [3,300 × 2.35)/52]. The model assumes a balanced, long-term parity distribution[4] and productivity measures obtained from the database mentioned above (See Table 1). The farm has an historical farrowing rate of 85% and the plan is to introduce an average of 32 gilts into the breeding herd each week to meet long-term productivity goals and replace unexpected breeding female losses/culls. For the conventional example, the target females to breed each week is estimated from the historical farrowing rate to be 175, (149/0.85). In the case of OvuGel®, we adjusted the breeding target upward to account for recycling among the small subset of females which would have been culled or returned to the breeding line under the conventional approach because of apparent anestrous by day seven post-weaning but are

Table 1: Parity distributions and productivity measures modeled

Parity structure	Parity distribution	Head	Mean BA	BA StDEV	BA mean weight	BA weight StDev	Mean PWM%	PWM% STDEV	Mean pigs weaned
Parity 0 breeding females	17.0%	561	11.91	3.31	3.37	0.45	13.00%	3.20%	10.36
Parity 1 breeding females	16.0%	528	12.58	3.48	3.70	0.60	10.97%	3.15%	11.20
Parity 2 breeding females	15.0%	495	13.44	3.55	3.72	0.62	14.36%	3.18%	11.51
Parity 3 breeding females	14.0%	462	13.90	3.65	3.70	0.66	18.42%	4.14%	11.34
Parity 4 breeding females	13.0%	429	13.85	3.69	3.61	0.59	19.28%	4.19%	11.18
Parity 5 breeding females	11.0%	363	13.54	3.75	3.61	0.65	19.20%	5.54%	10.94
Parity 6 breeding females	9.0%	297	13.08	3.71	3.52	0.66	17.89%	6.27%	10.74
Parity 7 and Up breeding females	5.0%	165	11.77	3.51	3.39	0.58	13.34%	6.23%	10.20

none-the-less bred when using OvuGel®. We estimated the size of this group to be about seven percent of the weekly breeding group or 10 females in our example ($0.07 \times 149 \approx 10$). Studies using OvuGel® suggest that on average about 30% of these animals will eventually produce a litter. The target breeding group for the OvuGel® administered animals is derived from a compound farrowing rate constructed in two parts, the historical one of 85% on 165 breeding females and 30% on the 10 females which would have been delayed or anestrus sows, ($[0.85 \times 165] + [0.30 \times 10]) = 81.9\%$. The conventional farrowing rate and this number are not comparable "farrowing rates". The OvuGel® adjustment value of 81.9% reflects the increased sow efficiency of successfully breeding the three sows (0.30×10) which would have been rejected for breeding under conventional protocols thereby increasing non-productive days. The target breeding group in the OvuGel® case is 182 (149/0.819). Each group includes 32 gilts.

Tables 2 and 3 illustrate the distribution of the wean-to-estrus period for the conventional group and day post-weaning the animals bred for the OvuGel® group. We assume that animals are bred for the first time in the conventional group when estrus is discovered. The gilts are introduced daily in equal numbers for both groups reflecting the available gilts in the pool in estrus on any given day. We use a separate stochastic model to estimate the interval between the first discovery of estrus and ovulation for individual females. The model randomly selects the number of hours after estrus begins until it is discovered, based on the number of hours during the day the animals are observed for estrus. Next, we randomly select for the duration of estrus which is inversely correlated with wean-to-estrus length and then randomly select the time of ovulation within estrus. A similar but simpler procedure is carried out for OvuGel® using published literature for the timing of ovulation post administration of OvuGel®. While these intervals are distributions of differences in the model, we note the means here as an increase of one day between breeding and the beginning of biological gestation in OvuGel® administered animals and two days for the conventionally bred group.

Table 4 gives the distribution of gestation intervals taken from the database mentioned previously. Actual gestational age is less certain using conventional breeding and may be up to two days off. This increases the risk of inducing farrowing too early or at the wrong time. OvuGel® administered animals, the variance between recorded and actual gestational age is much less, thereby reducing the risk of early termination of pregnancy.

Each pig generated from a litter is placed on an identical average growth curve (Gompertz type)[5] of the form:

W_t =Wm*exp(-a*exp(-Wm*b))

Where:

$W_{t=}$ Current pig live weight

Wm = weight of the pig at maturity (492.8),

a = 4.942076,

b = age of the pig in days

To generate an individual growth curve for each pig, the average growth curve was adjusted based on the weight of the pig at birth compared to the average weight of all pigs at birth on that day. The percent difference from average birth weight on a given day, determines the percent difference in growth rate from the average growth rate throughout the remainder of the grow out.

Daily feed intake is estimated from a polynomial consumption function of the form:

$c^* W_t^2 + b^*W_t + a$,

where W_t is the current pig live weight

c, b and a are parameter estimates (constants).

Feed ingredient costs are based on an average of recent year's prices ($0.071/lb. corn; $0.071/lb. DDGS; $0.20/lb. HiPro Soybean Meal). Nonfeed costs are estimated at a constant value of $60.00/head.

Table 2: WEI distribution table (conventional breeding)

	Percent in estrus and bred	Weaned BF	Percent of total gilts bred	Gilts/recyles	Percent of target group bred	Total bred
Day 1	0.00%	0	15%	6	3%	6
Day 2	0.00%	0	15%	6	3%	6
Day 3	5.00%	7	15%	6	7%	13
Day 4	49.00%	65	13%	5	40%	70
Day 5	38.00%	51	13%	6	33%	57
Day 6	6.00%	8	14%	6	8%	14
Day 7	2.00%	3	14%	6	5%	9
Total	**100%**	**134**	**99%**	**41**	**100%**	**175**

Table 3: WEI distribution table (OvuGel®)

	Percent of weaned sows bred	Weaned BF	Percent of total gilts bred	Gilts/recyles	Percent of target group bred	Total bred
Day 1	0%	0	15%	6	3%	6
Day 2	0%	0	15%	6	3%	6
Day 3	0%	0	15%	6	3%	6
Day 4	0%	0	13%	5	3%	5
Day 5	100%	144	13%	5	82%	149
Day 6	0%	0	14%	5	3%	5
Day 7	0%	0	14%	5	3%	5
Total	**100%**	**144**	**100%**	**38**	**1**	**182**

Table 4: Gestation distribution

Conventional and OvuGel® without Induction		**OvuGel® with Induction at Day 114**	
Gestation distribution		**Gestation distribution**	
Day	Percent	Day	Percent
113	2.8%	113	2.8%
114	8.9%	114	8.9%
115	26.5%	115	88.3%
116	34.6%	116	0.0%
117	17.3%	117	0.0%
118	7.0%	118	0.0%
119	2.9%	119	0.0%

Finished animals are valued using the public pricing scheme obtained from the website of Indiana Packers Corporation[6] at https://indianapackerscorp.com/files/procurement-program.pdf. The lean percent of each carcass is determined as a function of its current live weight by the simple linear equation:

Lean Percent = $0.6399-0.00035 * W_t$.

Results and discussion

The farm productivity results in an average of 11 pigs weaned per breeding group or about 1,640 pigs per week or 25.85 pigs weaned per breeding female per year. The mechanism by which OvuGel® creates value in finishing is by reorganizing the pregnancies of the breeding group causing them to happen earlier and with less variance across the day when biological gestation starts for each member of the target breeding group. In doing so, the breeding females on average, farrow earlier and in a tighter group which allows them to take advantage of the limited time allocated to them in the farrowing house. This occurs with or without farrowing induction but the greatest benefit is achieved utilizing farrowing induction on day 114 of gestation for each OvuGel® administered animal. We modelled each of these three scenarios, conventional breeding, the use of OvuGel® without farrowing induction and with farrowing induction to demonstrate the difference in profits created by each.

Figure 1 contains an illustration of the distribution of lactation days for the three protocols. You can observe the relative size of the standard deviation of each distribution which is a factor in the productivity change. By the time of weaning, the OvuGel® offspring, on average, record a higher number of lactation days and are heavier. In addition, their weaning weight has a lower standard deviation than the conventional group. See Table 5. The OvuGel® offspring are older at weaning than piglets bred with conventional means and therefore are heavier on average, all the way to harvest. At weaning, the weight distribution of these older, heavier pigs results in 1.87% fewer piglets below the standard cull weight of 8lbs for OvuGel® offspring and 3.17% fewer piglets below this threshold when compared to OvuGel® plus induced farrowing. When harvest nears, pigs that are 1-2 days older and near the mean of the growth curve will be, all things equal, 4-5 lbs. heavier as typical instantaneous growth rates at these weights average 2.2+lbs. per day. This value can be realized in additional turns of the finishing assets at a target finishing weight or by added weight in fixed time finishing systems.

Marketing programs are many and varied and the reader should estimate the value difference in finishing by the marketing method which they believe is best. However, as a baseline, to estimate the value of this additional weight, we divide the finishing animals into a topping load (first 16% of animals by weight), and a closeout (the remainder of the animals less the lightest 9-10% of animals which form a light closeout). We used a stochastic optimization routine to discover the day which each of these three groups returned the greatest total group profit for the OvuGel® offspring. OvuGel® offspring were marketed with the corresponding group of conventional pigs on that same day. Since the conventional pig weights were on average lighter for each group, we assume they would not have reached their maximum profit prior to the day the OvuGel® offspring were marketed.

Table 5 reveals that average weights in the closeout group were four pounds heavier for the OvuGel® offspring with induction vs. conventional farrowing. The weighted average profit per pig from each of the groups combined demonstrates a mean expected profit increase of \$0.29/head and \$0.79/head for the OvuGel®

Figure 1: Lactation length distributions

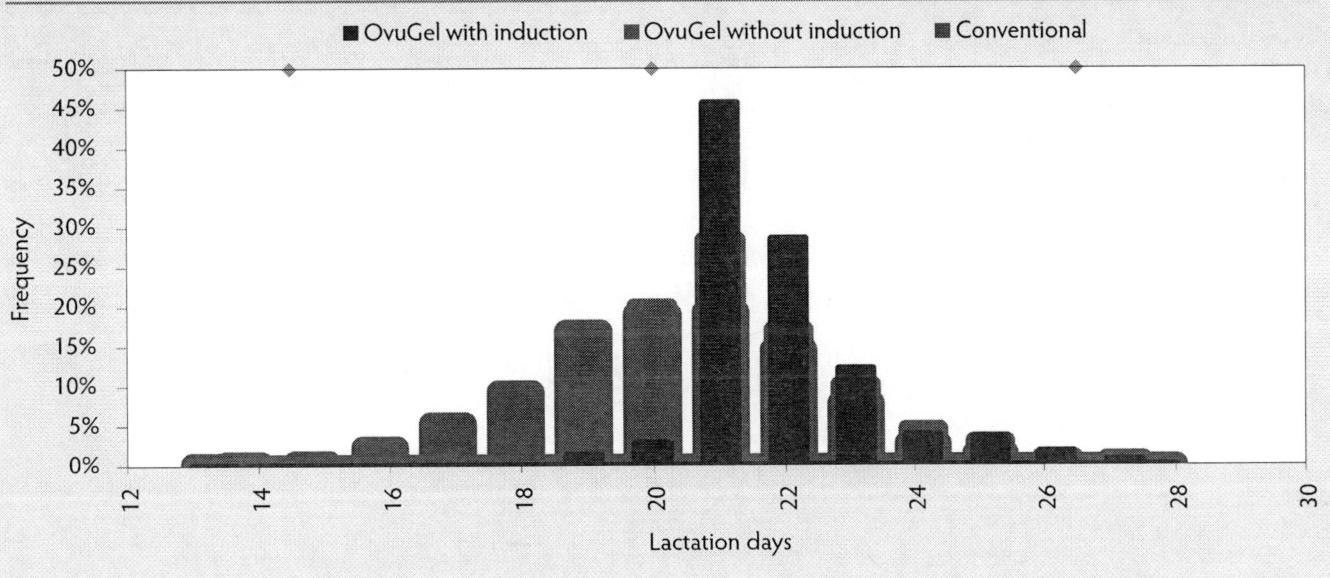

Table 5: Results of Stochastic Simulation

Conventional and OvuGel® with and without Induction			
Per head values	**Conventional**	**OvuGel®**	**OvuGel® with Induction**
Wean weight mean	11.67	12.21	12.64
Wean weight STDEV	2.49	2.54	2.51
Mean lacation days	20.24	21.14	21.86
Lactation days STDEV	1.96	1.73	1.27
% piglets below 8lbs at wean	5.67%	3.80%	2.50%
Increased standard weaned pigs		1.87%	3.17%
Mean top load mkg weight	298	299	299
Mean closeout load mktg wgt	284	286	288
Mean light close mktg wgt	266	266	267
Mean top load feed	$73.93	$74.11	$73.94
Mean closeout load feed	$68.93	$69.67	$70.36
Mean light close feed	$62.11	$62.64	$62.72
Mean top load profit	$36.65	$36.99	$36.84
Mean closeout load feed profit	$37.50	$38.10	$38.40
Mean light closeout feed profit	$33.37	$34.11	$33.57
Mean profit/head summary	$36.99	$37.29	$37.79
Increased profit per head		$0.29	$0.79
Increased profit per litter*		$3.16	$8.64
Increased profit/BF/year*		$7.43	$20.31

* Assumes 4% W-F death loss; Credit given for increase in standard weaned pigs

offspring; no induction and with induction groups compared to the conventional farrowing. Assuming a 4% wean-finish mortality, these values rise to $3.11/litter and $8.38/litter and $7.30 and $19.69 per breeding female per year.

Acknowledgement

The authors acknowledge with gratitude, the special help provided to this effort by Dr. Nathan Augspurger, Sr. Manager, Product Development, JBS United and Dr. Rick Tubbs, DVM, MBA, we accept their help but remain exclusively responsible for any errors herein.

References

1. B. A. Belstra, Flowers, W.L., Rozeboom, K.J., and See, M.T., Factors Contributing to Variation of Duration of Estrus and Time of Ovulation in a Commercial Sow Herd, NC State University Cooperative Extension, Annual Swine Report 2001.

2. Indiana Packers Corporation, Procurement Program, Available at: https://indianapackerscorp.com/files/procurement-program.pdf.

3. M.E. Johnston, G. Sconyers, R.D. Boyd, D. Dau, R.R. Kraeling, S. K. Webel, Use of OvuGel® Improves Pigs Born Alive per 100 Sows Weaned and Increases Average Age of Pigs Weaned, IPVS Abstract, Dublin Ireland, 2016.

4. PigChamp Data, PigChamp, Inc., 1531 Airport Road, Ames, IA, 50010.

5. Pinilla, Juan Carlos, Lecznieski, Luiz, Sow Removal and Parity Distribution Management, Technical Services, PIC, available at: http://www.pic.com/Images/Users/1/salesportal/technicalservices/paritycontrolinformation/ParityDistributionManagementJCP2010.pdf.

6. PIC Wean to Finish Manual, available at: http://www.pic.com/Images/Users/1/salesportal/literature/manuals/WeanToFinishManual_April2013_small%5B1%5D.pdf.

Systematic implementation of VFDs using FeedLINK electronic VFD solution

B. Leuwerke
Swine Vet Center, Saint Peter, Minnesota

Introduction

As a practice, Swine Vet Center (SVC) began to prepare its office staff, clients and veterinarians well in advance of the January 1, 2017 Veterinary Feed Directive (VFD) implementation date for all medically important feed grade antibiotics. SVC had experience with GlobalVetLINK (GVL®) as a service that had historically provided a user-friendly format to create valid VFDs, along with providing the capability to electronically sign, issue and file these VFDs for audit purposes. With this familiarity, in 2016, SVC chose to sign-on with GVL to be its provider of VFD services.

Preparation

Starting in 2015 and throughout all of 2016, SVC began working closely with its swine producers to review VFD requirements so that they were prepared to implement starting in 2017. Several of those producers had used Pulmotil, so they were familiar with the general concept of VFDs and what information was required to issue a VFD. More of the preparation revolved around understanding what feed grades were moving from over-the-counter (OTC) to VFD status, explaining that feed efficiency label claims were being discontinued and evaluating medicated feed budgets to continue to ensure that these matched labeled duration and dosage claims.

In addition to those discussions, SVC worked with producers to determine how they would best implement VFDs for each of their herds, flows, systems, etc. The scope of the Veterinary Feed Directive is a relationship of the medication being directed, the veterinarian signing the VFD, the client (producer) responsible for the care and feeding of the animals receiving the VFD feed, the feed distributor (individual, partnership, corporation or association) providing the VFD feed, and the site(s) associated with the VFD.

A VFD can be issued to cover multiple sites when fed out of the same feed mill, so producers with their veterinarian could choose to assign multiple sites that house the same stage of pig, fed out of the same feed mill to one VFD. Alternatively, producers could choose to have a VFD issued for each individual group of pigs on a site-by-site basis. Within our practice, we had producers that chose to manage VFDs one of these ways based on their comfort with managing those records or being able to accurately work through an FDA audit of the VFD process.

At the time of drafting this proceedings paper, SVC has yet to see how managing the sheer volume of VFDs will go, but expect with producer and feed mill preparation the transition should be relatively smooth.

Real-world VFD scenarios

As referenced earlier, the decision on how to manage VFDs, either by site, by feed mill or flow (if fed by same feed mill) is largely based on how comfortable producers have felt with both managing the number of VFDs required or how they have perceived being best prepared for FDA audits of their VFDs. The following are common scenarios that illustrate how VFDs can be managed within a farm or system.

Consider the scenario of one sow farm source that flows pigs across 12 individual wean to finish sites. All sites are fed by the same feed mill and within the nursery phase there are 4 separate rations, each containing a VFD-required antibiotic. In this scenario, a producer could choose to list all sites on one VFD and have a VFD for each medicated ration, for a minimum of 4 VFDs to manage for this producer. Alternatively, if this producer is more comfortable with each individual site being issued a VFD to medicate each of four diets, there would be a total of 4 VFDs issued each time the site filled for a total of 48 VFDs to manage.

If this same system switches four of their twelve sites to an alternative feed mill if listing multiple sites on a VFD, at minimum they would have 8 VFDs to manage, or if electing to manage a separate VFD for each site, they would still have 48 VFDs to manage.

What about a scenario where a producer has sick pigs and needs to feed a VFD feed for treatment? This is still acceptable to do, but will require that a VFD is issued before the medicated feed can be manufactured and delivered to the site. With the mobile electronic access and signing capabilities of GVL's FeedLINK system, quick ability to create and sign a VFD is possible to be able to treat pigs.

Accessing FeedLINK to create VFDs

Paramount to SVC was the ability to generate for its clients a highly accurate and valid VFD that took a relatively brief amount of time to complete and issue electronically to both the producer and the feed distributor.

GVL SmartEngine™ technology allows for generating an accurate, complete and compliant VFD. To create a VFD using the FeedLINK system, users create an online account with GVL. Once signed into the account, the user will select FeedLINK, giving them the ability to create, renew or void a VFD. With the ability to upload owner/client information, the producer name and address information is quickly available for entry into the VFD form. Similarly, feed distributor name and location information is entered and managed by GVL, allowing quick searches and entry of feed mill information for easier completion of the VFD. Animal site information can also be uploaded into the FeedLINK database to make entry of this information for single sites or multiple sites easier when completing the VFD.

Safeguards provided by GVL SmartEngine technology ensure that once an antimicrobial is chosen, only the compliant animal class, duration, indications, expiration, etc. are available to be chosen to complete the VFD.

Distributors who provide an email address can have completed VFDs sent directly to them after the veterinarian electronically signs it. Likewise, producers who provide an email address can be granted access to retrieve created VFDs to file for their records. The ability to electronically sign and send VFDs is acceptable and encouraged by the FDA, and GVL is 21 CFR part 11 compliant for the 2-year file storage requirement.

Changes to VFD format and additional requirements

VFDs are no longer required to have the estimated pounds of feed to medicate. In lieu of this, it is required that the number of animals to be fed with the medicated feed be listed on the issued VFD. Estimates of total number of animals fed under the timeline of the VFD can be difficult to make, especially if a site has multiple turns of animals go through from issuance to expiration of the VFD. From a record-keeping standpoint, VFDs are required to be stored for 2 years by the veterinarian, the feed distributor and the owner. A veterinarian signing a VFD must be licensed in the state that the VFD is being issued and must be operating under a valid Veterinary-Client-Patient-Relationship (VCPR) as defined by the state in which the VFD is being issued.

Frequently asked VFD questions

See Table 1.

References

1. Global Vet Link. Available at https://www.globalvetlink.com/. Accessed 9 November 2016.

2. Feedstuffs VFD Central. Available at http://feedstuffs.com/vfd.aspx. Accessed 9 November 2016.

3. FeedLINK Demo Video. Available at https://www.youtube.com/watch?v=-b5oJgR5DFc&feature=youtu.be. Accessed 9 November 2016.

Table 1: Frequently asked VFD questions

What happens if a producer places a feed order to be delivered on the last day of the VFD?	Producer would need to get a new VFD from their veterinarian in order to continue feeding that medicated feed after the VFD expiration date, and to ensure that he/she feeds it for the required duration.
Can a client still manufacture their own feed on-farm?	Yes! The client will need a VFD to purchase Type B medicated premix and will also need valid VFDs to feed to their animals. They do not need to register as a distributor unless they plan to sell/distribute feed to another producer.
What if a producer has multiple locations, can they all be on the same VFD?	Yes, a veterinarian can issue a VFD for multiple locations and each animal location must be specified on the VFD. The justification must be a common health status.
So what is the scope of a VFD?	One veterinarian (not clinic), one client (owner or agent for the owner), one feed distributor (can be multiple sites as long as it is one business entity), one or more animal locations, and one VFD medication (and applicable combination drugs).
How often do you need to visit a specific site to maintain a valid VCPR?	Find out whether state or federal VCPR guidelines apply to you. The FDA has provided a chart on their website that shows whether the state or federal VCPR definition applies to a lawful VFD in each state: www.fda.gov/AnimalVeterinary/DevelopmentApprovalProcess/ucm460406.htm.
Who has to keep a copy of the VFD?	Veterinarians are responsible for storage of every original VFD for two years. Feed distributors and producers/clients must retain a copy of the VFD for two years.
How will an FDA inspection process work?	At the feed distributor, an investigator will examine three randomly selected VFD forms. From these, they will pick one form to follow back to the vet and to the producer. Learn more at www.globalvetlink.com/news/fda-begins-vfd-inspections-pilot-project/.
If a VFD has multiple animal locations that are contract growers, do they need to have the VFD at their site?	No, as long as the owner/headquarter has a copy and can provide the VFD to the grower within 24 hours of an inspection.

Pre-slaughter stunning of swine – a scientific viewpoint

Steve Wotton MBE, HNC, MSc
University of Bristol, School of Veterinary Sciences, Langford, North Somerset, United Kingdom

Introduction

Pre-slaughter stunning is defined in European Legislation[1] as "any intentionally induced process, which causes loss of consciousness and sensibility without pain, including any process resulting in instantaneous death." In order for stunning to be humane, loss of consciousness must be induced before the animal can feel any pain associated with the stunning method, i.e. within 100 – 150 milliseconds. The regulation also requires that "the loss of consciousness and sensibility shall be maintained until the death of the animal." Simple, or recoverable, stunning must be followed as quickly as possible by a procedure ensuring death, such as bleeding. An animal "shall be regarded as continuing to live until the permanent cessation of circulation or the destruction of its brain".[2] Therefore, from slaughter or killing point of view, death can be recognised from the absence of cardiac activity (e.g. pulse or heartbeat) when bleeding has ceased or destruction, or dysfunction, of the brain.[2] Brain death in animals can be recognised from the absence of brain stem reflexes such as rhythmic breathing, pupillary light reflex, corneal reflex and gagging.[2] The heart will continue to beat for several minutes following stunning and slaughter, in an animal that is irreversibly brain dead, and has no bearing on the criteria for identifying the death of an animal.[2] Simple, or recoverable, stunning methods that are permitted by legislation[1] are mechanical, electrical and controlled atmosphere stunning.

Mechanical stunning

The use of a penetrating captive bolt gun is not a preferred method with pigs due to the excessive kicking (clonic activity) displayed immediately following the shot that can last for several minutes. In the majority of cases pigs will be killed (brain dead) by an accurate shot and the kicking is as a result of spinal reflexes 'out of control' of the brain. The cause of this excessive kicking post-stun is unknown but may be due to the predominance of white (fast twitch) muscle fibres in pigs, together with their large muscle bulk.

The usual methodology for the on-farm emergency killing, or culling of, neonate piglets is by manual blunt force trauma, carried out by administering a blow to the head, which is generally performed by swinging the young animal against the floor or a wall. Although widely used as a means of emergency killing, or culling, the effectiveness of this method is heavily dependent on the strength and skill of the operative and, consequently, the probability of achieving an immediate kill in all cases, is low. Furthermore, a lack of proper training and human error can lead to pain and distress to the animal. It is also a method of killing that is aesthetically unpleasant for both the operator and any bystanders.

Mechanical blunt force trauma is an alternative method that delivers a more controlled percussive blow to the head that can result in an immediate stun/kill. Research at Bristol University into the use of non-penetrating captive bolt devices (Zephyr EXL, Bock Industries) has shown that this method can be used successfully with neonate piglets (≤ 10.9 kg).[3] A further trial demonstrated that, under commercial production conditions, it took less than seven seconds to select, place and shoot a piglet.[4]

The use of a free-bullet is classed as a killing method but has Health and Safety implications when used within an abattoir setting and, in the UK, is only commonly applied to water buffalo. Information on the use of firearms with a free projectile (shotguns, rifles) is available through an on-line guide.[5]

Electrical stunning

The passage of sufficient alternating electrical current across the brain of a pig (≥ 1.3 Amps[1]) will initially disrupt neuronal function followed in approximately 200 milliseconds (from the start of the application) by tonic/clonic epilepsy.[6] Increasing the amplitude of the current or the time of application will not affect the period of unconsciousness produced. Increasing the current field strength or increasing the time will, however, affect the degree of post-stun movement. During the epileptic fit the animal is unconscious and its physical response is characterised by two distinct phases, the tonic and the clonic phases, which are followed by recovery if the animal is not exsanguinated. With electrical stunning systems, research[7] has shown that the minimum time to the first sign of recovery (return of rhythmic breathing) is 37 s. The important organ when determining death in an animal, following stunning and slaughter, is the brain and research[8] has shown that (with pigs) loss of cortical function will occur, on average, in 18 s (maximum 22 s) following an effective thoracic stick (exsanguination). Therefore, effective stunning must produce unconsciousness that lasts from the start of the application, through the sticking operation and for at least 22 s following sticking (Figure 1). Applying research data to the practice of stunning and slaughter results in a maximum stun-to-stick time of 15 s when head-only stunning is used with pigs. Longer times may result in the animal regaining consciousness during bleeding.

Research at Bristol University has demonstrated that good electrical contact between the electrodes and the animal is critical to maintain the resistance between the electrodes below the threshold necessary to deliver the required current (≥ 1.3 Amps[1]). Modern electrical stunners are constant current by design and will automatically adjust the applied voltage based on the resistance. However, health and safety requirements limit the maximum

Figure 1: Recommended minimum time between electrical stun start and stick with pigs.

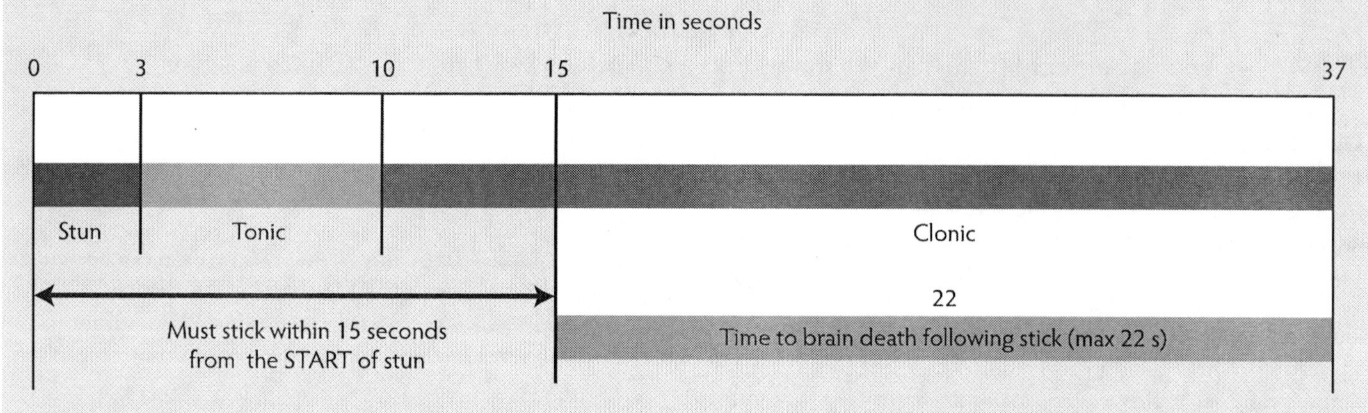

voltage that can be applied to 400-500 volts therefore, keeping the resistance as low as possible is advantageous. Electrodes should be kept clean and replaced when worn as this affects the ability to maintain contact. Operators should be rotated as fatigue affects contact resistance and the application should be quick, to allow the applied voltage to overcome the inherent high resistance to produce an immediate stun.

Some electrical systems have been designed to initiate an electrically induced cardiac arrest either at the point of the stun or immediately afterwards. These systems promote the start of death to the point of the stun and pigs will not recover from an effective 'cardiac arrest stun'. It is essential that a low frequency current (50 or 60 Hz) be applied across the heart to induce ventricular fibrillation. Unfortunately, the application of low frequency current is likely to cause damage to blood vessels torn by the force of contraction (blood splash) and/or broken bones or dislocated joints.[9] Automated and semi-automated systems have been developed in Europe (the Stork 'Midas' Stunner and the BANNS semi-automatic pig trap), which apply a higher frequency current to the head to stun and a secondary low frequency current across the heart to induce ventricular fibrillation. These 'split-stun' systems allow the damaging low frequency current to be applied across the heart at a lower amplitude than that required to induce an effective stun as the heart is not protected by the insulating properties of bone, i.e. the skull.

Controlled atmospheres

Exposure of pigs to high concentrations of CO_2 (> 80%) was introduced to Europe from the US in the early 1950's. Consciousness is lost in approximately 10 – 15 s due to the anaesthetic effect of the gas. In the UK we categorize the use of controlled atmospheres as a killing method because the induction of unconsciousness is not immediate in action. However, there are welfare concerns over the aversiveness of the induction of anaesthesia with high concentrations of CO_2 and other gases (or gas mixtures) may offer a welfare improvement.[10,11] Emphasis has been placed on the development of good pig handling facilities, e.g. the Butina Backloader system which, when combined with moving gates to minimise the handling of animals in the lairage, has helped to counterbalance the welfare problems with the use of CO_2.

The use of anoxia ($\leq$ 2% O_2) with an inert gas has been demonstrated to offer welfare advantages, however, the exposure times required to kill pigs are approximately 7 minutes[12] which is not commercially viable. Further research is necessary to investigate the combination of exposure to anoxia to induce unconsciousness followed by immersion in a high concentration of CO_2 to kill.

Discussion

Modern technology has resulted in improvements in animal welfare at pre-slaughter stunning but has resulted in the need for managers, supervisors and operators to have a good understanding of the procedures together with the underlying science, that they apply. This information, and more, is covered in the Animal Welfare Officer training course that is offered by the University of Bristol to the meat industry in the UK and elsewhere.[13]

References

1. Council Regulation (EC) No 1099/2009. On the Protection of Animals at the Time of Killing. 2013.

2. European Food Safety Authority (EFSA). Scientific opinion on monitoring procedures at slaughterhouses for pigs. EFSA Journal 2. 2013; 11(12):3523.

3. Grist A et al. Humane Euthanasia of Neonates I: Validation of the effectiveness of the Zephyr EXL Non-penetrating Captive Bolt system for euthanasia of new-born and weaned piglets up to 10.9 kg. Animal Welfare 2016a; (In Press).

4. Grist A. et al. The assessment of the effectiveness of the Bock Industries Zephyr EXL non-penetrating captive Bolt system for euthanasia of new-born and weaned piglets up to 10.9 Kg. Proceedings of the AASV Annual Meeting. 2016b.

5. Humane Slaughter Association (HSA). Humane killing of livestock using firearms. On-line Guide. `http://www.hsa.org.uk/methods/free-bullet-firearms`. Accessed 7 November 2016.

6. Cook CJ et al. The effect of electrical head-only stun duration on electroencephalographic-measured seizure and brain amino-acid neurotransmitter release. Meat Science. 1995; 40 (2): 137-147.

7. Anil MH et al. Electrical stunning and slaughter of pigs - Guidelines for good welfare assurance. Fleischwirtschaft. 1997; 77 (5) 473-476.

8. Wotton SB, et al. Pig slaughtering procedures: time to loss of brain responsiveness after exsanguination or cardiac arrest. Research in Veterinary Science. 1986; 40, 148-151.

9. Wotton SB et al. Pig slaughtering procedures: Head-to-back stunning. Meat Science. 1992: 32, 245-255.

10. Raj ABM et al. Welfare implications of gas stunning pigs: 2. Stress of induction of anaesthesia. Animal Welfare. 1996; 5, 71-78.

11. Raj ABM et al. Welfare implications of gas stunning pigs: 3. The time to loss of somatosensory evoked potentials and spontaneous electroencephalogram of pigs during exposure to gases. Veterinary Journal. 1997; 153, 329-340.

12. Raj ABM. Behaviour of pigs exposed to mixtures of gases and the time required to stun and kill them: welfare implications. Veterinary Record. 1999; 144, 165-168.

13. Animal Welfare Training, University of Bristol; available at: **http://www.awotraining.com**. Accessed 7 November 2016.

An interferon inducing porcine reproductive and respiratory syndrome virus vaccine candidate elicits protection against challenge with a heterologous virulent type 2 strain

E. Fontanella[1], MS; Z. Ma[2], PhD; Y. Zhang[2], PhD; A. Castro[1], DVM, PhD; H. Shen[1], DVM, PhD; P. Halbur[1], MS, DVM, PhD; T. Opriessnig[1,3], Dr med vet, PhD
[1]VDPAM, Iowa State University, Ames, Iowa; [2]University of Maryland, College Park, Maryland;
[3]The Roslin Institute, University of Edinburgh, Midlothian, United Kingdom

Introduction

Achieving consistent protection by vaccinating pigs against porcine reproductive and respiratory syndrome virus (PRRSV) remains difficult. It has been demonstrated that PRRSV inhibits synthesis of type 1 interferons (IFNs) which are important factors in the innate immunity against viral infections and in activation of the adaptive immune response.[1,2] Type I IFNs have a central role in the differentiation of CD4+ and CD8+ T cells. Signaling of type I IFNs is required for IFN-γ production during viral infection. Recently, an interferon inducing PRRSV vaccine candidate strain A2MC2 was demonstrated to be attenuated and induce neutralizing antibodies.[1] Moreover, when A2MC2 was passaged 90 times *in vitro*, the resulting strain A2P90 was found to be avirulent in pigs while inducing type I IFNs in both MARC-145 and pulmonary alveolar macrophages similar to its parent strain, and eliciting higher levels of neutralizing antibodies against VR-2332 compared to the Ingelvac PRRS® MLV virus vaccine strain.[3] The objective of this study was to determine the efficacy A2P90 to protect pigs against challenge with moderately virulent PRRSV strain VR-2385 (92.3% nucleic acid identity with A2MC2) and highly virulent atypical PRRSV MN184 (84.5% nucleic acid identity with A2MC2).

Materials and methods

Forty 3-week-old pigs were randomly assigned to five groups including a NEG-CONTROL group (non-vaccinated, non-challenged), VAC-VR2385 (vaccinated, challenged with strain VR-2385), VR2385 (challenged with strain VR-2385), VAC-MN184 (vaccinated, challenged with strain MN184) and MN184 group (challenged with MN184 virus). Vaccination was done at 3 weeks of age followed by challenge at 8 weeks of age. Serum samples and nasal swabs were collected in regular intervals and tested for presence of PRRSV RNA by an in-house real-time RT-PCR assay and for presence of antibodies including neutralizing antibodies by a commercial ELISA (IDEXX PRRS X3 Ab Test; IDEXX Inc., Westbrook, MA, USA) and a fluorescence focus neutralization (FFN) assay using strains ISU-P (97.7% homology), VR-2385 (92.3% homology) and MN184 (84.5% homology). All pigs were necropsied 10 days after challenge and macroscopic and microscopic lesions were assessed by a pathologist.

Results and discussion

No viremia was detectable in any of the vaccinated pigs; however, by the time of challenge 15/16 vaccinated pigs had seroconverted based on ELISA and had neutralizing antibodies against the homologous strain ISU-P. Infection with VR-2385 resulted in mild-to-moderate clinical disease and lesions. For VR-2385 infected pigs, vaccination significantly decreased PRRSV viremia and nasal shedding by 9 days post challenge (dpc), significantly reduced macroscopic lung lesions, and significantly increased the average daily weight gain compared to the non-vaccinated pigs. Infection with MN184 resulted in moderate-to-severe clinical disease and lesions regardless of vaccination status; however, vaccinated pigs had significantly less nasal shedding by dpc 5 compared to non-vaccinated pigs. Under the study conditions, the A2P90 vaccine strain was attenuated without detectable shedding, improved weight gain, and offered protection to the pigs challenged with VR-2385 by reduction of virus load and macroscopic lung lesions. Further work is needed to investigate different vaccination and challenge protocols, including routes, doses, timing and strains.

References

1. Albina E, Carrat C, Charley B. J Interferon Cytokine Res 1998 Jul;18(7):485-90.
2. Takaoka A, Yanai H. Cell Microbiol 2006 Jun;8(6):907-22.
3. Ma X, Yu Y, Xiao Y, et al. Sci Rep 2016;6:36312.

Determining the efficacy of pooling different fecal swab types in the field versus the lab for detection of porcine epidemic diarrhea virus by PCR

Sara Hamlett, MS; Katie Woodard, DVM; Rodger Main, DVM, PhD
Iowa State University Veterinary Diagnostic Laboratory, Ames, Iowa

Introduction

Surveillance testing for porcine epidemic diarrhea virus (PEDV) is a common practice in the swine industry; however, running routine diagnostics is costly for a producer. Swine researchers have already established that pooled testing is cost effective and helpful to determine herd status with other diseases. However, the methodology of pooling samples for PEDV testing has not been fully explored. The objective of this study was to evaluate whether there was a difference in the detection of PEDV using fecal swabs pooled in the field vs. fecal swabs pooled in the laboratory and to compare the efficacy of four different types of fecal swabs.

Materials and methods

Seven serial dilutions (10^{-1} to 10^{-7}) were made from stock PEDV. The serial dilutions were tested via quantitative PCR (qPCR) to distinguish viral load. Feces were collected from two PEDV negative sows and softened with 250 ml of PBS. Eighteen grams of the fecal mixture were weighed into 50 mL conical tubes. For each serial dilution, there were four negative fecal samples and one positive sample. The positive sample was made by mixing 18 g of PEDV negative feces with 2 mL of virus dilution. Fecal swabs were collected in a 4 × 2 factorial fashion with four swab types used: 1) Copan FecalSwab, 2) Copan ESwab, 3) Copan Universal Viral Transport and 4) Starswab II. Swabs were pooled using a lab or field method. Fecal matter collections using the swabs were conducted as directed by the manufacturer. All samples were refrigerated for 48 hours before pooling to mimic shipment from the field to a laboratory. Sample pooling via the lab method was the same for swab types 1, 2, and 3. An equal volume, 150 µl, of each sample was combined into a 5 mL snap cap tube, which was then submitted for testing. The lab method for swab 4 was completed by swirling the swab into a 5 ml snap cap tube filled with 2.5 ml of PBS and then taking 150 µl combined into a 5 mL snap cap tube. The field pooling method for swabs1, 2, and 3 was conducted by pouring all of the media (1 ml) in each of the five tubes for a dilution into a single tube. A smaller amount of the pool was submitted for testing. The Starswabs were prepared the same as the lab method but for the field method all five samples for the dilution were poured into the same tube. Again, a smaller amount of the pool was submitted for testing. Each sampling was repeated for five replicates. The finished pooled sample was submitted to the ISU VDL for qPCR.

Results and discussion

The stock dilution had positive amounts of PEDV from dilution 10^{-1} to 10^{-6}. When the stock dilutions were added to negative feces, virus was detected in samples 10^{-1} to 10^{-4}. Pooling diluted the virus to a detectable level at 10^{-3}; however, detection was not consistent in all pools past 10^{-1}. The lab method of pooling appears to be more sensitive than the field method when using swab 1 or 2. The method of pooling did not make a difference with swab 3 or 4. Type 2 appeared to be the most sensitive compared to the other three, with 4 being the second most sensitive. The decreased numbers of positive PEDV samples after diluting the stock with the feces and after pooling the swabs was not unexpected. After fecal collection on swab 1, the mixture of media and feces created a gel like consistency which may have prevented the sample from becoming homogenous. Thus, when the small sample was taken to pool, it may not have included the virus. This resulted in type 1 detecting fewer positives in both the lab and field methods compared to the other swabs. In conclusion, the Copan ESwab (swab 3) was most effective at detecting PEDV. The method of pooling did not matter when using the most effective swab. Therefore this study suggests the method of pooling is dependent on the swab used to collect fecal material.

Evaluation of autogenous vaccine efficacy in western Canadian swine herds

L. Redies; J.C.S Harding, DVM, MS, Dipl. ABVP; S.E. Detmer, DVM, PhD
Western College of Veterinary Medicine, University of Saskatchewan, Saskatoon, Saskatchewan, Canada

Introduction

Influenza A viruses in swine (IAV-S) are a cause of severe respiratory disease and are highly contagious. Endemic influenza in a growing pig population can result in significant production costs related to disease treatment and reduced growth rates. As a result, vaccination is a commonly used control strategy to limit influenza infections on swine farms. However, due to the diversity of IAV-S strains circulating in North America, commercial vaccines are not always effective.[1] Autogenous vaccines are utilized when commercial vaccine strains differ significantly from field strains, although the efficacy of these vaccines against IAV-S infection has not been well evaluated.

Materials and methods

In order to determine autogenous vaccine effectiveness, 23 swine farms from Alberta, Saskatchewan and Manitoba were evaluated based on routine diagnostic submissions submitted to Prairie Diagnostic Services, Inc. and the University of Minnesota Veterinary Diagnostic Lab from 2012-2016. Autogenous vaccines had been utilized in each farm. Ten swine farms that had experienced influenza but had not utilized commercial or autogenous vaccines were also included in the analysis as a comparative group. Strains isolated pre- and post-vaccination were sequenced and antigenic sites in the hemagglutinin protein were compared between strains.[2,3] Each farm was classified according to the strains identified post-vaccination as: 1) strain disappeared, 2) new subtype emerged, 3) virus did not change, 4) an antigenic change occurred. A Fisher's Exact test was performed to determine if outcomes differed between the autogenously vaccinated and non-vaccinated farms.

Results and discussion

Our results showed that in the non-vaccinated farms, the virus was unchanged in 8/10, but there were antigenic changes in 2/10. (Table 1). This differed significantly ($P < 0.001$) from the outcomes in autogenous farms where: 12/23 farms did not identify any influenza post-vaccination, 7/23 farms had a new subtype emerge, and 4/23 showed evidence of antigenic change (Table 1). In these four autogenous farms, the strains experienced significant changes in antigenicity such that the autogenous vaccine was no longer effective. This increased evolutionary rate was likely driven by pressure from the vaccine that can allow the virus to escape vaccine-mediated immune protection. Our research showed that autogenous vaccines are largely successful at eliminating the specific IAV-S strain targeted, but more work is needed to evaluate farm or virus-specific variables resulting in increased antigenic evolution seen on these farms. Understanding factors that can impact autogenous vaccine efficacy is critical for practitioners in terms of herd health management and influenza prevention. To support our ongoing research in this area, we have developed a proactive strategy for farm surveillance and vaccine monitoring.

Acknowledgements

The authors wish to thank the interprovincial undergraduate student summer research fund at the Western College of Veterinary Medicine for financing this research. This research was supported by the Alberta Livestock and Meat Agency, Saskatchewan Ministry of Agriculture and the Saskatchewan Agricultural Development Fund.

References

1. Anderson TK, Campbell BA, Nelson MI, Lewis NS, Janas-Martindale A, Killian ML, Vincent AL: Characterization of co-circulating swine influenza A viruses in North America and the identification of a novel H1 genetic clade with antigenic significance. Virus Res 201:24-31, 2015.

2. Nobusawa E, Aoyama T, Kato H, Suzuki Y, Tateno Y, Nakalima K: Comparison of complete amino acid sequences and receptor-binding properties among 13 serotypes of hemagglutinins of influenza A viruses. J Virol 182:475-485, 1991.

3. Lewis NS, Anderson TK, Kitikoon P, Skepner E, Burke DF, Vincent AL: Substitutions near the hemagglutinin receptor-binding site determine the antigenic evolution of in uenza A H3N2 viruses in U.S. swine. J Virol 88:4752–4763, 2014.

Table 1: Timeline of influenza sampling and isolation from Farms 1-33. Farms 1-23 represent the autogenous vaccine group as shown by "A" while Farms 24-33 represent the control vaccine group as shown by "C." T0 represents the original strain with each interval equivalent to one month. Strains shown in blue represent one antigenic change in the hemagglutinin protein while strains shown in pink represent two antigenic changes. Strains shown in green indicate three or more antigenic changes and strains shown in black indicate no antigenic changes.

Farm	Group	T0	T1	T2	T3	T4	T6	T7	T8	T9	T10	T11	T12	T13	T14	T15	T16	T17	T18
1	A	H1N2	Gone																
2	A	H1N2	Gone																
3	A	H1N2	-	-	H1N2	-	H1N2	-	-	-	H1N2	H1N2	-	-	H1N2	-	-	-	-
4	A	H1N2	Gone																
5	A	H1N2	Gone																
6	A	H1N2	Gone																
7	A	H1N1	-	-	-	-	-	H3N2	-	-	-	-	-	-	-	-	-	-	-
8	A	H3N2	-	-	H3N2	-	-	-	-	-	-	-	H3N2	H3N2	H3N2	-	H3N2	-	-
9	A	H3N2	Gone																
10	A	H3N2	Gone																
11	A	H1N2	-	-	-	-	-	-		-	-	-	-	-	-	H1N2	-	-	-
12	A	H3N2	-	-	-	-	-	-	H1N2	-	-	-	-	-	-	-	-	-	-
13	A	H3N2	-	-	-	-	-	-	-	-	-	-	-	-	-	-	H1N2	-	-
14	A	H3N2	Gone																
15	A	H3N2	-	H3N2	-	H3N2	-	-	-	-	H3N2	-	-	-	-	-	-	-	-
16	A	H3N2	Gone																
17	A	H3N2	-	-	-	-	-	-	-	-	-	H1N1/ N2	-	-	-	-	-	-	-
18	A	H1N2	Gone																
19	A	H1N2	-	-	-	-	-	-	-	-	-	-	H3N2	-	-	-	-	-	-
20	A	H1N2	Gone																
21	A	H3N2	-	H1N2	H1N2	H1N2	H1N2	-	H1N2	-	-	-	H1N2	-	H1N2	-	-	-	H1N2
22	A	H3N2	Gone																
23	A	H3N2	H3N2	-	-	-	-	-	H3N2	H3N2	-	-	H3N2	H1N2	-	-	-	-	-
24	C	H3N2	-	-	-	-	-	-	-	-	-	-	-	H3N2	-	-	-	-	-
25	C	H3N2	-	-	-	-	-	H3N2	H3N2	H3N2	-	-	-	-	H3N2	-	-	-	-
26	C	H3N2	-	-	-	-	-	-	-	-	-	H3N2	-	-	-	-	-	-	-
27	C	H1N2	H1N2	H1N2	H1N2	-	-	-	-	-	-	-	-	-	-	-	-	-	-
28	C	H1N1	-	-	-	-	-	-	-	-	-	-	-	-	-	-	H1N1	-	-
29	C	H1N2	H1N2	-	-	-	H1N2	-	-	-	-	-	-	-	-	-	-	-	-
30	C	H1N2	-	-	-	-	H1N2	-	-	-	-	-	-	-	-	-	-	-	-
31	C	H1N2	-	H1N2	-	-	-	-	-	-	-	-	-	-	-	-	-	-	-
32	C	H3N2	-	-	-	-	H3N2	-	-	-	-	-	-	-	-	-	-	-	-
33	C	H3N2	H3N2	-	H3N2	-	-	H3N2	-	-	-	-	-	-	-	-	-	-	-

Trends of *Lawsonia intracellularis* polymerase chain reaction submissions at the University of Minnesota Veterinary Diagnostic Laboratory over a 10-year period

Donna Drebes, BS; Connie Gebhart, PhD; Julio Alvarez, DVM
College of Veterinary Medicine, University of Minnesota

Introduction

Lawsonia intracellularis, an obligatory intracellular gram negative bacterium, is the etiological agent of porcine proliferative enteropathy (PPE). PPE is endemic in the United States swine herd, though no recent national/regional incidence estimates are available. Therefore, it is difficult to assess its impact and evaluate whether changes in management (such as the upcoming January 2017 ban on the use of growth promoters) may influence its distribution. The objective of this study was to determine whether patterns could be detected in diagnostic submissions in order to guide future diagnostic sampling strategies.

Material and methods

The Minnesota Veterinary Diagnostic Laboratory (MVDL) database was screened. All cases submitted for diagnosis of *L. intracellularis* by conventional PCR (PCR) and real-time PCR (rt-PCR) from 2006 to 2015 were selected. Seven thousand forty-seven accessions (a unique number generated by the MVDL for each submission) containing 25,894 samples were found. Non-diagnostic samples and samples from countries other than the United States and Canada (n = 376 accessions containing 6,697 samples) were removed leaving 6,671 accessions containing 19,197 samples for the analysis.

Variables considered at the sample level included accession, year and month of submission, owner information, animal location, test type, specimen and result (with Ct value for rt-PCR). Samples were grouped by type as intestinal, fecal, or oral fluid samples. Trends in submission patterns (number of samples and accessions, number of samples per accession, type of samples submitted) and test results (percentage of positive/negative samples, Ct values for rt-PCR overall and per sample type) were assessed visually and tested using Chi square, ANOVA, and non-parametric tests. Statistically significant results were defined as those having p values ≤ 0.05. Analyses were carried out using IBM SPSS V.23.

Results and discussion

The number of diagnostic samples submitted increased throughout the 2009-2015 period, with an annual mean of 751 samples in 2006-2009 compared to 4,853 samples in 2015. The proportion of each sample type submitted changed over time, with oral fluids increasing as a percentage of the total samples (10% in 2013; 56% in 2015). No seasonal pattern in diagnostic submissions were present. However, significant increases in the number of samples received compared to the expected were observed during certain periods suggesting occurrence of outbreaks. No statistically significant differences in the proportion of positive samples regardless of year or technique used (PCR vs. rt-PCR) were observed. Among sample types testing positive by rt-PCR there was a difference in the Ct value, with oral fluid samples showing significantly higher Ct values when compared to fecal and intestinal samples (Figure 1) (*P* value < 0.01).

Our results support the theory that the swine industry understands the value of including *Lawsonia* PCR for disease diagnosis and surveillance. A potential increasing awareness of its importance may be reflected by the increasing number of sample submissions. Oral fluids are becoming more frequently utilized for diagnosis. This may be due to increased surveillance, although the significantly higher Ct (suggestive of a lower bacterial load) may indicate bacterial destruction by saliva enzymes prior to arrival at the MVDL. Potential follow-up studies include reviewing the updated database after the implementation of the veterinary feed directive, comparing other *Lawsonia* diagnostic test type results (serology or immunohistochemistry) and comparing *Lawsonia* PCR and rt-PCR results to those of other enteric pathogens such as *Salmonella*.

Figure 1: Boxplot of the Ct values obtained among positive (Ct<35) and suspect (Ct between 35 and 40) samples by the analysis of 214 intestinal, 904 fecal and 1,133 oral fluid samples using rt-PCR in years 2013-2015. Ct value shown on y axis while specimen type shown on x axis.

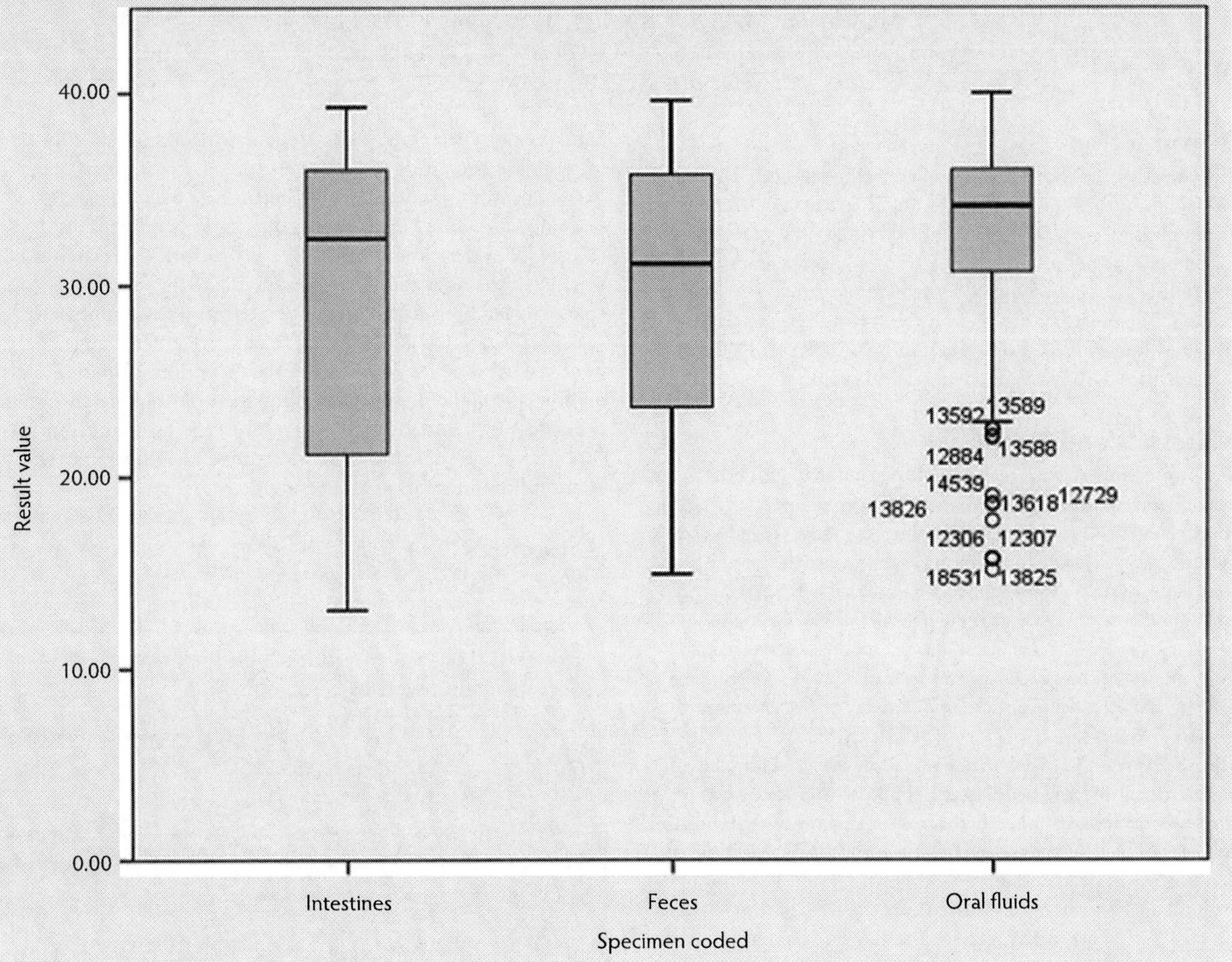

Sand as a bedding for loose-housed gestating sows fed via ESF

J. Applebaum; M. Pierdon, VMD; T. Parsons, VMD, PhD
Swine Teaching and Research Center, School of Veterinary Medicine, University of Pennsylvania, Kennett Square, Pennsylvania

Introduction

The use of pens to house groups of gestating sows is becoming more common. Some studies point to an increase in lameness in group housed sows.[1] The dairy industry often uses sand as a preferred bedding because it is thought to be sanitary, comfortable and reduce lameness. Sand bedding has not yet been investigated in sows. The objective of this study was to understand the use of sand bedding by loose-housed gestating sows fed via ESF and any relationships with parity, feed rank and lameness.

Materials and methods

Sows were housed in a large dynamic pen and fed with ESF. Sows were observed morning and afternoon, approximately seven hours apart, daily for nine weeks. At that time, each sow's location was recorded as well as the temperature and whether the bedding areas were clean or dirty. Sows had free access to a 4 × 8.5m straw area and a sand area the same size (Figure 1). Halfway through the study, the locations of the straw and the sand were reversed. At three week intervals, each sow was walked down a solid concrete hallway, scored as lame or not lame, and day of gestation was recorded. 206 sows were included in the study. Statistics were done using STATA v.13.1. Spearman correlation was used for feed rank, barn temperature and bedding use. A point bi-serial correlation was used to look at parity and whether or not an animal used the sand. A Kruskal Wallis test was used to look at the difference between bedding use in clean and dirty conditions. A multi-level linear regression analysis was done to investigate the relationship between lameness, feed rank, parity, stage of gestation and sand usage across each three-week interval. Feed rank was calculated as the average order in which a sow ate over the two weeks prior to lameness scoring.

Results and discussion

The majority of sows in the herd were observed in the straw or sand areas at least once with 66% of sows using the straw and 52% of the herd using the sand. Younger parity sows were initially more likely to use the sand ($r = -.18$, $P = .015$), as were sows that ate later in the day ($r = .20$, $P = .018$), though both of these relationships were not significant after the first time period. The percent of the herd observed in the straw was significantly greater when it was clean than when it was dirty ($P = .002$) whereas this effect was not seen with the sand ($P = .35$). Using a multi-level regression model, there was no effect of lameness, parity, day of gestation, or feed rank on the percentage of time an animal spent in the sand ($P > .05$).

Sand seems to be a viable and inexpensive bedding choice for loose-housed gestating sows. Because no relationship was seen between cleanliness of sand and percent of sows using sand, sand bedding may require less cleaning than straw bedding. This study took place in the summer and suggests that further research may be needed to understand the use of sand bedding over more seasons, in other housing systems, and in environments of varying size. Further studies could also determine what percentage of a herd will use sand if not given straw as an option. Which animals use the sand could be based on individual animal preference since the sand was eventually used across parity and feed rank, but future studies would be needed to determine the influence of social relationships on location and bedding preference.

References

1. Kronemen et al. 1993, J. Vet Med 704-712.

Figure 1: (Left) Straw-bedded area; (Right) Sand-bedded area after reversing bedding type.

Evaluation of the sensitivity of Envirobooties™ and Swiffer™ kits for viral detection of PRRSV and PEDV in livestock trailers

T. Engle[1], MS; R. Jones[2], DVM; D. Johnson[3], BS; A. Scheidt[3], DVM, MS
[1]Virginia-Maryland College of Veterinary Medicine; [2]Livestock Veterinary Services; [3]Boehringer Ingelheim Vetmedica Inc.

Introduction

Disease transmission by transport vehicles is a major economic concern for the swine industry. Improving detection methods for viral pathogens on transport trailers is essential for monitoring this risk, and research involving environmental sampling with Envirobooties™ is limited for swine viral pathogens. Therefore, the purpose of this study was to evaluate the sensitivity of Envirobooties™ and Swiffer™ pads for detection of two economically important viral pathogens. This study consisted of two trials to evaluate the sensitivity of using Envirobooties™ and Swiffer™ pads for detecting PRRSV and PEDV. The first trial evaluated both detection methods using commercially available Ingelvac PRRS® MLV. The second trial also evaluated both detection methods using live PEDV.

Materials and methods

An aluminum stock trailer was used for experimental inoculation. The trailer was power washed, disinfected with Synergize™, and allowed to dry before each trial. To determine sensitivity, 6 dilutions and 1 negative control were made serving as treatments for PRRSV and PEDV trials. Each dilution was quantified to determine the starting concentration. The trailer was divided into 70 experimental squares (1ft × 1ft) with 10 rows and 7 columns. Each row was randomly assigned a sampling method, either Envirobooties™ or Boehringer Ingelheim Vetmedica (BI) Swiffer™ kits, and each column represented each dilution treatment(n = 35 Envirobooties, n = 35 BI Swiffer™ kit). A 5 ml sample was used to inoculate assigned sample squares. After a minimum of 5 minutes to dry, each square was sampled with its designated sampling method. All PEDV samples were sent to the Boehringer Ingelheim Health Management Center, and all PRRSV samples were sent to Iowa State Veterinary Diagnostic Laboratory. The results of this study indicate that viral RNA cannot be detected for PCR for PRRSV and PEDV samples from 10^{-4} to 10^{-5} diluted virus, regardless of detection method. Both Envirobooties™ and BI Swiffer™ kits were able to detect PRRSV and PEDV from 10^{0} to 10^{-3} diluted virus.

Results and discussion

PRRSV Swiffer™ and Envirobooties™ results can be seen in Table 1A. PRRSV Swiffer™ results are as follows 10^{0} (5/5 positives; CT ranges 24.3-30.0), 10^{-1} (5/5 positives; CT 31.4-34.2), 10^{-2} (5/5 positives; CT ranges 31.0-31.7), and 10^{-3} (1/5positives; CT value = 34.2). PRRSV Envirobooties™ results are as follows 10^{0} (5/5 positives; CT ranges 21.5 – 25.3), 10^{-1} (4/5 positives; CT ranges 27 – 31.4), 10^{-2} (5/5 positives; C ranges 29.9 – 33.3), and 10^{-3} (3/5 positives; CT ranges = 32.5 – 35.7). PEDV Swiffer™ and Envirobooties™ results can be seen in Table 1B. PED Swiffer™ results are as follows 0^{0} (5/5 positives; CT value ranges 24.2 – 26.5), 10^{-1} (5/5 positives; CT ranges 26.6 – 29.1), 10^{-2} (4/5 positives; CT ranges 31.2 – 35.8), and 10^{-3} (1/5positives; CT value = 35.7). PEDV Envirobooties™ results are as follows 10^{0} (4/5 positives; CT ranges 30.1 – 34.3), 10^{-1} (5/5 positives; CT ranges 29.2 – 33.8), 10^{-2} (5/5 positives; CT ranges 30.2 – 34.5), 10^{-} (1/5positives; CT value = 34.3), and 10^{-4} (1/5positives; CT value = 33.3). The results from the PEDV trial indicate that the BI Swiffer™ kits were more sensitive at detecting PEDV based on lower CT values compared to Envirobooties™. In addition, Envirobooties™ were more sensitive for detecting PRRSV based on lower CT values. These results suggest that Swiffer™ pads should be used for environmental sampling and monitoring of PEDV, and Envirobooties™ may be implemented as a new diagnostic tool for PRRSV environmental sampling.

Table 1: PRRSV and PEDV Swiffer™ and Envirobooties™ results

A — PRRSV

Dilution	Sample	Number positive	Mean	Median	Min	Max
Control	Envirobooties™	(0/5)	NA	NA	NA	NA
	Swiffer™	(0/5)	NA	NA	NA	NA
10^-5	Envirobooties™	(0/5)	NA	NA	NA	NA
	Swiffer™	(0/5)	NA	NA	NA	NA
10^-4	Envirobooties™	(0/5)	NA	NA	NA	NA
	Swiffer™	(0/5)	NA	NA	NA	NA
10^-3	Envirobooties™	(3/5)	34.13	34.20	32.50	35.70
	Swiffer™	(1/5)	34.20	34.20	34.20	34.20
10^-2	Envirobooties™	(5/5)	32.54	33.20	29.90	33.30
	Swiffer™	(5/5)	31.30	31.30	31.00	31.70
10^-1	Envirobooties™	(4/5)	29.75	30.30	27.00	31.40
	Swiffer™	(5/5)	32.98	33.10	31.40	34.20
10^0	Envirobooties™	(5/5)	23.70	24.00	21.50	25.30
	Swiffer™	(5/5)	26.64	25.90	24.30	30.00

B — PEDV

Dilution	Sample	Number positive	Mean	Median	Min	Max
Control	Envirobooties™	(0/5)	NA	NA	NA	NA
	Swiffer™	(0/5)	NA	NA	NA	NA
10^-5	Envirobooties™	(0/5)	NA	NA	NA	NA
	Swiffer™	(0/5)	NA	NA	NA	NA
10^-4	Envirobooties™	(1/4)	33.30	33.30	33.30	33.30
	Swiffer™	(0/5)	NA	NA	NA	NA
10^-3	Envirobooties™	(1/5)	34.40	34.40	34.40	34.40
	Swiffer™	(1/5)	35.70	35.70	35.70	35.70
10^-2	Envirobooties™	(5/5)	32.06	32.20	30.20	34.50
	Swiffer™	(4/5)	34.08	34.65	31.20	35.80
10^-1	Envirobooties™	(5/5)	31.84	32.00	29.20	33.80
	Swiffer™	(5/5)	28.06	28.60	26.60	29.10
10^0	Envirobooties™	(4/5)	32.33	32.50	30.10	34.20
	Swiffer™	(5/5)	25.00	24.60	24.20	26.50

Comparison of morbidity and mortality after challenge with two North American PRRS virus isolates shows marked variation in time course and prevalence of clinical disease between isolates

L.A. Constance[1]; B. Bloomberg[3]; J.K. Lunney[4]; J.C.M. Dekkers[5]; R.R.R. Rowland[1]; M.C. Niederwerder[1,2]
[1]Department of Diagnostic Medicine/Pathobiology, College of Veterinary Medicine, Kansas State University, Manhattan, Kansas; [2]Kansas State Veterinary Diagnostic Laboratory, Kansas State University, Manhattan, Kansas; [3]Comparative Medicine Group, Kansas State University, Manhattan, Kansas; [4]United States Department of Agriculture, Agricultural Research Services, Beltsville Agricultural Research Center, Beltsville, Maryland; [5]Department of Animal Science, Iowa State University, Ames, Iowa

Introduction

Porcine reproductive and respiratory syndrome (PRRS) is the most economically devastating disease of swine worldwide.[1] Significant genetic variation among PRRS virus isolates can correlate to differences in clinical signs and disease progression.[2] The objective of this study was to compare morbidity and mortality of two heterologous North American (Type 2) PRRSV isolates after experimental challenge in nursery pigs.

Materials and methods

Two experimental populations of approximately 200 commercial crossbred pigs (Pietrain × LW) each were infected with either NVSL-97-7985 (NVSL) or KS-2006-72109 (KS06) and followed for 42 days post-infection (dpi).

Results

Overall morbidity after challenge with NVSL or KS06 was 39.2 and 21.6%, respectively ($P = 0.0002$). Mortality rates were also higher after infection with NVSL when compared to KS06 (11.5 and 9.0%, respectively); however, this difference was not statistically significant ($P = 0.51$). The time course of clinical disease post-infection with NVSL was chronic, with clinical signs occurring at a similar rate throughout the course of the trial. In contrast, clinical disease post-infection with KS06 was acute, with clinical signs primarily occurring between 4 and 10 dpi. The mean duration of clinical disease also reflected the chronic and acute disease progression of the two isolates; 11.5 ± 6.0 dpi with NVSL and 4.0 ± 2.1 dpi with KS06 ($P < 0.0001$). Clinical signs of respiratory disease, such as open mouth breathing, dyspnea, tachypnea and nasal discharge, were 11.3 times more likely to occur after challenge with NVSL than after challenge with KS06 ($P < 0.0001$). Infection with NVSL was 3.9 times more likely to reduce body condition during the course of the trial compared to infection with KS06 ($P = 0.001$). Severity of clinical disease correlated with an overall increase in PRRSV replication after challenge with NVSL; total virus load was 153.7 ± 20.3 and 140.3 ± 17.2 for NVSL and KS06, respectively ($P < 0.0001$). Average daily gain (ADG) during the 42 day trial was also significantly lower in pigs challenged with NVSL; 0.43 ± 0.09 kg and 0.45 ± 0.10 kg in pigs challenged with NVSL and KS06, respectively ($P = 0.03$).

Discussion

In this model, challenge with NVSL had significantly higher morbidity, greater virus replication, and decreased ADG due to the chronic duration of disease. However, challenge with KS06 resulted in similar mortality due to acute and severe disease within the first week post-infection. These results provide evidence for the acute and chronic forms of clinical disease that can occur with different PRRSV isolates.

Acknowledgements

This research was supported by the National Pork Board Project #10-156, Genome Canada, and members of the PRRS Host Genomics Consortium. LAC was supported by The State of Kansas National Bio and Agro-Defense Facility fund.

References

1. Holtkamp et al. Assessment of the economic impact of porcine reproductive and respiratory syndrome virus on United States pork producers. *J. Swine Heal. Prod.* 2013;21:72–84.

2. Kimman et al. Challenges for porcine reproductive and respiratory syndrome virus (PRRSV) vaccinology. *Vaccine* 2009;27:3704–3718.

Detection of influenza A virus on inanimate fair surfaces

Courtney M. Wright; Michele M. Zentkovich; Jacqueline M. Nolting, MS;
Andrew S. Bowman, MS, DVM, PhD, DACVPM
The Ohio State University, College of Veterinary Medicine Department of Veterinary Preventive Medicine

Introduction

Week-long livestock exhibitions have a multitude of swine-swine and swine-human interactions, which can facilitate influenza A virus (IAV) transmission between species resulting in the development of novel viral strains.[1] IAV can survive days to weeks outside a host,[2] meaning the movement of portable objects between livestock exhibitions may amplify rapid virus dissemination amongst livestock and human populations. The goal of this study was to document the frequency of IAV contamination of portable fomites at agricultural fairs to identify mitigation strategies for pathogen spread.

Materials and methods

The present study sampled inanimate surfaces at 20 agricultural fairs during the summer of 2016. Cotton gauze was used to sample objects in the swine barn (i.e. feeders, waterers, tack boxes, sort boards, and chairs), which were then placed into brain heart infusion broth and frozen until RRT-PCR could be performed. Viral isolates were cultured and subtyped according to standard protocols.

Results

The completed study detected viral RNA on fomites from 10 fairs. In total, 75/400 (18.75%) of environmental surfaces were PCR positive, of which 64/75 (85.33%) were in direct contact with swine during time of sampling. Viable virus was recovered from 7/75 (9.33%) samples.

Discussion

Agricultural fairs provide a unique comingling of swine from many farms. The viable virus detected in this study highlights the potential role of fomites in viral dissemination from a fair where there were IAV-infected pigs. These results indicate the importance of disinfecting inanimate objects in contact with swine and highlight the need for swine exhibitors to have effective disinfection plans to minimize viral transmission via fomites.

References

1. Ma, Wenjun, Robert E. Kahn, and Juergen A. Richt. "The pig as a mixing vessel for influenza viruses: human and veterinary implications." *J Mol Genet Med* 3.1 (2008): 158-166.

2. Bean, B., et al. "Survival of influenza viruses on environmental surfaces." *Journal of Infectious Diseases* 146.1 (1982): 47-51.

Evaluation of the effects of flushing feed manufacturing equipment with chemically treated rice hulls on porcine epidemic diarrhea virus (PEDV) cross contamination during feed manufacturing

Jordan Gebhardt[1], BS; Steve Dritz[1], DVM, PhD; Jason Woodworth[1], PhD; Cassandra Jones[1], PhD; Phil Gauger[2], DVM, PhD; Mike Tokach[1], PhD; Charles Stark[1], PhD; Mary Muckey[1], MS; Jianqiang Zhang,[2] PhD; Rodger Main,[2] DVM, PhD;
[1]Kansas State University, Manhattan, Kansas; [2]Iowa State University, Ames, Iowa

Introduction

Feed contaminated with porcine epidemic diarrhea virus (PEDV) has been demonstrated to contaminate subsequent batches of feed due to carryover on feed contact surfaces. Wet decontamination is the most effective method of equipment surface decontamination, but this method is not practical in most commercial feed production settings. A potential alternative, practical decontamination strategy is to remove surface contamination with chemically-treated rice hull flushes. Rice hulls were selected as a carrier for the chemical due to their high level of abrasiveness, excellent adsorption characteristics, and relatively low cost. Therefore, our objective was to determine the impact of medium chain fatty acids (MCFA) or formaldehyde-treated rice hull flushes on residual PEDV contamination during feed manufacturing, as determined by PEDV RNA detection and 10-day old pig bioassay. Additionally, dust collected from non-feed contact surfaces was evaluated for infectivity.

Materials and methods

Flush treatments were constructed using untreated rice hulls (control) or rice hulls treated with 0.325% formaldehyde, 2% MCFA blend, or 10% MCFA blend (1:1:1 ratio of caproic, caprylic, and capric acid). In 8 separate laboratory-scale mixers, a PEDV-contaminated (10^4 $TCID_{50}$/g) batch of feed was mixed, followed by one of the four treatment flushes. There were two replicates per treatment. To test the applicability at larger scale, PEDV-contaminated batch of feed was manufactured in a single 50-kg mixer and discharged through a bucket elevator, followed by a 10% MCFA-treated rice hull flush through the system and then a subsequent batch of PEDV-free feed. Contaminated feed and flush samples were analyzed for presence of PEDV RNA by qRT-PCR, and samples were stored at -80°C until determination of infectivity via 10-day old pig bioassay using rectal swab and cecal content qRT-PCR analysis, immunohistochemistry, and histopathology. Data were analyzed using PROC GLIMMIX (SAS Institute, Inc., Cary, NC) to determine differences in qRT-PCR cycle threshold (Ct) values.

Results and discussion

The untreated rice hull flush yielded 3 of 6 samples with detectable PEDV RNA, while 1 of 6 formaldehyde-treated rice hull flushes and 2 of 6 of the 2% MCFA rice hull flushes had detectable PEDV RNA. Genetic material was not detected in any 10% MCFA rice hull flush samples. In the production scale system, no flush or subsequent feed samples from the mixer had detectible PEDV RNA. However, one 10% MCFA rice hull sample collected from the bucket elevator discharge spout had detectible PEDV RNA. The PEDV-contaminated control feed resulted in infectivity in bioassay, but the chemically-treated rice hulls and feed collected after the flush did not result in infectivity. Dust collected after mixing PEDV-contaminated feed had a large quantity of PEDV RNA (Avg. Ct = 29.4), while dust collected immediately after the 10% MCFA rice hull flush had a reduced quantity of viral RNA (Ct = 33.7). The subsequently manufactured feed had no detectable PEDV RNA. The PEDV-contaminated dust resulted in infectivity in bioassay, while the MCFA rice hull flush or subsequent feed did not. Rice hull flushes effectively reduce PEDV infectivity and chemical treatment of the flushes reduces the quantity of detectable PEDV RNA. Additionally, flushing a feed manufacturing system with a rice hull flush treated with 10% MCFA eliminated dust infectivity.

Impact of sow parity on the influenza A (IAV-S) infection timing of the suckling pig in an endemic herd

Megan E. Bloemer[1], BS; Eva Jablonski[2], DVM, MS; Luis Ochoa[3], PhD; Aaron Lower[3], DVM; Laura Greiner[3], PhD, MS
[1]University of Illinois College of Veterinary Medicine, Urbana, Illinois; [2]Zoetis, Parsippany, New Jersey;
[3]Carthage Veterinary Service, Carthage, Illinois

Introduction

Influenza A viruses commonly infect sow herds and can become endemic in the suckling pig population. Understanding the timing of infection in suckling piglets could enable producers to implement different management strategies to decrease the percentage of positive piglets or eradicate the virus. To the authors' knowledge, timing of infection by age and parity has not been evaluated.

Objective

The objective of this study was to determine the effect of sow parity [i.e. P0 (gilts) vs. P4+ (sows)] on the age at which suckling piglets become infected with influenza virus on an endemic H1N1 sow farm. We hypothesized that piglets from primiparous sows would be more likely to be infected with influenza and at a younger age due to lower maternal antibody titers as compared to multiparous sows.

Materials and methods

In a 6,000 head influenza-vaccinated commercial sow farm, 15 gilts and 15 sows were randomly selected from three farrowing rooms, with 10 test litters (five per parity group) in each room. Cross fostering was not allowed. Nasal samples were collected from both nostrils of six tagged piglets per litter. Swabs were collected on 3, 10, 16, and 22 days of age using synthetic calcium-alginate tipped swabs (Puritan Medical Products, Guilford, ME) and placed into viral media tubes (Uni-Tranz RT, Puritan Medical Products) for storage and transport to ISU VDL for influenza PCR testing. Nasal swabs were pooled by three, yielding two pools per litter with the same pigs in each pool at each collection point. Litter was the experimental unit. Litter was considered positive when at least one of the two pools was PCR positive. Data were analyzed as a randomized controlled block design (block = room) and the model included parity group and piglet age as fixed effects and room as a random effect.

Results

Both parity groups had seven total positive litters over the course of the study (Table 1). Parity had no effect on the incidence and timing of influenza infection in piglets ($P \geq 0.8$). Age was highly significant ($P < 0.05$) with the number of positive litters increasing as age increased, starting at 10 days of age onward. The incidence by age was the same for both parity groups.

Discussion

This study does not support managing populations according to parity for influenza. Practitioners should continue to vaccinate the whole sow population since pigs from primiparous litters do not appear to be more likely to be infected or infected at a younger age with influenza than multiparous sows.

Table 1: Effect of parity on the percentage of IAV-S PCR positive litters at 3, 10, 16, and 22 days of age*

	Day of age			
	3	10	16	22
Gilts (P0)				
Number of litters on test	15	15	15	14
Number of positive litters	0^a	1.0^b	3.0^c	7.0^d
Percentage of positive litters, %	0^a	6.7^b	20.0^c	50.0^d
Sows (P4+)				
Number of litters on test	15	14	13	12
Number of positive litters	0^a	1.0^b	3.0^c	7.0^d
Percentage of positive litters, %	0^a	7.1^b	23.1^c	58.3^d

abcd Effect of day of age: Means with differing superscript within a row differ $P < 0.05$.
Effect of parity: No significant effect ($P \geq 0.8$; Chi square analysis)
* The number of sows/litters on test changed over the course of the trial due to unexpected sow deaths and accidental commingling of piglets.

Evaluation of a novel antibiotic-free extender utilizing short-term hypothermic storage

R. Warner[1], MS; J. Brown[1], DVM; T. Dohlman[2], DVM, MS, DACT; A. Krull[2], DVM, PhD; J. Ji[3], MS; K. Skoland[1], BS; M. Breuer[1], V. Thompson[1], M. Nickel[1], BS; K. O'Brien[1], BS; A. Forseth[1], DVM; P. Canning[1], DVM; L. Karriker[1], DVM, MS, DACVPM

1Swine Medicine Education Center, Iowa State University, Ames, Iowa; [2]Veterinary Diagnostic and Production Animal Medicine, Iowa State University, Ames, Iowa; [3]Department of Statistics, Iowa State University, Ames, Iowa

Introduction

Swine veterinarians are committed to judicious use of antimicrobials, which necessitates objective consideration of alternatives to common uses. The goal of this study was to evaluate two commercially available semen extenders by assessing semen quality and bacterial growth in a novel antibiotic-free hypothermic semen extender (novel) and an extender containing gentamicin (conventional).

Materials and methods

Semen collection: Sperm-rich ejaculate fractions were collected from three Yorkshire boars using the gloved-hand technique. The semen was pooled and an aliquot was obtained for bacterial culture and analysis. Analysis of the aliquot was completed 24 and 48-hours post-collection. The remaining semen was divided into two treatment groups. Group one was diluted (1:1, vol/vol) with the novel extender. Group two was diluted (1:1, vol/vol) in the conventional extender containing 250 mg gentamicin. The samples were cooled utilizing cooling curves specific to the extenders and following manufacture recommendations (novel @ 5° C, conventional @ 17° C).[1] The samples were further diluted to a concentration of 3.0×10^9 spermatozoa per dose using the NucleoCounter® SP-100™ (ChemoMetec, Denmark). The samples were then divided into six doses for analysis (12 total). Three doses of each extender type were inoculated with *Serratia* spp. (5.0×10^8 CFU/mL). Semen quality analysis: The extended semen samples were evaluated using computer-assisted sperm analysis (CASA), Sperm Vision® SAR software. A small aliquot of each dose was loaded in a Leja standard count four-chambered slide (IMV Technologies) for sperm motility characteristic evaluation, at 24-hour intervals, for six days. The following parameters were chosen for analysis based on correlation to fertility:[2] progressive forward motility (FPM), motility % of sperm (MOT), velocity straight line (VSL), velocity average path (VAP), distance average path (DAP), and amplitude of lateral head displacement (ALH), as well as, individual and group motility. Morphological characteristics were also evaluated at 24-hour intervals for six days using the eosin-nigrosin[3] and modified acrosome staining techniques.[4] Semen bacterial analysis: Standard bacterial plate counts using blood agar media were completed for all samples at 48, 72, and 120-hours post-collection.

Results and discussion

Unlike traditional extenders, the novel extender utilizes storage at 5° C to reduce bacterial growth. The unique formulation and properties of this extender allows for storage at hypothermic temperatures. Over six days, there was no significant change in bacterial growth between the conventional and novel extenders in non-challenged doses. At all time points, challenged doses regardless of treatment contained more bacteria than non-challenged doses as expected. Standard plate counts for novel extender doses challenged with *Serratia* spp. displayed statistically higher growth than challenged conventional extender doses after day-3 ($P < 0.05$). When motility was assessed FPM for the novel extender doses was significantly lower than the conventional extender doses beginning at 72-hours ($P < 0.05$). MOT, VSL, VAP, DAP, and ALH were also significantly lower ($P < 0.05$). However, this conclusion was potentially biased by insufficient rewarming of the novel extender doses that were stored at the lower temperature (5°C). Morphological characteristics did not vary statistically between treatment groups, but transition from acrosome intact to non-intact increased significantly across the six-day time period ($P < 0.05$) as expected. These results indicate that viable sperm can be maintained using antibiotic-free extension when compared to traditional extension for short-term storage. Care must be taken when evaluating motility parameters so that storage temperature does not bias assessments. No fertility endpoints were evaluated in this study. Innovative developments in the field of boar semen extension allows for further judicious use of antimicrobials by practitioners.

References

1. Grossfeld, R., et al. (2016) Antibiotic-free hypothermic storage of boar semen in Androstar +5 extender results in similar fertility results compared to semen stored at 17°C in extender with antibiotic content. *Animal Reproduction Science*. Vol. 169. Page 125.

2. Didion, BA. (2008) Computer-assisted semen analysis and its utility for profiling boar semen samples. *Theriogenology*. Vol 70. Pages 1374-1376.

3. Blom, E. (1950) A one-minute live-dead sperm stain by means of eosin-nigrosin. *American Journal of Obstetrics & Gynecology*. Vol 61. Issue 3. Page 711.

4. Larson, JL., Miller, DJ. (1999) Simple histochemical stain for acrosomes on sperm from several species. *Molecular Reproduction and Development*. Vol. 52, Issue 4. Pages 445-449.

Timing differences of *Mycoplasma hyopneumoniae* colonization in growing replacement gilts with and without tulathromycin administration at weaning

B. Burton[1], BS; K. Crawford[2], DVM, MS
[1]College of Veterinary Medicine, University of Illinois, Urbana, Illinois; [2]Merck Animal Health, De Soto, Kansas

Introduction

Mycoplasma hyopneumoniae (Mhp), the primary agent of enzootic pneumonia, is an economically-significant pathogen in the swine industry. The sow has been implicated as a potential source of colonization during the lactation period.[1] Because an animal sheds Mhp for approximately 240 days, it is essential that colonization begins early to eliminate the infection before farrowing.- For farms supplying breeding stock, treatment with antibiotics at weaning may postpone bacterial colonization, thereby increasing the risk of Mhp shedding from first litter gilts to their offspring in farrowing.[2]

Objective

The objective of this study was to estimate the time of initial detection of Mhp colonization via PCR in pigs treated with 15 mg/kg tulathromycin (Draxxin®, Zoetis) at weaning compared with littermates that did not receive tulathromycin treatment.

Materials and methods

Pigs (n = 120) from two multiplication sow farms (Farm A and Farm B) with separate nursery-GDU barns were evaluated. On each farm, two gilt pigs per litter were selected from 30 sows; 10 sows each from parity 1, parity 2-3, or parity 4 or greater. Pigs in the treatment group received 15mg/kg of tulathromycin IM at weaning; control pigs were not treated. Tracheobronchial mucous was collected from each pig via tracheal swabbing with a sterile urethral catheter inserted deeply into the trachea. A section of the catheter tip was placed in a sterile snap-cap tube containing 1cc of sterile saline. Samples were assayed individually via PCR for Mhp. Pigs were tested at 3, 9, 13, 17 and 20 weeks of age. No further testing was done on pigs subsequent to a positive finding. All testing was done at the Iowa State University Veterinary Diagnostic Laboratory.

Results

PCR results are presented in Table 1. All samples were PCR-negative at 3, 9 and 13 weeks of age. Mhp colonization was first detected at 17 weeks of age.

Discussion

Treatment had no effect on timing of initial detection. The timing of initial Mhp detection was later than expected. Accordingly, the age at colonization was older than desired for replacement gilts, and we predict that the gilts in this study will shed Mhp to their offspring.

References

1. Calsamiglia M, Pijoan C et al. Colonization state and colostral immunity to *Mycoplasma hyopneumoniae* of different parity sows. *Vet Rec.* 2000;146:530–532.

2. Pieters, M et al. An assessment of the duration of *Mycoplasma hyopneumoniae* infection in an experimentally infected population. Vet Microbiology 2009;1324:261-266.

Table 1: Mhp PCR results

Age in weeks	Farm	Controls No. positive	Controls % Positive	Treated No. positive	Treated % Positive
3		0/30	0.0	0/30	0.0
9		0/29	0.0	0/30	0.0
13	A	0/27	0.0	0/30	0.0
17		9/27	33.0	9/29	31.0
20		13/17	76.5	17/19	89.5
3		0/30	0.0	0/30	0.0
9		0/27	0.0	0/30	0.0
13	B	0/26	0.0	0/28	0.0
17		11/12	42.0	13/28	46.0
20		14/14	100.0	14/14	100.0
3		0/60	0.0	0/60	0.0
9		0/56	0.0	0/60	0.0
13	A & B	0/53	0.0	0/58	0.0
17		20/53	37.7	22/57	38.6
20		27/31	87.1	31/33	93.9

Refinement of field anesthesia protocols for joint fluid collection in healthy swine

K. O'Brien[1], BS; P. Canning[1], DVM; K. Skoland[1], BS; D. Madson[2], DVM, PhD, DACVP; M. Nickel[1], BS; V. Thompson[1]; M. Breuer[1]; R.Warner[1], MS; A. Forseth[1], DVM; J. Brown[1], DVM; L. Karriker[1], DVM, MS, DACVCP
[1]Swine Medicine Education Center, Iowa State University, Ames, Iowa;
[2]Iowa State University Veterinary Diagnostic Laboratory, Ames, Iowa

Introduction

Lameness in the swine industry is a costly reoccuring problem for producers. Antemortem synovial fluid sampling is a useful tool to diagnose infectious causes of lameness. It facilitates expanded diagnostic sample sizes for evaluation of treatment success in sampled animals. The procedure has not been formally refined or evaluated in a research setting. Specifically, the comparison of anesthetic protocols and their utility for joint sampling have had no evaluation prior to this study. The objectives of this study were to compare the effectiveness of four anesthetic combinations for the collection of joint fluid and to assess lameness and sample site healing after the procedure.

Materials and methods

Twenty-one finisher pigs showing no clinical signs of lameness were divided into four treatment groups (n = 6 pigs per group): Telazol (4.4 mg/kg) ketamine (2.2 mg/kg) xylazine (4.4 mg/kg) (TKX), Telazol (4.4 mg/kg) ketamine (2.2 mg/kg) acepromazine (0.03 mg/kg) (TKA), ketamine (5 mg/kg) acepromazine (0.5mg/kg) with a 2% lidocaine epidural (2.2 mg/kg) (KAL), and Telazol (4.4 mg/kg) acepromazine (0.3mg/kg) with a 2% lidocaine epidural (2.2 mg/kg) (TAL). All anesthetics were given intramuscularly and dosed according to individual pig weights. Once the animals were anesthetized, one carpus (TKX and TKA only) and one hock joint were aseptically aspirated for synovial fluid. Each pig was assessed using the following parameters: time to surgical plane, time to sternal recumbency, and time to ambulation. Joint swelling and lameness assessment scoring occurred on day zero, two, four, and seven using an established scoring system for each. On day seven, pigs were euthanized and the shoulder, elbow, carpus, hip, stifle and hock joints were opened evaluated grossly in terms of fluid, cartilage and synovium. Synovium was collected for histology evaluation from each joint that had been sampled antemortem. The synovium was fixed in formalin and histological evaluation was performed by a diagnostician blinded to treatment allocation.

Results and discussion

TKX and TAL provided a sufficient plane of anesthesia for sample collection in 6/6 and 4/6 pigs in their respective treatment groups. TKA and KAL allowed for successful joint fluid collection in 1/6 pigs in each group. TKA and KAL did not provide sufficient muscle relaxation or depth of plane for the procedure. As seen in Table 1, TKX was the most cost effective treatment group, with a cost of $24.98 per animal and an average time from first injection to return to ambulation post-procedure of 205 minutes (standard deviation 73 min). Nine out of twelve joints were successfully tapped from all six pigs in the group with the average amount of synovial fluid collected as 1.3 mL (standard deviation 0.55 mL) (Table 1). When evaluated grossly, 5/9 samples contained some blood contamination. For TAL, the cost was $38.96 per pig and an average time from first injection to return to ambulation post-procedure for pigs of which a sample was obtained of 373 minutes (standard deviation 101 min) (Table 1). Four out of six hock joints were successfully tapped (Table 1). Three out of the four synovial fluid samples collected were red/serosanguinous when evaluated grossly. The average score for joint swelling and lameness for all four groups was zero during the trial (Table 1). TKX performed well during this antemortem procedure and allowed for the successful collection of joint fluid and prompt recovery post procedure. An anesthetic protocol for antemortem joint taps that is applicable to commercial settings and cost effective is a valuable tool to practitioners for diagnostic investigations of lameness. The antemortem technique allows practitioners to increase their diagnostic sample size, monitor treatment success in sampled pigs and can complement post mortem examinations in affected herds.

Table 1: Comparison of four anesthetic protocols for suitability in antemortem joint fluid sampling.

	TKX	KA + Lido	TA + Lido	TKA
Number of pigs in group	6	3	6	6
Number of pigs with at least one joint sample obtained	6	1	4	1
Average cost for all pigs in group (Standard deviation)	$24.98 ($4.16)	$22.37 ($0.82)	$38.96 ($4.65)	$50.99 ($6.35)
Average time to surgical plane for pigs where sampled was obtained (Standard deviation)	13 min (9 min)	71 min (N/A)	26 min* (15 min)	74 min** (N/A)
Average time to sternal for pigs where sampled was obtained (Standard deviation)	125 min (26 min)	151 min (N/A)	203 min (33 min)	180 min (N/A)
Average time to ambulatory for pigs where sampled was obtained (Standard deviation)	266 min (73 min)	317 min (N/A)	373 min (101 min)	287 min (N/A)
Average lameness score	0	0	0	0
Average joint swelling score	0	0	0	0
Average volume collected (Standard deviation)	1.3 mL (0.55 mL)	2 mL (N/A)	1 mL (0.71 mL)	1 mL (0 mL)

*** TKX refers to Telazol, Ketamine, and Xylazine
KA+Lido refers to Ketamine, Acepromazine, and Lidocaine epidural
TA + Lido refers to Telazol, Acepromazine, and Lidocaine epidural
TKA refers to Telazol, Ketamine, and Acepromazine

* Average is for 2 of the 4 pigs in the group because two of the start times were not recorded

** This is a rough estimate. The pig required multiple attempts at sample collection until a successful sample was obtained. Only the first attempt for sample collection was recorded, not the actual time. However, the end of collection time was recorded so an estimated start time was used based on the end of collection time.

Validation of snout wipes to detect influenza A virus in swine (IAV-S)

Joel Steckelberg[1]; Todd Wolff[2], DVM, MPH; Daren Miller[3], DVM; Andrew Bowman[4], MS, DVM, PhD, DACVPM; Deborah Amodie[2], MS; Daniel Hendrickson[3], DVM
[1]College of Vet Med, Iowa State University, Ames, Iowa; [2]Zoetis, Florham Park, New Jersey;
[3]Four Star Veterinary Services, Farmland, Indiana; [4]College of Vet Med, The Ohio State University, Columbus, Ohio

Introduction

Influenza A virus in swine (IAV-S) is a significant respiratory pathogen and the associated loss of performance can be severe. The economic impact of IAV-S has been estimated at a loss of $3.23 per head. In the presence of coinfection, production losses can be much higher.[1] A sensitive and easy-to-use testing method to determine the stability of a herd is important in monitoring IAV-S when implementing a control strategy. The objective of this study was to compare three different testing strategies in nursery-aged pigs to detect IAV-S.

Materials and methods

Recently placed nursery pigs that were exhibiting clinical symptoms of IAV-S (coughing, nasal discharge, un-thriftiness) were included in this study. At six different nurseries, 30 pigs were selected to sample. A snout wipe (SW), oropharyngeal (OP) swab, and nasal swab (NS) were collected on each pig and pooled with three samples per pool. The SW collection process consisted of using a sterile cotton gauze pad to wipe across the snout of the pig while also entering the external nares and collecting nasal secretions, if present. The wipe was placed into a vial containing brain heart infusion broth, penicillin, and streptomycin. To collect the NS, viral transport swabs (BD, UVT) were inserted in the nares. OP swabs (BD, BBL CultureSwab) were collected by swabbing the oropharynx. Each pig was sampled in a randomized order. RT-PCR was performed on the samples and a cycle time (CT) value less than 38 was considered positive for IAV-S. In total, an equal number of samples for each testing strategy were taken ($n = 60$). The variables of interest in this study included CT results (positive/negative) and CT values. CT results were categorically defined as binary and analyzed using a generalized linear mixed model approach. CT results were analyzed with a model that considered the fixed effect of sample type and the random effects of farm and the residual error. CT values were analyzed by a linear mixed model approach. Treatment Least Squares Means were calculated for each group and comparison of LSMeans were performed by the two-sided Student's t-test at the 5% level.

Results

No statistical differences were observed between sample types for categorical CT results (positive/negative). However, the analysis of CT values did indicate that the OP swab had significantly higher results when compared to both the NS and the SW (Table 1).

Discussion

Previously, the SW technique has been proven to be an effective and a non-invasive sampling technique in exhibition swine.[2] Our study validates in a production setting, the SW is a practical and efficient technique to adequately detect IAV-S in a nursery. To implement control strategies, the IAV-S status of recently weaned pigs needs to be known to make informed vaccination and protocol changes. The SW is a useful technique that can be used to accurately reflect the IAV-S status of a flow.

References

1. Dykhuis Haden C et al. Assessing production parameters and economic impact of swine influenza, PRRS, and *Mycoplasma hyopneumoniae* on finishing pigs in a large production system. *Proc AASV*. 2012;P75-76.

2. Edwards J et al. Utility of snout wipe samples for influenza A virus surveillance in exhibition swine populations. *Influenza and Other Respiratory Viruses*. 2014;8(5), 574-579.

Table 1: Comparison of CT results and CT values between nasal swab, oropharyngeal swab, and snout wipe

	Least square means		
Variable	Nasal	Oropharyngeal	Snout wipe
CT results (Pos./Neg.)	83.73%[a]	86.55%[a]	90.81%[a]
CT values	28.48[b]	31.89[a]	28.95[b]

[LS Means with the same letter grouping within a row are not significantly different][1]

Observations of sleep-related behaviors in piglets pre- and post-castration

Anna Martin[1]; Kristina Horback[2], PhD; Thomas Parsons[1], VMD, PhD
[1]Swine Teaching and Research Center, University of Pennsylvania, Kennett Square, Pennsylvania;
[2]Department of Animal Science, University of California, Davis, Davis, California

Introduction

Societal concerns regarding farm animal welfare are growing; therefore, new methods are needed for swine producers and veterinarians to readily evaluate the welfare of the animals under their care. Most piglets in the United States are castrated at an early age to reduce the incidence of boar taint at slaughter. This routine husbandry practice is commonly done without analgesia or anesthesia despite research suggesting that it is painful.[1] However, the extent and duration of this pain is not well understood nor is the use of analgesics to mitigate the pain.[2] Some studies report that painful practices and other stressors may affect the quantity and quality of sleep-related behavior in animals.[3] The purpose of this study was to determine if there are significant differences in the occurrence of sleep-related behaviors between castrated and sham-castrated piglets.

Materials and methods

At the University of Pennsylvania's Swine Teaching and Research Center, 242 piglets from 21 litters were observed for six 35min periods: 24, 3-4, and 1-2 hours prior to castration, and 1-2, 3-4, and 24 hours after castration. No piglets were tail-docked or teeth-clipped in this study. All boars (n = 123) were castrated between 7 and 9 days old. After physically restraining piglets, castration was performed by spraying 2% iodine with lidocaine on the scrotum, making two vertical incisions over the scrotum, and removing the testes by tearing the spermatic cord. Afterward, the incision was sprayed once more with 2% iodine with lidocaine. Gilts (n = 119) were used as a control group and sham-castration was performed by physically restraining animals and using an alcohol wipe on both sides lateral to the vulva. For every 35-min observation period the piglets' postures (standing; abnormal standing; statue standing; lateral; straight out; sternal; sternal, hind legs partially extended; sternal, hind legs fully extended; dog sit) and their behavioral states (walk, fight, rest, sleep, nurse) were recorded every 5 min. Behaviors during sleep were continuously recorded and included change of posture, snout twitch, ear twitch, limb paddle, myoclonus, shiver, tail twitch, and stretch. All data were analyzed using SPSS 23.0 with a significance level set at $P < 0.05$. Repeated measures ANOVAs were run on the post-castration data, with the within-subject factor being time of observation and the between-subjects factor being the treatment (castrated/ sham castrated).

Results and discussion

Prior to castration, there were no significant differences between control and treatment piglets in the occurrence of specific sleep behaviors, the proportion of observation time the piglets were in each body posture or behavioral state. After castration, there was a significant effect of treatment on certain sleep-related behaviors, with castrated piglets performing more ear twitches ($P = 0.02$), more shivers ($P = 0.006$), more tail twitches ($P < 0.001$), and spending a greater time in the sternal, hind limbs fully extended position ($P = 0.02$) than the sham-castrated piglets (Figure 1). In contrast, the sham-castrated piglets were observed to perform significantly more limb paddling ($P = 0.01$) and stretches ($P = .003$), and spent a greater proportion of time walking ($P = .04$) than the castrated piglets. This study shows that there is a significant difference in the sleep behavior between castrated and sham-castrated piglets. The welfare metrics described here were easy to implement and readily quantifiable. Taken together this work suggests an experimental paradigm for the study of putative analgesics in reducing castration associated sleep behaviors and presumable helping to mitigate post-operative pain in castrated piglets.

Figure 1: Time spent in sternal, hind legs fully extended posture after castration.

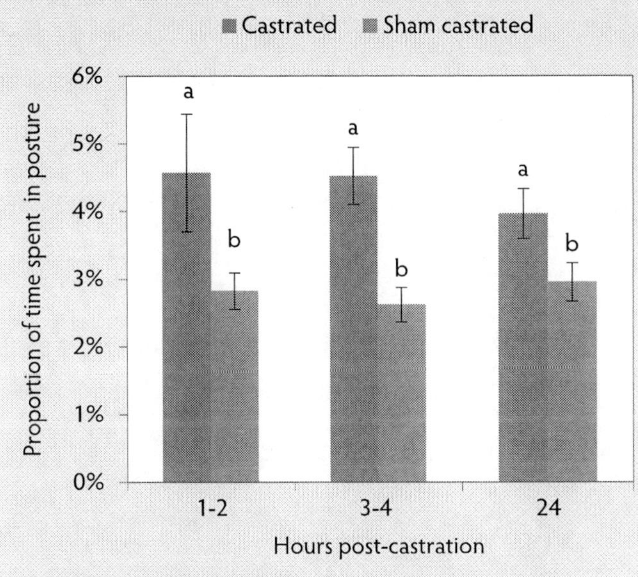

Note: a, b indicated $P < 0.05$ significant difference following repeated measures ANOVA.

References

1. White, RG et al. Vocalization and physiological response of pigs during castration with or without a local anesthetic, Journal of Animal Science. 1995; 73:381-386

2. Hay M et al. Assessment of pain induced by castration in piglets: behavioral and physiological responses over the subsequent 5 days. Applied Animal Behaviour Science. 2003; 82 : 201-218.

3. Langford, FM, Cockram, MS. Is sleep in animals affected by prior waking experiences? Review. *Animal Welfare.* 2010;19:215-222.

4. Kent JE, Molony V. Royal (Dick) School of Veterinary Studies, University of Edinburgh.Guidelines for the Recognition and Assessment of animal pain. 20 October 2008. Available at: http://www.vet.ed.ac.uk/animalpain/Default.htm Accessed 11 September 2016.

Survey of enteric disease pathogens within a swine production company

Stephanie Betbeze[1]; Seth Krantz[2], DVM; Jeffrey Luebbe[3], DVM; Jessica Seate[3], DVM
[1]Lincoln Memorial University-CVM; [2]Tosh Pork LLC; [3]Boehringer-Ingelheim Vetmedica, Inc.

Introduction

Enteric bacterial pathogens such as *Lawsonia intracellularis, Salmonella* spp., and *Brachyspira* spp. commonly cause clinical and subclinical infection that result in production losses in swine. Identification of the prevalence of these organisms can minimize future disease by determining if pigs should be vaccinated. Subclinical ileitis caused by *L. intracellularis* negatively impacts market weight and decreases pig throughput. *Salmonella* spp. can cause a variety of disease syndromes depending on the immune status of a pig such as high mortality. Surveillance of *Brachyspira* spp. aids in the prevention and early detection of swine dysentery. The objective of this study is to determine the prevalence of these major enteric pathogens in a swine production system through cross-sectional surveillance of wean-to-finish sites.

Materials and methods

Surveillance of 24 wean-to-finish sites targeted pigs ages 10, 15, and 20 weeks. Per site, 20 random pig serum samples were tested by SVANOVA ELISA for *L. intracellularis* and *Salmonella* spp. Also, 12 environmental samples were collected by placing an EnviroBootie™ on each foot upon entry to an individual pen. After a thorough walk through, each pair was removed and placed in individual Whirl-Pak® bags. Samples were tested for *L. intracellularis* PCR and cultured for *Salmonella* spp. and *Brachyspira* spp. To further compare serological status of *L. intracellularis* across the 24 wean-to-finish sites, 180 pigs ranging 6-21 weeks of age were sampled. Secondary diagnostics for *Salmonella* spp. included 12 full tissue necropsies on fall-out and sick-pen pigs ages 6-9 weeks. Samples were refrigerated and sent overnight to ISU-VDL and Boehringer Ingelheim Health Management Center in Ames, IA.

Results

Among the initial 24 wean-to-finish sites tested, *L. intracellularis* was highest in pigs ages 15- weeks with 75% positive ELISA and 82% positive PCR. Prevalence of *Salmonella* spp. was highest in pigs ages 20-weeks with 91% positive ELISA. *Brachyspira* spp. cultures were all negative. Upon secondary testing shown in Figure 1, *L. intracellularis* had the highest prevalence in single-sourced pigs at 11-weeks compared to the mixed-source group which peaked at 20-weeks of age. Prevalence for *L. intracellularis* for both groups increased significantly at 9-weeks of age then decreased over 50% between ages 11 to 14 weeks. Tissue diagnostics for *Salmonella* spp. identified clinical salmonellosis with necrotizing colitis in 1 of the 12 pigs submitted.

Conclusion and discussion

Variation in prevalence among set age groups in initial surveillance *Lawsonia* spp. and *Salmonella* spp. concluded that secondary testing was essential. Results in Figure 1 show the *L. intracellularis* prevalence in single-sourced pigs peaked 9 weeks' sooner than multi-sourced pigs. *Salmonella* spp. diagnostics revealed 1 pig with clinical salmonellosis, yet high exposure within the herd. This raises a large concern within a swine production company as it questions routine hygiene, biosecurity, and subclinical disease. Vaccination may reduce overall disease status within the herd, however, it is important to have a holistic all inclusive management and biosecurity protocols in place which is crucial to minimize risk and prevent exposure.

Figure 1: Profile of *Lawsonia* ELISA suspect/positives by age and flow.

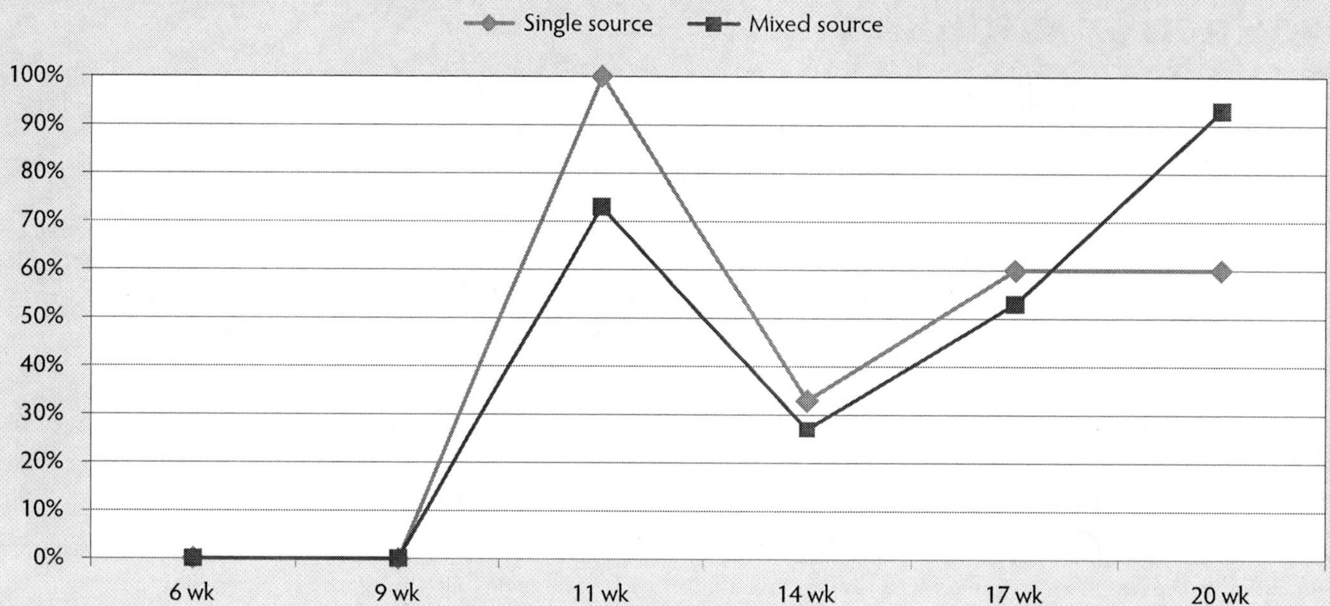

A field trial of an injectable *Lawsonia intracellularis* vaccine on an antibiotic-free farm

G. Chan; R. Friendship, DVM, MSc
Population Medicine, University of Guelph, Guelph, Ontario, Canada

Introduction

Porcine proliferative enteropathy (PPE), or ileitis, caused by *Lawsonia intracellularis* results in thickening of the intestinal wall leading to poor growth rate and feed conversion, diarrhea, and occasionally mortality.[1] The disease can be controlled by in-feed antibiotics[2] but recently developed vaccines may be an alternative for antimicrobial-free farms. The objectives of this pilot study are: (1) to assess the efficacy of an injectable *L. intracellularis* vaccine (Porcilis™ Ileitis, Merck) to control ileitis on a highly-challenged, antibiotic-free farm, (2) to evaluate the safety of the vaccine, and (3) to assess fecal shedding and antibody titers to *L. intracellularis* in vaccinated and non-vaccinated animals.

Materials and methods

A total of approximately 2500 pigs were included in this study on a 600-sow, multisite farrow-finish, antimicrobial-free, highly challenged, commercial farm. Each week, 90% of newly weaned pigs (3 weeks of age) were randomly selected for the treatment group and 10% for the control group and then vaccinated according to their group. Both groups were housed together. Control piglets were given a 2 cc intramuscular (IM) vaccination in the neck against porcine circovirus type 2 and *Mycoplasma hyopneumoniae* while pigs in the treatment group were vaccinated with 2 cc of Porcilis™ Ileitis on one side of the neck and the circovirus-Mycoplasma vaccine on the other side, both IM. During the first week, 30 of the pigs in the treatment group and 24 of the control pigs were weighed and samples taken (blood and fecal) for testing. The same pigs were sampled at the end of the nursery stage (11 weeks), 13 weeks, 17 weeks of age, and prior to slaughter. Sera were tested for the presence of *L. intracellularis* antibodies using indirect fluorescent antibody test (IFAT). Feces were tested for shedding of *L. intracellularis* using PCR. Positive fecal shedding with a Ct value >30 was considered low, 24-29 was considered medium, and 19-23 was considered high. Descriptive statistics and cross tabulations were used for the statistical analysis.

Results and discussion

Over the course of the experiment, 0.04% of piglets experienced an adverse reaction as a direct result of vaccination but recovered without complications. Small pigs (< 6 kg) at weaning were able to tolerate vaccination and were not more susceptible to adverse reaction than large pigs. There was no difference in average daily gain between vaccinated and control pigs at the end of nursery phase. Of the pigs sampled during the first week, 70% of pigs were seropositive for antibodies to *L. intracellularis*, presumably due to the remaining presence of maternal antibodies. When measured at the end of the nursery stage, all vaccinated pigs were seropositive for antibodies and remained positive at 13 and 17 weeks. Control pigs had no detectable antibody titers at the end of the nursery nor at 13 weeks, but at 17 weeks, 68.4% were seropositive in response to a natural challenge. All pigs were negative for fecal shedding at the end of the nursery. At 17 weeks, 63% of vaccinated pigs and 28% of control pigs were negative or low shedders, while 37% of vaccinated and 72% of control pigs were either medium or high shedders. As of 17.5 weeks of age, no pigs have shown clinical signs of ileitis. The results of this clinical trial show that the intramuscular *L. intracellularis* vaccine Porcilis™ Ileitis stimulates the production of protective antibodies and reduces the amount of fecal shedding. Based on the lack of clinical signs in control pigs, the decrease in shedding from vaccinated pigs may have provided some indirect protection by reducing the likelihood of exposure to *L. intracellularis*. Practitioners can use this information in making recommendations to producers regarding the control of ileitis caused by *L. intracellularis*.

Acknowledgements

Merck Animal Health for financial support; University of Montreal diagnostic services and the Animal Health Laboratory of the University of Guelph for testing.

References

1. Lawson GHK and Gebhart CJ. Proliferative enteropathy. *J Com Pathol*. 2000;122:2.

2. McOrist S et al. Treatment and prevention of porcine proliferative enteropathy with oral Tiamulin. *Vet Rec*. 1995;139:25.

Determining the effect of agitation of chlortetracycline stock solutions on chlortetracycline concentrations at different areas of water delivery systems

T. Girard[1], BS; A. Stricker[2], MS, DVM; J.Kelly[2], DVM; L. Karriker[1], DVM, MS, DACVPM
[1]Iowa State University, Ames, Iowa; [2]Suidae Health and Production, Algona, Iowa

Introduction

Medicating through a water delivery system is a convenient and commonly used method to treat populations of pigs in a barn. Chlortetracycline (CTC) is available in a water soluble form and is labeled to be mixed for a final concentration of one gram per gallon, which achieves a dose of 10 mg/lb daily, assuming water consumption of 10% body weight. The CTC label also requires that stock solutions are mixed every 24 hours. Some producers agitate stock solutions of CTC to achieve a consistently mixed solution.

Objective

The objective of this study was to determine whether agitation of stock solutions of CTC affects the concentration of CTC in both the stock solution and at the nipple drinker when administered through a proportioner and measured over time.

Materials and methods

Three separate swine facilities were selected for this study. At each barn a stock solution of PennChlor64 (PharmGate Animal Health Omaha, Nebraska) was mixed per label by dissolving 5 packets in 4 gallons of stocks solution and then proportioned through a Dosatron proportioner (Clearwater, Florida) at a dilution rate of 1:128 to produce a final concentration of 1 gram per gallon water delivered. New stock solutions were prepared every 24 hours for five days. Stock solutions were thoroughly mixed by hand after each packet was added. Stock solution remained un-agitated for the first three days in barns A and C and for the first two days in barn B. An agitator (Little giant submersible pump, Franklin Electric Company, Inc, Fort Wayne, Indiana) was added to the stock solution on day four in barns A and C and on day three in barn B. Water samples were collected from the proportioner intake tube with a needle and syringe immediately after the stock solution was prepared and again four hours later. Samples were also collected from the closest (nipple 1) and furthest (nipple 2) nipple drinker in the barn four hours after preparing the stock solution. Water quality was analyzed in each barn prior to the trial and water pH was measured at ten random nipple drinkers each day. High performance liquid chromatography was used to determine CTC concentrations in each sample.

Significant results

Analysis of results demonstrated the following: 1) agitation had an effect on stock solution chlortetracycline concentration, 2) there was a higher chlortetracycline concentration found in the closest nipple to the proportioner compared to the furthest nipple (Figure 1) and 3) the magnitude of difference in concentration due to agitation and nipple position was also impacted by site specific factors.

Discussion

The results in this study demonstrated high variation in chlortetracycline concentrations presented to the pigs through water delivery systems. Caretakers should be educated on the details of consistent oral drug administration and be prepared to support animals that do not respond with timely parenteral treatment.

Figure 1: Results of mean chlortetracycline concentration at nipple 1 (the closest nipple from the medicator) and nipple 2 (the furthest nipple from the medicator)

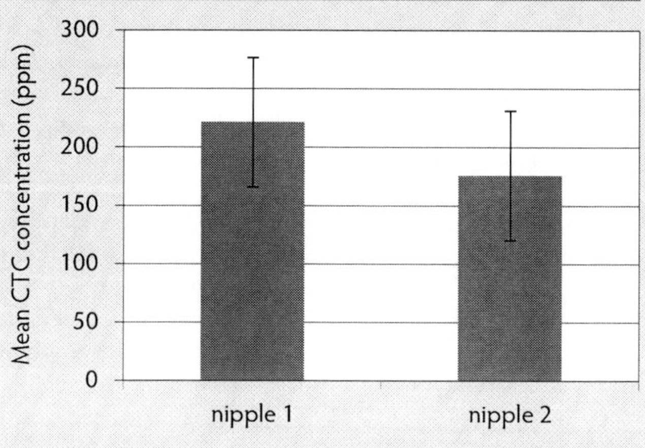

Effect of ketoprofen (Anafen®) on platelet aggregation in piglets

H. Golightly, HBSc; T.L. O'Sullivan, DVM, PhD; K. Reynolds, DVM; S. Enouri, BVSc, MSc, PhD; R. Johnson, DVM, PhD, DACVCP
Ontario Veterinary College, University of Guelph, Guelph, Ontario, Canada

Background

The current Canadian code of practice for the care and handling of pigs requires that piglets of any age receive an analgesic around the time of castration.[1] Injectable non-steroidal anti-inflammatory drugs (NSAIDs) are used to meet this requirement. However, the pharmacodynamic properties associated with the different drugs in this category should be considered before use peri-operatively. Specifically, the NSAID ketoprofen which is approved for use in swine in Canada, has been reported to have adverse affects on blood coagulation in dogs.[2] Very little work has been done to evaluate the effects of ketoprofen on piglet platelet function. The objective of this study was to determine the effect of ketoprofen on platelet aggregation in piglets, as a measure of overall blood-clotting ability, using multiple electrode platelet aggregometry (MEA) (Multiplate®; Roche, Switzerland).[3]

Materials and methods

On day 1 of the trial, eight 4-day old commercial Yorkshire/Landrace × Duroc piglets were individually housed at the University of Guelph. On day 3, jugular catheters were surgically placed on the right side of the neck and the piglets were randomly assigned to one of two treatment groups: ketoprofen alone (3 mg/kg) (TXT-A) and ketoprofen (3 mg/kg) compounded with iron dextran (52.82 mg/kg) (TXT-B). Control group data from a previous study were used for reference (TXT-C, iron dextran alone at approximately 58.8 mg/kg). Piglets then received an intramuscular injection of their assigned treatment on day 5. Whole blood was collected into heparinized tubes (BD, Canada) at pre-treatment, 6 and 24 hours post-treatment. Within 30-90 minutes of collection, platelet aggregation was assessed using MEA (Multiplate®; Roche, Switzerland) with adenosine di-phosphate (ADP, 5 µM) and collagen (COL, 2.5 µg/mL) as agonists.[3] A general linear mixed model was developed using the Proc Mixed procedure (SAS 9.2, Carry, North Carolina, USA) and was used to assess effects of treatment, agonist type, time and sex as fixed effects. Agonist type, treatment and time were modeled as split plot factors.

Results and discussion

No significant difference was found for sex (male/female) on platelet aggregation ($P = .330$). Effect of treatment over time was significant ($P = .002$) and agonist used to elicit platelet aggregation had a significant effect on the magnitude of aggregation ($P = .023$). Treatment differences over time are presented in Table 1. There were no differences between any

Table 1: Within treatment differences over time of platelet aggregation as assessed by the use of ADP and COL quantified as area under the curve (AU*min or U) for piglets receiving TXT-A, TXT-B or TXT-C measured at different time points.

ADP	Baseline	6 hours	24 hours
TXT-A (Ketoprofen)	78.63 ± 9.01 (8)	86.06 ± 9.01 (8)†	65.08 ± 9.41 (7)§
TXT-B (Ketoprofen + iron dextran)	65.38 ± 9.67 (8)	51.50 ± 9.67 (8)	50.25 ± 9.67 (8)
TXT-C (Iron dextran)	85.25 ± 7.23 (8)§	62.75 ± 7.23 (8)*†	79.25 ± 7.23 (8)§
COL	**Baseline**	**6 hours**	**24 hours**
TXT-A (Ketoprofen)	61.31 ± 9.01 (8)	73.63 ± 9.01 (8)	66.58 ± 9.41 (7)
TXT-B (Ketoprofen + iron dextran)	63.19 ± 9.67 (8)†	46.88 ± 9.67 (8)	28.50 ± 9.67 (8)*

Data are expressed as mean ± SE ($P < .05$), the number of animals in each group is indicated in parenthesis.
Significant differences within treatment: * vs. baseline, § vs. 6 hours, † vs. 24 hours.
TXT-A = ketoprofen (3 mg/kg;100 mg/mL)
TXT-B = compounded ketoprofen (3 mg/kg) + iron dextran (52.82 mg/kg) (solutions mixed so that a 3.4kg pig would receive 1mL of solution)
TXT-C = iron dextran alone (~58.8 mg/kg) (control group data that were carried over from a previous study evaluating a different NSAID in 2013 and is only available for ADP test).
ADP = Adenosine diphosphate (5 µM) used as aggregation stimulant
COL = Collagen (2.5 µg/mL) used as aggregation stimulant

treatment group area under the curve (AUC) values at baseline, regardless of agonist used. AUC values represent aggregation response over time (AU*min), with increased values indicating greater platelet attachment to machine electrodes and thus a greater clotting response.[3] At 6 hours post-treatment, TXT-A piglets had significantly higher AUC values compared to TXT-B or control piglets when aggregation was stimulated with ADP ($P = .011$ and $P = .047$, respectively). The same pattern was observed in the COL test at 6 hours ($P = .046$). At 24 hours post-injection, TXT-C AUC values had increased higher than TXT-B ($P = 0.019$, ADP agonist). The COL test at 24 hours showed that TXT-A piglets continued to have significantly higher AUC values than TXT-B piglets ($P = .006$), suggesting a prolonged decrease in clotting ability for TXT-B piglets. The significantly lower AUC values observed for TXT-B should be assessed for biological translation of a clotting deficit in a pig model. Methodological error related to the machine used or confounding error may explain the unexpected decrease in platelet aggregation observed in TXT-C (ADP agonist) at 6 hours post-injection (Table 1). Future research should aim to optimize methodological variables for the Multiplate® analysis for swine including anticoagulant used and agonist type and concentration which have not been investigated to the authors' knowledge and may improve test accuracy.

References

1. National Farm Animal Care Council. Code of Practice for the Care and Handling of Pigs. 2014. Available at: `https://www.nfacc.ca/pdfs/codes/pig_code_of_practice.pdf`. Accessed 11 August 2016.

2. Plumb, Donald C. Plumb's Veterinary Drug Handbook. New York: Wiley-Blackwell, 2015. Print.

3. Toth O et al. Multiple electrode aggregometry: A new device to measure platelet aggregation in whole blood. *Thromb Haemost.* 2006;96.

B-cell tetramer to evaluate the memory immune response to PRRSV

K.L. Gustafson, BS; M.C. Rahe, MPH, DVM; M.P. Murtaugh, PhD
Department of Veterinary and Biomedical Sciences, University of Minnesota, Saint Paul

Introduction

Memory B cells are crucial for the body to prevent reinfection. These cells circulate throughout the body after a pathogen has been cleared, and continue to monitor for pathogen re-appearance long after infection. This sentinel activity is the basis for why vaccinations are able to prevent disease from reoccurring. Vaccine development requires clinical trials to ensure safety and efficacy which necessitates that animals be challenged with the pathogen after vaccination. Vaccination is particularly important for production animals which are consistently at risk from devastating viral infections such as Porcine reproductive and respiratory syndrome virus (PRRSV). This pig-specific pathogen is a major threat to porcine welfare and production worldwide and a pathogen lacking a vaccine capable of producing sterilizing immunity. Despite memory B cells playing this large role in memory immunity, little research has been performed to evaluate antigen specific B cells for their molecular variability. To address this gap in knowledge, we have developed a B cell tetramer reagent which may be used to isolate memory B cells that are specific to specific PRRSV antigens so that these cells can be evaluated in vitro.

Materials and methods

The B cell tetramer our lab developed consists of 4 molecules of PRRSV nonstructural protein 7 (NSP7). The NSP7 molecules are bound via biotin to a streptavidin molecule. Streptavidin is conjugated to phycoerythrin (PE) to make a tetramer consisting of 4 conjugated NSP7 molecules. This results in a highly fluorescent reagent containing four sites for anti-NSP7 antibodies to bind. PRRSV-immune and naïve sera were incubated with the tetramer and anti-PE microbeads. Complexes were run over a magnetic column to select for NSP7-specific antibodies and then evaluated with an anti-porcine indirect IgG ELISA to determine recovery of antibodies. Splenocytes from PRRSV-immune and PRRSV-naïve pigs were incubated with the tetramer, and isolated using and anti-PE microbeads. Samples were analyzed by PE fluorescence as the readout to determine the variation in memory B cell response to PRRSV infection.

Results and discussion

ELISA results showed the NSP7-tetramer binds antibodies specifically in PRRSV immune serum, and not PRRSV naïve serum. The PRRSV immune serum produced an average OD_{450} 10 times that of the PRRSV naïve serum on indirect ELISA, indicating an increased presence of anti-NSP7 IgG antibodies (Figure 1). The data show that the tetramer has the ability to bind NSP7-specific IgG antibodies in PRRSV immune serum. Incubation of splenocytes from immune pig serum with tetramer, followed by magnetic selection and flow cytometry analysis, showed that the NSP7 B cell tetramer was able to bind memory B cells at a higher frequency than PRRSV naïve splenocytes. These results indicate that the NSP7 B cell tetramer will be a useful reagent to determine the immune status of, and variation in, pigs after vaccination or natural exposure to PRRSV.

Figure 1: ELISA results show the tetramer binds anti-PRRSV IgG antibodies only in pigs exposed to PRRSV.

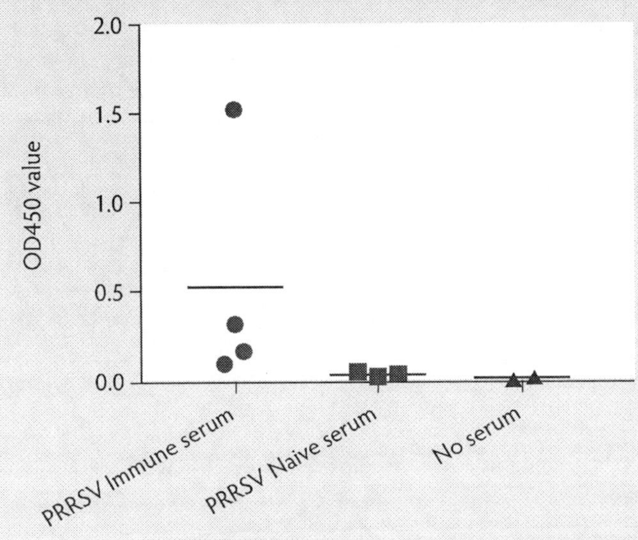

Determining *Lawsonia intracellularis* antibody status in pigs during three early stages of life

Kayla Henness[1], BS; John Waddell[2], DVM, MBA; Luis Ochoa[3], PhD; Megan Schnur[3], DVM
[1]University of Illinois CVM; [2]Boehringer Ingelheim Vetmedica Inc.; [3]Carthage Veterinary Service, Carthage, Illinois

Introduction

Ileitis costs the US swine industry more than $100 million per year.[1] Ileitis is caused by *Lawsonia intracellularis*, which is an obligate intracellular bacterium that grows within the cytoplasm of intestinal epithelial cells.[2] Ileitis causes significant financial losses in the swine industry as it is responsible for mortality, decreased average daily gain, reduced feed utilization as well as wider variation of market weights. The objective of this study was to establish the antibody status of piglets as an initial step in determining the potential for maternal antibody interference in pigs vaccinated with avirulent live *Lawsonia intracellularis* vaccine at a younger age.

Materials and methods

Thirty-one (31) sows and sixty-two (62) piglets on two different sow farms were selected based on parity for this study. At each farm, approximately five gilts, five parity 2, and five parity 3+ sows were selected and identified by tag number at farrowing for serum collection. Two robust piglets (male and female) were selected from each litter at birth and individually ear tagged for identification. The piglets were selected prior to cross-fostering so that the selected piglets remained with their own dam throughout lactation. Three serum collections from the jugular vein were obtained from the piglets on days 3, 18, and 42 of life. Three days after farrowing a serum sample was obtained from the dams. Enterisol® Ileitis vaccine (full dose of lyophilized presentation) was administered via oral gavage at three days prior to weaning (18 days old). A brief vaccination compliance assessment was completed at each of the two study sites to evaluate accuracy of vaccination rehydration protocol, dose, age at vaccination and other compliance parameters. The serum samples were sent to the Boehringer Ingelheim Vetmedica Inc. Health Management Center in Ames, Iowa and were tested for *Lawsonia intracellularis* antibody levels using the SVANOVA enzyme-linked immunosorbent assay (ELISA) test.

Results and discussion

The ELISA results were compiled and analyzed for comparison across litter, gender, parity and correlation with the antibody titers of the dam at the time of farrowing. The piglet ELISA PI values dropped from approximately 60% positive at 3 days of age to all negative by day 18 of age and remained in the negative range at day 42 (Figure 1). There was poor correlation of dam and piglet ELISA PI values ($R = 0.41$). There were no significant differences in ELISA values based on gender of the pigs.

Under the conditions of this survey, on these two farms, maternal antibodies (as measured by *L. intracellularis* ELISA) were negative by the age of 18 days in piglets which nursed exposed and positive dams. Therefore, it appears as though maternal antibody interference would not be expected in this situation.

Figure 1: *Lawsonia intracellularis* ELISA.

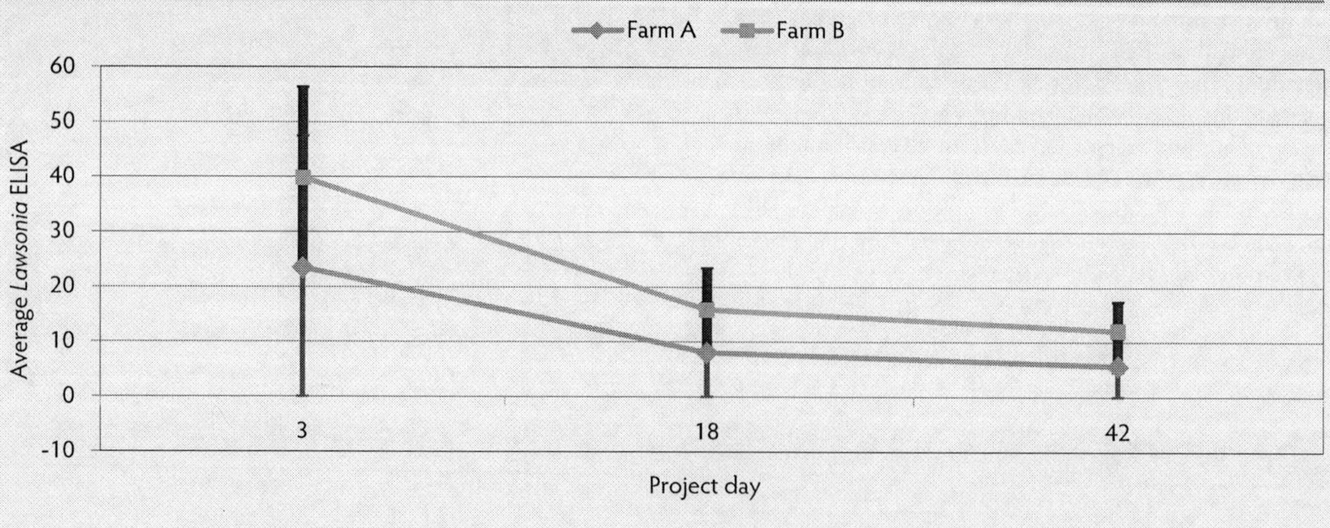

References

1. McOrist S. Defining the full costs of endemic porcine proliferative enteropathy. Vet J 2005;170(1):8–9.

2. McOrist, S., & Gebhart, C. J. (2006). *Diseases of Swine* (9th ed.). Ames, IA: Blackwell Publishing.

Characterizing piglet loss from PRRSV outbreak

T. Homann; C. Vilalta, DVM, PhD; B. Morrison, DVM, MBA, PhD
University of Minnesota, Population Veterinary Medicine

Introduction

PRRSv is a disease that causes high economic impact, particularly in sow farms, with the main implication being a decrease in pigs weaned. This decrease can come from one of three sources: a decrease in litters farrowed (i.e. decreased conception rates, increased sow death, and increased abortions), a decrease in live born per litter (i.e. increased stillborn and mummified fetuses) and an increase in pre wean mortality (i.e. low vitality piglets, decreased immunity, respiratory disease).1 The objective of this study was to determine percent attribution for a decrease in litters farrowed, decrease in live born per litter and increase in pre wean mortality for wean pig loss in a PRRSv break.

Method

24 farms were selected as a convenient sample of herds enrolled in the Swine Health Monitoring Project (SHMP). Inclusion criteria were that farms had to exhibit clinical signs and have reported to SHMP as a new break. Farms had to have broken and recovered to baseline between 2013 and 2016, according to the method reported by Linhares et al (2015).2 The Linhares method determined the timeframe with which to analyze farms' loss.

Data was used from a cohort farrowing report or equivalent to calculate: average sows farrowed, average live born per litter, and average pre-wean mortality for 21 weeks prior to break. This was used as a baseline to compare pre-break performance and isolate change in performance due to PRRSv. Baseline was compared the each factor for each week of the break to determine amount pigs lost that week per factor. The sum total loss from each category was then used to determine overall percent loss attributed to each factor. Note: if one factor outperformed the average prior to the break, it was characterized as not contributing to overall performance loss due to PRRSv.

Results and discussion

Average attribution of total loss across all farms was: 38% from reduced litters farrowed, 24% from decreased live born and 38% from increased pre wean mortality. (Figure 1) While the percent of each type of loss is approximately equally distributed, individual farms do not exhibit this distribution. This variation in loss attribution may be associated with management, at system and farm levels, or virus strain. This has practical implications in assessing the economic impact of a break and in predicting how certain management practices or strains might behave in a break. Further studies are needed to further quantify and predict how to better manage breaks and the economic impact of different attributions.

References

1. Zimmerman JL, et al, Diseases of Swine, 10th Edition, Ch 31, p 467-9.

2. Linhares DCL. Comparison of time to PRRSv-stability and productive losses between two exposure programs to control PRRSv in sow herds. PVM. 2014; 116:111-119.

Figure 1: Percent attribution of piglet weaning loss in all participant farms.

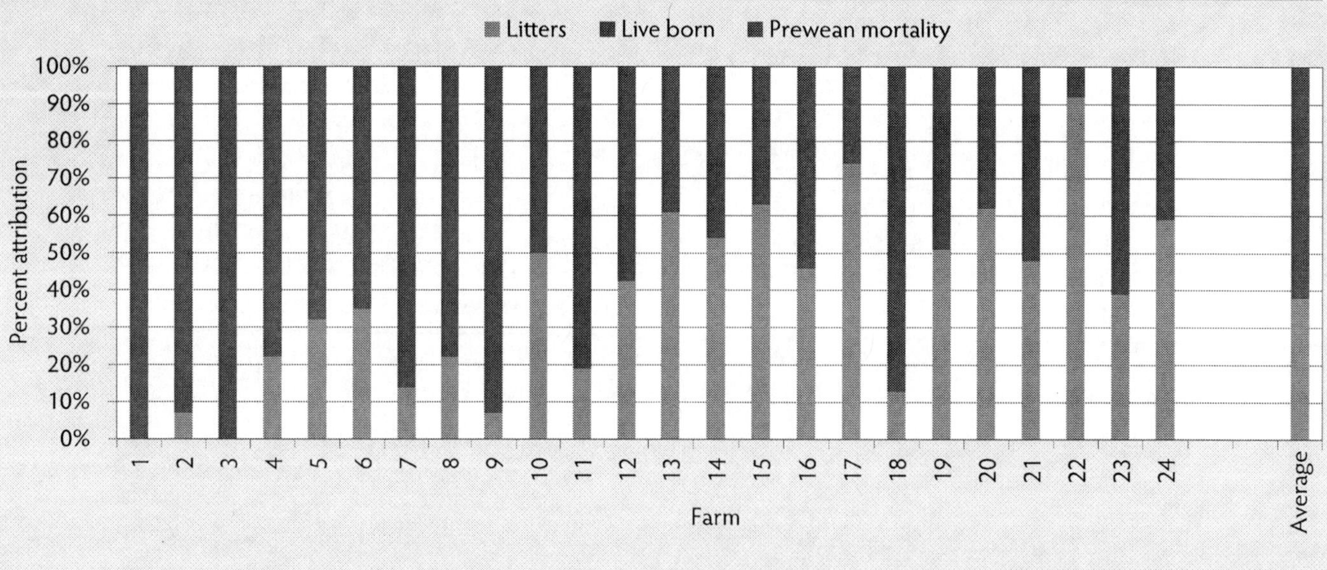

The sow fecal microbiome across parity and hierarchy

John Hurst; Meghann Pierdon, VMD; Thomas Parsons, VMD, PhD
Swine Teaching and Research Center, University of Pennsylvania School of Veterinary Medicine, Philadelphia, Pennsylvania

Introduction

Considerable resources have been expended investigating the fecal microbiome of growing pigs; however, little is known of the microbiome of gestating sows. Additionally, researchers have demonstrated that lower parity sows experience more aggressive encounters and have higher scratch scores and salivary cortisol levels when mixed with high parity sows.[1] We hypothesize, given observations in mice of a decrease in fecal bacterial diversity when subjected to chronic social defeat,[2] that lower parity or lower ranking sows would exhibit decreased fecal microbiome diversity.

Materials and methods

Sows were housed in a large dynamic pen and fed via ESF. Feed hierarchy was determined as the mean order in which the sow ate for seven days prior to sample collection. From three consecutive breeding groups 53 sows were confirmed pregnant by ultrasound and were enrolled for fecal collection. With a freshly gloved hand, rectal contents were collected in individual specimen containers and stored at -80°C. DNA was isolated from fecal samples using the PowerSoil® DNA Isolation Kit. The V4 hypervariable region of bacterial 16S rRNA genes was amplified via PCR and the resulting amplicons were sequenced on the Illumina MiSeq Genome Sequencer. Thus far, 32 randomly selected samples have completed sequencing and have been analyzed by parity and feed hierarchy. Analysis was done using QIIME program with R for statistical testing. Parities were grouped for analysis. A weighted UniFrac was done for parity and feed rank with an Anosim to test for significance. Alpha diversity statistics done include a Shannon, Observed species, Chao1 and PD whole tree with a Wilcoxin Rank sum test used to test for significance. FDR adjustment was used to account for multiple comparisons. The LEfSe program was used for comparison of taxa.

Results and discussion

Bacterial communities varied significantly with parity ($R = 0.17$, $P = .035$) but not feed rank ($P = .78$). With increasing parity there was a downward trend in all measures of microbiome diversity. A significant difference in diversity was observed between sows over fourth parity when compared to either parities 0 and 1 ($P = .002$), or parity 2 and 3 sows ($P = .044$). Specifically a decrease in Prevotella and Ruminococcacea species was observed in parity 4 and older sows compared to sows in parity 0 and 1, and a decrease in Ruminococcaceae and Lactobacillus when compared to parity 2 and 3 sows.

Research in other species indicates that the microbiome may be sensitive to chronic stress and therefore an objective measure of stress. Despite our hypothesis that younger animals may be more stressed in a novel environment compared to older animals, measures of bacterial diversity decreased in older sows at a community and an individual level. There were also changes in the taxa present in older compared to younger sows. Considering feed rank as an indicator of hierarchy, there was no relationship found between diversity and hierarchy. Ongoing study is needed to characterize the fecal microbiome of gestating sows and the potential consequences of microbiome on clinical and production outcomes but in this case we did not find that the microbiome was indicative of stress as measured by low parity or low hierarchy.

References

1. Strawford et al. 2008. The effect of management strategies and parity on the behaviour and physiology of gestating sows housed in an electronic sow feeding system. Can. J. Anim. Sci. 88:559–567.

2. Bharwani, et al. 2016. Structural & functional consequences of chronic psychosocial stress on the microbiome & host. Psychoneuroendocrinology 63:217–227.

Investigating the vertical transmission of *Erysipelothrix rhusiopathiae* by real-time polymerase chain reaction (RT-PCR) and enrichment culture

E. Koep[1], BS; D. Murray[2], DVM; A. Sponheim[3], DVM
[1]Iowa State University, Ames, Iowa; [2]New Fashion Pork, Jackson, Minnesota;
[3]Boehringer Ingelheim Vetmedica, Saint Joseph, Missouri

Introduction

Arthritis from the chronic form of erysipelas is the most economically significant form of the disease. The resulting lameness from arthritis leads to decreased performance, sudden death due to proliferative endocarditis, and possible carcass condemnation.[1] Real-time polymerase chain reaction (RT-PCR) and enrichment broth culture for *Erysipelothrix* spp. have been shown to be an important tool for erysipelas surveillance.[2] Erysipelas has been managed successfully in the past due to effective vaccination strategies and its susceptibility to penicillin; however, recent withdrawal time increases on penicillin have caused a decrease in the use of the antibiotic within breeding herds, which may result in an increased incidence of the disease downstream.[3] Coincidently, an increase in chronic lameness has been seen in vaccinated finishing pigs. For this reason, this study looks at the timing and route of piglet colonization and infection with *E. rhusiopathiae* to identify sounder intervention strategies and prevent chronic lameness in the grower phase.

Materials and methods

Ninety-nine (99) sows from three sow farms with chronic lameness downstream were selected for sampling with an even distribution among two parity groups of $P < 2$ and $P \geq 2$. Sows were observed upon entering the farrowing barn for lethargy, poor appetite, and fever. Copan flocculated swabs with aimes media were used to collect four swab samples from each sow pre-farrow; two swab samples were collected from approximately 4 to 6 inches within the vagina, and two swab samples were collected from the soft palate tonsils. One vaginal swab and one tonsil swab were submitted for RT-PCR for *Erysipelothrix* spp., and the other vaginal swab and tonsil swab were submitted for *E. rhusiopathiae* enrichment culture. Offspring from positive sows were sampled prior to processing with two swabs of the soft palate tonsils and one fecal sample. One tonsil swab and the fecal swab were submitted for RT-PCR for *E. rhusiopathiae*, and the other tonsil swab was submitted for *Erysipelothrix* spp. enrichment culture. Additionally, nine-week old pigs showing signs of lameness were sampled at the end of the nursery phase for *E. rhusiopathiae* shedding. Sixteen oral fluid samples were collected on 4,200 pigs, and soft palate tonsil swabs were collected from sixteen individual pigs. Samples were sent for *Erysipelothrix* spp. RT-PCR and enrichment culture.

Results

All 99 sows were negative for *Erysipelothrix* spp., thus, no piglets were sampled on the sow. Oral fluids from the nursery were negative for *Erysipelothrix* spp. The tonsil swabs from the sixteen individual pigs submitted for RT-PCR were also negative, however, two of sixteen (12.5%) submitted for *E. rhusiopathiae* enrichment culture were positive for *E. rhusiopathiae*.

Discussion

The negative samples from the sows, coupled with continued observed lameness in growing pigs suggests that there may be other factors that explain why *E. rhusiopathiae* was not detected. The sows may be shedding it during a different phase of the reproductive cycle, the sampling in this study occurred too early or too late due to a short period of detection by PCR, or growing pigs are being colonized after leaving the sow farm.[2] The environmental viability of *E. rhusiopathiae* should also be assessed in the farrowing rooms, staging rooms, nurseries, and finishing barns of historical erysipelas infected flows.

References

1. Wang et al. Review: *Erysipelothrix rhusiopathiae. Veterinary microbiology* 2010;140:405-417.

2. Opriessnig et al. Novel Diagnostic Tools for *E. rhusiopathiae* Infection. IPVS. 2014.

3. Behnke et al. Use of Oral Fluids and Fecal Samples in Sow Herds for Detection of *Erysipelothrix rhusiopathiae* by Polymerase Chain Reaction (PCR). AASV. 2016.

Developing urine catheterization in sows for calcium and phosphorus analysis

N. Lauterbach[1], BS; M. Hoogland[2], DVM, MS; G. Cline[3], DVM; M. Nickel[1], BS; L. Karriker[4], DVM, MS, DACVPM
[1]Iowa State University College of Veterinary Medicine, Ames, Iowa; [2]Smithfield hog production division, Algona, Iowa; [3]Boehringer Ingelheim Vetmedica, Inc., St. Joseph, Missouri; [4]Swine Medicine Education Center, Ames, Iowa

Problem and objective

Increasing sow mortality and decreased sow longevity are emerging issues facing the swine industry. Nutritional imbalances have been theorized as a major cause for these problems. Better tools to assess nutritional imbalances are needed. Blood analysis is problematic, as the body aims to maintain homeostasis. Urine could indicate nutritional deficiencies within a sow and provide a quick and easy diagnostic tool to assess any imbalances. The objective of this study was to develop a consistent and repeatable method for placing urinary catheters in sows and to begin using urine to monitor various mineral levels across different stages of gestation.

Materials and methods

Several different devices were used to visualize the external urethral orifice (EUO), ranging from commercial vaginal speculums to handmade devices. Various lengths and sizes of red rubber, Foley, and intermittent catheters were tested, along with post-cervical artificial insemination rods. A 60 milliliter catheter tip syringe was used to draw urine out of the bladder. Urine was submitted to Iowa State Diagnostic Lab in 15 milliliter conical tubes for calcium, phosphorus, and creatinine analysis.

Results and discussion

It was important to visualize the EUO in order to place the catheter. Location of the EUO was variable from sow to sow, often two to six inches past the vulva tip. A two prong vaginal speculum with a pen light was most efficient at locating the EUO. The speculum was lubricated with a general insemination lube and inserted into the vagina at a dorsal angle. The speculum was opened slowly and steadily while directing the tip ventrally, at which point the EUO was visualized. A 16 inch long, 14 French, coude tipped intermittent catheter was optimal. Cold catheters passed more easily through the urinary diverticulum, which was found to be the most time consuming step of the procedure. Syringes less than 60 milliliters were avoided, as they lacked adequate pressure to draw urine out of the bladder. Urinary catheterization was performed without restraining the sows. Average time to place the catheter, collect the urine, and separate it into aliquots to be shipped was three minutes per sow, while the quickest placement and collection took less than 20 seconds. Fifteen to twenty samples could be comfortably collected per person per hour. We demonstrated that urine collection via catheterization was a repeatable and efficient process to obtain individual diagnostic samples, as 220 sows were sampled from three different sow farms. Further analysis of the calcium, phosphorus, and creatinine levels are forthcoming and will be presented at the AASV conference. Urine collection as a diagnostic tool should be further developed to assess nutritional deficiencies in sows. Future diagnostic capabilities of urine include detection of viruses, mycotoxin residues, estrogenic compounds, oxidative stress, metabolomics, and antibiotic residues. Reliably collecting sow urine will advance the study of these issues.

Testing the association between multiple influenza A infections and average daily gain in nursery aged piglets under field conditions

Andrea Patterson, BScH; Helena Grgic, DVM, MSc, PhD; Zvonimir Poljak, DVM, MSc, PhD
Department of Population Medicine, Ontario Veterinary College, Guelph, Ontario, Canada

Introduction

Influenza A virus (IAV) has been a challenge facing pork producers for a number of years. It has been confirmed that piglets can be infected multiple times within the nursery phase with multiple strains of the virus. This research was undertaken as a pilot project to determine if multiple infections with IAV were associated with decrease in average daily gain (ADG).

Materials and methods

Nursery barn selected for the study was a 2000-head facility managed as all-in all-out facility. 15 piglets were selected from each of four sow sources (N = 60), on weaning day. Piglets were housed in 3 separate rooms with shared airflow. Each piglet was weighted on weaning day, and 1 week prior to shipment to the finishing barn. Average daily gain was calculated based on these two weights. Nasal swabs were taken weekly for determination of IAV infection status. Virus isolation was used to identify IAV-positive samples, using MDCK cells. Statistical analysis was completed using Stata 13 Intercooled (Stata corp LP, College Station, Texas). Piglets were grouped according to the number of positive virological samples during nursery phase into 3 distinct groups: 0+ve, 1-2+ve, and > 3 positive results. Simple linear regression was used to determine statistical significance ($P < 0.05$). The final model contained nominal variable reflective of groupings based on the number of infections as the primary exposure of interest, and initial weight as a potential confounding variable. Interaction between the two variables was evaluated.

Significant results

Descriptively, pigs with 0 +ve detections had mean average daily gain of 0.47Kg/day (SD = 0.08), pigs with 1-2 detections had mean average daily gain of 0.45 Kg/day (SD = 0.7), and pigs with > 3+ve detections had mean average daily gain of 0.50 Kg/day (SD = 0.07), these results are shown in Table 1. The only statistically significant variable found was the initial weight ($P < 0.01$), both in the univariable and multivariable models. However, number of infections was not found to be statistically significant, before or after adjustment for the initial weight. Interaction terms were not found to be statistically significant either.

Discussion

A previous study showed that nursery aged piglets are susceptible to multiple infection with influenza A virus. This study once again proved that to be correct with piglets ranging from 0-4 infections through out the 8-week nursery period. Association between number of infections and average daily gain could not be established, which could be a consequence of several factors: (1) weekly sampling was perhaps to sparse and pigs shedding for shorter duration could not be detected, which could have contributed to misclassification of the exposure status, (2) interplay between maternally derived antibodies and circulating viruses could have created conditions for multiple infections but not clinical disease, (3) insufficient power to detect the difference, or (4) confounding effect of variables that were not measured.

Table 1: Summary of average daily gain, initial weight, gender and source farm separated by number of infections from 3 weeks of age to 11 weeks of age on a commercial Ontario nursery farm

Parameter	Number of influenza infections			All animals
	0	1	> 1	
ADG mean (Kg)*	0.47 ± 0.08	0.45 ± 0.7	0.50 ± 0.07	0.45 ± 0.07
Initial weight mean (Kg)†	5.22 ± 0.26	5.68 ± 0.65	6.30 ± 0.88	5.70 ± 0.69
Female pigs (%)	60.00	55.56	50.00	51.72
Source 1 N = 15 (%)	0.00	100	0.00	100
Source 2 N = 15 (%)	6.67	60.00	33.33	100
Source 3 N = 14 (%)	14.29	78.57	7.14	100
Source 4 N = 14 (%)	14.29	85.71	0.00	100
Number of observations	5	47	6	58

* ADG shown as mean (±) SD
† Initial Weight shown as mean (±) SD
ADG= average daily gain; SD= standard deviation

Individual differences in social behavior and affective state in group-housed gestating sows

K. Pruett; K.M. Horback, PhD; T.D. Parsons VMD/PhD, DACAW
University of Pennsylvania School of Veterinary Medicine, Kennett Square, Pennsylvania

Introduction

Societal concern about animal welfare is driving the transition of gestating sows from individual stalls to group housing. In the absence of gestation stalls, animals have greater mobility but the establishment of social hierarchies can result in injuries and competition for resources. Common methods of measuring swine welfare focus on the physical attributes of the animal, but typically neglect their psychological welfare or affective state. The affective state of an animal can alter cognitive processes such as appraising ambiguous stimuli. Animals experiencing positive affective states (relaxation, contentment) are more likely to judge ambiguous stimuli positively, while those experiencing negative affective states (anxiety, fear) are more likely to interpret information as threatening. Here the relationships between social rank, parity, and both physical and psychological measures of welfare in group-housed gestating sows were assessed. An additional objective was to see whether a computer screen-based operant task was possible with sows.

Materials and methods

At the Swine Teaching and Research Center of the University of Pennsylvania, approximately 150 gestating sows are housed in dynamic groups and fed via electronic sow feeding stations. Eighteen sows were trained a computer based go/no-go task with a joystick movement as a conditioned response. Sows were rewarded (corn-soy feed) for responding to a positive stimulus (small red square) and punished (whistle blow) for responding to a negative stimulus (large white square). Of the sows trained, 7 met the training criteria and were assessed for cognitive bias. Each sow's response toward 10 trials of an ambiguous stimulus (medium pink square) was recorded as an assessment of the sow's cognitive bias. This information, combined with feed rank, aggression at pen mixing, and body lesion scoring, was used to determine how sow affective state might be influenced by social rank in group pens. A cognitive bias index (CBI) was calculated for each sow. These indices can range from 0 to 1, with 0 reflecting a response to the ambiguous stimulus analogous to a negative conditioned response, and 1 being analogous to a positive conditioned response. Spearman's rank-order correlations were conducted to investigate relationships among the variables. Wilcoxon signed-rank tests were run to assess difference in average lesion scores per pen for each region between pre-mixing day and post-mixing days.

Results and discussion

Feed rank remained significantly ($P < 0.05$) consistent for each sow throughout the study. Body lesion scores significantly increased 4 days after pen mixing ($Z = -3.7, P < 0.001$), followed by a significant decrease in anterior lesions 10 weeks later by the day of the cognitive bias test ($Z = -2.0, P < 0.05$). Five of the seven sows had a positive CBI (> 0.5) for the first ambiguous trial, and five of the seven sows had a positive CBI for the average response to all 10 ambiguous trials. Significant correlations were found between the average CBI and feed rank during first week in pen ($r_s = 0.86, P < 0.05$), third week in pen ($r_s = 0.78, P < 0.05$), fourth week in pen ($r_s = 0.78, P < 0.05$) and ninth week in pen ($r_s = 0.82, P < 0.05$). This means that sows that ate earlier in the pen responded to fewer ambiguous trials and displayed more negative biases. There was a significantly positive correlation between CBI and aggression received during pen mixing ($r_s= 0.96$, $P < 0.001$) (Figure 1). As a result, sows with more positive biases also had higher lesion scores 4 days post-mixing ($r_s= 0.96, P < 0.01$). The sows that received more aggression were faster at hitting the joystick on ambiguous trials ($r_s= -0.93, P < 0.01$). Repeated exposure to the ambiguous stimulus changed each sow's CBI over time. This research suggests that judgment bias may be a better indicator of personality traits, learning ability, and coping styles in sows. Recognizing individual differences in sows is important for improved stockmanship of group housed systems; however a better measure of affective state should be developed.

Acknowledgements

Project funding from the NIH/Merial Summer Scholars program and Kraft Foods/The Oscar Mayer Company.

Figure 1: Relationships between feed rank during week 1 of testing, aggression received during pen mixing, and cognitive bias index (CBI) in 7 group housed sows. Significant correlations were found between the average CBI and feed rank during the first week in the pen ($r_s = 0.86$, $P < 0.05$). There was a significantly positive correlation between CBI and aggression received during pen mixing ($r_s = 0.96$, $P < 0.001$).

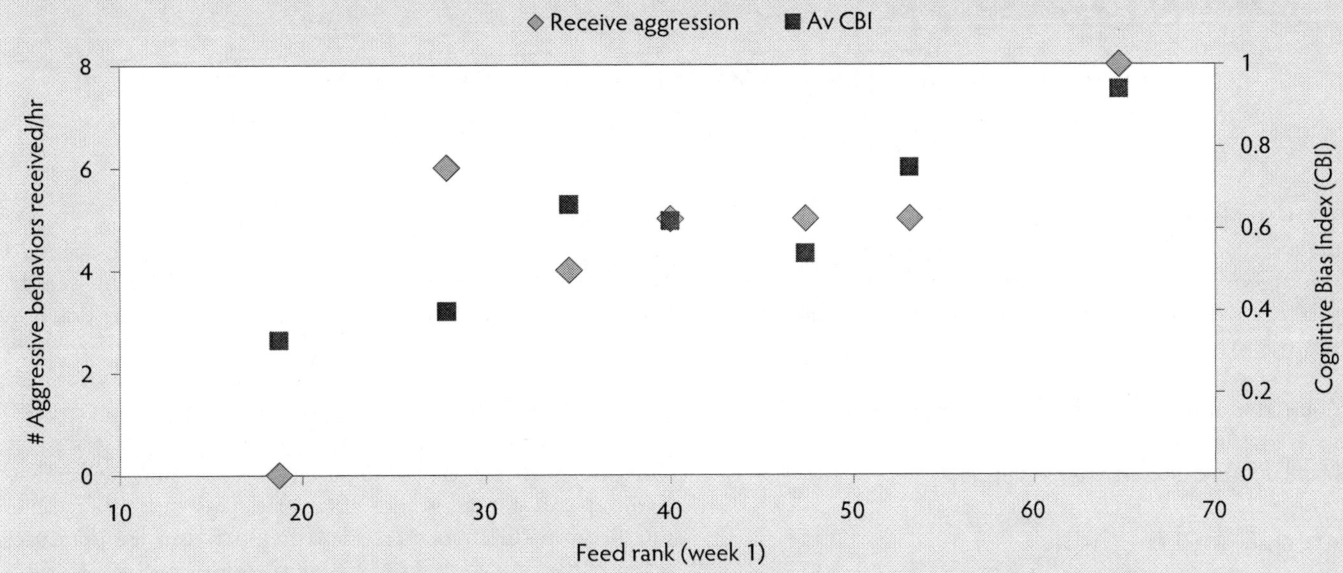

Determination of behavioral and physiological pain indicators associated with castration and tail docking in young piglets

Brent Sexton[1], BS; Brett O'Brien[2], VMD,MS; Sam Holst[3], DVM; Deb Amodie[2], MS
[1]Iowa State University, College of Veterinary Medicine, Ames, Iowa; [2]Zoetis, Florham Park, New Jersey;
[3]Swine Vet Center, Saint Peter, Minnesota

Introduction

Swine in the United States are routinely processed prior to weaning, allowing producers to administer supplemental iron; dock tails; and surgically castrate male piglets. Due to mounting consumer interest in livestock welfare, there is a growing need for an effective and consistent way to measure the pain associated with these procedures.[1] The objective of this study was to identify and discern differences in behavioral and physiological parameters associated with the pain from tail docking and castrating young piglets.

Materials and methods

58 gilts and 62 boars between 3 and 6 days of age, blocked by sex within litter, were randomly allocated to 1 of 4 treatment groups: 1) female, not tail docked [T01]; 2) female, tail docked [T02]; 3)male, not tail docked or castrated [T03]; and 4) male, tail docked and castrated [T04]. All piglets received an iron injection at the time of treatment. Piglets were assessed 2 hours before and 2 hours after treatment was applied. Activity scans were taken every 2 minutes for 20 minutes, then every 5 minutes for the remainder of 2 hours. Recorded activities included: walking, standing, sitting, lying, huddling, nursing, playing/fighting, urinating, defecating, and scooching. Nursing frequency and duration were also recorded during this period. Rectal temperatures were recorded immediately prior to [M-1], immediately after [M+1], and 20 minutes after [M+20] treatment. SAS was used to analyze variables with a general linear mixed model approach for repeated measures. Statistical significance was attributed to differences with $P \leq 0.05$.

Results

Hunched posture, lethargy, stiff gait, and trembling occurred more often after treatment ($P \leq 0.05$). T04 had the highest incidence of each of these activities after treatment, as shown in Table 1. Activity scan results revealed a significant difference in the occurrence of huddling, isolation, and lying after treatment ($P \leq 0.01$), but with no differences between treatment groups. There was also no significant difference found between groups for either nursing frequency or duration, however; nursing frequency did decrease in all groups after treatments were applied. No significant differences were identified between temperatures taken at M-1 or M+1. At M+20, LSMean rectal temperatures of castrated males were significantly lower than either female group (T01: 102.44°F; TO2: 102.31°F; TO3: 102.03°F; T04: 101.77°F).

Discussion

In this study, hunched posture, lethargy, stiff gait, and trembling were identified as parameters that occurred more frequently in castrated and tail docked boars than in any other treatment group. Therefore, these behaviors may be better indicators of acute pain in neonatal piglets. Huddling, lying, isolation, nursing frequency, and nursing duration were parameters with no observable differences between groups and appear to be poor indicators of acute pain in piglets, yet showed to be indicators of handling stress. Additionally, castrated and tail docked boars appeared to be a better model of pain, as little differentiation was seen between

Table 1: LSMean comparision for the four parameters found to be significantly different between treatment groups.

	LSMeans			
Behavior	**T01 After Tx**	**T02 After Tx**	**T03 After Tx**	**T04 After Tx**
Hunched	0.13^b	0.31^b	0.45^b	1.03^a
Lethargic	0.26^b	0.26^b	0.22^b	1.58^a
Stiff	0.17^{bc}	0.51^b	0.13^c	0.91^a
Trembling	0.06^b	0.03^b	0.12^b	0.74^a

Superscripts within a row represent a significant ($P \leq 0.05$) difference female groups. Overall, these results identify specific behaviors that should be assessed in future comparisons of pain mitigation strategies for processed piglets.

References

1. O'Connor, A. M., Anthony, R., Bergamasco, L., Coetzee, J. F., Gould, S. A., Johnson, A. K., Dzikamunhenga, R. S. (2014, June). Pain management in the neonatal piglet during routine management procedures. Part 2:Grading the quality of evidence and the strength of recommendations. Animal Health Research Reviews, 15(01), 39-62.

Soy isoflavones and porcine respiratory and reproductive syndrome viral infection in weanling pigs

B.N. Smith; A. Morris; M.L. Oelschlager; R.N. Dilger
Department of Animal Sciences, College of Agriculture, Consumer, and Environmental Sciences, University of Illinois, Urbana, Illinois

Introduction

Porcine reproductive and respiratory syndrome virus (PRRSV) is the most prevalent disease of swine globally. Infection of weanling pigs with PRRSV leads to a complex immune response resulting in significant illness, reduced feed intake, and decreased growth performance. Previous experimental evidence from our lab suggests that increasing the concentration of soybean meal in the diet of young pigs confers benefits in terms of growth performance and immune parameters. The objective of this experiment was to identify a potential mode of action for this benefit, specifically the ability for soy-derived isoflavones to confer immunological benefits to young pigs infected with PRRSV.

Material and methods

Four dietary treatments differed in their soy protein source (soy protein isolate vs. HP 300®; Hamlet Protein, Findlay, OH) and isoflavone supplementation (none vs. 1,500 mg total isoflavones/kg diet from Novasoy 400®; ADM, Decatur, IL); the control diet (CON) contained soy protein isolate and no supplemental isoflavones. All diets were isonitrogenous, isocaloric, and contained identical standardized ileal digestible amino acid concentrations. Weanling pigs (60 barrows, 21 d of age, 5.71 ± 0.44 kg) were individually housed in disease containment chambers and provided ad libitum access to experimental diets for 7 d before receiving either a sham inoculation (sterile PBS) or a 9.28 × 103 50% tissue culture infective dose of PRRSV at 28 d of age (0 d post-inoculation, DPI). A total of 5 experimental treatments included an uninfected group receiving the CON diet, plus four infected groups each receiving a different dietary treatment. Growth performance and rectal temperatures were recorded throughout the study, and blood was collected for quantification of serum PRRSV load, differential complete blood counts, cytokine concentrations, ex vivo stimulation of immune cells, and T-cell immunophenotyping. All data were analyzed as a 1-way ANOVA.

Results and discussion

Preliminary results indicate no differences ($P > 0.05$) within infected pigs receiving experimental diets for growth performance, rectal temperatures, or presence or titer of circulating PRRSV antibodies, confirmed by IDEXX-ELISA and indirect fluorescent antibody detection, respectively, at any time-point. PRRSV-infected pigs receiving diet A had a larger proportion of Helper T Cells (CD3+CD4+) in their total T Cell populations (CD3+) compared to other diets ($P < 0.05$) (Table 1). All other immunophenotypes were not affected ($P > 0.05$) by dietary treatment. Within infected groups, differences ($P < 0.05$) were noted for PRRSV rtPCR Ct values at 3 DPI, but not at 6 or 14 DPI. Infected groups maintained lower white blood cells counts (WBC) at 3 and 6 DPI compared to uninfected pigs, though Diet C did have a higher count ($P < 0.05$) on 6 DPI compared to other infected treatment groups. Analysis and interpretation of immune parameters including serum cytokine analysis is ongoing, but preliminary results indicate supplemental soy isoflavones may have an effect on the immune response to PRRSV infection, but not on growth parameters, under these conditions.

Table 1: Effects of dietary soy isoflavones level and porcine reproductive and respiratory virus (PRRSV) infection on peripheral blood T cell immunophenotypes of weanlings pigs[1]

Item	Uninfected	PRRSV-infected				
	Control Diet	Diet A	Diet B	Diet C	Diet D	SEM
T Cell, % (CD3+)[2]	55.30	59.02	53.01	53.89	53.21	6.66
Helper T Cell, % (CD3+CD4+)[3]	22.03[ab]	26.41[b]	23.09[ab]	18.03[a]	20.18[a]	2.13
Cytotoxic T Cell, % (CD3+CD8+)[3]	12.44	19.20	22.43	22.71	22.78	3.21
Memory T Cell, % (CD3+CD4+CD8+)[3]	4.30	5.13	6.32	6.15	6.66	1.19
Memory T Cell Expressing IFNγ, % (CD3+CD4+CD8+IFNγ+)[3]	90.10	83.41	79.25	77.94	79.72	4.06
Memory T Cell Expressing IFNγ MFI (CD3+CD4+CD8+IFNγ+)[4]	764.1	577.3	515.4	490.6	501.0	88.5

ab Means without a common superscript letter differ ($P < 0.05$).

1 Values represent least square means of 10 to 12 pigs with collection of blood occurring at 12 DPI. All pigs received allotted treatment diet starting -7 DPI. Abbreviations: DPI = days post-inoculation.

2 Percent of total lymphocytes that are positive for cell-surface marker CD3.

3 Percent of CD3-positive lymphocytes that are also positive for cell-surface markers CD4, CD8, CD4/CD8, or CD4/CD8 and intercellular IFN γ

4 Median fluorescence intensity (MFI) is a measure of the fluorescence intensity in a fluorescence channel being measured. It provides an alternative measurement to percent positive for comparison of cell populations between individuals. It is less sensitive to outliers, which is important for very small cell populations like memory T cells.

Utility of antemortem joint fluid collection for diagnostic investigations of infectious lameness

V. Thompson[1]; P. Canning[1], DVM; E. Koep[3], BS; D. Murray[3], DVM; K. O'Brien[1], BS; M. Nickel[1], BS; A. Forseth[1], DVM; J. Brown[1], DVM; R. Warner[1], MS; M. Breuer[1]; K. Skoland[1], BS; D. Madson[2], DVM, PhD, DACVP; A. Viall[2], DVM, MS, DACVP; L. Karriker[1], DVM, MS, DACVPM
[1]Swine Medicine Education Center, Iowa State University, Ames, Iowa;
[2]Iowa State University Veterinary Diagnostic Lab, Ames, Iowa; [3]New Fashion Pork, Jackson, Minnesota

Introduction

Lameness in growing pigs is a significant welfare and economical issue for United States (US) hog production companies. In the US, farms experiencing lameness reported a prevalence of up to 30% and these cases accounted for 5.4% of all mortalities. In 2015, it was estimated that the production costs of lameness were $2.00 to $10.00 per affected pig. Current diagnostic approaches rely on postmortem sample collection, whereas antemortem techniques would allow practitioners to evaluate a larger, more representative number of pigs that could ultimately lead to development of a more focused treatment plan. The objective of this study was to evaluate the utility of antemortem synovial fluid for the diagnostic investigation of lameness cases in a field setting.

Case history

Two 4,300 head finisher sites with a common sow source were investigated. The prevalence of lameness at Site 1 and 2 was 12% and 7% respectively. Pigs were between 14 and 16 weeks of age. The clinical presentation at both sites was as follows: difficulty rising from a dog-sitting position, stiff gait, shortened length of stride, and variable degrees of swelling associated with the hocks and elbows.

Materials and methods

Seven pigs were selected at Site 1 and six pigs were selected at Site 2 for synovial fluid sampling. Selection criteria included: observation of dog-sitting, gait abnormalities, and/or swelling of one or both hock joints. Oral fluid samples were collected from the home pens of the pigs chosen for evaluation. Antemortem arthrocenteses were performed on four pigs at Site 1 and three pigs at Site 2 after intramuscular injection of telazol (4.4 mg/kg), ketamine (2.2 mg/kg), and xylazine (4.4 mg/kg)(TKX). Synovial fluid was collected from one hock and one carpus from each pig. Synovial fluid samples were submitted for clinical pathology, bacterial culture, and *Mycoplasma hyosynoviae*, *Mycoplasma hyorhinis*, and *Erysipelothrix* PCR. The pigs were recovered and moved to a sick pen for continued monitoring for seven days. Antemortem pigs were scored daily for lameness and joint swelling to determine whether the antemortem arthrocenteses had any effects on locomotion. The remaining three pigs at each site were humanely euthanized and necropsied. Sample collection included: full tissue set, second rib, synovial fluid, and synovium from the carpi and hocks. Grossly abnormal joints were also sampled for synovial fluid and synovium._

Results and discussion

Both antemortem and postmortem synovial samples were *Mycoplasma hyosynoviae, Mycoplasma hyorhinis,* and *Erysipelothrix* PCR negative and had no significant growth on culture. Both antemortem and postmortem synovial fluid samples had clinical pathology parameters within the normal ranges. Forty-five percent of postmortem grossly abnormal joints had mild synovitis and 22% of all postmortem limb joints had mild cartilage defects or thinning. All oral fluid samples tested negative for *Mycoplasma hyosynoviae*, *Mycoplasma hyorhinis*, and *Erysipelothrix* on PCR. The pigs selected for antemortem sample collection did not show changes in locomotion during the seven-day observation period. Pigs were fully recovered and ambulatory about one to two hours after the initial injection of TKX. The results of this study demonstrated that for clinical pathology, bacterial culture and PCR, there were no differences between the two methods. Also, the antemortem sampled pigs did not suffer morbidity due to the procedure as their lameness and joint swelling scores decreased or remained constant throughout the observation period. Therefore, antemortem arthrocentesis is a practical tool for practitioners to complete diagnostic work-ups in situations where euthanasia is not an option or ideal. It is best suited for the identification of infectious pathogens by means of PCR, culture, and clinical pathology.

We can't answer that: The importance of careful interpretation of swine production data

J. Tubbs[1], MS; M. Pieters[2], DVM, PhD; J. Menard[3], DVM; C. Surprenant[3], DVM; B. Morrison[2], DVM, PhD, MBA
[1]Auburn University College of Veterinary Medicine, Auburn, Alabama; [2]University of Minnesota College of Veterinary Medicine, Saint Paul, Minnesota; [3]F. Ménard, Ange-Gardien, Quebec, Canada

Introduction

Health management and disease mitigation are key factors in deciding which system to use in swine production. The batch farrowing system is built around sow reproduction and mitigation of disease. This system has been used for several decades across the Midwest.[1] In this system farrowing occurs every 3 or 4 weeks (although variations may be used) in an all-in all-out approach. Batch farrowing allows for simultaneous down time in all farrowing rooms for cleaning, disinfection, and drying and is a tool for disease management. Some of the major swine diseases that could be mitigated by use of batch farrowing include Influenza A virus, Porcine Circovirus type 2, and Porcine Reproductive and Respiratory Syndrome virus. *Mycoplasma hyopneumoniae* (*M. hyopneumoniae*) is endemic in most swine herds worldwide and is the principle cause of enzootic pneumonia.[2] The effect of batch farrowing on prevalence and clinical disease suggestive of *M. hyopneumoniae* infection has not been investigated. Therefore, the objective of this study was to analyze diagnostic and production data from a swine production system to identify whether the prevalence of *Mycoplasma hyopneumoniae* and associated respiratory disease had changed post-conversion to batch farrowing.

Materials and methods

The research question was developed through discussions with representatives of a swine production system of approximately 20,000 sows who reported a perceived increase in respiratory disease post conversion to batch farrowing of approximately 50% of the system. A working hypothesis and research goals were generated and a timeline established. The hypothesis was that conversion to batch farrowing increased respiratory disease associated with *M. hyopneumoniae* infection. Initial data review to summarize finisher close-outs and ELISA for *M. hyopneumoniae* (ELISA Oxoid) from finishers through descriptive statistics was conducted. Production flows in the data were classified as either batch or control (non-converted) groups with pre and post conversion data for each. Further analysis prompted a comprehensive investigation using sow farm production data divided into the same classifications. All data sets included 6 years from June 2010 to June 2016 and ELISA data extended to June of 2006.

Results and discussion

A comparison of means of key production outcomes pre and post conversion follows: Feed Efficiency – Batch 2.68 to 2.7, control 2.59 to 2.65; finish viability % – Batch 92.5 to 94.0, control 94.7 to 95.3; average daily gain (kg) – Batch 0.87 to 0.89, control 0.92 to 0.91; average wean age – Batch 20.3 to 20.2, control 19.5 to 20.2. Detailed discussion and analysis of these data revealed several confounders present in the data sets. Conversion to batch farrowing required a change in breeding timing affecting wean age by parity and extended nursery time. Confounders also included changes to the vaccination program and other health challenges. Confounders could not be accounted for in data analysis and made further statistical analyses and interpretation difficult. It is noted therefore, that careful interpretation of swine production data with extensive knowledge of the system is needed by researchers and practitioners before recommendations can be made and questions can be answered. If significant confounders exist, production data analysis should be carefully performed to avoid misleading conclusions. The definition of a research question prior to data collection is ideal to obtain meaningful answers.

References

1. Pijoan, C., Torremorell, M. 2000. Band production: A high health production system for small swine farms. Retrieved from University of Minnesota, College of Veterinary Medicine.

2. Thacker, E. L. 2004. Diagnosis of *Mycoplasma hyopneumoniae*. *Journal of Swine Health and Production*. 2004, 12(5), 252-254.

Identifying disease agents contributing to porcine respiratory disease complex (PRDC) within a large production system

Shannon Ure[1]; Jana Morgan[2], DVM; Thomas Wetzell[2], DVM; Tyler Te Grotenhuis[3], DVM
[1]University of Illinois CVM; [2]Boehringer Ingelheim Vetmedica Inc., [3]Christensen Farms

Introduction

Porcine respiratory disease complex (PRDC) presents a large concern for the swine industry. One of the primary components is influenza A virus in swine (IAV-S). IAV-S can account for over $10 loss per market pig.[2] IAV-S is a ubiquitous virus with a high speed of replication. Shedding of virus can begin one-day post infection and continue up to six days, leaving minimal time for detection. Diagnostic sampling to detect IAV-S should be repeated and multiple sample types should be used. This investigation was performed to identify if IAV-S was the primary cause of coughing and if not what other viruses could be contributing in pre-wean and post-wean pigs, the evaluation of vertical transmission of IAV-S from sow farm to nursery, and the specific IAV-S strains that could be causing cough in these pigs. For this system coughing began in piglets around 10 days of age and increases in severity closer to weaning. The concern is that early nursery aged pigs show difficulty starting on feed, thereby inhibiting average daily gain, feed efficiency and increasing mortality.

Materials and methods

Specific flows were identified based on coughing in farrowing and nurseries. Six flows with a total of seven sow farms and 11 nursery barns were enrolled. Twenty intranasal swabs (INS) and three tissue sample sets (TS) were collected from coughing pigs near 21 days of age at each sow farm. Ten INS and 3 TS from coughing pigs, and 2 oral fluid samples per airspace were collected at each nursery. Diagnostic samples were submitted to Iowa State University-VDL and Boehringer Ingelheim HMC (Ames, IA). Aliquots of the INS samples (120) were sent to Kansas State University where Next Generation sequencing was performed for virus detection.

Results/discussion

In four of the six flows investigated, IAV-S sequences appeared similar vertically when compared on a phylogenetic tree. This suggests IAV-S originates in farrowing and is vertically transmitted to pigs prior to weaning (Figure 1). Results also suggested lateral spread in the nursery in two of six flows (additional sequences pending). Metagenomic sequencing revealed many other viruses within the system that could potentially be contributing to respiratory clinical signs, further investigation necessary. When looking to explain respiratory illness in swine, once the primary pathogens of PRDC are eliminated as possibilities, it is important to begin looking for alternate causes.

Figure 1: Percent IAV-S positive nasal swabs organized by flow including sow farm, early stage nursery and late stage nursery.

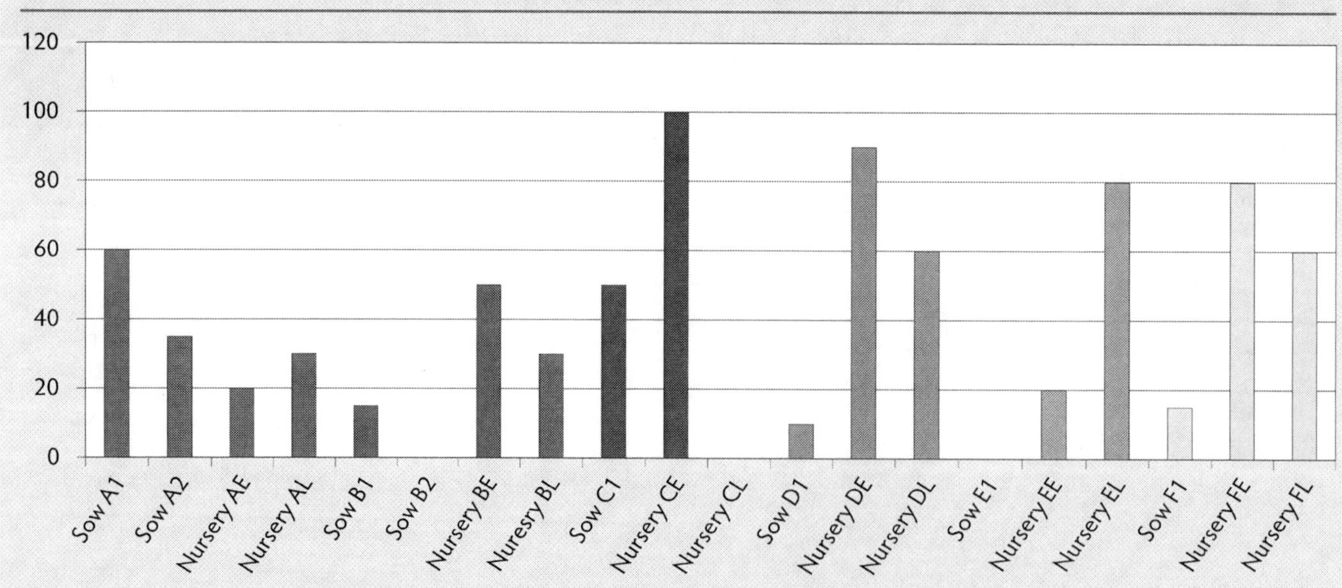

References

1. Donovan TS. Influenza isolate selection methodology for timely autogenous vaccine use. American Association of Swine Veterinarians 2008:557-561

2. Gauger, P. (16, July 6). Influenza A Virus in Swine. Lecture presented at BIVI DVM Swine Academy in Iowa State University, Ames.

3. Hause, B., PhD. (2016, July 20). Metagenomic Sequencing for virus discovery and characterization. Lecture presented at PRRSV Symposium in InterContinental Chicago Magnificent Mile, Chicago.

4. Janke, B.H. "Influenza A Virus Infections in Swine: Pathogenesis and Diagnosis. "Veterinary Pathology 51.2 (2013): 410-26.

Evaluating different methods of antibody detection for porcine epidemic diarrhea virus in animals exposed at various intervals

Kathleen Wood[1], BS; Brad Heins[2], DVM; Randy Jones[2], DVM; Cary Sexton[2], DVM; Alan Scheidt[3], DVM
[1]North Carolina State University College of Veterinary Medicine, Raleigh, North Carolina; [2]Livestock Veterinary Services, Kinston, North Carolina; [3] Boehringer Ingelheim Vetmedica, Inc, Saint Joseph, Missouri

Introduction

Porcine epidemic diarrhea virus (PEDV) is a highly contagious, rapidly spreading enteric coronavirus that affects all phases of swine production. Infection with PEDV results in severe diarrhea and dehydration, as well as high mortality in suckling piglets. Establishing herd immunity is critical for controlling PEDV; protective immune response on clinical farms is short-lived, and recovered animals may be re-infected months later. Current commercial vaccines produce an antibody titer response, yet fail to fully protect the herd from clinical disease. The objectives of this study were to assess humoral immunity at various intervals following PEDV exposure, and to evaluate different methods of antibody detection including placental umbilical cord serum (PUCS), colostrum, and piglet serum.

Materials and methods

This study was conducted on a 2400 head sow farm where PEDV was first detected in January 2014. A total of 19 sows of various parities (parity 0-7) were selected for evaluation based on herd entry dates. On-site sows (parity 3-7) were naturally exposed during the PEDV break, while subsequent replacement gilts (parity 0) were exposed during GDU phase after farm placement. PUCS (n = 19) and colostrum (n = 19) samples were collected after farrowing. Piglet serum (n = 36) was collected 2-3 post-farrowing from 2 pigs per litter. All samples were submitted for PEDv Fluorescent Focus Neutralization (FFN) assay to determine anti-PEDV neutralizing antibody titers.

Results and discussion

All colostrum FFN results were positive (Table 1). All PUCS and piglet serum FFN results were negative (titer results < 1:20). There was a tendency for increased titer variability in gilts compared to sows, in spite of a more recent PEDV exposure. Determining protection through sow colostral transfer is still difficult, and there is variation in transfer of maternal antibodies to colostrum. Further research is needed to evaluate herd protection against PEDV.

Table 1: Colostrum PEDV FFN titer results

Sow ID	Parity	1:640	1:1280	1:2560	> 1:2560
3099	0			X	
3104	0				X
3086	0	X			
3087	0			X	
3093	0		X		
1563	3			X	
1870	3		X		
1583	4		X		
1600	4				X
1589	4			X	
1634	4				X
1630	4				X
1042	5				X
930	6		X		
923	6			X	
941	6				X
1756	6	X			
556	7			X	
677	7				X

Increased frequency of isolation of multi-drug resistant *Salmonella* I 4,[5],12:i: from swine with histologic lesions consistent with salmonellosis

B. Arruda, DVM, PhD; A. Krull, DVM, PhD; A. Kreuder, DVM, PhD; E. Burrough, DVM, PhD, DACVP; K. Schwartz, DVM, MS; O. Sahin, DVM, PhD
Department of Veterinary Diagnostic and Production Animal Medicine, College of Veterinary Medicine, Iowa State University, Ames, Iowa

Introduction

Salmonella serotype I 4,[5],12:i:- has emerged over the last decade as a leading cause of human disease in Europe and the United States.[1,2] Resistance to four common classes of antibiotics deemed ASSuT (ampicillin, streptomycin, sulfonamide and tetracycline) resistance has been rapidly increasing in human isolates over the last several years.[2] Concurrently, an increase in the isolation of *Salmonella* serotype I 4,[5],12:i:- from swine clinical samples submitted to the Iowa State University Veterinary Diagnostic Laboratory (ISU-VDL) has been noted. Although there have been reports of isolation of this organism from swine, documentation of its prevalence, resistance in swine, and association with disease has not been well defined (published).

Methods

All data was compiled from the ISU-VDL from samples submitted from swine between 2007 and 2016. Histopathology results from a subset of cases that met qualification criteria (pigs between 3 and 13 weeks of age, history of diarrhea, loss of body condition and/or colitis noted grossly, enteric culture with isolation of Salmonella serotype I 4,[5],12:i:-, Typhimurium, or Derby, and histopathology of the large intestine) between January and March of 2016 were summarized. The information within diagnostic reports was compiled from multiple diagnostic pathologists, and lesions were considered consistent with salmonellosis when they included multifocal superficial to deep ulceration of the large and less commonly the small intestinal mucosa, expansion of the lamina propria by neutrophils, and luminal accumulation of fibrin and cellular debris. Antimicrobial susceptibility testing was completed on the TREK Sensititre™ system and ASSuT resistance was determined using ampicillin, gentamicin as a substitute for streptomycin, sulfadimethoxine as the sulfonamide, and oxytetracycline as the tetracycline.

Results

A substantial increase in the relative frequency of isolation *Salmonella* serotype I 4,[5],12:i:- was observed over the past 5 years (< 3% in 2011 to > 15% of all *Salmonella* isolated in 2015). In a review of case data from clinical submissions to the ISU-VDL, there is a consistent association between enteric disease with concurrent histologic lesions consistent with salmonellosis and the isolation of *Salmonella* I 4,[5],12:i:-, particularly during the post-weaning nursery phase of production. Of the serotypes identified in 2015, 76% of the *Salmonella* I 4,[5],12:i:- isolates were associated with microscopic lesions consistent with salmonellosis, compared to 94% of Typhimurium and 14% of Derby isolates. Antibiotic susceptibility testing data on the ISU-VDL swine isolates of I 4,[5],12:i- also confirms a rapid increase in the identification of ASSuT MDR isolates (from 3.5% in 2012 to 36.8% of isolates identified thus far in 2016). In comparison, the percentage of Typhimurium isolates exhibiting the same resistance profile has remained relatively steady at approximately 10-11% since 2007. Importantly, 11 of 663 isolates of I 4,[5],12:i:- had resistance to all antibiotic classes tested including beta-lactams, cephalosporins, pleuromutilins, tetracyclines, aminoglycosides, amphenicols, fluoroquinolones, lincosomides, macrolides, and folate pathway inhibitors. In contrast, however, no isolates of Typhimurium with this profile have been identified to date.

Discussion

The increased isolation of *Salmonella* I 4,[5],12:i:- in association with clinical disease in swine along with increasing identification of the ASSuT MDR resistance profile in this serotype may represent an emerging threat to the swine industry and warrants increased awareness in diagnostic labs nationwide.

References

1. EFSA. The European Union summary report on trends and sources of zoonoses, zoonotic agents and food-bourne outbreaks in 2012. EFSA Journal 12, 3547 (2014).
2. CDC. National Antimicrobial Resistance Monitoring System (NARMS): Enteric bacteria. Human Isolates Final Report., (2013).

Influenza A virus prevalence and seasonality in midwestern US breeding herds

F. Chamba[1], DVM, MS; J. Nerem[2], DVM; A. Alba[1], DVM, PhD; M. Torremorell[1], DVM, PhD
[1]Department of Veterinary Population Medicine, College of Veterinary Medicine, University of Minnesota, Saint Paul, Minnesota; [2]Pipestone Veterinary Services, Pipestone, Minnesota

Introduction

Influenza is a costly disease for US pig producers. There is interest in understanding influenza epidemiology and control in pig populations, in particular in understanding influenza seasonality to target timing of intervention strategies.[1]

In this study, we aim to: 1) estimate the herd-level prevalence of influenza in breeding herds over time and 2) evaluate the seasonality of influenza in breeding herds.

Materials and methods

Influenza testing data was compiled from 2011 to 2016. There were 40 farms conveniently and voluntarily enrolled in a retrospective-prospective cohort study started last year. Piglets within the breeding herds were monitored on a monthly basis using oral fluids. Oral fluids were tested individually to detect IAV matrix gene by RT-PCR. A farm was considered positive if at least one oral fluid tested positive.

Monthly herd prevalence was calculated as the ratio between the numbers of positive farms in each month divided by the total number of farms submitting in that month. Trend and seasonality was evaluated using a regression model that includes herd-level prevalence as the outcome in each month and sine, cosine and autoregressive functions as predictors. Herd-level prevalence data was modeled under a logit transformation in R 3.3 (R Foundation for statistical computing). Herd-level prevalence was compared between months calculating a year-adjusted prevalence ratio (PR) using a generalized estimating equations (GEE) log-binomial model in SAS 9.4® (SAS Foundation, Cary, NC).

Results

There were 57 months of data (almost 5 years) and 40 farms. All farms were positive at least once across the study period. There were a total of 1801 submissions of which 504 were positive (28%). There were 45 submissions per farm on average (range 29-56) of which an average of 13 submissions per farm were positive (range 3-32). Finally, average monthly herd prevalence was 28%, ranging from 9 to 57%.

Herd-level prevalence in this cohort of breeding herds had a seasonal pattern composed of 2 peaks of detection in the colder months and 1 deep in the hotter months. This pattern was detected every year. Based on the monthly herd-level prevalence ratios, influenza herd-level prevalence in this cohort rose during fall, peaked in both winter and spring, and decreased in summer months.

Conclusions and discussion

Under the conditions of this study, influenza was shown to have different herd-level prevalence across the year with a seasonal pattern that included higher herd-level prevalence during winter and spring months, and lower herd-level prevalence in summer months.

Results from this study indicate the benefits of having year round surveillance because it allows producers and veterinarians to identify high risk periods, providing guidance on when to focus their control strategies.

Acknowledgements

This study was funded by the Minnesota Rapid Agricultural Response Fund and producers participating in the influenza surveillance program.

References

1. Poljak, Z., Carman, S. & McEwen, B. Assessment of seasonality of influenza in swine using field submissions to a diagnostic laboratory in Ontario between 2007 and 2012. Influenza and other respiratory viruses 8, 482-492, doi:10.1111/irv.12248 (2014).

Identification of multiple viruses in swine fecal samples using next-generation sequencing technology coupled with Kraken algorithm-based bioinformatics analysis pipeline

Qi Chen; Ying Zheng; Baoqing Guo; Jianqiang Zhang; Kyoung-Jin Yoon; Phillip C. Gauger; Sarah A. Bade; Wendy Stensland; Karen M. Harmon; Rodger G. Main; Ganwu Li
Department of Veterinary Diagnostic and Production Animal Medicine, College of Veterinary Medicine, Iowa State University, Ames, Iowa

Introduction

The primary advantage of next-generation sequencing (NGS) technology is that the hypothesis-free metagenomics strategy enables NGS to simultaneously detect the presence of multiple microorganisms and to identify previously unrecognized pathogens directly from clinical samples without prior knowledge. However, identifying '*a viral needle in a metagenomics haystack*' extensively relies on bioinformatics to tackle the huge amounts of sequence data involved. The objective of this study is to develop a bioinformatics pipeline and utilize this pipeline to identify the presence of multiple microorganisms including both known and unknown pathogens in clinical samples.

Materials and methods

Metagenomics analysis using NGS was performed on 217 swine fecal samples from various geographic locations in the United States between April of 2014 and February of 2015. All samples were obtained from pigs with acute diarrhea and were positive for porcine epidemic diarrhea virus (PEDV) RNA by PCR testing. A bioinformatics analysis pipeline based on Kraken algorithm was established and used to analyze the sequencing data.

Results and discussion

Only 4 fecal samples were positive for PEDV alone, while multiple viral agents (up to 9) were concurrently identified with PEDV in the remaining 213 samples (98.2%). In addition to PEDV, 93% of the samples were detected with at least 3 more viruses mixed infection, and over half (65%) of the tested samples had mixed infection with more than 5 viruses. Additionally, 28 samples (13%) were detected with all 9 viruses detected in this study, along with PEDV. Those 9 viruses detected in the current study were porcine astrovirus (53.9%, 117/217), enterovirus G (32.7%, 71/217), posavirus (32.3%, 70/217), porcine sapelovirus (PSV; 31.8%, 69/217), kobuvirus (28.5%, 62/217), sapovirus (12.9%, 28/217), porcine deltacoronavirus (8.8%, 19/217), and pasivirus (6.9%, 15/217) (Figure 1).

Among these viruses, the complete genome sequence of PSV type 1 (PSV-1), now named as sapelovirus A, was obtained for the first time in US swine. This US PSV-1 strain (USA/IA33375/2015) had the genome of 7565 nucleotides in length and shared 87.8%-83.9% nucleotide identities with 7 additional PSV-1 strains reported in other countries whose whole genomic sequences are available in GenBank to date. Phylogenetic analysis revealed that global PSV-1 isolates formed two clusters and the US PSV-1 strain was closer to Korean and Chinese PSV-1 strains but distinct from UK PSV-1 V13 strain.

In summary, we have established a Kraken algorithm-based bioinformatics analysis pipeline and by which we successfully identified the presence of multiple viruses in fecal samples from diarrheic pigs. Mixed viral infections in the intestinal tract appeared to be highly frequent in pigs undergoing PEDV infection although the clinical significance of such a concurrent infection remains to be further studied. Additionally, complete genomes of previously not well-characterized viruses were obtained and further characterized in this study. Our method offers a powerful and cost-effective tool to identify multiple pathogens in a single sequencing run, and even can be used to estimate the prevalence rate of each pathogen in a mixed infection in animals.

Figure 1: Prevalence of detected RNA virus in examined 217 fecal samples.

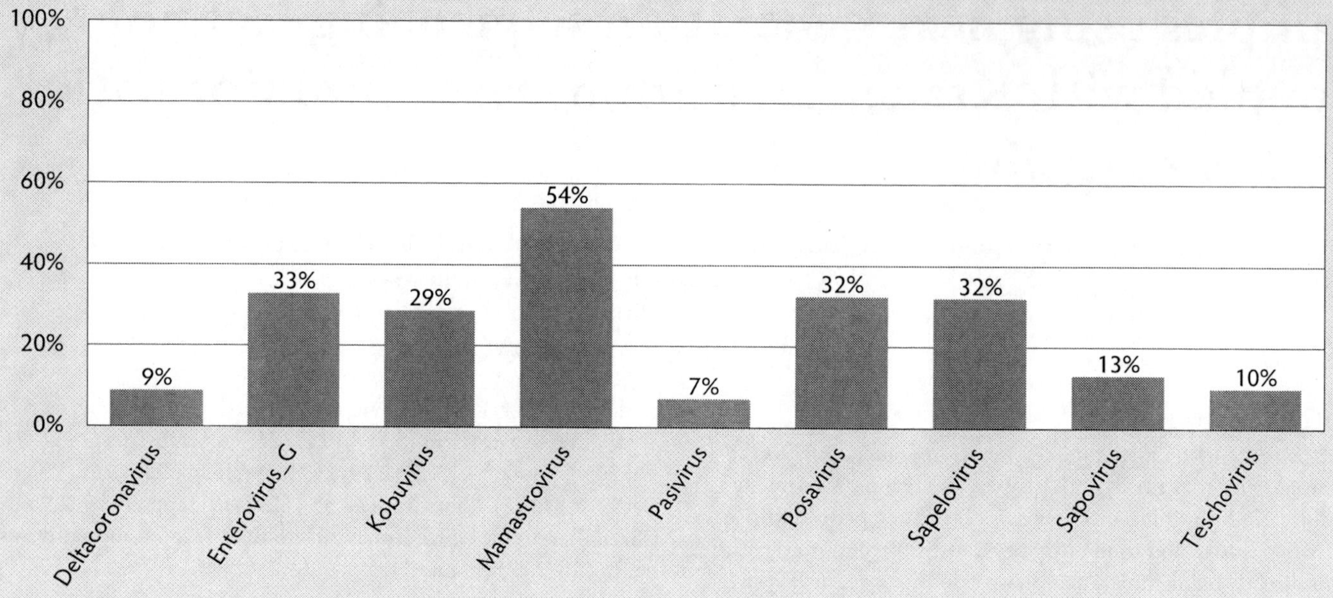

Evaluation of pathogenesis and antibody responses of porcine deltacoronavirus in 3-week-old weaned pigs

Qi Chen[1]; Phillip Gauger[1]; Joseph Thomas[1]; Michael Welch[1]; Priscilla Gerber[2]; Tanja Opriessnig[1,2]; Paulo Arruda[1]; Drew Magstadt[1]; Luis Giménez-Lirola[1]; and Jianqiang Zhang[1]

[1]Department of Veterinary Diagnostic and Production Animal Medicine, College of Veterinary Medicine, Iowa State University, Ames, Iowa; [2]The Roslin Institute and the Royal (Dick) School of Veterinary Studies, University of Edinburgh, Roslin, Midlothian, United Kingdom

Introduction

Pathogenicity and kinetics of antibody response of porcine deltacoronavirus (PDCoV) in weaned pigs have not been investigated. In this study, we assessed the pathogenicity of PDCoV in 21-day-old weaned pigs and the effect of re-challenge. Additionally, we evaluated the appropriateness of different specimen types for PDCoV nucleic acid and antibody (Ab) detection.

Materials and methods

Thirty PDCoV negative, 3-week-old weaned pigs were randomly divided into 2 groups with 15 pigs in each group. On Day 0 (D0), group 1 (G1) were orogastrically inoculated with PDCoV USA/IL/2014 isolate (10^5 $TCID_{50}$/pig) and group 2 (G2) were inoculated with virus-negative medium. Five pigs per group were euthanized and necropsied on D7 for pathogenesis evaluation. The remaining 10 pigs of each group were kept through D42. On D35, when pigs were 8 weeks old, both groups were challenged with the same PDCoV isolate (10^5 $TCID_{50}$/pig). Clinical observations were recorded. Individual fecal swabs and sera, and pen-based feces and oral fluids (OF) were collected. Intestinal tissues were evaluated for gross and microscopic lesions, and also tested by PDCoV specific immunohistochemistry (IHC). Fecal swabs, feces, and OF were tested by PDCoV real-time RT-PCR to determine the amount of virus genomic material. PDCoV Ab in sera was tested by indirect fluorescent antibody (IFA) and virus neutralization (VN) assays. Sera, feces and OF were tested for PDCoV Ab by a S1-based ELISA.

Results and discussion

Negative control pigs (G2) did not have diarrhea and virus shedding from D0-D35. PDCoV-inoculated pigs developed soft to semi-watery diarrhea during D4-D10. Virus shedding peaked on D7 and gradually decreased thereafter in fecal swabs, feces, and oral fluids (Figure 1A). Specifically, PDCoV RNA was detected through at least D14 (negative on D21) in pen-based feces, at least D28 (negative on D35) in some of the individual fecal swabs, and at least D35 (negative on D42) in pen-based OF. After 2nd challenge on D35, no diarrhea was observed, and PDCoV was not detected in fecal swabs, feces and OF during D36-D42 from either group.

On D7, 3 out of 5 necropsied pigs from G1 had soft to semi-watery intestine content, and 2 of them had thin-walled intestines. G1 pigs also had significantly lower villus-height-to-crypt-depth ratio in middle- and distal-jejunums and ileum compared to G2 (Figure 1B). PDCoV IHC staining was positive in distal jejunum and ileum of all 5 G1 pigs, but negative in all 5 G2 pigs. On D42, no remarkable gross or histologic lesions were observed from either group and IHC was consistently negative in both groups.

In G1, serum IFA IgG Ab was detected from D14, peaked on D21-D28 and started to decrease from D35. Serum VN Ab appeared on D7 and maintained at high level during D14-D42 (Figure 1C). As tested by ELISA, both IgG and IgA Abs were detected in sera from D14 through D42, in pen-based feces from D21-D42, and in pen-based oral fluids from D14-D42.

In summary, PDCoV can cause infection in 3-week-old weaned pigs and result in diarrhea, gross and microscopic lesions. Virus shedding in individual fecal swabs, pen-based feces and pen-based oral fluids were characterized. Ab responses in sera, feces, and oral fluids were also characterized. Notably, it was observed that PDCoV naïve 8-week-old pigs appear to be less susceptible to PDCoV infection than 3-week-old pigs under the conditions of this study and this warrants further investigation.

Figure 1: Pathogenicity and kinetics of antibody response of PDCoV in weaned pigs. **(A)** Virus shedding in fecal swabs, feces, and oral fluids as determined by a PDCoV quantitative real-time RT-PCR. Serial 10-fold dilutions of a PDCoV isolate with known titers ($TCID_{50}$/ml) were used to generate the standard curve. No virus shedding was detected in any sample type from Neg/PDCoV groups throughout the study. No OF samples were obtained on D7 and D28. **(B)** Villus-height-to-crypt-depth ratios in small intestines of pigs necropsied on D7 (5 pigs per group). *: $P < 0.05$ between 2 groups; **: $P < 0.01$. **(C)** Virus neutralizaing (VN) antibody titers in serum samples.

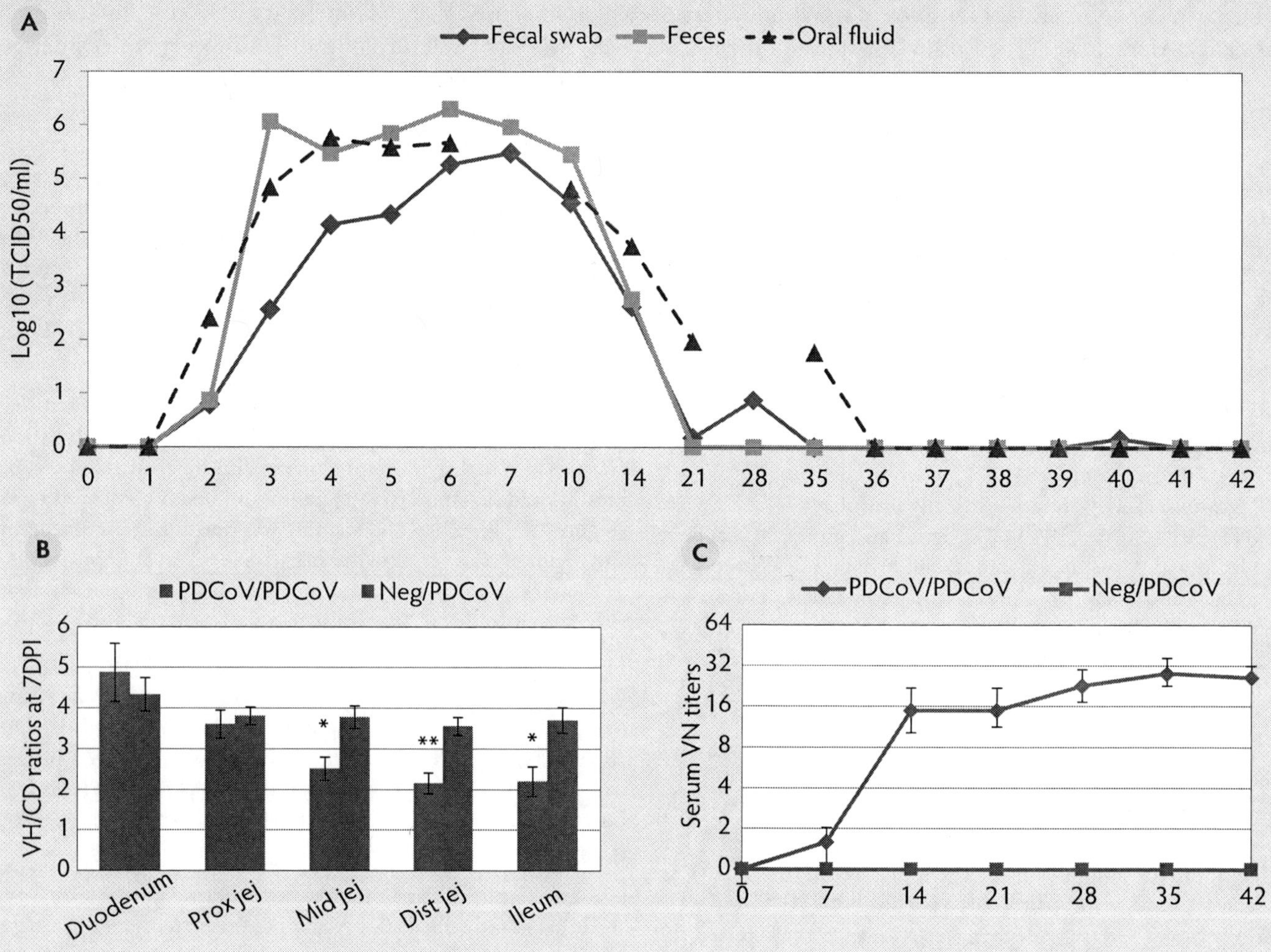

Microbiome associations in pigs with the best and worst clinical outcomes following co-infection with porcine reproductive and respiratory syndrome virus (PRRSV) and porcine circovirus type 2 (PCV2)

M. Niederwerder[1,2], DVM, PhD; C. Jaing[3], PhD; J. Thissen[3], BS; A. Cino-Ozuna[1,2], DVM, PhD, DACVP; K. McLoughlin[4], PhD; R. Rowland[1], PhD
[1]Department of Diagnostic Medicine/Pathobiology, Kansas State University, Manhattan, Kansas; [2]Kansas State Veterinary Diagnostic Laboratory, Manhattan, Kansas; [3]Physical & Life Sciences Directorate, Lawrence Livermore National Laboratory, Livermore, California; [4]Computations Directorate, Lawrence Livermore National Laboratory, Livermore, California

Introduction

On a world-wide basis, co-infections involving porcine reproductive and respiratory syndrome virus (PRRSV) and porcine circovirus type 2 (PCV2) are common and contribute to a range of polymicrobial disease syndromes in swine. Both viruses compromise host defenses, resulting in increased susceptibility to infections by primary and secondary pathogens that can affect growth performance as well as increase morbidity and mortality.

Materials and methods

An experimental population of 95 nursery pigs was co-infected with PRRSV and PCV2.[1] At 70 days post-infection (dpi), 20 representative pigs were selected as having the best or worst clinical outcome based on average daily gain (ADG) as well as the presence and severity of clinical disease. Serum, lung and fecal samples collected at 70 dpi were analyzed using a comprehensive DNA microarray technology, the Lawrence Livermore Microbial Detection Array, which can detect over 8,000 microbes.

Results and discussion

Worst clinical outcome pigs had prolonged and greater levels of both PRRSV and PCV2 viremia as measured by qPCR. Bacterial species, such as *Bacillus cereus*, were detected at a significantly higher rate in the serum of worst performing pigs. At the level of the fecal microbiome, the overall microbial diversity was significantly lower in the worst clinical outcome group. Members of the *Proteobacteria* phylum, specifically *Escherichia coli*, were detected at a significantly higher rate in the feces of best performing pigs.[1] The results reinforce the importance of pathogen load in determining clinical outcome and suggest an important role of microbial diversity as a contributing factor in disease. With growing pressure to eliminate antibiotic use and innovate new management tools for infectious diseases in food animals, this work provides foundational knowledge about the microbiome characteristics that may contribute to performance enhancement in the presence of common viral pathogens.

Acknowledgements

This work was supported by the USDA NIFA Award #2013-68004-20362, the State of Kansas National Bio and Agro-Defense Facility Fund, and by the Lawrence Livermore National Laboratory Derived Research and Development effort (14ERD081).

References

1. Niederwerder, M.C., Jaing, C.J., Thissen, J.B., Cino-Ozuna, A.G., McLoughlin, K.S., Rowland, R.R., 2016. Microbiome associations in pigs with the best and worst clinical outcomes following co-infection with porcine reproductive and respiratory syndrome virus (PRRSV) and porcine circovirus type 2 (PCV2). Veterinary microbiology 188, 1-11.

Evaluation of nutrient availability when using feed disinfection in nursery diets

L. Greiner[1], PhD; J. Connor[1], DVM; B. Knopf[1]; A. Graham[1], MS; L. Ochoa[1], PhD; R.J. Harrell[2], PhD
[1]Carthage Innovative Swine Solutions, LLC, Carthage, Illinois; [2]Professional Swine Management, LLC, Carthage, Illinois

Introduction

Contaminated feed has been shown to be a potential vehicle for the introduction of pathogens into farms.[1] As a result, a number of hog producers have opted to use different protocols for disinfection of feed, including the use of formaldehyde-based antimicrobials. However, it has been suggested that the chemical properties of formaldehyde may interact with dietary nutrients, such as lysine, making them less bioavailable for the pig, therefore, affecting their growth performance. Nevertheless, there is a lack of research results to validate these claims. Moreover, if such effect exists, it will be important to determine its extent and evaluate strategies to overcome such effects. Therefore, the objective of this study was to evaluate the effects of using a formaldehyde based feed disinfectant (Sal CURB®, Kemin Industries, Des Moines, IA USA) on the growth performance of nursery pigs fed diets with or without increasing levels of amino acid (AA) or phytase/vitamins.

Materials and methods

This study was conducted in a commercial wean-to-finish facility during a 44-day study period. A total of 1,125 barrows and gilts (average BW = 6.10, SD = 0.594 kg) were used in a randomized complete block design (Block = BW) with the following dietary treatments: Trt 1) Std/NoSC: diet formulated at 100% of nutrient requirements/feed not disinfected with Sal CURB; Trt 2) Std/SC: formulated as Trt 1/feed disinfected with Sal CURB; Trt 3) AA/SC: diet formulated at 115% of amino acid requirements/feed treated with Sal CURB; Trt 4) V&P/SC: diet formulated at 115% of vitamin and phytase requirements/feed treated with Sal CURB; Trt 5) AA+V&P/SC: diet formulated at 115% of amino acids and vitamin and phytase requirements/feed was treated with Sal CURB. A three-phase dietary nursery program was utilized, with phases changing on days 7, 22, and 44 post-wean. Phase-1 diet formulations were identical for treatments 2-5, due to practical limitations on batch size. Therefore, experimental treatments included only two test diets until d 7. Pen was the experimental unit (25 pigs per pen), and growth performance of pigs was measured every time diet phases changed. Data were analyzed using the MIXED procedure of SAS to evaluate the main effects of treatment and the random effect of block.

Results and discussion

Results are shown in Table 1. Adding Sal CURB to feed formulated to 100% of nutrient requirements had no effect ($P > 0.05$) on any of the measured variables on any of the interim periods (i.e., Trt 1 vs. Trt 2 on d 0 to 7, 7 to 22, or 22 to 44). Moreover, final BW (28.53 vs. 28.03 kg), overall (d 0 to 44) ADFI (0.735 vs. 0.708 kg) and overall G:F (0.654 vs. 0.656) were not different ($P \geq 0.21$) between Trt 1 and 2, respectively. However, pigs on Trt1 grew faster ($P < 0.05$; 0.517kg/day) than pigs in Trt 2 (0.499 kg/day) for the overall ADG study period. Interestingly, the single formulation to 115% requirements of AA (Trt 3, 0.503 kg) or V&P (Trt 4, 0.494 kg) failed to improve ($P < 0.05$) overall ADG to the levels observed in Trt 1. However, no differences ($P = 0.60$) were found between Trt 1 and 5 (0.517 and 0.522 kg/day, respectively) for overall ADG.

Results from this study suggest that disinfecting nursery feed with a formaldehyde-based product may decrease the growth rate of nursery pigs by as much as 3.5%. However, this effect was relatively small, as seen on the lack of effects on other performance parameters, including final BW. Moreover, the impact of using feed disinfected with Sal CURB on ADG was overcome by the combined addition of AA and vitamins & phytase above the requirements, as seen in Treatment 5. Further research is needed to determine if nutrient additions < 115% are able to recover the ADG of pigs, and identify the extent to which the availability of specific nutrients are affected by Sal CURB.

References

1. Dee, S., Clement, T., Schelkopf, A., Nerem, J., Knudsen, D., Christopher-Hennings, J., &Nelson, E. (2014). An evaluation of contaminated complete feed as a vehicle for porcine epidemic diarrhea virus infection of naive pigs following consumption via natural feeding behavior: proof of concept. *BMC Vet Res*, *10*(1), 176.

Table 1: Least square means for the effects of feed disinfection (Sal CURB®) on the growth performance of nursery pigs fed diets with or without increasing levels of amino acid (AA) or phytase/vitamins.[1]

	Treatment						
	1	2	3	4	5		
Feed disinfection	**No**	**Yes**	**Yes**	**Yes**	**Yes**		
AA needs	**100%**	**100%**	**115%**	**100%**	**115%**		
Phytase/vitamins needs	**100%**	**100%**	**100%**	**115%**	**115%**	**SEM**	***P*-value**
Body weight, kg							
Start (d 0)[2]	6.10	6.10	6.10	6.10	6.10	-	-
Day 7[2]	6.97	7.00	6.86	6.90	6.99	0.095	0.78
Day 22	13.35	13.17	13.29	13.01	13.57	0.177	0.25
End (d 44)	28.53[ab]	28.04[b]	28.27[ab]	27.85[b]	29.03[a]	0.277	0.04
Average daily gain, kg							
Day 0 to 7[2]	0.127	0.127	0.109	0.127	0.127	0.0136	0.78
Day 7 to 22	0.426[ab]	0.404[b]	0.426[ab]	0.404[b]	0.440[a]	0.0091	0.02
Day 22 to 44	0.685[ab]	0.676[b]	0.680[ab]	0.667[b]	0.699[a]	0.0068	0.03
Overall (d0 to 44)	0.517[a]	0.499[b]	0.504[ab]	0.494[b]	0.522[a]	0.0066	0.02
Average daily feed intake, kg							
Day 0 to 7[2]	0.132	0.141	0.122	0.127	0.141	0.0091	0.61
Day 7 to 22	0.513	0.508	0.526	0.481	0.522	0.0240	0.62
Day 22 to 44	1.048	1.029	1.052	1.025	1.057	0.0123	0.24
Overall (d0 to 44)	0.735	0.708	0.730	0.703	0.739	0.0131	0.25
Gain:feed							
Day 0 to 7[2]	0.959	0.875	0.896	0.959	0.915	0.0800	0.93
Day 7 to 22	0.837	0.804	0.831	0.840	0.851	0.0280	0.81
Day 22 to 44	0.654	0.658	0.645	0.649	0.658	0.0060	0.65
Overall (d0 to 44)	0.704	0.699	0.690	0.704	0.704	0.0097	0.85
Removals, %	4.39	3.59	2.24	3.06	3.61	0.85	0.19

[1] Data adjusted to a common start weight of 6.1 kg.

[2] Feed from dietary treatment 2 was fed to treatments 2 to 5 from day 0 to 7, therefore, only comparison can be made between Treatment 1 and Treatment 2 (i.e., feed disinfection vs. not feed disinfection).

[ab] Means with different superscript within a row differ $P \leq 0.05$.

Comparison of the efficacy of a commercial inactivated pandemic (p) H1N1 influenza A virus vaccine and two experimental M2e-based vaccines against pH1N1 challenge in the growing pig

Tanja Opriessnig[1,2], Dr med vet, PhD; Philip Gauger[2], DVM, MS, PhD; Xi Jiang[3], PhD; Ming Tan[3], PhD; Patrick Halbur[2], DVM, MS, PhD

[1]The Roslin Institute, University of Edinburgh, Midlothian, United Kingdom; [2]Department of Veterinary Diagnostic and Production Animal Medicine, Iowa State University, Ames, Iowa; [3]Division of Infectious Diseases, Cincinnati Children's Hospital Medical Center and Department of Pediatrics, University of Cincinnati, College of Medicine, Cincinnati , Ohio

Introduction

Influenza A virus (IAV) is considered one of the most important respiratory diseases in US swine herds.[1] Vaccination, while often used today, is commonly not effective. The populations of IAV's found in North American pigs are currently very diverse and the lack of cross-protection among heterologous IAV strains is a concern. The objective of this study was to compare a commercial inactivated pandemic (p) H1N1 vaccine and two novel subunit vaccines,[2,3] using the IAV M2e epitope as antigen, in a growing pig model.

Materials and methods

Thirty-nine 2-week-old IAV-negative pigs were randomly assigned to five groups and rooms. At 3 weeks of age and again at 6 weeks of age, pigs were vaccinated intranasally with an experimental subunit particle vaccine (EXP-PARTICLE- IAV) or a subunit complex-based vaccine (EXP-COMPLEX-IAV) or intramuscularly with a commercial inactivated vaccine (COM-pH1N1-IAV; FluSure® Pandemic; Zoetis). At 7 weeks of age the pigs were challenged with pH1N1 virus intranasally and intratracheally or were sham-inoculated. Necropsy was conducted 5 days post pH1N1 challenge (dpc).

Results and discussion

At the time of challenge one of the COM-pH1N1- IAV pigs had seroconverted based on IAV nucleoprotein-based blocking ELISA. COM-pH1N1 pigs had significantly higher pH1N1 hemagglutination inhibition (HI) titers compared to all other groups. After challenge, nasal IAV RNA shedding was significantly reduced in COM-pH1N1 pigs compared to all other pH1N1 infected groups and this group also had reduced IAV RNA in oral fluids. The mean macroscopic lung lesions were 15.4 ± 3.7 for EXP-PARTICLE-IAV, 14.3 ± 3.7 for the EXP-COMPLEX-IAV group, 11.8 ± 3.2 for the POS-CONTROL pigs, 0.1 ± 0.1 for the COM-pH1N1-IAV pigs and 0.1 ± 0.1 for the NEG-CONTROL pigs. Lesions were significantly less severe in the NEG-CONTROL and the COM-pH1N1-IAV pigs. Under the conditions of this study a commercial pH1N1 specific vaccine effectively protected pigs against homologous challenge as evidenced by reduced clinical signs, virus shedding in nasal secretions and oral fluids and reduced macroscopic and microscopic lesions whereas intranasal vaccination with experimental M2e epitope-based subunit vaccines did not. The results further highlight the importance using IAV type specific vaccines in pigs.

References

1. Janke BH. Clinicopathological features of Swine influenza. Curr Top Microbiol Immunol 2013;370:69-83.

2. Tan M, Huang P, Xia M, et al. Norovirus P particle, a novel platform for vaccine development and antibody production. J Virol 2011;85(2):753-64.

3. Wang L, Cao D, Wei C, Meng XJ, Jiang X, Tan M. A dual vaccine candidate against norovirus and hepatitis E virus. Vaccine2014;32:445-52.

Mycoplasma hyopneumoniae detection *in vivo*: investigating the diagnostic sensitivity of laryngeal swabs

Luiza R. Roos[1]; Nitipong Homwong[1]; Eduardo Fano[2]; Maria Pieters[1]
[1]Department of Veterinary Population Medicine, College of Veterinary Medicine, University of Minnesota, Saint Paul, Minnesota; [2]Boehringer Ingelheim VetMedica Inc., Saint Joseph, Missouri

Introduction

Mycoplasma hyopneumoniae contributes to major losses in the swine industry, as primary agent of enzootic pneumonia and associated with the porcine respiratory disease complex. After experimental inoculation, *M. hyopneumoniae* infection has been reported to last up 7 months, with infected pigs becoming chronic carriers and being capable of transmitting the pathogen to susceptible pigs.[1] This chronic characteristic of *M. hyopneumoniae* infection increases the challenge of disease control and elimination, however, the identification of an *in vivo* diagnostic tool that incorporates accuracy and practicality for correspondence to the actual *M. hyopneumoniae* infection status of the herd aids swine practitioners in field settings. Currently, laryngeal swabs are used for early *M. hyopneumoniae* detection and have been suggested to have higher sensitivity in comparison to oral fluids and nasal swabs.[2] However, the diagnostic sensitivity of laryngeal swabs has yet to be reported in the literature. Therefore, the aim of this investigation was to assess the diagnostic sensitivity of laryngeal swab by Bayesian estimation.

Materials and methods

Results of a previous study, in which laryngeal and bronchial swabs were tested real-time PCR,[3] were employed in this investigation. Laryngeal swabs (LS) were collected from 21 gilts experimentally inoculated with *M. hyopneumoniae* at 0, 14, 28, 42 and 56 days post inoculation (dpi), and from 39 naturally exposed gilts at 0, 14 and 28 days post natural exposure (dpe). Bronchial swabs (BS) were collected from all gilts at 56 dpi or 28 dpe. All swabs were tested by species-specific *M. hyopneumoniae* real time PCR. Bayesian estimation was used to assess the diagnostic sensitivity of LS in comparison to BS, in the absence of a gold standard and assuming specificity ≤ 1 for LS and specificity = 1 for BS. Three models with conditional dependent assumptions were run: Model 1 compared LS collected at 14 dpi/dpe with BS collected at 56 dpi/28 dpe. Model 2 compared LS collected at 28 dpi/dpe with BS collected at 56 dpi/28 dpe. Model 3 comprised a concurrent analysis of Models 1 and 2.

Results and discussion

The estimated median sensitivity (Se) for model 1 was 88.5% for LS. Model 2 resulted in 82.1% sensitivity for LS. Lastly, model 3 resulted in 86.6% sensitivity for LS. The credible interval of all three models overlapped, implying no statistical significance. The Deviance Information Criteria (DIC) estimated was 27.7, 34.1 and 46.8 for Models 1, 2 and 3, respectively. The deviance is a measure of how well the model fits the data and was used to determinate the fittest model in this analysis. Under the conditions of this analysis, the diagnostic sensitivity of LS collected at 14 dpi or dpe, when compared to BS collected at 56 dpi and 28 dpe, was highly sensitive for early *M. hyopneumoniae* detection in infected gilts in experimental settings. This analysis, and its assumptions, is suggested as the best model for biological and statistical significance of the diagnostic sensitivity of LS as an accurate alternative to BS for early *M. hyopneumoniae* identification in live pigs.

References

1. Pieters et al., 2009. Vet. Microbiol. 134, 261 – 266.
2. Pieters and Rovira, 2013. Allen D. Leman Swine Conference, pp. 75–76.
3. Roos et al., 2016. Vet Microbiol. 184, 51–58.

Adaptation of porcine epidemic diarrhea virus (PEDV) to Vero cells

T. Sato, PhD, DVM; K. Oroku; A. Nagao; Y. Furuya; O. Taira, PhD, DVM; N. Tsutsumi, PhD
Nippon Institute for Biological Science, Tokyo, Japan

Introduction

Since 1990s, the modified-live and/or killed porcine epidemic diarrhea virus (PEDV) vaccines have been commercially available in the countries of East Asia including China, Japan and South Korea. These vaccine strains correspond to genogroup 1 (classical genotype), such as CV777 and isolated using Vero cells in culture media supplemented with trypsin.[1] Subsequently, the virus was then attenuated by further serial passages in the Vero cells.[2,3] Previously, we reported that the 100th passaged 83P-5 strain, which was corresponding to genogroup 1, in Vero cells resulted in a growth adaptation in vitro and attenuation in vivo.[3] Since 2010, variant PEDVs (genogroup 2) have been the etiological agent responsible for the outbreak of PED worldwide. In this study, we have attempted to develop a Vero cell-adapted strain of genogroup 2 PEDV and characterized the in vitro growth.

Materials and methods

Vero cells were cultured in Dulbecco's modified Eagle's medium, supplemented with 10% fetal bovine serum (FBS), 0.02% yeast extract and 0.3% tryptose phosphate broth. In 2013, two field strains were isolated from the intestine of suckling piglets with acute diarrhea. The spike gene sequences form the isolated viruses were subjected to the phylogenetic analysis. These strains were cultured with above medium without FBS supplemented 10 µg/ml trypsin and decreased stepwise trypsin concentration as serial passage progresses. The virus stocks were 10-fold serial diluting and virus titration was performed in 24-well plates in triplicate per dilution. At 2-3 days post inoculation, the plates were daily observed and the virus titration was determined by observation of cytopathic effect (CPE) according the Reed and Muench method and expressed as 50% tissue culture infective doses ($TCID_{50}$)/ml.

Results

Two PEDV isolates designated MZ0116-2/2013 and IK1217/2013 were successfully obtained from the small intestines of affected piglets in Miyazaki (collected on December 2013) and Ibaraki (collected on November 2013) prefectures, respectively. They were classified in genogroup 2 by phylogenetic analysis of spike gene sequences. A distinct CPE, characterized by cell fusion and syncytium formation, was observed for both isolates. Serial passage of isolated strains was performed 130 times in Vero cell cultures. Above CPE was observed in the cultures as far as remaining trypsin in culture media. When the virus could be maintained without trypsin, a changed CPE was observed as rounded cells and eventual cell detachment.

Discussion

It has been reported that the attenuated 83P-5 is able to culture in Vero cells without trypsin in culture media.[3] IK1217/2013 and MZ0116-2/2013 strains might have been attenuated during the serial passages as well as 83P-5. Further investigation will be needed to investigate the pathogenicity and immunogenicity of each passaged viruses in vivo.

References

1. Kusanagi K et al. Isolation and serial propagation of porcine epidemic diarrhea virus in cell cultures and partial characterization of the isolate. *J Vet Med Sci.* 1992;54:313-318.

2. Song DS et al. Oral efficacy of Vero cell attenuated porcine epidemic diarrhea virus DR13 strain. *Res Vet Sci.* 2007;82:134-140.

3. Sato T et al. Mutations in the spike gene of porcine epidemic diarrhea virus associated with growth adaptation in vitro and attenuation of virulence in vivo. *Virus Genes.* 2011;43:72-78.

Identification of neuropathogenic *Teschovirus A* serotype 2 and serotype 11 and experimental reproduction of disease

F. Matias Ferreyra, DVM; B. Arruda, DVM, PhD; D. Madson, DVM, PhD; K. Schwartz, DVM, MS; G. Stevenson, DVM, PhD; J. Zhang, MD, PhD; Q. Chen, DVM, PhD; K. Yoon, DVM, PhD; P. Arruda, DVM, PhD
Department of Veterinary Diagnostic and Production Animal Medicine, College of Veterinary Medicine, Iowa State University, Ames, Iowa

Introduction

Teschovirus A (PTV) infections in swine are often asymptomatic, although in some cases infections result in paralysis due to poliomyelitis or polioencephalomyelitis. Teschovirus encephalomyelitis is a neurologic condition of pigs commonly characterized by locomotor disturbances including posterior ataxia, paresis, and/or paralysis. The disease was first recognized in the Czech Republic in 1929 as a highly fatal encephalomyelitis of pigs caused by strains of PTV serogroup 1 (PTV-1) of the genus *Teschovirus*, family *Picornaviridae*. Initially, the PTVs were classified within the genus *Enterovirus*. However, PTVs have been reclassified based on genomic organization and phylogenic analysis. PEV-1 to -7 have been renamed PTV-1 to -7 and PEV-11 to -13 were renamed PTV-8 to -10. PTV-11 to -13 are the serotypes most recently identified. Although multiple serotypes have been identified and genetically characterized, the neuropathogenicity of some of these serotypes has not been fully elucidated.

Material and methods

PTV-11 and PTV-2 field strains were isolated from two separate diagnostic cases with a history of pigs with neurologic disease submitted to the Iowa State University Veterinary Diagnostic Lab (ISU VDL) in 2013 and 2014, respectively. The viruses were initially identified by PCR and final corroboration was achieved by virus isolation and sequencing of the VP1 capsid protein. Eighteen cesarean-derived colostrum-deprived pigs (CDCD) were purchased from a commercial source. Fecal swabs were collected 5 days prior to inoculation and were PCR negative for PTV (CT 0.0; 35 cycles). Animals were housed in separate rooms based on inoculum at ISU Livestock Infectious Disease Isolation Facility. CDCD pigs were randomly assigned to three treatment groups: negative control (n = 4); PTV-2-inoculated (n = 7); and PTV-11-inoculated (n = 7). Pigs were sedated and inoculum was administrated intravenously using a 25-gauge butterfly catheter. Negative control animals were inoculated with 3 ml of cell culture media; animals in PTV-2- and PTV-11-inoculated groups were inoculated with 3 ml of 10^6 TCID50/ml of PTV-2 or PTV-11. Animals were evaluated daily for consistent clinical signs. Eight pigs were necropsied at DPI 14 (negative control: n = 2; PTV-2-inoculation: n = 2; PTV-11-inoculated: n = 4). At DPI 16 a PTV-2-inoculated pig was necropsied due to progression of clinical signs. The remaining piglets were necropsied at DPI 21. Macroscopic lesions were evaluated by diagnostic pathologists and a complete set of tissues were collected. Sections of each sample were frozen at -80°C and placed in 10% formalin for histopathologic examination.

Results and discussion

Negative control animals developed no clinical signs. Three out of seven animals in the PTV-2-inoculated group developed diarrhea at DPI 7 and mild incoordination of hind limbs at DPI 11. Five out of seven animals in the PTV-11-inoculated group developed diarrhea at DPI 7, mild to moderate depression at DPI 10, and posterior ataxia to paralysis at DPI 11. Macroscopic examination of all animals was unremarkable. Microscopic examination of negative control animals was unremarkable. Histologic lesions in clinical and non-clinical PTV-2- and PTV-11-inoculated animals consisted of mild to severe lymphoplasmacytic myelitis with multifocal areas of gliosis, neuron degeneration and satellitosis (Figure 1 and 2). To our knowledge this is the first experimental study demonstrating the neuropathogenicity of PTV-2 and PTV-11.

Figure 1: Pig #67, inoculated with PTV-2. Lymphoplasmacytic myelitis with expansion of the Virchow-Robin spaces (*) and multifocal areas of gliosis (^) in the cervical (1A), thoracic (1B), and lumbar (2C) spinal cord. Lymphoplasmacytic ganglioneuritis (1D). Pig #75, inoculated with PTV-11. Lymphoplasmacytic myelitis with expansion of the Virchow-Robin spaces (*) and multifocal areas of gliosis (^) in the cervical (2A), thoracic (2B), and lumbar (2C) spinal cord. Lymphoplasma-cytic ganglioneuritis (2D).

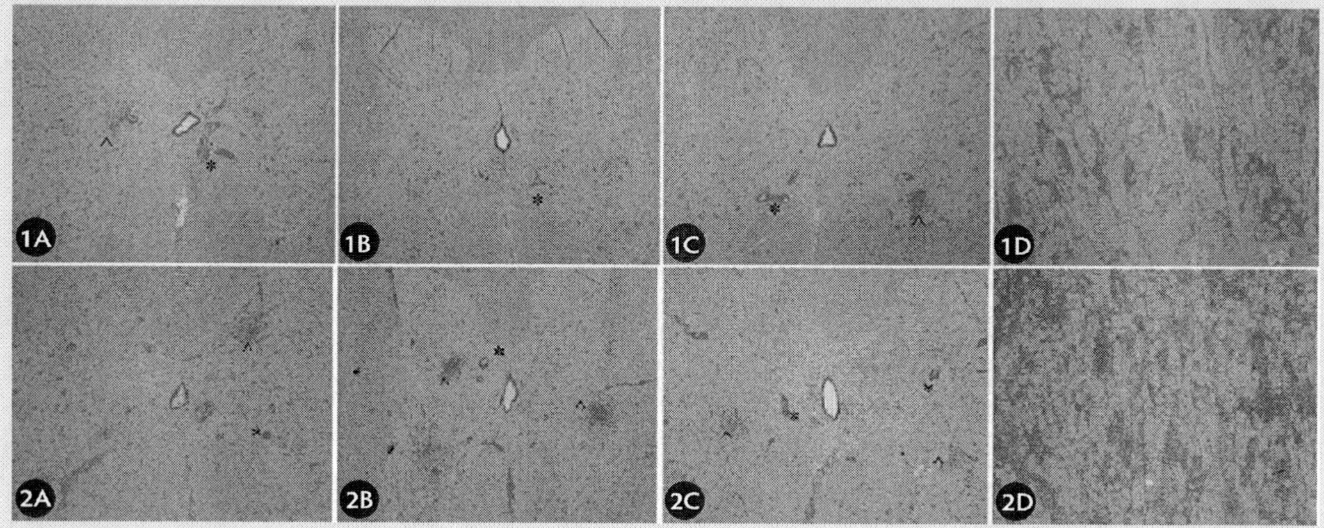

Identification of a divergent strain of Sapelovirus associated with a severe polioencephalomyelitis outbreak in the US

Paulo Arruda[1]; Kent Schwartz[1]; Bailey Arruda[1]; Jerome Nietfeld[2]; Fabio Vannucci[3]; Talita Resende[3]; Albert Rovira[3]; Paul Sundberg[4]; Ben Hause[2]
[1]Iowa State University Veterinary Diagnostic Laboratory, Iowa State University, Ames, Iowa;
[2]Kansas State Veterinary Diagnostic Laboratory, Kansas State University, Manhattan, Kansas;
[3]University of Minnesota Diagnostic Laboratory, University of Minnesota, Saint Paul, Minnesota;
[4]Swine Health Information Center, Perry, Iowa

An acute outbreak of atypical neurologic disease in 11-week-old pigs in an approximately 3,000 finishing swine operation in the United States was recently investigated by the Iowa State University Veterinary Diagnostic laboratory (ISUVDL). Affected animals originated from a single nursery and were placed in two different finishers at 9 weeks-of-age. Over the course of 3 weeks, the clinical onset and progression of signs included decrease of water and feed consumption, compromised ambulation with ataxia, incoordination of limbs, mental dullness, paresis, paralysis and decreased response to environmental stimuli. Clinical examination revealed that affected animals had discernible ataxia however still had deep pain perception and withdrawal reflexes in hind limbs. Despite mental dullness and aimless wandering, pigs often had central awareness with no nystagmus or cranial nerve deficits observed. Overall morbidity of 20% and case fatality rate of 30% was reported by the field veterinarian involved in the case investigation. Laboratory Findings: Histopathologic examination of central nervous system revealed severe lymphoplasmacytic and necrotizing encephalomyelitis with multifocal areas of gliosis and neuron satelliosis, suggestive of a neurotropic viral infection. Cultures did not demonstrate bacterial pathogens in multiple tissues from affected pigs. Immunohistochemistry and PCR for PRRS and porcine circovirus were negative. Sections of spinal cord submitted for PCR for enterovirus, teschovirus and sapelovirus were found consistently positive for porcine sapelovirus. Next Generation Sequencing of brain stem and spinal cord samples investigated the possible presence of other and/or novel pathogens as well as to confirm previous sapelovirus PCR results. Interestingly, a genetically (perhaps antigenically) novel sapelovirus was identified within CNS tissues of affected animals. The 2,323 amino acid polyprotein sequence has an overall 94% amino acid identity and 86% nucleotide identity to recent Korean sapelovirus. No other viral agents were identified within examined samples. Additionally, in situ hybridization assay was developed and results demonstrated positive staining within glial fibrillary acidic protein positive cells and neurons from tissues of affected animals. Between January 1, 2014, and June 21, 2016, a total of 65 cases were submitted for neurologic disease, had histopathology examination of neurologic tissue and had nested PCR for PSV, PTV, and PEV performed. Of these 65 cases, 30 cases had histologic lesions attributed to a viral infection. One or more viruses were detected in 19 of those 30 cases. PSV was detected in 14 of the 19 cases, either alone (6 cases) or with PTV or PEV (four and three cases, respectively). All three viruses were detected in one case. The biologic relevance of this finding is not clear at this point as there is a significant gap of knowledge concerning the pathophysiology and the potential role of this particular virus in cases of encephalomyelitis in swine. To date, sapelovirus from other species have not been reported to be associated with nervous disease. In this case, novel sapelovirus was the only agent detected associated with a unique clinical presentation of CNS disease. At least one previous case report documents the neuroinvasive potential of porcine sapelovirus in swine [1] (Alex Schock, *et al.* 2014). However, knowledge gaps remain in disease causation (i.e. Koch's postulates as yet unfulfilled), epidemiology, pathogenesis and biologic relevance of this potential pathogen.

References

1. Schock, A., Gurrala, R., Fuller, H., Foyle, L., Dauber, M., Martelli, F., Scholes, S., Roberts, L., Steinbach, F., Dastjerdi, A., 2014: Investigation into an outbreak of encephalomyelitis caused by a neuroinvasive porcine sapelovirus in the United Kingdom. Vet Microbiol. 172, 381-389.

Piglet feed intake during the end of the lactation period impacts post-weaning diarrhea incidence and survival rate

Frédéric Vangroenweghe[1]; Hinnerk Kraneburg[2]; Diedrich Hendriks[3]; Tim van Sprang[4]; Ellen Van Driessche[5]
[1]Elanco Animal Health, Brussels, Belgium; [2]De Varkenspraktijk – locatie Oss, Oss, The Netherlands;
[3]DAC Zuid-Oost, Deurne, The Netherlands; [4]Slingeland Dierenartsen, Silvolde, The Netherlands;
[5]Animal Health Care – Flanders, Torhout, Belgium

Introduction

Post-weaning *Escherichia coli* diarrhea (PWD), also called post-weaning (pw) enteric colibacillosis, in pigs remains a major cause of economic losses for the pig industry.[1,2] PWD typically causes mild to severe watery diarrhea between 5 and 10 days after weaning and is caused primarily by enterotoxigenic *Escherichia coli* (ETEC). The most common adhesins found on ETEC from PWD in pigs are associated with fimbriae F4 (previously called K88) and F18, while the predominant enterotoxins are heat-labile toxin (LT), heat-stable toxin a (STa), and heat-stable toxin b (STb). In addition to F4 and F18, other fimbrial adhesins, such as F5 (K99), F6 (987P), and F7 (F41), have been associated with PWD, but less frequently.[3-6] Besides the impact of the presence of the pathogenic bacteria, other influencing factors can impact the pathogenesis of the disease, such as feed intake before and after weaning. The objective of the present study was to study the relationship between piglet feed intake at the end of the lactation period with the subsequent mortality or survival in the first 2 weeks post-weaning.

Materials and methods

A 600-sow farms with diagnosed problems of PWD was selected to perform the study. Piglets were followed from 7 days before weaning till 14 days post-weaning. In order to be able to quantify the uptake of pre-weaning creep feed during the lactation period, AviBlue® (Elanco Animal Health) added into the feed of the piglets during the lactation period. In total, piglets of 20 litters, evenly distributed over different sow parities were included into the study. Feed uptake was quantified through spectrophotometric measurement of blue color in the feces of the piglets at the day of weaning. All piglets were individually identified, weighed at weaning and subsequently followed for the next 14 days post-weaning. All diarrhea symptoms and mortality was registered. Relation between the feed intake, post-weaning clinical diarrhea presence and survival rate was statistically analyzed.

Results

Results show that heavy piglets tend to have a lower feed intake pre-weaning, subsequently resulting in a higher risk for diarrhea occurrence and mortality.

Discussion

The results clearly show the importance of a sufficient feed intake pre-weaning for all piglets in order to perform well during the immediate post-weaning period under the stress of pathogenic *E. coli* bacteria. Large piglets pre-weaning are mostly positioned at the best producing teat of the sow's udder, and therefore these piglets tend to eat less pre-weaning creep feed, resulting in less adaptation of their intestinal flora and digestive capacity to the diets provided during the post-weaning period. The ultimate goal for the swine farmer will be to stimulate the heaviest piglets for sufficient feed intake in order to overcome the connected post-weaning problems.

References

1. van Beers-Schreurs et al., 1992. *Veterinary Quarterly* 14(1), 29-34.
2. Fairbrother et al., 2005. *Animal Health Research Reviews* 6(1), 17-39.
3. Kwon et al., 2002. *The Veterinary Record* 150, 35-37.
4. Frydendahl, 2002. *Veterinary Microbiology* 85, 169-182.
5. Chen et al., 2004. *Veterinary Microbiology* 103, 13-20.
6. Vu Khac et al., 2006. *BMC Veterinary Research* 2, 10

Mycoplasma hyopneumoniae vaccination check using quantification of IFN-gamma levels following T-lymphocyte stimulation of *M. hyopneumoniae*-vaccinated piglets

Frédéric Vangroenweghe[1]; Hinnerk Kraneburg[2]; Diedrich Hendriks[3]; Tim van Sprang[4]; Ellen Van Driessche[5]
[1]Elanco Animal Health, Brussels, Belgium; [2] De Varkenspraktijk – locatie Oss, Oss, The Netherlands;
[3]DAC Zuid-Oost, Deurne, The Netherlands; [4]Slingeland Dierenartsen, Silvolde, The Netherlands;
[5]Animal Health Care – Flanders, Torhout, Belgium

Introduction

Mycoplasma hyopneumoniae (M. hyo) is the causative agent of swine mycoplasmal pneumonia, a disease is wide-spread and causes huge economic losses in the swine industry through growth retardation and susceptibility to infection with other primary and secondary pathogenic organisms.[1] The infection is transmitted vertically from sow to piglets during the lactation period[2] and further horizontal transmission occurs during the post-weaning period.[3,4] Recent studies have shown difference in humoral and cellular immune response following M. hyo vaccination in terms of locally produced sIgA antibodies[5] and production of IFN-gamma by the peripheral blood T-lymphocytes.[6] The objective of the present study was to develop an M. hyo vaccination check using quantification of IFN-γ levels following T-lymphocyte stimulation of M. hyo-vaccinated piglets.

Materials and methods

Whole blood from 20 piglets, vaccinated at 3 days of age with an early single dose M. hyo vaccine, were used in the study. The piglets (n = 5 per age category) were 3 and 73 days of age at bleeding. Blood was collected in a lithium-heparine vacutainer tube and transported under controlled temperature conditions (+18-25°C) to the laboratory. Stimulating antigen was prepared through different methods: pure M. hyo vaccine antigen, oil fraction of M. hyo vaccine, osmotic lysis, ultrasonic lysis,[7] a combination of both osmotic and subsequent ultrasonic lysis. Phospate buffered saline and pokeweed was used respectively as negative and positive control. A standard antigen load of 20 µL on a total volume of 200 µL were tested during the incubation at 37°C for 24h. All samples were activated in duplicate. Subsequently, the whole blood was centrifuged and supernatant fluid was collected. IFN-γ levels were analysed in supernatant fluid using a commercially available ELISA test (Thermo Scientific – Pierce Biotechnology, Rockford, IL, US) according to the manufacturer's instructions.

Results

Summary of the results is given in Table 1.

Discussion

This study shows that M. hyo vaccination can be checked using whole blood from vaccinated piglets and subsequent in vitro stimulation with M. hyo antigen. Results show an increase in IFN-gamma at 73 days of age, with variation based on the antigen

Table 1: IFN-γ levels (pg/ml) in piglets bled at d3 (day of vaccination) and d73 (10 weeks post-vaccination). NEG – negative control; OL – osmotic lysis; SON – sonication; POS – positive control

Antigen code	Decription	IFN-γ at d3	IFN-γ at d 73
1 – negative control	PBS	0.0	0.0
2 – antigen (Ag)	Vaccine pellet	0.0	0.0
3 – OL-Ag	Ag following osmotic lysis	0.0	239.8
4 – SON-Ag	Ag following sonication	0.0	9.7
5 – OL+SON-Ag	Ag after combined osmotic lysis and sonication	0.0	6.7
6 – positive control	Pokeweed	50.0	1115.5
7 – vaccine oil fraction	Vaccine oil fraction following centrifugation	0	25.4
8 – entire O/W M. hyo vaccine	Entire oil in water M. hyo vaccine	6.3	182.4

used to stimulate the whole blood assay. Further optimization will be performed in order to be able to check the vaccination compliance in younger piglets and to obtain a reproducible result that can be compared between different swine operations. The test can also be valuable in case of any doubt whether piglets were vaccinated against M. hyo.

References

1. Maes et al., 2008. Veterinary Microbiology 126: 297–309.
2. Pieters et al. 2009. Veterinary Microbiology 134 (2009) 261–266.
3. Meyns et al., 2004. Preventive Veterinary Medicine 66: 265–275.
4. Meyns et al., 2006. Vaccine 24: 7081–7086.
5. Martelli et al., 2014. Veterinary Microbiology 168: 357–364.
6. Seo et al., 2013. J. Vet. Med. Sci. 75(2): 245–247.
7. Razin, 1963. J. Gen. Microbiol. 33, 471-475.

Evaluation of a vaccine prototype using a chimeric protein and a molecular enhancer produced by *Pichia pastoris* fermentation

A. Villamil[1]; E. Lamazares[2,3]; A. Ruíz[1]; O. Sánchez[3]; J.R. Toledo[2]
[1]Departamento de Patología y Medicina Preventiva, Facultad de Ciencias Veterinarias, Universidad de Concepción, Chile;
[2]Departmento de Fisiología, Facultad de Ciencias Biológicas, Universidad de Concepción, Chile;
[3]Departmento de Farmacología, Facultad de Ciencias Biológicas, Universidad de Concepción, Chile

Introduction

The post-weaning multisystemic wasting syndrome (PMWS) is a viral disease with a high prevalence worldwide[1] affecting pigs between 6 and 15 weeks old.[1,2] Nowadays, four vaccines are used to control porcine circovirus type 2 (PCV2), which has been described as the virus that triggers this wasting condition.[3,4] The aim of this work was to evaluate a capsid chimeric protein and alpha interpheron as a molecular enhancer of immune response, both produced by *Pichia pastoris* fermentation, and used as a PCV2 vaccine in order to see protection against PCV2, in three weeks old swine.

Materials and methods

A genetically engineered *P.pastoris* expressing a chimeric PCV2 antigen and alpha interferon as a molecular enhancer was developed, using its products as a vaccine prototype on a preliminary assay. Thirty-four animals were randomly selected from a commercial farm to be used within four treatment groups and two control groups. Animals were vaccinated at three weeks of age with 2mL of emulsion containing 600g/mL of total protein, administered intramuscularly, and followed by a booster three weeks later. Intranasal challenge with 10 mL of PCV2 ($1 \times 10^{4.7}$ TCID50) was performed two weeks after the booster and animals were taken to necropsy four weeks after challenge. An indirect ELISA (Biocheck) was used to measure titters of serum antibodies against Cap protein and DNA was extracted from serum samples using QIAamp DNA MiniKit in order to quantify PCV2 nucleic acid by RT- PCR.

Results and discussion

Non of the pigs exhibited clinical signs or remarkable macroscopic lesions during the experiment, as well as any local reaction. All of the vaccinated and non-vaccinated pigs were positive for anti-Cap specific antibodies even before the experiment began. It should be taken into consideration that pigs came from a commercial farm where PMWS is not present but they are not free from PCV2 so maternal antibodies could be still present and be detected on serum samples. PCV2 DNA was detected in blood samples of all pigs during the first 35 days of experiment as shown in figure 1.

Pigs from groups 1 to 4 had an important decrease of viral load after challenge, which suggest that vaccinated animals had a better protection against virus exposure than negative control groups. We conclude The chimeric protein obtained from *P. pastoris* fermentation was able to reduce the viral load of experimentally challenge pigs with PCV2 when is used as a vaccine emulsion on three weeks old pigs.

References

1. Liu, Q., et.al., (2000). Quantitative, competitive PCR analysis of porcine circovirus DNA in serum from pigs with post-weaning multisystemic wasting syndrome. Journal of Clinical Microbiology 38, 3474–3477.

2. Olvera, A., et. al. (2004). Comparison of porcine circovirus type 2 load in serum quantified by a real time PCR in post-weaning multisystemic wasting syndrome and porcine dermatitis and nephropathy syndrome naturally affected pigs. Journal of Virological Methods117, 75–80.

3. Opriessnig, T., et al. (2009). Difference in severity of porcine circovirus type 2 (PCV2)-induced pathological lesions between Landrace and Pietran in pigs. Journal of Animal Science.87: 1582-1590.

4. Grau-Roma, L., et al. (2009) Infection, excretion and seroconversion dynamics of porcine circovirus type 2 (PCV2) in pigs from post-weaning multysistemic wasting syndrome (PMWS) affected farms in Spain and Denmark. Veterinary Microbiology. 135: 272-282.

Figure 1: Mean values of the genomic copy number of PCV2 DNA in serum from group 1 to 6.

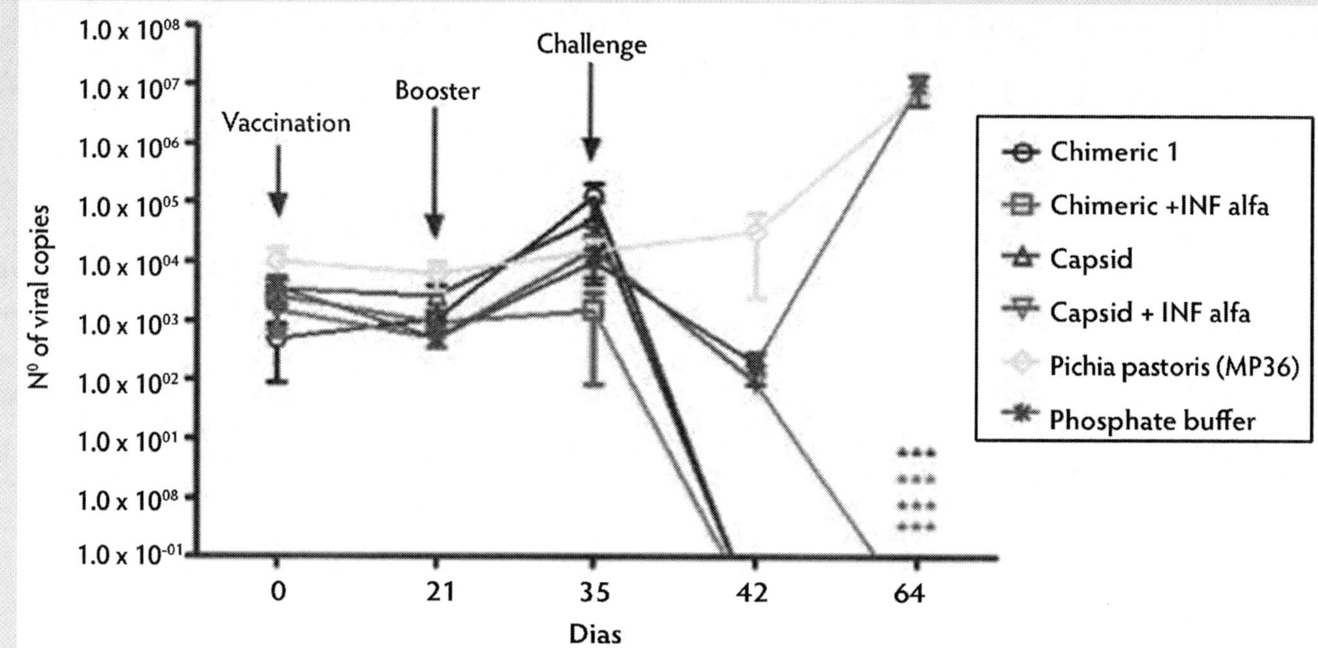

Grow-finish mortality in pigs consuming acidified water

J.A. Jendza, Ph.D
BASF Corporation, Florham Park, New Jersey

Introduction

Hemorrhagic Bowl Syndrome (HBS) is a disease that often kills some of the most valuable pigs in the barn. It is a source of chronic mortality at all ages, and the exact cause of the disease is unclear.[1] We hypothesized that HBS is ultimately the result of volvulus formation followed by exponential growth of microbes and gas formation rostral to the volvulus causing an intestinal rupture before the volvulus can reverse. Under this hypothesis, gut acidification would be predicted to reduce HBS mortality by

1. Reducing the level of bacteria present in the digesta
2. Slowing the rate of gas and toxin production by intestinal microbes during volvulus
3. Thereby granting the animal more time for the volvulus to reverse without rupturing

Materials and methods

Grow-finish pigs (1,260 pigs, 40 lbs) were assigned to 36 pens. Pens were assigned to treatments in a completely randomized design with 9 replicates per treatment. Treatments consisted of a 2 × 2 factorial of distillers dried grains plus solubles inclusion (0 or 35%; DDGS) and liquid feed grade sodium formate formulation (0 or 0.08% in the water; Na-formate). The level of Na-formate was chosen to achieve a final pH of between 4.3 and 4.0 in the drinking water. Feed intake, body weights, and mortality were recorded every 2 weeks. Resulting data were analyzed according to a 2 × 2 factorial with repeated measures. Model consisted of the main effects (DDGS, water acidification, and time period), and all 2 factor interactions. Mortality was analyzed as a generalized linear model with a poisson distribution.

Results and discussion

The DDGS treatment reduced ADG, ADFI and F:G by 3.0, 1.3, and 3.8 percent, respectively (Table 1). Whereas the Na-formate treatment improved by F:G by 1.0%, but did not affect ADG or ADFI. The negative effects from high DDGS on performance run counter to much of the published literature,[2] and are likely the result of the DDGS being of lower than expected quality. The benefit of Na-formate, however, is quite consistent with previous reviews.[3]

Total mortality was not improved by DDGS, but HBS mortality was reduced 76%. Na-formate reduced the total mortality by 34%, which is consistent with previous literature.[3] Interestingly, there was significant interaction between DDGS and Na-formate on HBS mortality, with pigs consuming acidified water having HBS mortality lower than pigs fed non-DDGS feed, but higher than those fed diets containing 35% DDGS. This suggests that the ability of Na-formate and DDGS to alter total mortality differ, depending on the cause of death.

References

1. Schwartz, K. J. Hemorrhagic Bowel Syndrome (HBS) in Swine. URL: `http://porkgateway.org/wp-content/uploads/2015/07/hemorrhagic-bowel-syndrome1.pdf` Access Date:11.01.2016.

2. Stein, H. H. and G. C. Shurson (2009). "Board-invited review: the use and application of distillers dried grains with solubles in swine diets." *J. Anim. Sci.* 87(4):1292-1303.

3. Tung, C. M. and J. E. Pettigrew (2006) *Critical Review of Acidifiers.* URL: `http://www.pork.org/ResearchDetail/541/CriticalRevieHwofAcid.aspx` Access date: 9.18.2014.

Table 1: Main effects of DDGS and water acidification with Na-formate on grow-finish performance and total mortality.

	ADG, lb/d	ADFI, lb/d	Feed:Gain, lb/lb	Gain:Feed, %	Mortality, pig/pen	HBS, pig/pen
DDGS*						
0%	1.90	3.83	2.00	51.1	1.05	0.36
35%	1.84	3.88	2.08	49.0	0.88	0.21
Na-formate**						
0.00%	1.87	3.87	2.05	49.8	1.16	0.29
0.08%	1.88	3.84	2.03	50.3	0.77	0.29
SEM	0.006	0.014	0.004	0.10	0.076	0.053
DDGS × Na-Formate						
0%	0%					0.46
35%	0%					0.11
0%	0.08%					0.25
35%	0.08%					0.32
SEM						0.076
P-values						
DDGS	< 0.0001	0.0295	< 0.0001	< 0.0001	0.5955	0.0274
Na-formate	0.3896	0.2261	0.0002	0.0024	0.0074	0.0192
Period	< 0.0001	< 0.0001	< 0.0001	< 0.0001	< 0.0001	< 0.0001
DDGS × Na-formate	0.6710	0.8584	0.7696	0.8337	0.7537	0.0012
DDGS × Period	0.7787	0.9066	0.1233	0.7779	0.9366	0.8237
Na-formate × Period	0.9151	0.9881	0.3973	0.4143	0.9717	0.3847

* Distillers dried grains with solubles.
** Feed grade sodium formate (61% formic acid, 20.5% sodium formate).

Development of a new swine *Mycoplasma* multiplex qPCR

Eric van Esch, DVM, Technical Director
BioChek BV, Reeuwijk, Netherlands

Introduction

In swine 3 Mycoplasma species are deemed to be important, *Mycoplasma hyopneumoniae, Mycoplasma hyosynoviae and Mycoplasma hyorhinis*. All three can be considered to be important in respiratory diseases, but in some cases also in lameness in swine (*M. hyosynoviae*). This paper is describing the development of a new multiplex Mycoplasma qPCR which detects all 3 species.

Materials and methods

Primer/Probe

Based on literature studies, blast in GenBank® and evaluation with post-blast software (in house product) specific primers for *M. hyopneumoniae, M. hyosynoviae* and *M. hyorhinis* were designed. To these specific primers Taqman probes were designed for detection of qPCR products. An internal control primer/probe set was used as well to validate negative results excluding inhibition.

Materials

For evaluation reference strains for the specific Mycoplasma's were used in combination with other Mycoplasma's species and other common agens in swine. Field samples with a known qPCR status were used. Also specific plasmids were used to determine the Lower Limit Of Detection (LLOD).

Methods

DNA extraction was done by using a spin column based method. qPCR was performed according to BioChek's RNA/DNA platform protocol (1 × 10 min 48°C, 1 × 3 min 95°C, 40 cycles 15 sec 95°C/60 sec 60°C, data collection at 60 sec 60°C). Three different cycler platforms were used for evaluation. Data was quantified by using Cq values.

Results

Specificity

In total 516 tests were performed to check specificity. All primer sets were found to be specific for their targets, no false positive results were obtained.

Sensitivity and Lower Limit Of Detection (LLOD)

68 Field positive samples were used for evaluating sensitivity. Based on these results sensitivity for *M. hyopneumoniae* was 98.2%, for *M. hyosynoviae* 100% and for *M. hyorhinis* 82.4% respectively. For the LLOD plasmids were used and LLOD's were for *M. hyopneumoniae* <100 copies/reaction, for *M. hyosynoviae* < 50 copies/reaction and for *M. hyrhinis* < 50 copies/reaction.

Discussion and conclusion

The new developed multiplex qPCR makes it possible to detect all 3 relevant Mycoplasma species in one single run. Both specificity and sensitivity are high. Sensitivity itself will depend on extraction methods used, where bead based methods seem to be improving sensitivity over column spin based methods. With the low LLOD values the qPCR will start detection below clinical relevant levels. Combining the 3 Mycoplasma species in one qPCR will save time and effort for the full evaluation of Mycoplasma in swine.

Annual occurrence of mycotoxins in US corn harvested from 2012 to 2015

E.G. Hendel[1], VMD, PhD; T. Jenkins[2], PhD; S.M. Mendoza[1], PhD; G. R. Murugesan[1], DVM, PhD
[1]BIOMIN America, Inc., San Antonio, Texas; [2]BIOMIN Holding GmbH, Getzersdorf, Austria

Introduction

Mycotoxins are harmful secondary metabolites produced by fungal species capable of infesting various commercial crops. These fungal species are roughly divided into two groups: those predominantly producing mycotoxins on the field (e.g. *Fusarium* spp.) and those predominantly occurring in storage (e.g. *Aspergillus* and *Penicillium* spp.). Mycotoxin contamination of feed materials is a global concern, as exposure to mycotoxins significantly impacts animal health and productivity.

BIOMIN has been conducting an annual global survey of commodity crops and completed feeds since 2004, including an annual corn survey in the United States since 2012. This presents an analysis of annual corn crops in the United States from 2012 to 2015.

Materials and methods

The number of corn samples from varied from year to year, peaking in 2014. A total of 994 samples were analyzed from the 2012 to 2015 corn harvest, including corn and corn silage. Corn DDGS were not included in the data-set. State information was not available for all samples. Overall, the majority of the samples originated from the Midwestern states.

Samples were analyzed utilizing HPLC and LC-MS/MS methodologies. The six major mycotoxin groups analyzed comprised of aflatoxins, type B trichothecenes such as deoxynivalenol (vomitoxin), ochratoxin-A, fumonisins, type A trichothecenes such as T-2 toxin, and zearalenone derivatives.

Results and discussion

A high proportion of samples contained at least one detected mycotoxin type (84% ± 5; mean % positive over all years ± SE), with co-occurrence of mycotoxins in many samples (≥ 2 mycotoxins detected in 42% ± 6 of samples annually; mean ± SE). The most prevalent mycotoxins found within samples from 2012 and 2013 included fumonisins (72% in 2012; 54% in 2013) and type B trichothecenes (33% in 2012; 27% in 2013). In the 2014 and 2015 crop, type B trichothecenes became most prevalent (62% in 2014; 75% in 2015), followed by fumonisins (56% in 2014; 53% in 2015). The occurrence of aflatoxins in the tested samples had a decreasing trend over the years (2012-15; 30%, 14%, 19%, 1%), while zearalenone remained steady (overall mean of 17% ± 3; mean ± SE).

The level of contamination detected in samples had fairly variable ranges (Table 1), and type B trichothecense and fumonisin contaminationed samples had the widest range of detected levels of mycotoxins. The percent of positive samples of fumonisins over 1000 ppb (2012-15: 20%, 35%, 49%, 40%) and percent of positive samples of type B trichothecenes over 1000 ppb (2012-15: 32%, 14%, 35%, 16%) varied annually, emphasizing the importance of comprehensive monitoring programs for mycotoxin risk assessment.

Table 1: Summary of mycotoxin contamination

State	Samples	State	Samples	State	Samples
AL	4	KS	3	OH	18
AR	1	KY	5	OK	2
AZ	8	MI	5	PA	1
CA	6	MN	24	SC	3
CO	1	MO	22	SD	35
CT	1	MS	19	RI	2
FL	43	MT	4	TX	21
GA	28	NC	12	VT	4
IA	100	NE	11	WI	120
IL	57	NM	8		
IN	48	NY	71		

Field efficacy study of RHINISENG® in Europe on a farm with problems associated with *Bordetella bronchiseptica*

A. Sánchez-Matamoros[1], PhD; A. Martos[1], MSc; O. Boix[1], MSc; A. Martínez-Abarca[2], DVM; F.J. Pallarés[3], PhD
[1]Laboratorios HIPRA S.A, Amer, Spain; [2]Dalland Hybrid España S.A., Murcia, Spain;
[3]Department of Anatomy and Comparative Pathology, Faculty of Veterinary Medicine, Murcia University, Murcia, Spain

Introduction

Bordetella bronchiseptica (Bb) is a bacterium found worldwide for which a high prevalence is reported in the swine industry.[1] This pathogen is the primary agent of non-progressive atrophic rhinitis (NPAR) and can also cause progressive atrophic rhinitis in mixed infection with toxigenic strains of *Pasteurella multocida* (PMT). Nowadays, EU legislation limits the metaphylactic use of antimicrobials and promotes the development of alternative prevention methods. To this end, vaccination against Bb seems to be a suitable choice. The aim of this study was to investigate the efficacy of the only available vaccine for preventing NPAR in Europe (RHINISENG®) under field conditions on a Spanish farm with Bb problems.

Material and methods

A randomized, controlled clinical trial with a parallel-group design was conducted in a Spanish farrow-to-nursery pig herd, which had consistently suffered from respiratory problems in nurseries due to NPAR. Eight pregnant sows with fewer than two parturitions were allocated to the two groups. The vaccinated group (Group V) was inoculated with RHINISENG®, while the control group (Group NV) received PBS. The parameters evaluated in the piglets of these sows were the detection of Bb and PMT, serum antibodies against Bb and PMT, and clinical signs of respiratory disease. In addition, the NPAR lesions (macroscopic and microscopic signs)[2,3] were evaluated at 6.5 weeks of age. At 9.5 weeks, productive parameters were also monitored.

Results and discussion

All sows were seropositive for Bb and seronegative for PMT prior to vaccination, as the Bb agent was present in the herd. Nonetheless, the antibodies against Bb in piglets in Group V were significantly higher than in Group NV until 46 days post-farrowing (dpf) (P value < 0.001). Moreover, piglets in Group V were seropositive for PMT, while those in Group NV remained seronegative. These results indicated that antibodies against Bb and PMT were effectively transferred via colostrum to protect their offspring, as published in previous experimental studies.[1,4]

Regarding the prevalence of Bb in piglets, the piglets in Group V showed a significant reduction in Bb prevalence at 21 and 46 dpf (P value < 0.001). This result showed that vaccination could decrease the Bb burden and prevent clinical outbreaks.[1] In addition, no clinical respiratory signs were observed in any of the piglets studied. However, the piglets in Group NV presented a higher percentage of animals with NPAR macroscopic signs, mean nasal lesion scores and more evident bone alterations than in Group V (Table 1). Furthermore, the mean body weight per piglet was higher in Group V (20.46 kg vs. 18.56 kg, respectively).

From the results of this study, we can conclude that immunization of pregnant sows with RHINISENG® prior to farrowing induced a high concentration of colostral antibodies in piglets. This protection enabled us to reduce the prevalence of Bb and NPAR lesions in piglets. It is also likely that implementation of vaccination on this farm has a beneficial effect on productive parameters during the nursery and weaning periods. These results confirm similar experiences with this vaccine in experimental trials.[4]

Acknowledgements

The authors would like to acknowledge the invaluable help of the technical services of HIPRA (Ramón Riquelme and Ramon Jordà) and UCAM staff for their technical assistance (DIAGNOS and M. Blanch).

References

1. Brockmeier SL *et al.* Bordetellosis. Diseases of swine, 10th edition, 2012;49:670-679.
2. European Pharmacopoeia 8.0. Monograph 04/2013:1361.
3. Brockmeier SL *et al.* Role of the Dermonecrotic Toxin of *Bordetella bronchiseptica* in the Pathogenesis of Respiratory Disease in Swine. *Infect. Immun.* 2002;70(2):481–490
4. Montañé J *et al.* RHINISENG®: efficacy of the basic vaccination scheme (vaccination and revaccination) and the booster vaccination. IPVS 2012; p.743.

Table 1: Macroscopic and histological lesions in the evaluation of turbinate and nasal septum at 6.5 weeks of age in groups studied (Group V and Group NV).

	Group V	Group NV
Macroscopic lesions		
Percentage of piglets with nasal lesions	30%	60%
Average nasal lesion scores	0.4/18	3.4/18
Maximum nasal lesion scores	2/18	10/18
Histological lesions		
Percentage of piglets with bone alterations	10%	60%
Average score of bone alterations	0.1/3	0.7/3
Maximum score of bone alterations	1/3	2/3

Evaluation of an inactivated vaccine against virus of porcine epidemic diarrhea virus (PEDV)

J. Hernandez[1]; Luis G. Giménez[2]; A. Massa[3]; A. Flores[3]
[1]Technical Service Lapisa Animal Health - La Piedad- Michoacán-México; [2]College of Veterinary Medicine, Iowa State University, Ames, Iowa; [3]Lapisa – Diagnostic Laboratory - La Piedad- Michoacán México

Introduction

PEDv was first detected in Mexico in July 2013 with piglets' mortality up to 100%. The aim of this study was to assess the ability of an experimental PEDV inactivated vaccine (LAPISA Animal Health) to induce an humoral response in vaccinated females (gilts and sows) measured by using a PEDV whole virus (WV)-based ELISA developed and validated at the Iowa State University Veterinary Diagnostic Laboratory (ISU-VDL) and adapted for antibody detection (IgG, IgA) in serum and colostrum /milk specimens.

Materials and methods

15 pregnant females were selected in the last third of gestation according to parity. Specifically, 4 negative gilts (A), 7 sows between 2-4 farrows (B), and 4 females with 4 or more farrows (C), were selected. Gilts had no previous contact with PEDV and were serologically negative, while sows were naturally exposed to PEDV around 100 days before this evaluation. Pregnant females were individually identified, and divided in two different treatment (Tx) groups defined as Tx1 (control group) and Tx2 (vaccinated group).

The vaccination protocol consisted on two doses (2 mL/dose) given intramuscular at 6 and 3 weeks pre-farrow (3 weeks between dose) for both groups: Tx1) saline solution (2 mL) intramuscular; Tx2) PEDV experimental inactivated vaccine (2 mL /dose) intramuscular.

Serum samples were collected from each female at 6, 3 and 1 weeks pre-farrow Colostrum/milk were collected at 12, 24 and 36 hours (h) after farrow, for each female in each group of treatment.

All samples were sent to ISU-VDL for testing (PEDV WV-based IgG/IgA ELISA IgG).

Results

The results were expressed as an average of the sample-to-positive (S/P) ratio for serum IgG and colostrum/milk IgA, identifying them as (basal sampling and final sampling) for both Treatments. A S/P of 0.8 and 0.6 were used as cut-off values for serum IgG and colostrum/milk IgA, respectively.

Conclusions

The humoral response, measured as IgG in serum and IgA in colostrum and expressed as S/P ratio, was higher in vaccinated females compared to unvaccinated females. Thus, based on preliminary results, the proposed PEDV experimental vaccine was able to stimulate a humoral response (serum IgG and milk IgA) when applied pre-farrow in both gilts and sows. However, these are preliminary results, and data related to transference of passive immunity to piglets are pending at this time.

Table 1:

PEDV WV IgG ELISA results on serum samples at 6th, 3rd and 1st weeks pre-farrow							
Tx1	6th	3rd	1st	Tx2	6th	3rd	1st
A	0,107	0,105	0,051	A	0,123	0,426	0,785
B	1,247	1.213	1,097	B	0,535	1,136	1,409
C	2,072	1,700	1,124	C	2,055	2,319	2,302
PEDV WV IgA ELISA results in colostrum samples at 12, 24 and 36 h post-farrow							
Tx1	12hr	24hr	36hr	Tx2	12hr	24hr	36hr
A	0,333	0,059	0,059	A	1,021	0,034	0,045
B	2,060	0,368	1,103	B	2,470	0,476	0,041
C	3,175	0,234	0,052	C	4,100	1,895	0,039

A novel approach for using P.G. 600®: Impact on gilt performance and sow longevity

Derald Holtkamp[1], MS, DVM; Jennifer Patterson[2], MS
[1]Department of Veterinary Diagnostic and Production Animal Medicine, Iowa State University, Ames, Iowa;
[2]Swine Research and Technology Centre, University of Alberta, Edmonton, Canada

Introduction

P.G. 600® (PG600, Merck Animal Health, De Soto, Kansas) contains exogenous gonadotropins and is labeled for inducing estrus in pre-pubertal gilts. A previous study validated the use of PG600 in pre-pubertal gilts that failed to exhibit a natural estrus during intensive boar stimulation in a boar exposure area (BEAR).[1] The investigators concluded that PG600 and intensive boar stimulation in the GDU (gilt development unit) may be used to deliver predictable numbers of breeding-eligible gilts to the sow herd. The objective of the present study was to follow up on the previous study by conducting a retrospective analysis of production records from the herd used for the previous report (BEAR-PG600 program) along with a similar herd (Baseline program) within the same Midwestern swine production system that was not using the BEAR-PG600 program. The farm on the BEAR-PG600 program used an offsite gilt development unit where gilts that failed to exhibit estrus after at least 23 days of boar exposure and heat detection were treated with PG600. All BEAR-PG600 program gilts had a recorded estrus prior to entry into the breeding herd. The Gilts entered in the Baseline program farm were not subjected to a BEAR-PG600 program and had no prior recorded estrus.

Materials and methods

Reproductive data was retrieved from the production system's computerized record program (Porcitec, Agritec Software, Barcelona, Spain) and provided to the investigators in Excel (Microsoft, Redmond, Washington) format. Further analysis of the data was performed in Excel. An analysis of production outcomes was performed to compare the production outcomes of the farm on the BEAR-PG600 program to the farm on the Baseline program for the same period of time on both farms. The lack of replication precluded statistical analysis of the productivity differences between the farms. The economic analysis was performed using the "Economic Calculator" from ReproPig (Merck Animal Health, De Soto, Kansas) that was modified to more specifically address the direct intervention costs of the BEAR-PG600 program. The economic calculator is a breed-to-wean enterprise budgeting model that includes all costs and revenues related to the swine breeding herd. The increased costs of the BEAR-PG600 program compared to the Baseline program were due to facility modifications, deprecation and operation, labor and management for boar exposure and PG600 administration. The economic analysis addressed differences in gilt retention and reproductive performance but did not include the value of differences in sow longevity.

Results and conclusions

With the BEAR-PG600 program, 30.6% of the gilts in the GDU were injected with PG600 and 80% of the gilts that entered the GDU were transferred to the sow farm. Summary reproductive data and the economic analysis comparing the BEAR-PG600 and Baseline programs are presented in Table 1. Compared to the Baseline program, the BEAR-PG600 program improved the retention rate of gilts following entry into the sow farm, reduced the number of days from gilt entry to first service, reduced the number of days from gilt entry to removal, increased the first service farrowing rate and increased the percentage of gilts that remained in the herd for at least three parities. The economic analysis indicated that the BEAR-PG600 program provided a benefit:cost ratio of 5.26. This study revealed both the biological and economic benefits of the BEAR-PG600 program in this Midwestern swine production system. The economic benefit of the BEAR-PG600 program depends on the price of a weaned pig and the overall intervention costs. However, a sensitivity analysis indicated that even in a worst case scenario where the weaned pig price was reduced by 22% and intervention costs were increased by 42%, the benefit:cost ratio was 2.84.

Acknowledgement

Funding provided, in part, by the National Pork Checkoff (Project 14-149) and PIC. Economic analysis completed with support from Merck Animal Health.

References

1. Paterson, J., Triemert, E., Gustafson, B., Werner, T., Holden, N., Pinilla J.C., Foxcroft, G. 2016. Validation of the use of exogenous gonadotropins (PG600) to increase the efficiency of gilt development programs without affecting lifetime productivity in the breeding herd. J. Anim. Sci. 94:805-815.

Table 1: Summary of production record data and economic analysis

Parameter		Baseline	BEAR-PG600 Program
Gilt retention rate (% of gilts entered into sow farm)		94.5%	96.6%
Average gilt entry-to-first service interval (days / female)		35.5	23.8
Average gilt entry-to-removal interval (no service) (days / female)		83.2	61.8
Farrowing rate of gilts - first service only		86.3%	93.0%
Removal rate by parity	P0	9.20%	3.50%
	P1	8.30%	6.00%
	P2	10.40%	8.00%
	P3+	72.10%	82.50%
Net profit per breeding female ($/female/year, no intervention costs)		$53.99	$83.62
Net profit improvement ($/female/year)			$29.62
Intervention cost ($/female/year)			$5.63
Benefit cost ratio			5.26
Return on investment (ROI)			426%

The bactericidal action of gamithromycin against *P. multocida*, *A. pleuropneumoniae*, *H. parasuis* and *B. bronchiseptica*

Olivier Merdy[1], Agronomist, A. Richard-Mazet[1], DVM, PhD, Rose Huang[2], PhD;
Paul Dorr[3], DVM, PhD, MACE; François Joisel[1], DVM, MSc, Univ. Doc.
[1]Merial SAS, Lyon, France; [2]Merial Inc., North Brunswick, New Jersey;[3]Merial Inc., Fulton, Missouri

Introduction

Azalides are similar to macrolides in that both inhibit protein synthesis by reversibly binding to 50S subunits of the ribosome and are thus generally assumed to be bacteriostatic based on this mechanism of action. Gamithromycin, the active pharmaceutical ingredient of ZACTRAN® (Merial, Lyon, France), is a novel azalide antimicrobial which has been reported to exhibit a strong and potent bactericidal action against *P. multocida* (Pm), *A. pleuropneumoniae* (App), *H. parasuis* (Hps) and *B. bronchiseptica* (Bb).[1] The purpose of this study was to investigate *in vitro* kill kinetic of gamithromycin against these gram-negative bacteria involved in swine respiratory disease.

Material and methods

The *in vitro* gamithromycin kill kinetics against 3 strains of each of species were determined. The 12 isolates were selected from a European collection tested in a previous study.[2] Under standard CLSI conditions, gamithromycin minimum inhibitory concentration (MIC) for these isolates corresponded to the published MIC_{50} of the species *i.e.* 0.5, 4, 0.25 and 1 µg/ml for Pm, App, Hps and Bb respectively.[1] Kill kinetics were assessed at 2, 6 and 10 µg/ml gamithromycin, i.e. concentrations that are achieved in target tissues in pigs after one IM administration at 6mg/kg.[2] Broths with no gamithromycin were included as growth controls. At each time point, each culture was sampled and plated in triplicate for further enumeration. Antibacterial substances are considered to be bactericidal if there is a decrease of at least 3 log (> 99.9% kill), which is consistent over time, with no return of the bacterial count at later time points within the 3 log cfu/mL kill area.

Results

The bacterial count in the starting *inocula* ranged between 4.76 and 6.23 $\log_{10}$ CFU/mL.

P. multocida: all three isolates demonstrated a > 3-log reduction at 24 hours, whatever the gamithromycin dose. The 3-log reduction was achieved between 6 and 24 hours. For two isolates, a reduction in colony count was observed as early at 1 hour whatever the dose tested. For the third isolate, killing was observed starting from 6 hours at 2 µg/mL and earlier at higher doses. A definite enhancement of the killing rate was observed with increasing gamithromycin dose.

A. pleuropneumoniae: all three isolates demonstrated a > 3-log reduction at 24 hours for gamithromycin at 6 and 10 µg/mL. A > 3-log reduction as early as at 6 hours was also observed in tests for 2 out of the 3 isolates at a gamithromycin concentration of 6 µg/mL and for 1 out of the 3 isolates at 10 µg/mL.

H. parasuis: all three isolates demonstrated a > 3-log reduction at 24 hours whatever the gamithromycin dose. For one particular isolate, a > 3-log reduction was also observed as early as at 1 hour at 10 µg/mL and at 3 hours whatever the tested gamithromycin concentration. A reduction of the bacterial count for the 3 tested isolates was also observed as early as at 1 hour whatever the gamithromycin dose. A definite enhancement of the killing rate was observed with increasing gamithromycin dose.

B. bronchiseptica: all three isolates demonstrated a > 3-log reduction at 24 hours whatever the gamithromycin dose. A reduction of the bacterial count was also observed for the 3 tested isolates as early as at 1 to 3 whatever the gamithromycin dose.

Discussion and conclusion

This study confirmed a definite bactericidal effect of gamithromycin against isolates of Pm, App, Hps and Bb. This effect was observed at drug concentrations sustained in the target tissues for at least 24 hours.[2] The minimal gamithromycin concentrations showing a bactericidal effect were close to the MIC values for the tested isolates. Under the conditions of the study, this effect could be observed as early as 6 hours after the exposure of bacteria to gamithromycin for the Pm and App isolates depending on the gamithromycin level tested. The bactericidal activity against Pm and Hps appeared to increase with increasing concentration above MIC as well, indicating a concentration-dependant bactericidal profile against these species.

® *ZACTRAN is a registered trademark of MERIAL Limited.* ZACTRAN is indicated in the U.S. for the treatment of bovine respiratory disease (BRD) associated with *Mannheimia haemolytica*, *Pasteurella multocida*, *Histophilus somni* and *Mycoplasma bovis* in beef and non-lactating dairy cattle. ZACTRAN is also indicated for the control of respiratory disease in beef and non-lactating dairy cattle at high risk of developing BRD associated with *Mannheimia haemolytica* and *Pasteurella multocida*.

References

1. Richard-Mazet, A., Hayward, J., Siegwart, E., Pfefferkorn, A., Dumont, P., Minimum inhibitory and minimum bactericidal concentrations of gamithromycin against European swine respiratory bacterial field isolates. *Proc. 24th IPVS congress/8th ESPHM*, 192.

2. Huang, R., Hamel, D., Tessman, R., Mullins, A., Malinski T., Rehbein, S., Letendre, L., Pharmacokinetics of gamithromycin in swine plasma and bronchoalveolar cells. *Proc. 24th IPVS congress/8th ESPHM*, 193.

Comparative study to evaluate the performance in grow-finish pigs treated with Aivlosin® 17% or Denagard® 10 following an experimental challenge with *Lawsonia intracellularis*

R. Kaptur[1], DVM; D. Rosener[1], DVM; A. Mueller[2], DVM; N. Winkelman[2], DVM; R. Domangue[3], PhD
[1]Pharmgate Animal Health, Wilmington, North Carolina; [2]Swine Services Unlimited, Inc., Rice, Minnesota;
[3]Statistical Consulting Services, Broadway, Virginia

Introduction

Porcine proliferative enteropathy (PPE, or ileitis) is an important enteric disease of pigs caused by the obligate intracellular bacterium *Lawsonia intracellularis* (*Li*). Reports indicate that 96% of US swine are sero-positive for *Li* with 28% having clinical disease.[1] The chronic form of PPE typically occurs in growing pigs at 6 to 20 weeks of age, causing reduced rates of weight gain and increased variability in body size. It has been estimated that ileitis can cost between $2.73 and $19.76 per pig annually.[2] AIVLOSIN® 17% Medicated Article, from Pharmgate Animal Health, is a new formulation of tylvalosin, a novel macrolide antibiotic with potent activity against *Li*.[3] AIVLOSIN® 17% is a granulated premix containing 17% tylvalosin (TVN) and is indicated for control of PPE in groups of swine in buildings experiencing an outbreak of this disease. The goal of this study was to compare the treatment benefits of Aivlosin® 17% and Denagard® 10 (labelled for the control of PPE) following a challenge with *Li*.

Materials and methods

Enrolled in the study were 216 commercial barrows and gilts averaging 9 weeks of age on arrival. They were sourced from a farm that was PPE, PEDV, PRRS, and APP negative, but was positive for *Mycoplasma hyopneumoniae*, *Mycoplasma hyosynoviae*, and *Mycoplasma hyorhinis*. Pigs were randomly assigned to pens based on gender and body weight (9 pigs per pen) to either: 4 pens non-medicated control, 10 pens AIVLOSIN® 17% (fed according to label); and 10 pens Denagard® 10 (fed according to label). Following acclimation, all pigs were orally challenged with an intestinal mucosal homogenate containing 1.6×10^9 *Li* organisms on study day -11. Treatments were initiated when at least 10% of all pigs were observed to have fecal scores greater than or equal to 1. The primary variables assessed were fecal shedding of *Li* (qPCR) and average daily gain (ADG). Three (3) pigs were tagged in each pen for fecal sampling. Pigs were maintained through to market.

Results and discussion

Following challenge and before treatment, all of the pigs tested in each treatment group were qPCR positive for *Li* with an average ct count of 19.5and 21.0 for Denagard® and Aivlosin®, respectively (Table 1). Following treatment, 93% of pigs in the Denagard' group and 83% in the Aivlosin® group tested positive with an average ct count of 26.5 and 29, respectively. Pigs treated with Aivlosin® 17% performed better versus pigs treated with Denagard® 10. During the treatment period Aivlosin® treated pigs had an ADG of 1.85 lbs/pig/day and Denagard® treated pigs had an ADG of 1.53 lbs/pig/day (Table 1). When pigs were followed to market, this resulted in an overall ADG of 1.84lb/pig/day for Aivosin® treated pigs bringing in $4/pig more value than Denagard® treated pigs (1.74lb/pig/day). Total value obtained for pigs was provided by the slaughter plant.

References

1. National Animal Health Monitoring Service (NAHMS)
2. Veenhuizen, M.F., Elam, T.E. and Soenksen N. The potential economic impact of porcine proliferative enteropathy on the US swine industry. IPVS. 1998; 64.
3. Rosener et al. The effectiveness of tylvalosin (Aivlosin® Type A Medicated Article) against porcine proliferative enteropathy (PPE; ileitis) in swine challenged with Lawsonia intracellularis. AASV. 2014. 283-284.

Table 1: Summary of variables

Treatment group	Aivlosin® 17%	Denagard® 10
Fecal shedding of *Lawsonia intracellularis* (PCR) at Day 0 pre-treatment	100% pigs positive (avg ct 21.0)	100% pigs positive (avg ct 19.5)
Fecal shedding of *Lawsonia intracellularis* (PCR) at Day 14 post-treatment	83% pigs positive (avg ct 29.0)	93% pigs positive (avg ct 26.5)
Pen average daily gain (ADG) Study Day 0-14	1.85[a] lbs/pig/day	1.53[b] lbs/pig/day
Slaughter plant sheets	$10,092.59 total value/85 market pigs = $118.73/pig	$9613.46 total value/84 market pigs = $114.45/pig

[ab] Rows with different letters are significantly different at $P < 0.05$.

Effect of Tonisity Px™ administration on intestinal morphology

Ava M. Firth[1], BS, DVM, MVS, MANZCVS, DACVECC, DECVECC; Guillermo Lopez Cano[2], DVM MSc; Alberto Morillo Alujas[2], BS DVM MS PhD MBA
[1]Tonisity International, Dublin, Ireland; [2]Tests and Trials S.A., Monzón, Huesca, Spain

Introduction

Intestinal morphology measurements, particularly increased villus height, are closely correlated with post-weaning weight gain.[1,2] The protein and amino acid profile of Px delivers key energy-producing substrates to the enterocyte which could in turn have an impact upon intestinal morphology. The aim of this study was to assess the impact of Px on intestinal morphology.

Materials and methods

This study was approved by the ethics committee of the University of Lleida, Spain. 52 sows and their litters (608 piglets) were enrolled in the study and randomized to one of two groups, control (group A) or supplementation with 3% Px solution (group B). From days 2-8 of age, group B litters were given 500 mL of 3% Px solution daily in an open pan. Pigs were individually weighed at 1, 9 and 19 days of age. At 9 and 19 days of age, 18 pigs from each group were euthanized for gut morphological histopathology. Selection of the piglets was done based on their ADG from initial weight to slaughtering time: SD1 to SD8 or SD1 to SD20. Within litters, quartiles (25%, 50% and 75%) of the ADG were used to select three piglets per litter, from six litters (three per treatment) at SD8 and another three piglets per litter from the other six litters (three per treatment) at SD20. Remaining pigs from the litters were then split into 3 treatment groups and given either dry creep feed (D), gruel made with creep feed and water (WG) or gruel made with 3% Px solution and creep feed (PG). A full description of that trial is published elsewhere in these proceedings. From those pigs, 6 pigs per treatment group were euthanized at 5 days after weaning (~24 days of age) and a further 6 pigs per treatment group were euthanized 9 days after weaning (~28 days of age). Selection of those pigs was based on quartiles of the BW at weaning.

Intestinal morphology measurements were performed as published by Berkeveld et al.[3] in which sections were cut from 10%, 50% and 90% along the length of the small intestine. Villus height, crypt depth, villus and crypt density, and intestinal mucus thickness were measured. The average measurements per slide were used as experimental observation.

Statistical analysis

The experimental unit for morphologic assay was the piglet. Gut morphology variables were analysing by a generalized linear model with the package GLM including the group, the intestine section and the study day as fixed effects. Interactions were explored but they were not significant so they were not included in the model. Tests were two-tailed and carried out with a risk $\alpha = 5$ %. P-values of $P \leq 0.05$ were considered statistically significant, while $0.05 < P \leq 0.10$ was considered a near-significant trend. All statistical analyses were performed using R software.†

Results

Histopathology analysis showed that pigs which received Px from day 2-8 of age had significantly greater ($P < 0.05$) villus height, villus density, and crypt density that persisted until at least 28 days of age, regardless of what creep feed they were given. Px pigs also tended to have greater intestinal mucus thickness ($P = 0.08$) (Table 1).

Discussion

The results of this study were consistent with previously published studies, in which villi height was reduced by 50% at 5 days after weaning and the villus height/crypt depth ratio was lower in weaned piglets than in unweaned piglets. 1,4,5 However, in this study, Tonisity Px, when given in the first week of life, had a significant and favourable impact upon villus height, villus height/crypt depth ratio and villus density that persisted until at least 9 days after weaning. This effect was seen across all creep feed groups and ages. This has favourable implications for both pre-weaning and post-weaning weight gain.

The Px pigs also had a significantly thicker intestinal mucus layer post-weaning. This has clinical relevance because the intestinal mucus layer has been shown to be critical to maintenance of normal gut permeability and restoration of gut barrier function in inflammation and ischaemic processes.6,7 Further studies in pigs are indicated to examine the relationship between Px, intestinal mucus thickness and the development of post-weaning diarrhea in pigs.

† R Core Team (2015). R: A language and environment for statistical computing. R Foundation for Statistical Computing, Vienna, Austria. ISBN 3-900051-07-0, URL http://www.R-project.org.

Table 1: Morphologic measurements of small intestinal sections.

	Villus height (µm)	Crypt depth (µm)	Villus height/ crypt depth ratio	Intestinal mucus thickness (µm)	Villus density (No./mm)	Crypt density (No./mm)
			Day 9 and 19 measurements combined			
Control	409 ± 16.5	139 ± 5.1	3.6 ± 0.16	536 ± 19.8	9.3 ± 0.22	20.6 ± 0.55
Px	443 ± 16.5	148 ± 5.1	3.6 ± 0.16	584 ± 19.8	8.7 ± 0.22	18.3 ± 0.55
P-value	**< 0.001**	0.199	0.962	0.087	**0.037**	**0.003**
			Day 24 and 28 measurements combined			
Control	249 ± 9.3	220 ± 6.4	1.2 ± 0.05	435 ± 14.2	8.0 ± 0.15	20.5 ± 0.71
Px	291 ± 10.4	226 ± 7.1	1.4 ± 0.06	481 ± 15.9	7.6 ± 0.17	21.8 ± 0.79
P-value	**0.003**	0.512	**0.018**	**0.033**	0.064	0.211

References

1. Al Masri, S., H. Hünigen, A. Al Aiyan, J. Rieger, J. Zentek, et al. 2015. Influence of age at weaning and feeding regimes on the postnatal morphology of the porcine small intestine. J Swine Health Prod. 23:186-203.

2. Tsukahara, T., E. Kishino, R. Inoue, N. Nakanishi, K. Nakayama, et al. 2013. Correlation between villous height and the disaccharidase activity in the small intestine of piglets from nursing to growing. Anim. Sci. J. 84:54-9.

3. Berkeveld, M., P. Langendijk, N.M. Soede, B. Kemp, M.A.M. Taverne, et al. 2009. Improving adaptation to weaning: effect of intermittent suckling regimens on piglet feed intake, growth, and gut characteristics. J. Anim. Sci. 87:3156-66.

4. Pluske, J.R., D.J. Hampson, I.H. Williams. 1997. Factors influencing the structure and function of the small intestine in the weaned pig: a review. Livest. Prod. Sci. 51:215-236.

5. Hampson, D.J. 1986 Alterations in piglet small intestinal structure at weaning. Res. Vet. Sci. 40:32-40.

6. Qin, X., S.U. Sheth, S.M. Sharpe, W. Dong, Q. Li, et al. 2011. The mucus layer is critical in protecting against ischemia-reperfusion-mediated gut injury and in the restitution of gut barrier function. Shock. 35:275-281.

7. Fishman, J.E., G. Levy, V. Alli, X. Zheng, D.J.Mole, E.A. Deitch. 2014. The intestinal mucus layer is a critical component of the gut barrier that is damaged during acute pancreatitis. Shock. 42:264-70.

Pig Atlas - enhancing phenotypic detail to increase genetic gain in pig breeding

L.E. Gangsei[1], PhD; J. Kongsro[2], PhD; E. Grindflek[2], PhD; J.M. Eggert[3], MS, MBA, PhD, Dipl. ACAG
[1]Animalia, Oslo, Norway; [2]Norsvin SA, Hamar, Norway; [3]Topigs Norsvin USA, Burnsville, Minnesota

The Pig Atlas

Topigs Norsvin has CT-scanned (Computed Tomography) approximately 3500 boars, annually, since 2008 to more accurately select boars with the most desirable characteristics. The information from CT provides valuable information about body composition (phenotype), which is used in the calculation of breeding values. Currently, Topigs Norsvin is implementing a virtual Pig Atlas[1,2] in an effort to effectively utilize even more information from the CT-scanned boars.

The Pig Atlas method offers an opportunity for "digital" segmentation of the CT-scanned pigs (Figure 1). In principle, there is no limit for the detail level in the segmentation. However, the precision of the segmentation depends on a non-rigid transformation based on approximately 2000 anatomical landmarks.

Applications

Currently, the Pig Atlas method is being applied to primal cuts (prediction of weight and fat/meat/bone content) and major organs. The goal is an automated method that quantifies a range of phenotypes not currently utilized in the calculation of the breeding index.

For some health factors, like osteochondrosis,[3] diagnostics are already being conducted for pigs via CT scanning. As in human medicine,[4] CT scanning of pigs can provide automated diagnoses of multiple diseases as well as quantification of skeletal, muscular and internal organ phenotypes. The Pig Atlas method will continue develop new, relevant phenotypes that were not previously possible to measure accurately in a live animal.

The phenomic gap

To take full advantage of the large amounts of genomic data now available (80K SNPs per boar), large volumes of accurate phenotypic data are also necessary; this practical problem for the utilization of genetic data, also evident in pig breeding, is known as the "phenomic gap".[5] As the detailed phenotypes from the Pig Atlas become available, Topigs Norsvin will develop a large data set that includes both known genotypes and detailed phenotypes. Together, these data will provide the foundation for accurately identifying the relationships between the genotypes and the phenotypes, which will increase the accuracy of breeding values, accelerate genetic progress, and increase the probability of identifying functional mutations.

References

1. Gangsei, L.E., et al. Building an in vivo anatomical atlas to close the phenomic gap in animal breeding. Computers and Electronics in Agriculture, 127: 739–743 (2016).
2. Gangsei, L.E., et al. Automatic segmentation of Computed Tomography (CT) images of domestic pig skeleton using a 3D expansion of Dijkstra's algorithm. Computers and Electronics in Agriculture, 121: 191–194 (2016).
3. Aasmundstad, T., et al. Osteochondrosis in pigs diagnosed with computed tomography: heritabilities and genetic correlations to weight gain in specific age intervals. Animal: An International Journal of Animal Bioscience, 7(10): 1576–1582. (2013).
4. Sotiras, A., et al. Deformable Medical Image Registration: A Survey. IEEE transactions on medical imaging, 32, 1153–1190 (2013).
5. Houle, D., et al. Phenomics: the next challenge. Nature reviews. Genetics, 11: 855–66 (2010).

Figure 1: One «slice» of a random CT-scanned boar in a sagittal view. The lines and text indicate the automatic Pig Atlas segmentation of this slice in 'head', 'shoulder', 'loin', 'belly', 'ham', 'viscera' and 'testicles'.

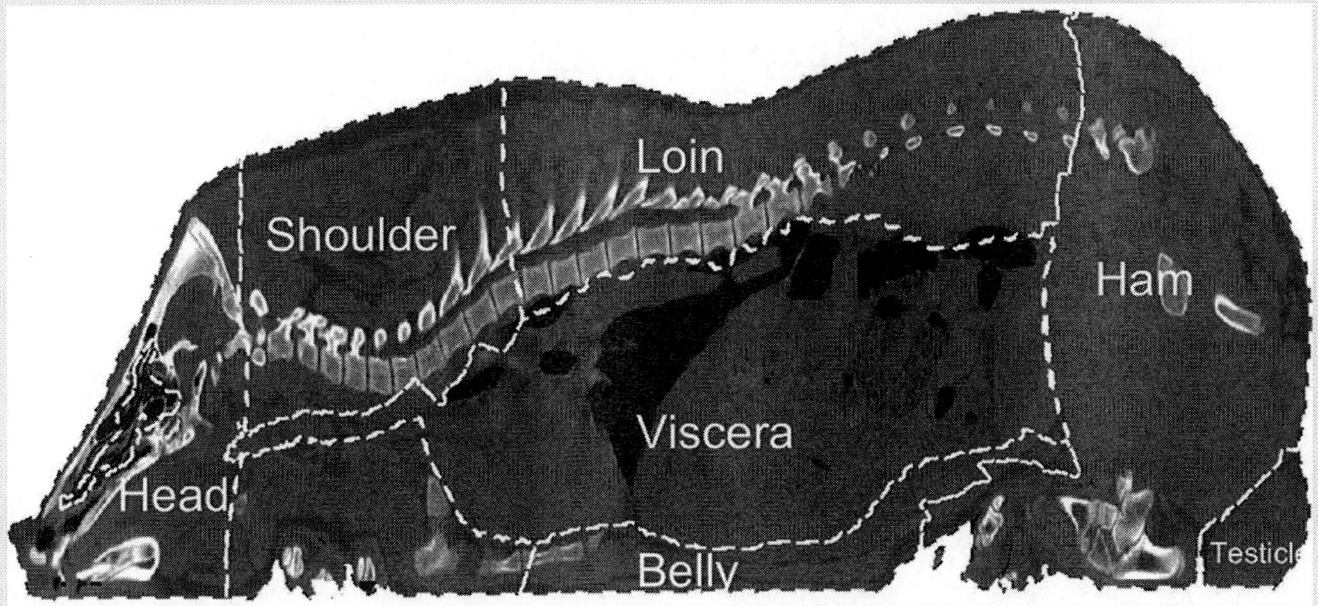

Disinfection in freezing temperatures: An experimental approach to identify best practices for applications in the animal health industry

F. Ahmadpour; S. van den Berg; G. Perez-Hernandez; L. Pantaleon, DVM, MS, DACVIM, MBA
Virox Animal Health, Oakville, Ontario, Canada

Introduction

Proper surface disinfection practices and environmental biosecurity parameters are necessary in order to ensure the health and welfare of livestock. During seasons with sub-zero temperatures, the practice of surface disinfection using aqueous disinfectants can become problematic in areas where ambient temperatures are not regulated. Inherently lower speed of antimicrobial efficacy for all chemical disinfectants in cold temperatures occurs, thus requiring a longer contact time. Furthermore aqueous disinfectants are likely to freeze when applied onto surfaces under sub-zero temperature conditions. These two factors described above, often result in lack of proper surface disinfection and therefore increase the likelihood of infectious diseases transmission among livestock. The objective of this study was to investigate the effect of using propylene glycol and common windshield washer fluids as diluents on the freezing point of different concentrate disinfectants.

Materials and methods

Seven disinfectant concentrate products were tested (Biosentry-905, Clidox-S, Farmphene, Prevail/Intervention Concentrate, Synergize, Thymox and Virkon). These were diluted with a propylene glycol (PG) solution (concentrations ranging from 0% to 30%) and two windshield washer (WW) fluids rated at -35°C/-31°F and -45°C/-49°F, according to their labels claims. Control experiments were performed with deionized (DI) water and the corresponding diluents.

Vials of diluted concentrates were cooled in a recirculating chiller set to -18°C/-0.4°F, and the freezing point was measured using a NIST-calibrated Digi-Sense thermocouple thermometer under agitation. The control experiments using were done under agitation and as a stagnant mixture. Overall change in pH was also measured.

Results and discussion

For the control experiments, freezing point decreased linearly with increasing PG concentration ($R^2 = 0.9475$ for agitated, $R^2 = 0.9424$ for non-agitated). Non-agitated solutions had freezing points about 8 degrees lower on average than the agitated solutions; nucleation sites from agitation or scratching the vial with the thermocouple were the likely cause of the difference in freezing temperatures, whereas the un-agitated mixtures were cooled further before freezing. This would not be seen with tap water, as there are impurities to act as nucleation sites. Different disinfectant compositions had a small effect on the exact freezing point at low PG concentrations (3.9°C/39°F), which became less noticeable as the PG concentration was increased (2.2°C/36°F). Only Clidox-S was an outlier at higher PG concentrations, leading to a variance of about 1°C (Figure 1).

When -45°C/-49°F WW fluid was used as the diluent, it resulted in significant upward shifts in the final pH of the diluted disinfectants. This can have a negative effect on microbial efficacy, as some disinfectants are more efficacious at lower pH. Diluting with either WW fluid gave a similar freezing temperature than when diluted with a 20% PG solution. Overall, dilution with a solution of 30% PG was required to ensure disinfectants remained in the liquid state below -10°C/14°F (Figure 1).

Liquid disinfectants used at sub-zero temperatures will require an antifreeze additive. PG is a practical antifreeze option and concentrations required for various sub-zero applications of disinfectants were determined in this study. Use of WW fluid has limited antifreeze function and caused a pH shift of the disinfectant, which in turn leads to changes in the chemistry and efficacy of disinfectants.

Figure 1: Freezing point of diluted concentrates vs. % propylene glycol as diluent.

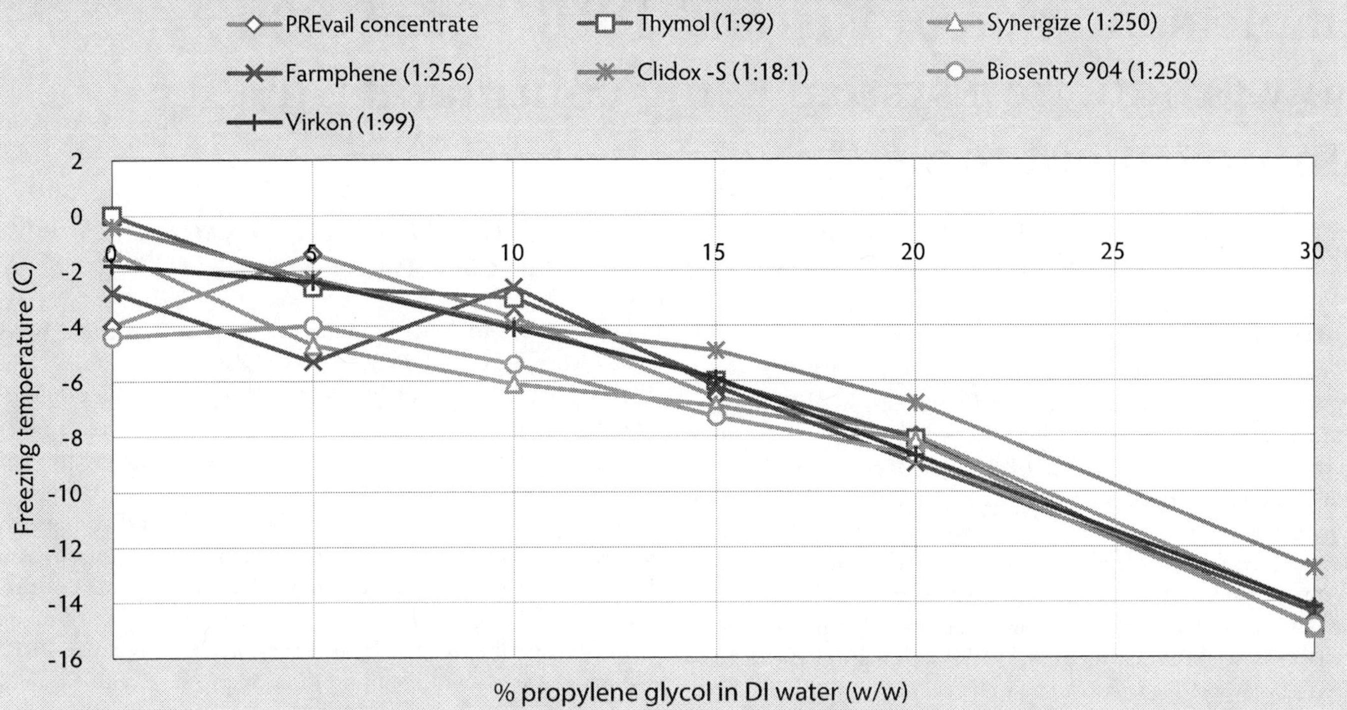

Evaluation of reproductive performance after implementation of Fostera® PRRS in a 9,000 sow production system using continuous quality improvement tools

Jose Angulo[1], DVM; Abigail Redalen[1], DVM; Larry Moore[2], DVM; Brent Pepin[2], DVM, MS
[1]Zoetis, Parsippany, New Jersey; [2]Postville Veterinary Clinic, Postville Iowa

Introduction

Porcine reproductive and respiratory syndrome (PRRS) causes significant negative financial impact in breeding herds. An acute outbreak can be very dramatic and costly having an estimated cost of $255 per sow and once PRRS virus is circulating in sows population as an endemic pathogen has an ongoing diminished reproductive performance.[1] The estimated cost of PRRS in a sow farm chronically infected is $74.16 per litter, this cost is calculated in loss of performance in reproductive parameters like farrowing rate, liveborn/litter and pre-wean mortality using margin-over-variable-cost (MOVC) method.[1] PRRS vaccination with Modified Live Vaccines is considered a preventive intervention to minimize PRRS-associated losses when pig populations become infected with Wild-Type PRRS virus (WT-PRRSv) and it helps to build PRRS herd immunity to achieve stability sooner and minimize production losses compared to farms without previous PRRS herd immunity.[2] Quality and process improvement is defined as a reduction of variability in processes and it uses statistical techniques like Statistical Process Control (SPC) and Capability Analysis.[3] A Control Chart is an enhanced time series plot designed to identify special patterns of variability in data called signals, and process capability is a statistical measure of inherent variation for a given characteristic in a stable process, in other words, it's the measure of the ability of a process to produce outputs that meet the specifications.[4] The objective of this project was to use quality improvement tools like Statistical Process Control charts and Capability Analysis as practical and reliable tools to evaluate PRRS control interventions implemented in a 9,000 sow production system.

Material and methods

The production system where this project was implemented has four breed-to-wean farms located in northeast Iowa. These farms utilized different strategies for controlling PRRS in the past, including a killed vaccine without clear success on controlling PRRS. A systematic approach based on Lean Six Sigma philosophy applying the DMAIC (Define-Measure-Analyze-Improve-Control) methodology was proposed to this production system working with whole herd immunity using Fostera® PRRS as the primary immunological tool. The project team defined the following goals for this project: Develop a PRRSv control strategy to maintain breeding herd stability; mitigate any potential Wild-Type PRRS virus (WT-PRRSv) re-circulation and/or new WT-PRRSv introduction as compared to baseline production; and establish standardized processes for measuring success in applied PRRS control interventions. After these goals were established, primary production interventions were discussed and agreed for managing whole breeding herd immunity and to improve Key Performance Indicators (KPIs) through the use of Fostera® PRRS. KPIs were compared to the previous 18 months of production data using Statistical Process Control (SPC) technique. A one-year period of evaluation is presented in this paper. PRRS stability was measured sampling 15 or 30 piglets around weaning age every other month at each sow farm. KPIs were evaluated using SPC I-Charts for each sow farm and X-bar-Charts for whole system evaluation (four sow farms together), and included the following parameters: Abortion rate per thousand sows; farrowing rate; average liveborn; pre-wean mortality and pigs weaned per sow weaned. In addition to SPC evaluation, a 2 sample t-Test was performed to detect statistical differences before and after in all KPIs. Additionally, a capability analysis was used to determine and quantify process improvement as compared to targets was implemented in weaned pigs per weaned sow (MINTAB® 17.1.0).

The primary interventions consisted in whole herd vaccination of the breeding herds with Fostera® PRRS on a quarterly basis and two doses of Fostera® PRRS in replacement gilts.

Results and discussion

Breeding herd stability was achieved in all four sow farms having PRRS PCR negative results in suckling pigs. Statistical Process Control (SPC) charts demonstrated an improvement in all reproductive Key Performance Indicators measured as compared to the before period. In a whole system analysis, statistical significant improvements (P value ≤ 0.05) were observed in abortion rate per thousand sows; average liveborn; pre-wean mortality and weaned pigs per weaned sow (Table 1). Capability analysis showed a reduction of 81% in out of specifications in pigs weaned per weaned sow, the production system went from having 41.89% out of system specs (# weaned pigs per sow: 9.5 to 12 having as Target 10.5) to only 7.85% (Table 1), this can be interpreted as consistently more piglets out of the door.

Table 1: Key Performance Indicators before and after PRRS control interventions and capability comparison for Weaned Pigs/Weaned Sow. (MINITAB® 17.1.0)

KPI	Period		*P* value
	Before avg. (StDev)	After avg. (StDev)	
Abortion rate/1,000 sows	2.322 (9.431)	0.746 (0.698)	**0.002**
Farrowing rate	81.191 (10.88)	82.156 (7.00)	0.216
Avg. live born	11.780 (1.717)	11.961 (0.546)	**0.001**
PWM%	11.677 (8.795)	7.396 (2.014)	**0.001**
Pigs weaned/sow weaned	10.236	10.881	**0.001**
Capability analysis:			
StDev	1.452	0.699	**0.001**
Pp index	0.29	0.60	
% out of spec (lower: 9.5, Target: 10.5, upper: 12)	**41.84**	**7.89**	

Conclusions

This year-round evaluation shows a significant increase of 0.645 weaned pigs per weaned sow after the implementation of PRRS control strategies (*P* value 0.001) along with a significant reduction of variability (*P* value 0.001). Outcomes of this project demonstrate the benefits of working in a systematic approach for controlling PRRS in breeding herds having Fostera® PRRS as primary tool for herd immunity. Also, it shows the value and practicality of using quality tools for measuring interventions for PRRS control in swine production systems helping practitioners lead data-driven decisions on health interventions to achieve long-term success for PRRS control.

References

1. Holck J.T., Polson D.D., The financial impact of PRRS virus. PRRS compendium producer edition. 2003. Chapter Six.
2. Linhares C.L., Johnson C., Morrison R.B. Economic analysis of vaccination strategies for PRRS control. 2015. PloS One 10(12).
3. Montgomery D.C., Introduction to Statistical Quality Control. Arizona State University. Seventh Edition. John Wiley & Sons Inc., Hoboken NJ. p. 7, 17.
4. Brue G., Howes R. Six Sigma. 2006.McGraw-Hill. New York, NY. p. 50-51.

Canadian perspective on porcine epidemic diarrhea

Egan Brockhoff, DVM
Prairie Swine Health Services, Red Deer, Alberta, Canada

Introduction

The first time I recognized porcine epidemic diarrhea (PED) causing severe clinical disease and high mortality in neonatal pigs was nearly a decade ago. I was in the Philippines for the first time, touring with veterinary friends and colleagues. We were working with a number of farms on disease prevention and mitigation strategies for various endemic diseases throughout the region. I had anticipated the on-farm discussion to focus largely on porcine circovirus, *mycoplasma hyopneumoniae*, classical swine fever and porcine reproductive and respiratory syndrome virus, all diseases common to the region and those that we had spent much of our pre-visit briefings discussing. Something else emerged though. Despite the farm, the discussion would invariably, and often rather quickly, turn to the topic of PED virus. These farms wanted to talk about neonatal scours, pre-weaning mortality and if I could provide some insight and recommendations for practical control programs. Their descriptions of the aggressive nature of the virus was puzzling to me, they very much sounded like they were describing transmissible gastroenteritis virus (TGE) and not PED. A review of their diagnostics quickly showed that indeed, PED was the culprit in these cases and TGE simply wasn't involved. Walking through affected barns was eye opening, I was seeing severe clinical diarrhea and depending on their history you could see that some of these herds had little to no immunity whatsoever. I walked away from these first visits with my eyes wide open. This disease had changed!

When it was announced to the world that the first case of PED had been found in the United States I recall pulling over from my drive, reading the release again closer, and then calling the Chief Veterinary Officer and the head of the Canadian Swine Health Board (CSHB). I remember sharing with them my experiences from Asia and my great concern about what this could mean for North America. This wasn't the PED of the past, it wasn't entirely similar to TGE and it had the potential to spread. Word moved quickly, people started talking and actions began. I am reminded of the words of one of my first mentors, Dr John Harding, "Never before has the risk of transboundary and emerging diseases been so real. The "perfect storm" caused by swine industry intensification, globalization, societal and environmental encroachment is having an insidious impact on the global swine industry." It appeared like that "perfect storm" was upon our doorstep.

In the weeks and months following the announcement of PED in the United States we could all see the frustration and concern growing as the virus spread from farm to farm and state to state. Various reports created various responses. Is the virus moving through the air? How is recycled water playing a role? How handle exposure at high traffic locations? Canada had a national biosecurity plan, industry lead disease surveillance, established area regional control and elimination projects and a fairly well framed disease response plan. There was good cooperation and communication between producers, industry and government. Would it be enough?

As a response to the porcine circovirus epidemic Agriculture and Agri-Food Canada funded the development of the Canadian Swine Health Board in 2008 with the mission to provide leadership and coordination in support of the management of the health of the Canadian swine herd. The CSHB was an independent multi-stakeholder board representing a broad cross-section of the pork industry. Four pillars were identified to support the establishment of a structured disease response plan for the Canadian Pork Sector: Biosecurity, Research, Long Term Disease Risk Management and Sustainability. Moreover, the CSHB facilitated and cultivated interprovincial communication through national cooperative projects, networking opportunities and health forums. This enhanced communication and cooperation set a necessary stage. This stage, along with the actions of the Biosecurity Pillar and the Long Term Disease Risk Management Pillar are what highly informed and shaped the Canadian response to PED in the United States and subsequently in Canada when PED arrived nine months later. it was that legacy, and the work of a lot of dedicated people within the industry, that lands us where we are today. Standing on the brink of being PED free once again.

Discussion

The series of events that lead Canada to where we are today began many years before the January 2014 PED outbreak in the Province of Ontario. Along with industry, biosecurity awareness, training and auditing had been identified by federal and provincial governments as a key funding priority in protecting the health of the Canadian Herd. In the spring of 2010 we started seeing coordinated projects established all across Canada using the Production Animal Disease Risk Assessment Program tool (PADRAP) as a mechanism for addressing and informing biosecurity around PRRS virus, but also, biosecurity in general. These early assessments were extremely valuable in showing the many GAPS we faced and they quickly came to the notice of the CSHB. It wasn't long thereafter that the CSHB launched their first national biosecurity benchmark project using a British biosecurity tool for the pilot. This project, far more broad in it's scope and nature than PADRAP and PRRS, again continued to inform the discussion and identify GAPS in on-farm and industry biosecurity and long term disease risk prevention and management. It was from that moment on that things began to rapidly accelerate. The CSHB immediately began work on a number of key projects including the development of a national biosecurity standard, a national biosecurity training and audit program, a PRRS free certification program, area regional control and elimination (ARC & E) projects for PRRS, and a disease surveillance system later to be called the Canadian Swine Health Intelligence Network. By the spring of 2011 tremendous effort had been expended on all of these projects and the National Biosecurity Swine Farm Training Program was launched simultaneously across Provinces with the goal of reaching every single pork producer in Canada.

The National Biosecurity Swine Farm Training Program initially began with training the trainers. Veterinarians from across Canada were invited and mobilized to participate in these programs with their clients while the provincial pork boards began organizing producers and producer/industry training events. Every pork producer in Canada was paid to participate in three events; A producer training event and bio-security self assessment, a veterinary audit of their site to identify GAPS and then a follow up veterinary audit to verify the noted GAPS had been addressed. Biosecurity became the core priority for the next two years as the program rolled out from coast to coast. The level of participation and action was inspiring. Producers, industry, government and veterinarians embraced this new found culture. You couldn't find a farm without an established and enforced Danish entrance! I recall one producer noting to me that it was nearly impossible to get in touch with their veterinarian, they were so busy doing biosecurity training and audit sessions. Transports were being washed and disinfected. Transporters were using boot covers and following Danish entry protocols to load and unload. The change was remarkable and compliance was good, not perfect, but good (Figure 1).

In the face of all of these activities two key seeds took hold that would ultimately shape everything that followed, including the response to PED when in arrived in January 2014. Those two seeds were cooperation and communication. Competitive farms and veterinary practices came together with industry and government to work together on CSHB projects, establish mechanisms, build tools and most importantly, act on priorities. Foresight is often a game of luck, and timing can be everything, when PED broke in the United States, Canada already had an established mechanism in place that quickly ramped up inside a culture of cooperation and communication in its efforts to keep PED at bay.

For nine months following the break of PED in the United States that national, provincial and regional cooperation and communication kept PED out of the country along with a culture of biosecurity and a willing attitude. Online and telephone town hall meetings were being hosted weekly and monthly to educate and inform anyone who wanted to listen in. Hundreds

Figure 1: One of the key slides from the National Biosecurity Swine Farm Training Program. If producers didn't embrace this slide the rest of the presentation really didn't matter.

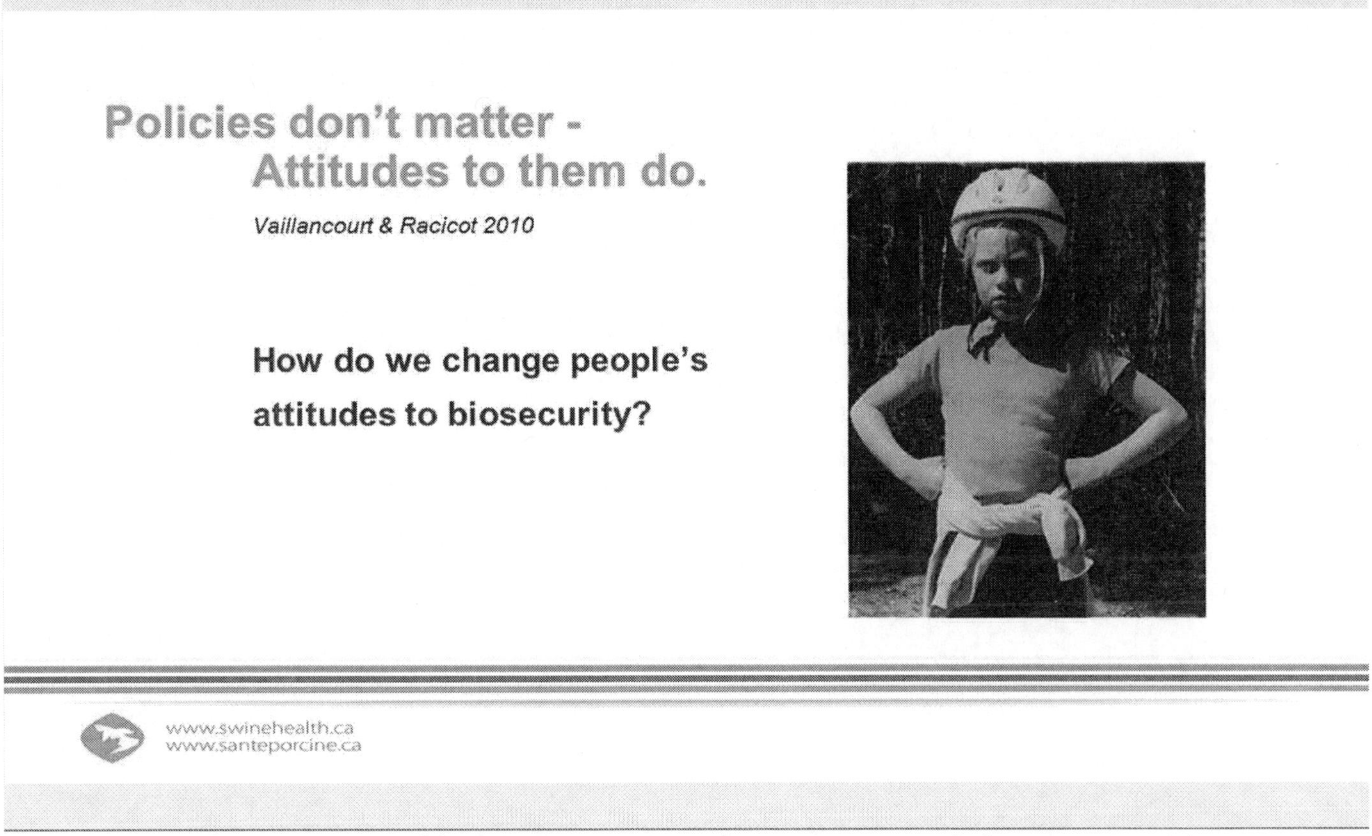

were participating in Provincial calls, thousands across the country. Producers, industry and government established weekly and monthly calls to keep the ball rolling on various biosecurity and surveillance initiatives. Each Province established weekly PED monitoring programs at high risk sites and response plans were established and practiced. Within weeks of PED arriving in Canada Provinces, under the recommendation of producers, veterinarians and industry placed it on their Notifiable disease list. Thus facilitating funding for diagnostics and monitoring. Huge effort was placed on transport biosecurity and transporter training. A new transporter certification program was launched along with truck wash certification and wash procedure certification. Compliance wasn't perfect but it was very good, the GAPS we recognized were either being filled or monitored. Feed truck drivers were wearing boot covers and respecting biosecurity, producers were disinfecting everything they could find to disinfect. Unfortunately, we missed one large GAP and that was feed and feed mill contamination.

In January of 2014 PED arrived in the Province of Ontario in central Canada. Within days over ten sites in Ontario with good biosecurity were found clinical for the disease. There was shock and surprise, some of these sites were considered to have impeccable biosecurity by their herd veterinarians. Initially, all eyes turned to transport biosecurity as the most likely culprit for letting the virus in. Canada exports millions of live pigs to the United States every year and returning transports had been identified as our single greatest risk of bringing PED back into Canada from the United States. Again, cooperation, communication and the hard work of a number of key people, including veterinary professionals, quickly showed that transport was not to blame for this outbreak. Within days all eyes were looking towards one feed mill, its transports and then quickly thereafter its feed and feedstuffs. Porcine plasma was identified as a potential risk factor and within weeks of the Ontario outbreak producers and feed mills across the country were pulling porcine plasma from their nursery diets. Unfortunately, the cat had been let out of the bag, PED virus had arrived.

The response to PED spreading in Canada was very similar to the response we saw prior to PED arriving. The focus on biosecurity remained strong and in many cases the increased awareness and recognition that the virus had arrived only served to inspire greater effort in prevention, control and then elimination. ARC & E mechanisms for PRRS had been well established and were well accepted by the pork industry. Veterinarians were experienced with control and elimination programs and there was a high level of confidence that PED ARC & E was attainable, not just on a local level but on a national level. The expectation that Canada could be PED free again had and has remained a realistic and attainable outcome. Without question the successful outcome of virus elimination has all come back to the evolution of those two key seeds, cooperation and communication in the face of a common goal.

Conclusion

As we move into to 2017, Canada stands on the doorstep of being PED free once again. Only a manageable number of farm premises across the country remain positive for the virus. Elimination programs have matured and have been successful. Monitoring programs remain in place as early detection and awareness tools and all stakeholders continue to work closely together including producers, industry and government. Cull sow and feeder pig assembly sites, the few that were positive, have established biocontainment protocols and are making gains in eliminating the virus. Efforts to mitigate the risk of transports spreading the virus continue to be well received and valued. For the most part, the never ending battle to mitigate complacency has succeeded. The work to push back on PED fatigue from setting in at all levels, has kept the control and elimination programs viable and the goal to eliminate the virus from Canada attainable. Regardless, the risk for further spread and infection remains real every single day of the year. Despite the many opportunities for the virus to move, the risk from positive transports and infected premises are not my greatest concerns. My greatest concern continues to be that attitudes will change, cooperation fails and communication breaks down. In 2014 all activities of the Canadian Swine Health Board ceased. A sustainable funding model could not be agreed upon and despite the recent introduction of PED into Canada it was dismantled. In its absence the vacuum has been filled by a lot of remarkable people doing remarkable things with few resources. Regardless of this tremendous effort, the very real risk that expectations of successful long term disease risk management outpace our current capabilities gives me pause.

Avian influenza: Lessons learned

Jill Nezworski, DVM
Blue House Veterinary, LLC, Hector, Minnesota

In early 2015 the Midwest US experienced an outbreak of HPAI. This outbreak started in turkeys and then spilling over into layers in MN, WI, SD, IA and NE. When it was all over approximately 40 million layers and associated pullets were depopulated due to the disease. After this traumatic event it is time to look back and assess what went well and what did not.

Several of the opportunities for improvement will focus on the front end of the disease. Some farms' delayed detection by only submitting samples due to a significant rise in mortality. Based on data from farms that were detected on surveillance this disease can simmer for a while in a caged layer facility. By the time mortality starts to increase the disease has reached critical mass and viral load increases exponentially. In some instances diagnosis was delayed due to long drive times and a lack of after-hours laboratory availability. MN tackled this issue by setting up couriers to deliver samples to the laboratories. These couriers provided excellent biosecurity information and managed the process well. Other states chose to recommend producers mail samples. This method lacked in both biosecurity and expediency.

In the midst of an on-farm outbreak some new opportunities came to light. First would be the overall lack of a farm specific catastrophe plan, especially revolving around depopulation and disposal. The outside crews used to perform these tasks added many extra people to a contaminated site. While farm staff seemed to only rarely challenge the clean/dirty line, outside contractors went back and forth with both people and equipment many times per day. This puts unnecessary strain on the decontamination process. This outbreak also occurred in an area of low human population density with caused comingling of depopulation contractors with normal farm employees at places like gas stations, restaurants, etc. Too many movements also kept occurring during the outbreak: propane, garbage, rendering, pullets. An immediate movement ban may be inconvenient but can provide value in a rapidly changing situation.

Now in the aftermath there still are producers out there that are not taking this serious. There is a perceived site immunity from having a truck wash, showers, audit, etc. Everyone need to focus on items the actually go in the barn. Pullet moving equipment, spent hen equipment, kill carts, CO2 tanks, employees, supplies. Implementation of a simple Danish entry is an effective way to reduce risk from employees and contractors. As much equipment as possible should be farm specific. Procedures need to be validated with tools such as GLOW GERM, salmonella testing of pullet/spent hen equipment and internal auditing.

Consumers, pigs, vets, and zoonoses: The critical role you play in earning trust with consumers

J.J. Jones
The Center for Food Integrity

It is not hard to see that today's consumers have many questions about agriculture and food. Whether we are reading an "old-fashioned" newspaper, logging onto social media or simply listening to conversations in restaurants, the curiosity and variety of topics being discussed is astounding. Consumers are asking, "What should I feed my family? Is this good for the environment? Is this pork chop from a humanely raised pig? Are farm workers paid enough? Why do vets us so many antibiotics?"

The list goes on for miles.

Curiosity is not a bad thing but how we, as food system leaders, respond is extremely crucial. For the past decade, The Center for Food Integrity has built a variety of tools and models based on its annual consumer trust research that allows food system leaders to better connect with consumers. While some influential voices would have you believe that the values of consumers and the values of food system leaders – like you, are vastly different when it comes to the food system, The Center for Food Integrity has found that our collective values are not that different. What is creating the perceived chasm is the way we communicate and engage.

Earning consumer trust

"People don't care how much you know until they know how much you care."

This quote, often attributed to Theodore Roosevelt, sums up what The Center for Food Integrity (CFI) learned in what is now a peer-reviewed and published model for earning trust in today's food system. Research sponsored by CFI and conducted in partnership with Iowa State University shows that confidence (shared values) is three to five times more important than competence (skill and expertise) in earning consumer trust. Specifically, CFI's study measured what drives consumer trust in the areas of food safety, nutrition, worker care, the humane treatment of farm animals and environmental protection.

In CFI's subsequent qualitative research CFI learned that consumers trust farmers because they believe farmers share their values. Unfortunately, consumers aren't sure today's agriculture still qualifies as farming. Why? Generational and geographic distance between farmers and consumers, technological advances in farming, and changes in farm size and structure. We see consumer alienation from agriculture and the food system expressed through concerns about nutrition, food safety, affordability, environmental sustainability, animal well-being and other issues.

Some argue that maintaining public trust is a worthy goal, but not relevant to success in business. This outdated notion fails to recognize the financial benefit of maintaining the trust of stakeholders who can determine the level of social license or social control an organization enjoys. A social license is the privilege of operating with minimal formalized restrictions (legislation, litigation, regulation or market mandates) based on maintaining public trust by doing what's right.

Every organization, no matter how large or small, operates with some level of social license. Organizations are granted a social license when they operate in a way that is consistent with the ethics, values and expectations of their stakeholders. Stakeholders include customers, employees, the local community, regulators, legislators and others who have an interest in how the organization impacts them. Maintaining the public trust that protects your social license to operate is not an act of altruism; it is enlightened self-interest.

Once lost, either through a single event or a series of events that reduce or eliminate stakeholder trust, social license is replaced with social control. Social control is regulation, legislation, litigation or market mandates designed to compel the organization to perform to the expectations of its stakeholders. Operating with a social license means more flexibility and lower cost. Operating with a high degree of social control increases costs, reduces operational flexibility and increases bureaucratic compliance.

Every sector of the food system, whether farmers, manufacturers, branded food companies, grocery stores or restaurants, is under ever-increasing pressure to demonstrate they are operating in a way that is consistent with stakeholder values and expectations. Groups opposed to today's food system are pursuing litigation, pressuring branded food companies, and initiating legislation to change how the system operates.

Historically when under pressure to change, the industry has responded by attacking the attackers and using science alone to justify current practices. Too frequently, the industry confuses scientific verification with ethical justification. Not only are these approaches ineffective in building stakeholder trust and support, they increase suspicion and skepticism that the food industry is worthy of public trust.

As consumer values change, the food system needs to evaluate and potentially modify current practices and fundamentally change the way it communicates in order to maintain consumer trust. Meaningful stakeholder engagement and effective values-based communication with consumers is essential to maintaining the trust that protects social license.

Building a truly sustainable food system requires balance. Maintaining public support requires our practices to be ethically grounded and consistent with the values of our stakeholders. Objective, independent data is essential to evaluate progress and support scientific claims of improvement. Reasonable profitability is essential to assure economic viability. It is only by balancing these sometimes competing interests that we can have a food system that is truly sustainable and supported by our stakeholders and the rational majority of consumers.

In order to better earn consumer trust, CFI research has taken deeper dives into social outrage factors, trusted messengers and trust-building transparency. When exploring social outrage factors, CFI sought to answer, "What takes an issue from a moderate level of concern to true social outrage?"

If the issue is a "concern" to the public and it has an "impact" on family, friends or groups perceived as vulnerable, it can trigger social outrage that drives societal change. Outrage factors identified and explored were:

- Lack of transparency
- Intentional wrongdoing
- Intentionally misleading
- Putting private interest ahead of public interest
- Insensitivity to public interest (tone deaf)
- Callous disregard for public
- Historical record of poor performance
- Failure or unwillingness to accept responsibility
- Impact on vulnerable populations or systems (people, animals, environment)
- Negligence in following industry best practices

CFI research has demonstrated that shared values drive trust. Confidence (shared values) is three to five times more important than demonstrating competency (skills and knowledge) in building trust. Research by CFI has also shown there is an inverse relationship between the size of a company or organization and the perception of shared values. The larger the food company or farm, the less likely the public is to believe they share their values and the more likely it is that they believe profit will be placed ahead of public interest.

While the food system has known intuitively for years that the American public has developed a "big is bad" mindset, CFI has now measured it. Whether its comparing local food companies to national food companies or small farms to large farms, the result is the same – many people think smaller operations are more likely to share their interests.

How do we overcome this bias? CFI focus groups and other secondary research tell us that transparency is important to building trust. But what exactly is transparency?

By looking at previous CFI research and examining other studies, CFI identified seven attributes that constitute trust-building transparency:

- Motivations – Actions are ethical and consistent with stakeholder interests
- Disclosure – All information, both positive and negative, is publicly shared
- Stakeholder Participation – Engaging those interested in your activities or impact
- Relevance – Sharing information stakeholders deem relevant
- Clarity – Sharing information that is easily understood and easily obtained
- Credibility – Sharing positive and negative information that supports informed stakeholder decision making and having a history of operating with integrity
- Accuracy – Sharing information that is truthful, objective, reliable and complete

In 2015, CFI took its research into trust-building transparency to another level – CFI explored transparency as it pertains to:

- Food safety
- Impact of food on health
- Environmental impact
- Labor and human rights
- Animal well-being
- Business ethics

Specifically, to identify the practices consumers associate with demonstrating trust-building transparency, survey participants were asked to rate a list of practices in each area. Here are some highlights:

CFI explored transparency as it pertains to:

- Food safety
- Impact of food on health
- Environmental impact
- Labor and human rights
- Animal well-being
- Business ethics

Food safety and impact of food on health

Consumers rate these two categories as the most important. For these issues, they want information on the product label. That includes all ingredients regardless of quantity, allergens, preservatives and whether ingredients were derived from GMO seed. For other issues, engagement and access to information are key themes. Consumers want to be able to engage via the company's website and they expect information to be provided in easy-to-understand language.

Environmental impact

Consumers want the opportunity to ask questions about environmental performance via the company website and they want answers provided in easy-to-understand language. When regulations are violated, corrective actions should be provided on the company website.

Labor and human rights

Consumers want the opportunity to ask questions about labor practices and human rights via the company website and they want answers provided in easy-to-understand language.

Animal well-being

Results of third-party audits on animal care should be shared on the company website. Consumers want the opportunity to ask questions via the company website and they want answers provided in easy-to-understand language.

Business ethics

Consumers want companies to accept responsibility on the company website for all business activity. They also expect whistleblowers to be protected.

Finally, a key learning from CFI's on-going research is in whom consumers trust as messengers to convey food system information and best practices for the introduction complex and, at time, controversial topics about food system. CFI's 2014 research explored consumers' complex decision-making processes to:

- Better understand how to introduce science and technical information into the public discussion on issues in agriculture and food so that they are considered in the social decision-making process.
- Better understand the communication channels and processes used by Moms, Millennials, Foodies and Early Adopters in today's environment when forming attitudes and opinions about issues in agriculture and food. The research provides a model for introducing and discussing complex, controversial issues. It's a model that can be applied when communicating and engaging with consumers to build trust around topics that are critical to our ability to meet the growing demand for food while preserving and protecting our natural resources.

CFI tested messaging from three different "voices":

- Mom Scientist (A mother with scientific educational and/or work experience)
- Federal Government Scientist(Self-explanatory)
- Peer (A person who shares your interest about food)

The data shows Mom Scientist was the most trusted source of information before respondents read any stimulus information. As respondents were exposed to more information from the three different voices, trust in Mom Scientist remained strong and the trust scores rose consistently for Government Scientist. The Peer voice ultimately dropped to the least-trusted source of information in most scenarios.

On-going research that builds on CFI's foundational research is key to earning consumer trust – and the important role swine veterinarians play in these very important conversations. To learn more about on-going research and for more in-depth research reports visit **www.foodintegrity.org**.

The Center for Food Integrity

CFI is a not-for-profit organization with members that represent every segment of the food system. CFI does not lobby or advocate for individual food companies or brands. Participating organizations represent the diversity of the food system, from farmers and ranchers to universities, NGOs, restaurants, food companies, retailers and food processors.

Opportunistic bacterial pathogens: Battles fought in daily practice

Cameron Schmitt, DVM, MS
Pipestone Veterinary Services, Independence, Iowa

Introduction

Opportunistic bacterial challenges occur in many situations in the swine industry. Most commonly they follow a primary viral challenge to the pig – such as PRRS – that weakens the hosts immune system effectively creating a defense void that that bacterial organisms exploit. Other situations beyond primary viral insults can result in opportunistic bacterial challenges as well – such as poorly managed ventilation, poor sanitation, poor diet or incorrect feed budget. One approach to categorizing opportunists is by body system affected and likely organisms that may be found. Specifically, breaking down opportunists into respiratory, enteric, and septic/lameness.

Respiratory

Commonly practitioners are dealing with primary viral challenges that do not have an effective treatment. Opportunistic pathogens secondary to PRRS and/or SIV infection commonly are *H. parasuis, S. suis, A. suis,* and *M. hyopneumoniae.* Practitioners end up utilizing antimicrobials that have a respiratory spectrum of activity just to battle the bacterial pathogens while the pig recovers from the viral agent. Antimicrobials may be administered injectibly, via the water or via the feed. Certain macrolides have been identified that have a tropism to concentrate in pulmonary alveolar macrophages (Draxxin® and Pulmotil®). Additionally, certain antimicrobials have higher volumes of distribution and will concentrate to a greater degree in lung and respiratory tissues. Beyond antimicrobial therapy, other management techniques may be used to improve the environment and air quality. Also, anti-inflammatories may be given to reduce fevers and increase feed and water intake. Data from challenge studies with antibiotic free pigs suggest that antimicrobial treatment reduces losses by approximately 50%. Prevention and/or vaccination against the primary viral pathogens are critical to reducing opportunistic bacterial respiratory disease.

Enteric

Enteric bacterial challenges during the nursery phase of production (the first ~8-9 weeks post weaning) most commonly are caused by hemolytic *E. coli*. K88 and F18 *E. coli's* make up the bulk of challenges and may cause very rapid mortalities of 10% or greater on groups of pigs. Tools to prevent/reduce losses with hemolytic *E. coli* include sanitation, vaccination, dietary modification, and antimicrobials. Gentamicin, neomycin, potentiated sulfa, amikacin, and enrofloxacin are the most commonly utilized antimicrobials. Resistant *E. coli's* are becoming more common in the author's experience and are driving the need for better preventative measures.

Salmonellosis is observed less frequently than *E. coli* infections and is commonly found in sick pens in pigs that became sick from some other malady and secondarily became infected due to concentration of immunocompromised pigs in a nosocomial fashion. Other times Salmonellosis is observed is when the incorrect feed phase is given to young pigs. Antimicrobials used for *E. coli* above are commonly used for *Salmonella* infections.

Septic/Lameness

Septic bacterial infections – similar to respiratory bacterial infections – are observed most frequently in PRRS infected pigs. While the pig's immune system is compromised by the viral insult, opportunists such as *S. suis, H. parasuis, A. suis,* and others gain access to the bloodstream and find immunologically conserved sites in the pigs body – meninges, pleura, pericardium, peritoneum, and synovia. These can be very difficult to treat as the practitioner is behind before they start. Clinical signs may be limited at the beginning of the infectious process. Additionally, there is little blood flow to these sites in the pig's body and delivering therapeutic levels of medications to these sites is difficult. Antimicrobials such as tetracyclines, sulfas, and various others are used to fight these infections. Ultimately, prevention of the primary challenge such as PRRS reduces/eliminates observation of sepsis.

Batch farrowing disease control benefits in a production system

Elise Toohill, DVM
The Maschhoffs, Carlyle, Illinois

Like many ideas and practices in the swine industry, batch farrowing is an old tool that is starting to regain momentum in today's world. For years the industry has struggled to break the cycle of PRRSV and batch farrowing may be the answer for farms that are able. The primary benefit of this system is the ability to run an all in all out farrowing house. Eliminating the presence of multiple ages of piglets prevents these virus factories from being able to continually shed and recirculate the virus. Due to the mass volume of animals weaned at one time, finishing sites are capable of being filled quickly thus minimizing the need to commingle various wean pig sources. Lest we forget, a benefit often overlooked is the production advantage gained due to the specialization of labor to perform specific tasks in shortened time frames.

It would be hard to find two farms that are in a better position to reap the benefits of batch farrowing such as these two within The Maschhoff system. These are two 2,500 head sow farms located within 2 miles of a large packing plant. These farms have an annualized break rate of every 6 months despite efforts to improve biosecurity. While these sow herds often experienced minimal production impact due to PRRS circulation, the PRRS virus, in conjunction with an assortment of secondary bacteria, proved detrimental to downstream performance. Despite exhaustive efforts from both veterinarians and production staff, it was not uncommon for downstream wean-to-market groups to close out at 75-80% Grade-A pigs. The in-process mortality (total dead pigs for the week divided by total pigs on feed) for this flow averaged 0.81% for the 3 months prior to converting to batch farrowing (Table 1.) The need for change was evident.

The aforementioned sow farms are located within one mile of each other and cumulatively possess one 800 head on-site GDU. Due to their proximity and equivalent sow capacity the ability to convert these two farms to a batch farrow system was completed with relative ease. The process required the movement of bred groups between sow farms, early weaning piglets to 10 days of age and occurred over the timeframe of one turn. These farms are now considered as a single sow source and each farm operates on a 4 week batch with one farm weaning every other week.

The primary health objective of this conversion was to minimize the impact of epidemic viruses. The action plan involved the removal of all shedding piglets and walking these viruses out of the farrowing house more quickly than a traditional load/close/homogenize and strict McREBEL procedure. In this situation, the farms were allowed to resume gilt entry and discontinue McREBEL procedures prior to batch conversion. Prior to conversion, these farms were circulating a 1-7-4 PRRS strain for approximately 8 months. Wean pig lots continually seroconverted to PRRSV at 2-4 weeks post placement.

Despite the continued seroconversion at 4 weeks post placement, subsequent months of batch farrowed pigs demonstrated a decreased severity of clinical signs in wean-to-market as finishing performance rebounded. At the conclusion of the first true batch or 6 months from the start of batch conversion, animals in wean-to-market were no longer seroconverting to PRRSV downstream. To date, the farms have not experienced a new PRRSV break or an alternative epidemic disease outbreak.

From a production standpoint, these farms are capable of weaning 5,000+ pigs per batch as mated inventory was not decreased in the conversion due to the ability to use the GDU and move bred animals between farms. This translates to an ability to fill wean to market sites within 2 days compared to 2-3 weeks previously. Take a moment to consider the advantages in the ability to manage these pigs more precisely in regards to nutrition,

Table 1: Average weekly in-process mortality (total dead pigs/total on feed) by phase (weeks post wean.) Weekly average represents maximum of 4 lots per phase and individual site variation may lead to over interpretation. Overall average of before and after at 0.81% vs 0.28% is key take away.

Phase	3 months prior to batch	6 month batch transition	Last 3 months on true batch
0-6 WPW	0.59%	0.75%	0.31%
7-12 WPW	1.08%	1.20%	0.40%
13-18 WPW	0.56%	0.52%	0.30%
19+ WPW	0.87%	0.37%	0.17%
All Lots	**0.81%**	**0.66%**	**0.28%**

mass medication via feed and/or water and vaccinations. Due to similar age within the population, the herd undergoes a similar timeline in regards to decay in maternal antibodies. For this reason the lots experience faster resolution of endemic viruses such as influenza and one would suspect, a decreased morbidity associated with endemic pathogens such as *Streptococcus suis* and *Haemophilus parasuis*.

Despite the obstacles encountered, the conversion to batch farrowing was well worth the effort as is evident by the profitability the farms now experience. This winter will provide the true test as PRRSV pressure intensifies. As evident in Figure 1, in-process mortality began to decline almost immediately during the batch conversion process. Previously affected lots still on feed resulted in in-process mortality remaining around 0.6-0.8% for the first 6 months. The most recent three months have averaged 0.28% in-process mortality, over half of what it was prior to the batch. Additionally, we are now consistently closing out wean to market lots at 90%+ Grade-A pigs. As a system there is a strong sense that this was the right decision for these farms and we will continue to seek opportunities to utilize the batch system elsewhere.

Figure 1: In-process mortality over one year batch conversion started week 12.

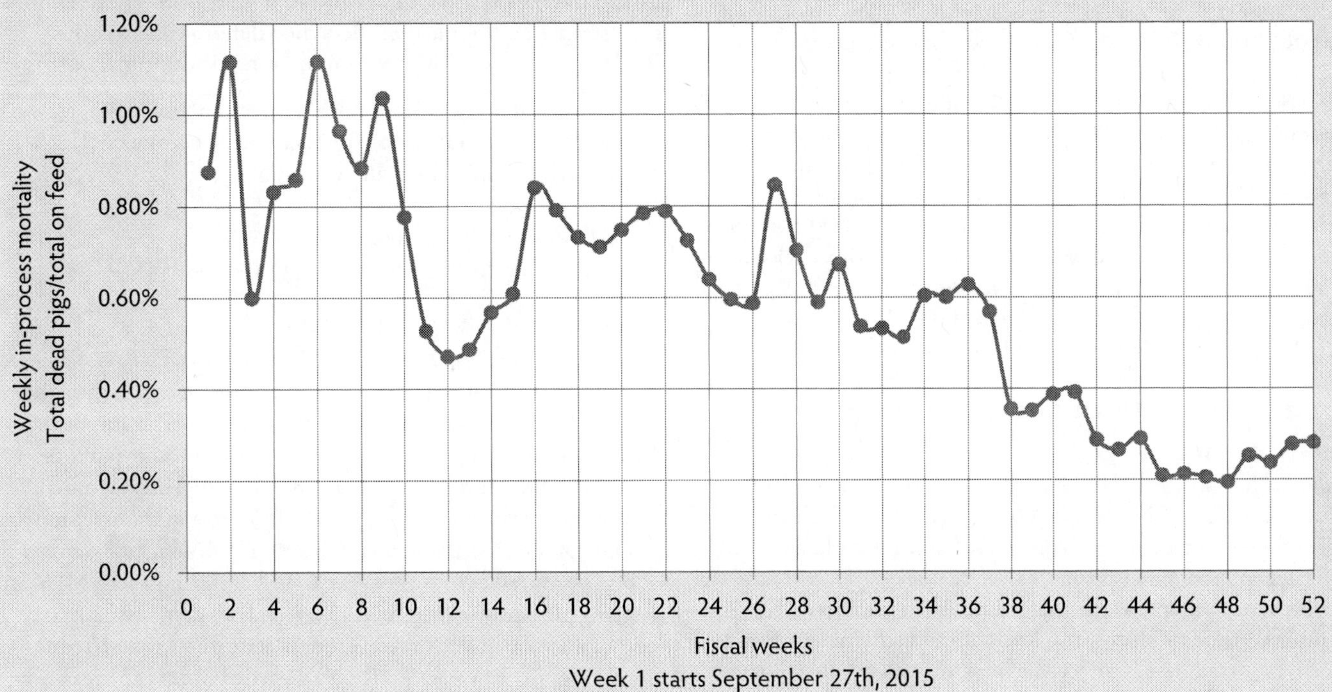

Determining optimum PRRSV management strategies

Clayton Johnson, DVM
Carthage Veterinary Service, Carthage, Illinois

Background

The swine veterinarian is charged with delivering optimal health management in a socially and economically sustainable manner. A novel porcine reproductive and respiratory syndrome virus (PRRSV) introduction on a breed to wean (BTW) farm results in dramatic performance losses that threaten farm and producer viability. In the face of this challenge the swine veterinarian must craft a PRRSV management plan which optimizes the client's biological and economic performance. To maximize BTW farm efficiency in the face of a novel PRRSV infection, the swine veterinarian must apply the best available evidence to their decision making process, and ultimately to their health management recommendations. This paper will communicate PRRSV control and eradication strategies with recommendations for optimal strategies based on farm specific attributes. PRRS control and eradication options presented below will focus on a Load, Close and Expose (LCE) model of PRRS management, with particular consideration given to two key decision points; the final goal of the PRRS management plan (PRRS Control vs. PRRS Eradication), and the method of BTW herd exposure (Live Virus Inoculation vs. Modified Live Virus Vaccination).

Time to stability and time to baseline performance

Time to Stability (TTS) is achieved when 4 consecutive negative PCRs for PRRSV RNA in a monthly basis testing of at least 30 piglet blood samples has been obtained. Time to baseline production (TTBP) is defined as time, in weeks, it takes to recover the levels of 'weaned pigs per week' that the herd had prior to PRRSV-detection (i.e. time to "in control" levels of productivity). TTS and TTBP estimates have been well defined through producer surveys and aid in modeling optimal producer outcomes.

TTS estimates contain tremendous variability regardless of PRRSV genotype or farm specific risk factors. Regardless of the variability in TTS, two significant associations can be made:

- BTW herds implementing Live Virus Inoculation (LVI) for exposure achieve TTS significantly faster than comparable herds achieving stabilization through Modified Live Virus (MLV)
- BTW herds with prior immunity (defined as PRRSV antibody positive at the time of novel PRRSV introduction) achieve TTS significantly faster than comparable herds without prior immunity (defined as PRRSV antibody negative at the time of novel PRRSV introduction)

When reviewing the associations with TTBP (defined as 21 week performance of the affected farm prior to novel PRRSV outbreak), four significant associations can be made:

- TTS is poorly correlated with TTBP
- TTBP is highly correlated with total pre-weaning production losses
- Prior herd immunity results in earlier TTBP, with a drastic increase in total weaned piglet production relative to PRRSV naïve herds and an apparent dose impact with herds exposed multiple times reaching TTBP sooner than herds exposed once
- BTW herds using MLV reach TTBP significantly faster than herds using LVI

Herd-specific considerations

The following is an incomplete list of herd specific considerations the veterinarian should consider when determining whether PRRSV Control or PRRSV Elimination is the optimal outcome:

- Historical BTW novel PRRSV infection rate
- BTW farm production type (genetic vs. commercial)
- BTW farm PRRSV piglet discount
- BTW farm gilt entry age
- BTW farm gilt ages and inventory
- PRRSV status of farms with shared ecosystem

Cost of PRRSV immunity

Immunizing breeding herds using Modified Live PRRSV Vaccination (MLV) impacts biologic and economic results though increased vaccine cost and the negative impact of the attenuated replicating virus on productivity levels. While vaccine purchase impact can be accurately estimated, the potential impact of MLV on sow and pig performance plays a critical role in this decision and is poorly understood. Potential impacts of MLV include number of pigs produced per sow per year as well as the impact of MLV on growing pig performance. Sensitivity analyses demonstrate that 'cost of attenuated PRRSV on growth performance' and 'reduction in pigs per sow per year (PSY)' heavily impact the break-even (number of years between PRRSV outbreaks) to justify preventive vaccination. Lower attenuated-PRRSV-impact on productivity (via growth performance or breeding herd productivity) reduces the break even for practicing preventive vaccination.

Duration of herd closure for PRRSV wild type eradication

Herd closure has long been considered a critical component to PRRSV Wild Type Eradication regardless of PRRSV Control or PRRSV Eradication being the objective. However, there is a growing body of evidence indicating PRRSV Wild Type Eradication is not only possible without herd closure but may not significantly impact TTS. The impact of herd closure on TTS is relevant to all BTW farms experiencing a novel PRRSV infection and is particularly critical for BTW farms which enter gilts at 'select' weights, raised at off-site gilt development units (GDUs).

Emerging genetic strategies for disease control

M. Culbertson; W.O. Herring
Genus PIC, Hendersonville, Tennessee

Introduction

The prevention, control and management of disease continues to be a primary driver for the operational efficiency and profitability of commercial swine production. Genetic strategies for disease control can widely be defined as including decisions related to animal intake and flow through the system, choice of genetic products to utilize and the on-going improvement initiatives that help create improved genetic potential into the future. The tools and programs utilized to create this improved genetic merit continue to progress at a rapid rate from utilization of field-based commercial data to genomic evaluation to emerging technologies such as gene editing.

Strategies and considerations

Genetic intake and dissemination

Although likely less dynamic in many mature systems today, the design of a genetic sourcing, multiplication and dissemination strategy remains a key consideration in many emerging regions as well as experiencing a continued evolution within other regions. Introduction of new animals remains a primary risk of disease spread and the acclimation of new animals to the specific system is an opportunity for control and management of disease and improvement of subsequent performance. The increased utilization of artificial insemination coupled with increasing capability of diagnostics has enabled the introduction and utilization of males with significantly reduced risk. Introduction of replacement gilts into systems positive for specific diseases remains an on-going area of evolution with a range of strategies including producing high health replacements for introduction and acclimation at a young age into the future population.

Utilization of commercial field data

Aligned with the discussion around introduction of live animals, primary genetic production today is typically done at a health status that is free of many of the production diseases which challenge commercial production around the globe. Although needed and beneficial for dissemination of elite genetic merit, the genetic industry must continually evaluate if the performance testing and selection of animals in this environment truly predicts commercial performance in the real-world Zumbach et al., 2007). Research has demonstrated relatively robust genetic correlations for a range of traits including those related to carcass lean yield and prolificacy; however, research has also clearly demonstrated that the genetic correlation for some traits, such as growth and efficiency, are low enough that maximum gain will not be achieved without accounting for the impact of environment (Zumbach et al., 2007). Finally, for traits specific to robustness and livability, selection practices that do not utilize commercially relevant data are likely to be ineffective.

Evolution of genetic evaluation and prediction

A significant break-through delivering the accelerated rates of genetic gain over the last 25 years was the implementation of Best Linear Unbiased Prediction (BLUP) (Henderson, 1975). An enhanced statistical tool for accurate combination of large sources of phenotypic data coupled with the genetic relationship between all animals, BLUP allowed for accurate utilization and weighting of these large data sets and subsequently dramatic progress for traits previously difficult to select for such as prolificacy. Continued advancement in the area of genomics has now allowed the genetic improvement industry to have tens of thousands of genetic markers on many of our selection candidates (Legarra, 2009). The utilization of these markers have also continued to evolve from attempting to find specific markers linked to desirable phenotypes to utilization of these markers to help more accurately define the degree of genetic relationships across the population. The utilization of genomic information via these new techniques has allowed a second dramatic increase in the rate of genetic improvement across all traits including those with primary impact on pig robustness and livability.

Emerging technologies for future impact

The continued development and evolution of genetic technologies appears to be accelerating. Genomic science has made dramatic progress in recent years resulting in more efficient high density genotyping and leading to advancements in genetic evaluation as described above. This development is expected to continue with projects actively underway capturing whole genome sequence information on thousands of animals (Hickey, 2013). In addition, breakthrough technologies such as gene editing offer dramatic opportunities to target improvement in animal robustness and disease resistance. A primary example of this progress being the development of a pig resistant to challenge with the PRRS virus (Whitworth et al., 2015).

Conclusion

Genetic gain has delivered improvement in many important traits over the 25 years. Robustness has historically proven difficult until large scale commercial field data was accurately captured and included in the evaluation process. Emerging technologies, coupled with these large commercial data sets, offer significant hope on accelerating the development of pigs with greater robustness and in some cases, potentially even resistance to common production diseases.

References

1. Henderson, C. R. 1975. Best linear unbiased estimation and prediction under a selection model. *Biometrics*. 31(2):423-447.

2. Hickey, J.M. Sequencing millions of animals for genomic selection 2.0. *J. Anim. Breed. Genet.* 130 (2013) 331-332.

3. Legarra A, Aguilar I, Misztal I: A relationship matrix including full pedigree and genomic information. *J Dairy Sci* 2009, 92:4656-4663.

4. Whitworth, K.M, R.R.R. Rowland, C.L. Ewen, B.R. Trible, M.A. Kerrigan, A.G. Cino-Ozuna, M.S. Samuel, J.E. Lightner, D.G. McLaren, A. Mileham, K.D. Wells, R.S. Prather. 2015. Gene edited pigs are protected from porcine reproductive and respiratory syndrome virus. *Nature Biotechnology* http://dx.doi.org/10.1038/nbt.3434.

5. Zumbach B., I. Misztal, S. Tsuruta, J. Holl, W. Herring and T. Long. Genetic correlations between two strains of Durocs and crossbreds from differing production environments for slaughter traits. *J. Animal Sci.* (2007) Apr,85(4):901-8.

Swine Health Information Center 2016 Progress

P. Sundberg, DVM, PhD, DACVPM
Swine Health Information Center, Ames, Iowa

Introduction

The Swine Health Information Center was funded by a National Pork Board grant in July of 2015. Its mission is to protect and enhance the health of the United States swine herd through coordinated global disease monitoring, targeted research investments that minimize the impact of future disease threats, and analysis of swine health data.

To give the Center strategic direction and set policy separate from the National Pork Board, a Board of Directors was named for the newly formed corporation. And, to help guide and give oversight to its programs and research, two working groups were formed with volunteer members from pork industry associations, pork production, veterinary medicine, academia, USDA and allied industry. Because of our linked production, SHIC also has engaged Canadian counterparts to help coordinate disease information and response with their domestic industry.

The Monitoring and Analysis Working Group is charged with assessing foreign, transboundary production disease risk using information from a variety of sources. The outcome of this assessment is the on-going prioritization of the Swine Disease Matrix. It is also responsible for improving the health of the nation's swine herd through the development and oversight for on-going projects. These include monitoring for domestic diseases affecting swine health and analyzing health and other data to support on-farm and prospective producer decision making.

The Preparedness and Response Working Group is charged with oversight of the Swine Disease Matrix research. It is responsible for funding decisions to fulfill the Matrix research objectives. It is also responsible for advising and oversight of SHIC's role in the emerging swine diseases response plan. That includes the appropriate SHIC response to an emerging swine disease and for the information and analysis necessary to support the proportional pork producer and pork industry response to these emerging diseases.

2016 Progress

Swine Disease Matrix

In collaboration with the American Association of Swine Veterinarians, the National Pork Board and the National Pork Producers Council, the Swine Health Information Center has prioritized a list of endemic and foreign swine pathogens, known as the Swine Disease Matrix. The SHIC Monitoring and Analysis Working Group reviewed the list and has prioritized it to help to focus SHIC activity on the highest risks to the US pork industry. The Working Group will be reviewing the prioritization as conditions change or new information is found.

Prioritization was done by assigning a 1 to 10 score to each of three criteria. The likelihood of introduction into the United States was included in the prioritization to account for introduction from foreign sources. Likewise, the likelihood of an endemic pathogen changing and causing a newly emerging disease or syndrome was also considered. The likely effect on production economics of disease from the pathogens was included as was the probable effect on domestic and/or international markets from an outbreak due to the pathogen.

Organizing and funding the research needed for diagnostic and informational preparedness for the prioritized, potential production diseases of the Swine Disease Matrix is a core activity of the Center. In 2016, SHIC Matrix research focused on the ability to detect the Matrix pathogens via nucleic acid detection, using platforms that are commonly available in the US major veterinary diagnostic laboratories, for example PCR testing. Given that some of these assays may already be produced and available in other countries, the research effort included an assessment of the fitness for purpose of these tests. And, since some of the pathogens in the Swine Disease Matrix are not currently in the US, some of the research will be conducted in collaboration with a foreign laboratory or in another manner to ensure that the pathogen will not be introduced to the US. Twenty-one research proposals from US universities, an allied industry biotechnology company and the Canadian national animal health laboratory were selected for funding.

Global disease assessment

SHIC has conducted a survey of swine diseases and related issues using the international network of the Monitoring and Analysis Working Group. The survey asked the respondents to list the top five pathogens circulating and causing production losses in their country or region. It then asked for a description of the morbidity, mortality and/or production losses from each pathogen and the control measures that are currently being used. Next was a query about pathogens that might not be among the top five but are appearing to be emerging and are deserving of further monitoring. The final question asked for an opinion about the possible consequences to the US pork industry should these pathogens gain entry here. Responses came from 13 foreign countries, including Japan, Korea, Philippines, China, Poland, Ukraine, Russia, United Kingdom, Spain, Brazil, Chile, Columbia and Canada.

VDL swine test standardization

SHIC is the primary source of funding for the veterinary diagnostic laboratories of Iowa State University, Kansas State University, University of Minnesota and South Dakota State University to work with Clemson University to standardize the way that they report their swine testing results. This project will improve the ability of the diagnostic labs to communicate test results and improve the ability for analysis of veterinary diagnostic lab test results that will offer real-time status of US disease trends and outbreaks. An important partner in the effort is the USDA. Their work with the National Animal Health Laboratory Network will help to ensure that, when completed, the results of the project can be offered to the other veterinary diagnostic labs that do swine work so there will be national coordination.

Disease risks research

SHIC is funding a project that will help define disease introduction risks to the US pork industry that come from importing feedstuffs and feed components. After PEDV was introduced into the United States in 2013, a USDA pathways analysis concluded that the most likely route of introduction was through the use of contaminated containers moving between countries and being used to import feed or feed ingredients. However, the pattern of the US outbreak was such that a more direct involvement of imported feed needed to be investigated. Because of the importation from China of high volumes of feed ingredients a model was developed to study whether PEDV harbored in imported ingredients could remain viable over the time and under the environmental conditions encountered during a "trans-Pacific" shipment from Asia to the USA. That model is being used to evaluate the survival of foreign viruses in feed ingredients using viral surrogates to represent the actual pathogens.

Rapid Response Corps

SHIC continues to make progress toward supporting nationwide operational disease preparedness with the Rapid Response Corps (RRC) project. The goal of this project is to develop a rapid response program for epidemiological investigations of emerging, transboundary and endemic swine diseases. Swine health experts who can be deployed on farms within 72 hours after a request will investigate possible introduction pathways and potential movements of the cause of the outbreak. An advisory group of swine veterinarians has been formed to help guide development and implementation of the program.

Diagnostic fee support

To help more producers and their veterinarians solve outbreaks from unknown causes, SHIC developed a system to help offset diagnostic fees for further investigation after the initial diagnostics are completed. To request support of diagnostic fees the diagnostician of the case must contact SHIC. In cases of unresolved high or ongoing morbidity or mortality a panel of diagnosticians will review the case and must concur that the most likely differential diagnoses have been addressed. To ensure that a foreign animal disease is not being missed, the SAHO of the premises' state must be aware of the case and the lack of a satisfactory diagnosis and has considered if a foreign animal disease investigation should be initiated. A novel Sapelovirus, causing morbidity and mortality via CSN infection, has been identified through this process.

Emerging disease response

SHIC responded to the emergence of Senecavirus A (Seneca Valley Virus or SVV) in 2015 with $270,000 worth of research designed to learn more about the virus and how to combat it on the farm. Projects funded included initial investigations into the breadth and distribution of SVV around the United States, the duration of viral shedding in finishing pigs and sows, sequencing and characterization of the virus, improving diagnostic tests and the efficacy of disinfectants against the virus. Fact sheets, recorded video webinars that include information about the USDA guidance for investigating outbreaks involving vesicular lesions and research reports are available on `www.swinehealth.org`.

The mission of the Swine Health Information Center is to protect and enhance the health of the United States swine herd through coordinated global disease monitoring, targeted research investments that minimize the impact of future disease threats, and analysis of swine health data. For more information, visit `www.swinehealth.org` or contact Dr. Paul Sundberg at `psundberg@swinehealth.org`.

Table 1: Swine Disease Matrix

Representative virus affecting swine	Economic impact	Market access/ trade impacts	Likelihood of introduction in the US or emergence of a domestic disease	Average
foot and mouth disease	9	9	9	8.4
classical swine fever	7	9	5	8.3
African swine fever	7	7	5	7.8
influenza A	8	4	8	6.2
pseudorabies -- Chinese strain	8	8	5	6.1
porcine reproductive and respiratory syndrome (PRRS)	6	3	6	6.0
Nipah	8	9	2	5.9
PRRS (Chinese high path)	6	5	5	5.5
porcine epidemic diarrhea	6	4	7	5.3
swine vesicular disease	4	3	2	5.1
Japanese encephalitis	5	5	4	4.8
vesicular stomatitis	1	1	1	4.5
porcine teschovirus (Teschen/PTV1)	5	5	5	4.3
Ebola-Restin	8	9	2	4.2
porcine circovirus	4	2	5	4.1
vesicular exanthema of swine	3	4	2	4.0
circovirus 3	4	2	5	3.9
porcine rubulavirus	5	2	2	3.9
transmissible gastroenteritis	5	4	4	3.9
Senecavirus A	3	3	3	3.9
Menangle virus	4	4	4	3.9
porcine deltacoronavirus	3	2	3	3.8
porcine rotavirus	4	1	5	3.6
influenza C	2	2	2	3.5
porcine parainfluenza 1	3	2	3	3.3
atypical swine pestivirus	3	2	3	3.2
hepatitis E virus	1	3	1	3.2
Getah virus	5	5	4	2.8
porcine parvovirus	3	1	5	2.7
orthoreovirus	2	1	1	2.2
porcine respiratory coronavirus	2	2	2	2.1
hemagglutinating encephalomyelitis	2	2	2	2.0
encephalomyocarditis virus	3	1	2	2.0
Chikungunya virus	1	1	1	2.0
rabies	1	1	1	1.7
porcine bocavirus	1	1	1	1.7

Table 1 continued on next page

Table 1: Continued

Representative virus affecting swine	Economic impact	Market access/ trade impacts	Likelihood of introduction in the US or emergence of a domestic disease	Average
porcine astrovirus	1	1	1	1.7
porcine adenovirus	2	1	2	1.6
porcine kobuvirus	2	1	2	1.5
porcine sapovirus	2	1	2	1.5
Sendai virus	1	1	2	1.5
porcine cytomegalovirus	2	1	1	1.4
swine pox	1	1	1	1.3
porcine sapelovirus	1	1	1	1.3
porcine torovirus	1	1	1	1.3
swine papillomavirus	1	1	1	1.1

Biosecurity disaster: Foot and mouth disease and classical swine fever outbreaks in Korea

W-I Kim, DVM, MS, PhD, DACVM
College of Veterinary Medicine, Chonbuk National University, Iksan, Korea

Introduction

Prior to November 2010, there were several outbreaks of foot and mouth disease (FMD) and classical swine fever (CSF) in Korea but those outbreaks were successfully controlled and eradicated without nationwide spread. Because intensive biosecurity and disease control strategies established based on the lessons learned from previous outbreaks still operated to prevent the introduction of FMD and CSF, the recent outbreaks resulting in nationwide spread of FMD and CSF are so discouraging and hard to explain. There is also concern that FMD and CSF might become endemic because it seems to occur throughout the year even under the current vaccination campaign. Therefore, many veterinary experts feel frustrated and depressed more than ever in the current FMD and CSF situation. Factors such as climate change, increased number of foreign farm workers, and international travel have been suggested as the cause of unexpected failure of biosecurity and disease control. In this talk, the history of FMD and CSF outbreaks in Korea will be reviewed to discuss differences between current and past biosecurity risk factors and future directions for disease eradication.

Materials and methods

Spatiotemporal analyses used data obtained from the World Organization for Animal Health Information Database (`http://www.oie.int`), the Food and Agriculture Organization World Reference Laboratory for foot-and-mouth disease (`http://www.wrlfmd.org`), and the Southeast Asia and China Foot-and-Mouth Disease Campaign (`http://www.seafmd-rcu.oie.int`). Statistical analysis was performed by using paired or unpaired t-tests, and correlation were made by using GraphPad InStat version 3.05 (Graph Pad Software, La Jolla, CA, USA). A phylogenetic tree was inferred by using the neighbor-joining method, and analysis was conducted by using MEGA version 6 (`http://www.megasoftware.net/`).

Results and discussion

FMD outbreaks were reported in 2000 (15 cases), 2002 (16 cases), and 2010 (20 cases), but all of the outbreaks were limited in only several provinces and successfully eradicated within a month. Another FMD outbreak, however, was reported in November 2010 and lasted for 5 months after spreading throughout the country except for Jeolla Province (153 cases).[1] FMD eradication was achieved in May 2014 under the condition of nationwide vaccination. Unfortunately, just 7 months later, in December 2014, a new outbreak was reported and continues through this point.[2]

CSF eradication was declared in December 2001 as a result of the national vaccination program that began in 1996. However, since a new outbreak was reported in April 2002, five months after the eradication declaration, and 85 cases occurred until 2003, vaccination was resumed nationwide except for Jeju Island. Since then, CSF were reported on mainland where maintains the vaccination policy in 2004 (9 cases), 2005 (5 cases), 2006 (2 cases), 2007 (5 cases), 2008 (7 cases), 2009 (2 cases), and 2013 (1 case). One case was also reported in 2016 and presumed to be derived from wild boars. In 2014, 20 farms in Jeju Island accidentally used an attenuated vaccine (LOM strain[3]) and were exposed to the vaccine virus. It becomes a major concern as the number of farms exposed to LOM strain continues to increase until 2016.

References

1. Park J-H et al. Reemergence of foot-and-mouth disease, Soth Korea, 2000-2011. Emerging Infectious Diseases. 2014;20:2158-2161.

2. Park J-H et al. Novel foot-and mouth disease virus in Korea, July-August 2014. Clinical and Experimental Vaccine Research. 2016;5:83-87.

3. Lim S-I et al. Safety of classical swine fever virus vaccine strain LOM in pregnant sows and their offspring. 2016;34:2021-2026.

Antibiotic resistance mechanisms

Randall S. Singer
Department of Veterinary and Biomedical Sciences, University of Minnesota, Saint Paul, Minnesota

Introduction

Bacterial infections in humans that are resistant to antibiotics pose a significant and growing public health challenge globally. Agricultural antibiotic use (AAU) can influence resistance in specific bacterial populations, and it is now well-documented that non-antibiotic compounds such as disinfectants and metals can also significantly impact the prevalence of resistant bacteria.[2] Within the poultry production system, which will be the focus of this paper, there is a complex web of possible pressures affecting resistance in bacteria of importance to human and animal health. The real challenge is determining which practices are making the greatest contribution to the emergence, amplification, persistence and dissemination of antimicrobial resistance (AMR) and then accurately predicting the net benefit to human health that modification or elimination of these practices would have.

The broiler chicken and turkey systems have been linked to many antibiotic resistant infections in humans and are therefore critically important systems in which to develop and implement strategic programs for mitigating antibiotic resistance. For example, *Salmonella enterica* and *Escherichia coli* are both prevalent in the broiler production system and have become increasingly resistant to antibiotics. Almost 50% of human salmonellosis illnesses in the U.S. have been attributed to chicken,[3] and as highlighted by recent multistate outbreaks, many of these *Salmonella* strains are multidrug resistant (MDR).[1] While poultry only accounts for a small portion of foodborne-associated *E. coli* illnesses, it importantly might serve as a reservoir of resistance genes that can be transferred to other human pathogens. There is also the possibility that a small fraction of antibiotic resistant *E. coli* extraintestinal illness in humans can be attributed to chicken.[4]

Linking AAU to human health

There are multiple pathways by which AAU can cause resistance to increase above background levels and then lead to a negative outcome (human health harm) (Figure 1, taken from Singer and Williams-Nguyen, 2014).[5] In Pathway A, AAU leads to an increase in resistant pathogens which are then transmitted to humans via the food chain or the environment. The selection pressure exerted by the AAU in this scenario occurs on the farm. In Pathway B, AAU selects for resistance in non-pathogens, such as commensals or environmental microbes, which then transfer resistance genes to pathogens leading to more resistant infections in humans. Again selection in this case occurs on the farm. Finally, Pathway C involves the release of active antimicrobial compounds into the environment where selection occurs predominantly in non-pathogens, such as soil microbes, and resistance is transferred horizontally to pathogens as in B. Pathways B and C might be referred to as forces increasing the size of the resistance gene pool.

To estimate the risk to human health in each of these scenarios (or combinations of them), we must establish quantifiable, causal links and determine what the current state of evidence tells us about the relative importance and interconnectedness of these links. For example, knowing that a human illness was caused by a resistant bacterium that originated from a specific poultry operation (the proverbial smoking gun) does not necessarily help us understand WHY the bacterium was resistant. The resistance in this bacterium was unlikely to have been created *de novo* by the AAU on the source farm. Many of the resistance genotypes, particularly when associated with multidrug resistance plasmids, have a global distribution.[6-9]

Figure 1: Pathways by which AAU can cause increased resistant infections in humans.[5]

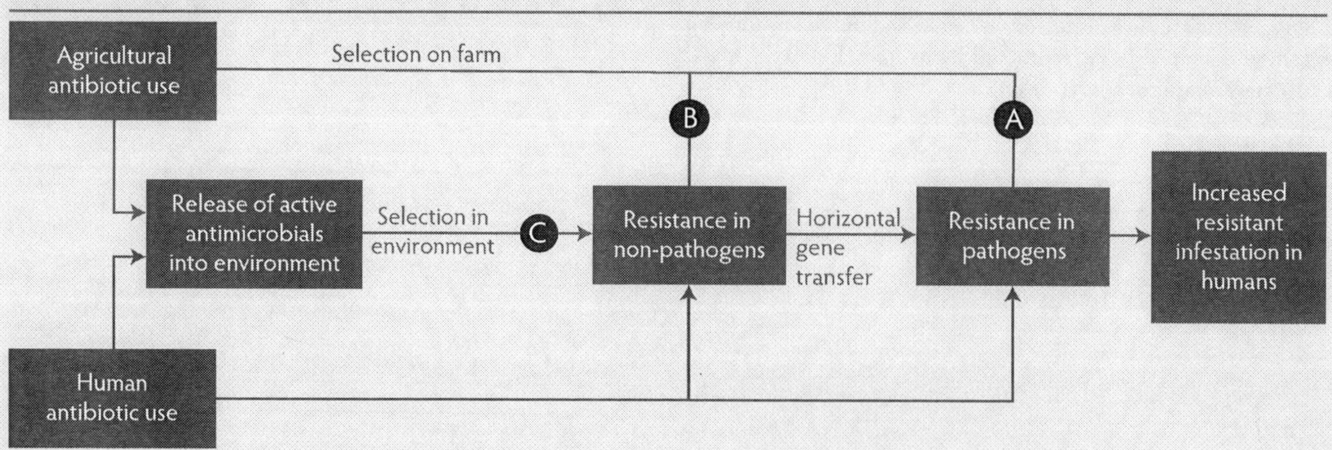

Antibiotic use in poultry production

Most studies of the relationship between antibiotic use and resistance in the poultry production system have focused on antimicrobial growth promoters (AGP).[10-12] Results of these studies have varied, but in general there has been an observed decrease in resistance to individual antibiotics in specific bacteria when AGPs were removed. In Europe, where the AGP have been banned, the commensal strains of *E. coli* from broilers in Denmark had reduced prevalence of resistance to sulfonamides and ampicillin but negligible changes to tetracycline and streptomycin[13,14]. With the recent implementation of Guidance for Industry #213 and the Veterinary Feed Directive by the US-FDA, the administration of medically-important AGPs are now phased out.[15,16] Consequently, it is more important to address the consequences, both positive and negative, of removing AGP from poultry production than to continue evaluating the potential for these uses to select for resistance.

Following the removal of AGPs in Europe, broiler chickens experienced a significant increase in necrotic enteritis (NE),[17-19] an extremely costly disease caused by *C. perfringens.*[18] AGP dosing regimes with antibiotics such as bacitracin, virginiamycin, avoparcin, lincomycin, tylosin, and penicillin were able to limit morbidity associated with NE,[20] but the increased incidence resulted in increases in antibiotic usage for NE disease prevention, treatment and control.[17,21] Infection with coccidian parasites has been shown to increase the likelihood of NE due to improved growth conditions for *C. perfringens*, and therefore, products that reduce coccidia such as ionophores[18,19,22-24] have been shown to decrease the incidence of NE.[25] What remains uncertain is whether the ionophores and the antibiotics used to prevent NE will have an impact on antibiotic resistance in poultry.

Given that there are at least 7 antibiotic compounds used in the US for the prevention of NE in broilers, studies are needed to examine the impact that these antibiotic administrations may have on resistance. Furthermore, antibiotics are commonly used in the hatchery for the control of early mortality associated with bacteria such as *E. coli* and *Salmonella*. The only two approved antibiotics in the US, gentamicin and ceftiofur sodium, are given subcutaneously in the day-old chick as labeled. Gentamicin though is most commonly administered extra-label *in ovo*; following the ban of extra-label cephalosporin administration,[26] ceftiofur cannot be used in this manner. In summary, very few of the non-AGP antibiotics approved for use in poultry have been evaluated directly for their impact on resistance, and therefore, quantifying the effect of AAU on human health through Pathways A, B or C, as shown in Figure 1, is problematic.

Antibiotic resistance and human health

Very few of the most important antibiotic resistant infections of humans have a direct relationship to agriculture.[1] Some scientists have begun linking *E. coli* in chicken with human illness.[4] As depicted in Figure 1, *E. coli* can serve as commensal bacteria that potentially transfer resistance to human pathogens via Scenario B. However, recent studies suggest that *E. coli* that cause human disease outside of the GI tract, known as Extraintestinal Pathogenic *E. coli* (ExPEC), have considerable genetic similarity to *E. coli* in animals. *E. coli* that cause disease in chickens, broadly known as avian pathogenic *E. coli* (APEC), have many of the same virulence factors as the human ExPEC strains. AMR in ExPEC and APEC strains is common, and thus there is the potential for resistant *E. coli* to infect humans through Scenario A of Figure 1 as well. The role of APEC in human disease, however, is uncertain.

The antimicrobial resistance genes found among *E. coli* and *Salmonella* are numerous and diverse.[27-35] In addition to harboring antibiotic resistance genes, plasmids of *E. coli* and *Salmonella* may also harbor genes encoding tolerance to heavy metals and disinfectants. Resistance towards a particular antibiotic can thus be a result of direct or indirect selection. For example, the use of gentamicin *in ovo* or in day-old chicks could directly select for *E. coli* and *Salmonella* that possess plasmids such as the IncHI2 plasmid pAPEC-O1-R (Figure 2a) and the IncA/C plasmid p199061_160 (Figure 2b). Selection of these plasmids through use of gentamicin would result in the co-acquisition of genes conferring resistance towards tetracycline, sulfisoxazole, copper compounds, silver compounds, and quaternary ammonium compounds. The reverse might also be true; use of heavy metals or disinfectants could directly select for *E. coli* and *Salmonella* harboring these resistance plasmids, thus co-selecting for AMR. Importantly, strains of *E. coli* and *Salmonella* within the poultry production system possess resistance to medically-important antibiotics, including the third-generation cephalosporins.

Antibiotic resistance is a critically important issue affecting health of human and animal populations. Although the role of poultry production practices in augmenting levels of antimicrobial resistance in key bacteria is uncertain, there are many practices in poultry production that have the potential to exacerbate the problem and aid in the selection, persistence and dissemination of resistance. In addition to using antibiotics in a responsible manner, we must also determine the effects that non-antibiotic uses, such as metals and disinfectants, have on resistant bacteria in the poultry production system.

References

1. Centers for Disease Control and Prevention (CDC). 2013. Antibiotic Resistance Threats in the United States, 2013. http://www.cdc.gov/drugresistance/threat-report-2013/.

2. Singer, R. S., M. P. Ward, and G. Maldonado. 2006. Can landscape ecology untangle the complexity of antibiotic resistance? Nat. Rev. Microbiol. 4: 943-952.

3. Guo, C., R. M. Hoekstra, C. M. Schroeder, S. M. Pires, K. L. Ong, E. Hartnett, A. Naugle, J. Harman, P. Bennett, P. Cieslak, et al. 2011. Application of Bayesian techniques to model the burden of human salmonellosis attributable to U.S. food commodities at the point of processing: adaptation of a Danish model. Foodborne Pathog. Dis. 8: 509-516.

4. Nordstrom, L., C. M. Liu, and L. B. Price. 2013. Foodborne urinary tract infections: a new paradigm for antimicrobial-resistant foodborne illness. Front. Microbiol. 4: 29.

5. Singer, R. S., and J. Williams-Nguyen. 2014. Human health impacts of antibiotic use in agriculture: A push for improved causal inference. Curr. Opin. Microbiol. 19: 1-8.

Figure 2A: pAPEC-O1-R, IncHI2 plasmid from APEC.[35] **Figure 2B:** p199061_160, IncA/C plasmid from APEC7.

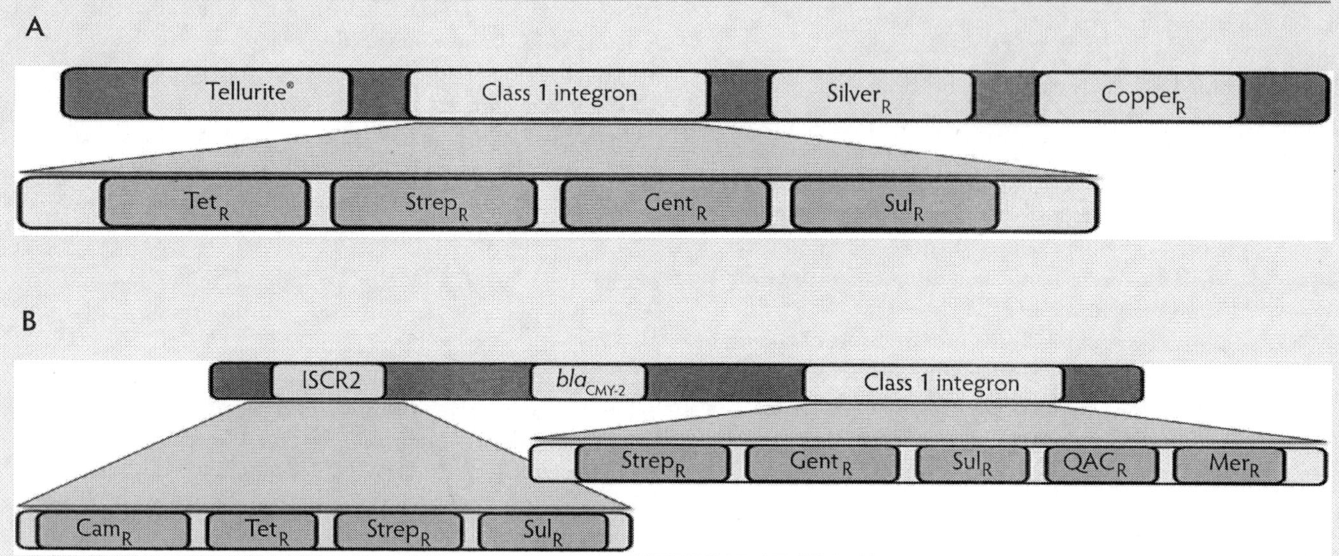

6. Call, D. R., R. S. Singer, D. Meng, S. L. Broschat, L. H. Orfe, J. M. Anderson, D. R. Herndon, L. S. Kappmeyer, J. B. Daniels, and T. E. Besser. 2010. blaCMY-2 -positive IncA/C plasmids from *Escherichia coli* and *Salmonella enterica* are a distinct component of a larger lineage of plasmids. Antimicrob. Agents Chemother. 54: 590-596.

7. Fernandez-Alarcon, C., R. S. Singer, and T. J. Johnson. 2011. Comparative genomics of multidrug resistance-encoding IncA/C plasmids from commensal and pathogenic *Escherichia coli* from multiple animal sources. PLoS ONE. 6: e23415.

8. Baquero, F., A. P. Tedim, and T. M. Coque. 2013. Antibiotic resistance shaping multi-level population biology of bacteria. Front. Microbiol. 4: 15.

9. Johnson, T. J. and K. S. Lang. 2012. IncA/C plasmids: An emerging threat to human and animal health? Mob. Genet. Elements 2: 55-58.

10. Fairchild, A. S., J. L. Smith, U. Idris, J. Lu, S. Sanchez, L. B. Purvis, C. Hofacre, and M. D. Lee: Effects of orally administered tetracycline on the intestinal community structure of chickens and on tet determinant carriage by commensal bacteria and *Campylobacter jejuni*. Appl. Environ. Microbiol. 71: 5865-5872.

11. Danzeisen, J. L., H. B. Kim, R. E. Isaacson, Z. J. Tu, and T. J. Johnson. 2011. Modulations of the chicken cecal microbiome and metagenome in response to anticoccidial and growth promoter treatment. PLoS ONE. 6: e27949.

12. Debnam, A. L., C. R. Jackson, G. E. Avellaneda, J. B. Barrett, and C. L. Hofacre. 2005. Effect of growth promotant usage on enterococci species on a poultry farm. Avian Dis. 49: 361-365.

13. DANMAP. 2009. DANMAP 2008. Use of antimicrobial agents and occurrence of antimicrobial resistance in bacteria from food animals, foods and humans in Denmark.

14. DANMAP. 2013. DANMAP 2012. Use of antimicrobial agents and occurrence of antimicrobial resistance in bacteria from food animals, food and humans in Denmark.

15. US Drug Administration Center for Veterinary Medicine. 2012. Guidance for Industry #209: The Judicious Use of Medically Important Antimicrobial Drugs in Food-Producing Animals, vol. Document No. 209 Available at: http://www.fda.gov/downloads/AnimalVeterinary/GuidanceComplianceEnforcement/GuidanceforIndustry/UCM216936.pdf.

16. US Food and Drug Administration Center for Veterinary Medicine. 2012. Guidance for Industry #213: New Animal Drugs and New Animal Drug Combination Products Administered in or on Medicated Feed or Drinking Water of Food-Producing Animals: Recommendations for Drug Sponsors for Voluntarily Aligning Product Use Conditions with GFI #209 vol. Document No. 209 Available at: http://www.fda.gov/downloads/AnimalVeterinary/GuidanceComplianceEnforcement/GuidanceforIndustry/UCM216936.pdf.

17. Casewell, M., C. Friis, E. Marco, P. McMullin, and I. Phillips. 2003. The European ban on growth-promoting antibiotics and emerging consequences for human and animal health. J. Antimicrob. Chemother. 52: 159-161.

18. Shojadoost, B., A. R. Vince, and J. F. Prescott. 2012. The successful experimental induction of necrotic enteritis in chickens by *Clostridium perfringens*: a critical review. Vet. Res. 43: 74.

19. Van Immerseel, F., J. De Buck, F. Pasmans, G. Huyghebaert, F. Haesebrouck, and R. Ducatelle. 2004. *Clostridium perfringens* in poultry: an emerging threat for animal and public health. Avian Pathol. 33: 537-549.

20. Butaye, P., L. A. Devriese, and F. Haesebrouck. 2003. Antimicrobial growth promoters used in animal feed: effects of less well known antibiotics on gram-positive bacteria. Clin. Microbiol. Rev. 16: 175-188.

21. Wierup, M. 2001. The Swedish experience of the 1986 year ban of antimicrobial growth promoters, with special reference to animal health, disease prevention, productivity, and usage of antimicrobials. Microb. Drug Resist. 7: 183-190.

22. Grave, K., M. C. Kaldhusdal, H. Kruse, L. M. Harr, and K. Flatlandsmo. 2004. What has happened in norway after the ban of avoparcin? Consumption of antimicrobials by poultry. Prev. Vet. Med. 62: 59-72.

23. Grave, K., V. F. Jensen, K. Odensvik, M. Wierup, and M. Bangen. 2006. Usage of veterinary therapeutic antimicrobials in Denmark, Norway and Sweden following termination of antimicrobial growth promoter use. Prev. Vet. Med. 75: 123-132.

24. Van Immerseel, F., J. I. Rood, R. J. Moore, and R. W. Titball. 2009. Rethinking our understanding of the pathogenesis of necrotic enteritis in chickens. Trends Microbiol. 17: 32-36.

25. Collier, C. T., C. L. Hofacre, A. M. Payne, D. B. Anderson, P. Kaiser, R. I. Mackie, and H. R. Gaskins. 2008. Coccidia-induced mucogenesis promotes the onset of necrotic enteritis by supporting *Clostridium perfringens* growth. Vet. Immunol. Immunopathol. 122: 104-115.

26. Anonymous. 2012. New Animal Drugs; Cephalosporin Drugs; Extralabel Animal Drug Use; Order of Prohibition, p. 735-745. In Food and Drug Administration H (ed.), vol. 77.

27. Antunes, P., J. Mourao, J. Machado, and L. Peixe. 2011. First description of *qnrS1*-IncN plasmid in a ST11 *Salmonella* Enteritidis clinical isolate from Portugal. Diagn. Microbiol. Infect. Dis. 69: 463-465.

28. Carattoli, A., L. Villa, L. Poirel, R. A. Bonnin, and P. Nordmann. 2012. Evolution of IncA/C *bla*CMY-(2)-carrying plasmids by acquisition of the *bla*NDM-(1) carbapenemase gene. Antimicrob. Agents Chemother. 56: 783-786.

29. Foley, S. L. and A. M. Lynne. 2008. Food animal-associated *Salmonella* challenges: pathogenicity and antimicrobial resistance. J. Anim. Sci. 86: E173-187.

30. Garcia Fernandez, A., A. Cloeckaert, A. Bertini, K. Praud, B. Doublet, F. X. Weill, and A. Carattoli. 2007. Comparative analysis of IncHI2 plasmids carrying blaCTX-M-2 or blaCTX-M-9 from *Escherichia coli* and *Salmonella enterica* strains isolated from poultry and humans. Antimicrob. Agents Chemother. 51: 4177-4180.

31. Phuc Nguyen, M. C., P. L. Woerther, M. Bouvet, A. Andremont, R. Leclercq, A. Canu. 2009. *Escherichia coli* as reservoir for macrolide resistance genes. Emerg. Infect. Dis., 15:1648-1650.

32. Poirel, L., R. A. Bonnin, and P. Nordmann. 2011. Analysis of the resistome of a multidrug-resistant NDM-1-producing *Escherichia coli* strain by high-throughput genome sequencing. Antimicrob. Agents Chemother. 55: 4224-4229.

33. Yong, D., M. A. Toleman, C. G. Giske, H. S. Cho, K. Sundman, K. Lee, and T. R. Walsh. 2009. Characterization of a new metallo-beta-lactamase gene, *bla*(NDM-1), and a novel erythromycin esterase gene carried on a unique genetic structure in *Klebsiella pneumoniae* sequence type 14 from India. Antimicrob. Agents Chemother. 53: 5046-5054.

34. Johnson, T. J., K. E. Siek, S. J. Johnson, and L. K. Nolan. 2005. DNA sequence and comparative genomics of pAPEC-O2-R, an avian pathogenic *Escherichia coli* transmissible R plasmid. Antimicrob. Agents Chemother. 49: 4681-4688.

35. Johnson, T. J., Y. M. Wannemeuhler, J. A. Scaccianoce, S. J. Johnson, and L. K. Nolan. 2006. Complete DNA sequence, comparative genomics, and prevalence of an IncHI2 plasmid occurring among extraintestinal pathogenic *Escherichia coli* isolates. Antimicrob. Agents Chemother., 50: 3929-3933.

Metrics for antibiotic use in food animals

P.R. Davies, BVSc, PhD
College of Veterinary Medicine, University of Minnesota, Saint Paul, Minnesota

Introduction

The urgency to address antibiotic resistance in human medicine is echoed among health oriented institutions from local to international levels. Bacterial genes that confer resistance to antibiotics are an ancient phenomenon and are widely distributed in nature, but it is generally accepted that antibiotic use is the primary selective force behind the occurrence of resistant bacteria that are clinically problematic.[15] There is little dispute that antibiotic use in any arena (human, food animal, companion animal, aquaculture, agriculture,...) will make some contribution to the global phenomenon of resistance, albeit with variable clinical consequences for human health. Access to antibiotics for both human and animal use varies enormously among countries, from completely unregulated access in many developing countries, to variably restricted access in more developed countries. Therefore, improving the processes of antibiotic use ('stewardship') must be a cornerstone of strategic efforts to combat the emergence of clinical resistance. Optimization of antibiotic use (ABU) requires measurement, the following discussion focuses on some of the challenges in defining appropriate metrics for antibiotic use.

Approaches to measurement of antibiotic use in food animals

Antibiotics have been available for use in animals for over 60 years, but systematic efforts to quantify antibiotic use at the national level have mostly evolved over the last 20 years. Analysis of national sales data has been, and remains, the predominant approach, but has recognized limitations that have driven efforts to obtain more granular data. Meaningful metrics of ABU requires consideration of both the numerators (measures of quantifying use) and the denominators (specifying the relevant populations and time). General approaches and metrics are summarized in Table 1, indicating key references.

Gross weight of ABU

Currently, weight of drugs is the most broadly used numerator, particularly for sales data, as it is relatively simple to obtain.[6,9] Beyond the limitation that sales data indicate distribution rather than actual use, measures of ABU based on weight alone are essentially meaningless without consideration of their relative potency, and definition of the population(s) at risk.

For example, a French study reported that total ABU in veterinary medicine (1,320 tonnes) was 1.7 times higher than in human medicine (760 tonnes), but when expressed on a demographic basis using the estimated weight of the animal and human populations, ABU was almost 3 times higher in humans (220 mg/kg/yr) than in animals (80 mg/kg/year), and there were substantial differences in the main classes of antibiotics used in animals (mostly tetracyclines and sulfonamides) and human medicine.[13]Furthermore, bundling or medically important and non-medically important antibiotics is uninformative, and in 2014 about 38% of the weight of ABU reported under ADUFA in the USA was for products deemed not to be medically important.[6]Another limitation of sales data relates to the distribution of products across animal species. In most countries including the USA, many antibiotics are labeled for use in multiple animal species, and sponsors are not required (or indeed able) to provide sales data by animal species. Furthermore, in the USA, the AMDUCA and MUMS legislation provide a legal basis for extra-label drug use in food animals, whereby products labeled for use in one species (including human products) can be used in other species when specific conditions are met.

A detailed analysis of ABU data from Denmark and the Netherlands concluded that simple country comparisons based on total sales figures 'entails the risk of serious misinterpretations'.[1] The analysis found estimates were strongly influenced by animal

Table 1: Overview of approaches to quantify antibiotic use in animals

Approach	Key references
Financial cost	Chauvin et al. (2001, 2008)[3,4]
Weight (active compound)	Grave et al. (2010)[8]
Animal daily doses	Jensen et al. (2004)[10]
Animal course doses	European Medicines Agency (2015)[5]
Prescribed daily dose	Pardon et al. (2012)[14]
Used daily doses/kg-pig	Callens et al. (2012)[2]
Used course doses	Menéndez González et al. (2010)[7]
% of herds using specific antibiotics	Jordan et al. (2009)[11]

demographics and gave 'very inaccurate indication of the true differences in exposure per animal species'. It was concluded that it is absolutely necessary to have species specific data to assess variation in antibiotic exposure between countries. However, accurate species level data is difficult and expensive to obtain. Given that different countries often face very different animal health challenges (e.g., type 1 PRRS in Europe vs. type 2 PRRS in North America), measurement systems should be designed to inform stewardship efforts within species and within countries, rather than for comparisons across species or ecological settings.

Potency (dose) adjusted measures of ABU

Variability in potency among antibiotics largely invalidates gross weight as a meaningful measure of antibiotic use with respect to improving stewardship. Substitution of a less potent compound (e.g., tetracyclines) by a more potent compound (e.g., cephalosporin) would result in an apparent reduction in ABU based on weight, when in reality the intensity of ABU (in terms of doses) could be unaltered or increased.[10] The accepted approach (devised by the WHO) to this problem is to use a divisor to convert the weight of compound to a dose related measure termed the defined daily dose (DDD). This converts the crude mass of product used in a setting to a standardized number of daily doses which can be expressed with the denominator of choice (e.g., population at risk, or hospital-bed-days) to quantify ABU. Use of DDD in practice has some caveats listed by the WHO, including

- DDD is a 'unit of measurement' and does not necessarily reflect the recommended or prescribed daily dose.
- Doses for individual patients and patient groups often differ from the DDD and will necessarily have to be based on individual characteristics
- Drug consumption data presented in DDDs only give a rough estimate of consumption and not an exact picture of actual use.
- DDD is often identical for various dosage forms of the same drug. Different DDDs may be established when the bioavailability is substantially different for various routes of administration (e.g. oral and parenteral administration) or if the dosage forms are used for different indications.
- The DDD is nearly always a compromise based on a review of the available information including doses used in various countries when this information is available.
- The DDD is sometimes a dose that is rarely if ever prescribed, because it is an average of two or more commonly used dose sizes.

The DDD approach has also been adopted in animal populations, but not necessarily in a uniform manner [10, 12] and the nomenclature used is inconsistent and confusing. In principle, daily dose conversions should be defined for each product, formulation, species, and route of administration, and compromise values will be required for products that have more than one label dose for different diseases within a species. The specifics can also vary over time within a country. In Denmark, the "animal daily dose" originally defined in 2003 was updated after 2013 with the "Defined Animal Daily Dose" due to refinement based on routes of administration. The Dutch values are available on line and are 'regularly updated' (http://www.autoriteitdiergeneesmiddelen.nl/en/dg-standard).

A study of the impact of the methods used between Denmark and the Netherlands concluded that the differences had substantial influence on the results of antibiotic consumption in pigs. Therefore harmonized units of measurement, animal weights and animal (sub) categories would be necessary to enable international comparison of ABU data.[16]

The variability in terminology and abbreviations in publications from various groups is confusing. However, although varying in details, the underlying calculations are rather generic with the following key components:

- For the numerator, weights are converted to account for potency based on consensus values derived from product labels (ADD, DADD, on either actual doses used, or less commonly based on actual use or based on doses indicated in written prescriptions.
- The number of 'daily doses' is expressed over a defined period (typically one year) in relation to some denominator representing the biomass of animals treated.
- Typically, the proxy measures for biomass are purposively chosen as reasonable estimates for animals in the respective stages of production (sows, piglets, nursery, finishing). The values used are not necessarily standardized among countries.
- A further consideration is the duration of action of single treatments together with the recommended duration of treatments using multiple administrations. This may be addressed by converting long-acting treatments to daily dose equivalents or comparing treatments in terms of courses rather than daily doses.

Defining the population at risk

Meaningful measurement of any population parameter is dependent on clear definition of a relevant denominator(s). The relationship between the inventory (and biomass) of food animals in a population (i.e., population existing at a point in time) and the number of animals raised over a defined period varies widely among animal species. For example, the US swine population is comprised of approximately 6 million sows and 60 million growing pigs but approximately 110 million animals are marketed annually due to the growing period of about 6 months. These relationships vary enormously from beef cattle (around 20-22 months growing period) and broilers (6-9 weeks growing period). The apparent intensity of ABU among species would therefore vary markedly depending on whether the choice of the 'biomass' denominator was based on the animal inventory or the annual production.

ESVAC has attempted to address this problem by developing a standardized approach to defining the population denominators, termed the population corrective unit (PCU.[9] The PCU for each 'animal category', acknowledged as a 'proxy for biomass', is calculated by multiplying numbers of live animals (dairy cows, sheep, sows and horses), and slaughtered animals (cattle, pigs, lambs, poultry and turkeys) using a standardized weight being "the time most likely for treatment". However, due to data restrictions, for farmed fish the biomass slaughtered in each country was used for the PCU.[9] This approach may constitute a reasonable compromise to achieve standardization of the population denominator among countries, but has inherent limitations in comparing ABU among species (as different proxies for biomass are used) and even within species (sow vs. growing pigs). Again, the differing needs for ABU among different species (due to different biology and pathogen profiles) means that species demographics will confound overall measures of ABU at the national level whenever species specific data are not available.

References

1. Bondt N et al. Comparing antimicrobial exposure based on sales data. *Prev Vet Med* 2013;108:10.

2. Callens B et al: 2012, Prophylactic and metaphylactic antimicrobial use in Belgian fattening pig herds. *Prev Vet Med* 106:53.

3. Chauvin C et al. The crucial question of standardisation when measuring drug consumption. *Vet Res* 2001; 32:533.

4. Chauvin C et al. Impact of antimicrobial drug usage measures on the identification of heavy users, patterns of usage of the different antimicrobial classes and time-trends evolution. J Vet Pharmacol Ther 2008;31:301-311.

5. European Medicines Agency: Principles on assignment of defined daily dose for animals (DDDvet) and defined course dose for animals (DCDvet). EMA/710019/2014.

6. Food and Drug Administration: 2014 Summary Report on Antimicrobials Sold or Distributed for Use in Food Producing Animals.

7. Gonzalez SM et al. Antimicrobial use in Swiss dairy farms: quantification and evaluation of data quality. Prev Vet Med 2010;95:50.

8. Grave K et al. Comparison of the sales of veterinary antibacterial agents between 10 European countries. J Antimicrob Chemother 2010;65:2037.

9. Grave K et al. Variations in the sales and sales patterns of veterinary antimicrobial agents in 25 European countries. J Antimicrob Chemother 2014;69:2284.

10. Jensen VF. et al. Veterinary antimicrobial-usage statistics based on standardized measures of dosage. Prev Vet Med 2004;64:201.

11. Jordan D et al. Antimicrobial use in the Australian pig industry: results of a national survey. Aust Vet J 2009;87:222.

12. Merle R et al. Feasibility study of veterinary antibiotic consumption in Germany--comparison of ADDs and UDDs by animal production type, antimicrobial class and indication. BMC Vet Res 2014;10:7-6148-10-7.

13. Moulin G et al. A comparison of antimicrobial usage in human and veterinary medicine in France from 1999 to 2005. J Antimicrob Chemother 2008;62:617.

14. Pardon B et al. Prospective study on quantitative and qualitative antimicrobial and anti-inflammatory drug use in white veal calves. J Antimicrob Chemother 2012;67:1027.

15. Perry J et al. The Prehistory of Antibiotic Resistance. Cold Spring Harb Perspect Med 2016;6:10.1101/cshperspect.a025197.

16. Taverne FJ et al. Influence of applying different units of measurement on reporting antimicrobial consumption data for pig farms. BMC Veterinary Research 2015;11:250.

Human models for reducing antibiotic use

Michael J. Sadowsky[1,2], PhD; Alexander Khoruts[1,3], MD
[1]BioTechnology Institute; [2]Department of Soil, Water and Climate, University of Minnesota, Saint Paul, Minnesota; [3]Division of Gastroenterology, Department of Medicine, Center for Immunology University of Minnesota, Minneapolis, Minnesota

Our society is at a crossroad. Antibiotics, perhaps one of the greatest medical success stories of the 20th century, are now proving to be ineffective in controlling growth of an increasing variety of pathogenic microorganism. This problem is not unique to humans, but also is problematic for large- and small-animal operations. Moreover, our current dilemma is not only due to antibiotic use, but also to misuse that leads to tremendous selection pressure on microorganisms to develop resistance. This has not only resulted in diseases that can no longer be treated with some antibiotics, but also to the development of antibiotic-induced dysbiosis in humans and animals. Since humans, and many animals, are composed of more microbial cells than those of animal origin, this has led to a variety of severe maladies that are difficult to treat. Since some practitioners try to treat these diseases with even more antibiotics, this often leads to chronic antibiotic use further exacerbating the problem.

Dysbiosis refers to disruption of microbial flora in or on the body. While nearly any body site can display antibiotic-induced dysbiosis, it is often used to refer to the gastrointestinal tract, a hollow organ containing a stunning variety of species (700-1000) and number of bacteria (up to 10^{12}/g feces). Antibiotic-induced gut dysbiosis is increasing in humans and has resulted in the increased occurrence of *Clostridium difficile* infection (CDI), in part due to loss of colonization resistance. *Clostridium difficile* has grown in incidence and mortality over the last several decades. When *C. difficile* spores reach the colons of individuals with dysbiotic GI tracts, spores germinate into vegetative cells that produce enterotoxins cause inflammation and resulting in often debilitating diarrhea. The disease spectrum caused by *C. difficile*, ranges from mild to severe cases, with the latter resulting in high mortality rates. The disease is typically treated with antibiotics (metronidazole and vancomycin). Despite this, the recurrence rate for the disease is 20-30%. While it is not known if this is due to our inability to rid the host of *C. difficile* or reinfection, we do know that there are environmental reservoirs for *C. difficile* spores in household, long term care, and hospital facilities. Moreover, patients cured of *C. difficile* often get it again following unrelated antibiotic treatment, such as for bladder infection.

Fecal microbiota transplantation (FMT), previously called bacteriotherapy, has now become widely accepted as a highly successful treatment for recurrent CDI (R-CDI). While this was originally done with blended or macerated fecal slurries from a donor (often a relative) and transferred into the recipient via nasogastric tube or enema. While this crude practice often worked, it had risks and had a negative esthetic factor that limited its practice among mainstream physicians. To this end, over the last 8 years we have developed FMT at the University of Minnesota as part of our Microbiota Therapeutics Program. FMT is almost exclusively used to treat R-CDI in humans that have failed conventional antibiotic therapy. Although, we have used it several time on monkeys.

The UMN Microbiota Therapeutics Program uses a pool of standardized donors that are rigorously screened and tested for communicable disease, have low body mass indices, and are regarded as healthy in all other aspects. Donors contribute feces in a clean and monitored restroom facility on campus, and standardized preparations of fecal microbiota are produced via wet-sieving and differential centrifugation. Final microbiota preparations contain 2.5×10^{11} to 2.5×10^{12} total bacteria and are stored frozen in cryobags at -80°C. The final material has limited odor and color and reduced esthetic barriers to treatment. Once thawed, the fecal microbiota can be instilled into the GI tract of recipients via multiple methods, although in practice we apply microorganisms to the colon via colonoscopy. The procedure is highly effective with cure rates approaching 96%. Most patients receive one preparation and this results in cure, but for patients with severe R-CDI we recommend two doses, one to get them into the intensive care unit and another for long-lasting cure. We have now developed encapsulated freeze-dried microbiota, and one capsule has been shown to result in cure with success rates approaching 90%. Thus, FMT is a highly successful alternative to treat R-CDI in humans without use of antibiotics. The same procedure has been used for large and small animals for quite some time and coprophagic animals likely "self-medicate" themselves repeatedly.

It was previously assumed, and rightly so, that FMT functions by restoring gut microbial structure and ecology that were originally made non-functioning due to toxin production and antibiotic use. Genome sequence analysis has shown that the feces of *C. difficile* patients are dominated by Proteobacteria, whereas a healthy human gut contains primarily members of the phyla *Firmucutes* (containing clostridia) and *Bacteriodetes* (containing *Bacterioides* among other genera. We now know that this hypothesis is partially correct and FMT results in rapid recolonization of the gut by fecal microbiota and reestablishment of *Firmucutes* and *Bacteriodetes*, with current reduction in the Proteobacteria. We now know, however, that FMT also functions by restoring bile acid composition in patients. The composition of the bile acid pool in humans is controlled by bacterial metabolism in the intestinal tract. In the liver, primary bile acids are conjugated to either glycine (predominantly in humans) or taurine. Bile acids that are not reabsorbed via hepatic circulation act as substrates for microbial metabolism in the gut via deconjugation, dehydroxylation, and epimerization reactions resulting in the formation of secondary bile acids. In the colon, nearly all bile acids are transformed by 7α-dehydroxylation into secondary bile acids (lithocholic acid or deoxycholic acid) and are either reabsorbed via portal circulation or excreted into feces.

We have long known that certain bile acids control the growth and germination of *C. difficile*. The primary bile acids cholic acid and taurocholic acid have been known to stimulate germination of *C. difficile* spores (Wilson 1983). In contrast, secondary bile acids produced by gut bacteria, including chenodeoxycholic acid, lithocholic acid, and ursodeoxycholic acid inhibit taurocholate-induced spore germination. Thus, the composition of bile acids in the colon has a significant impact on CDI, with primary bile acids, dominant after antibiotic treatment, and secondary bile acids, produced by colon bacteria restored after FMT, inhibiting germination.

These results indicate that a combination of restoration of gut microbiota and their ecology and changes to the composition of colonic bile acids lead to cure of R-CDI among most patients, and that this unprecedented and remarkable cure rate can be achieved in the absence of antibiotic therapy. Moreover, while the success of curing diseases in the 20th century was due to the development and use of antibiotics, the 21st century will likely see the use of microbiota to cure a variety of diseases and R-CDI is likely only the tip of the iceberg. That said, this technology is not a cure-all, magic-bullet, and must be used in a logical and thoughtful manner to reestablish a biotic state in humans and animals.

Veterinary Feed Directive: Feed industry perspective

Richard Sellers
American Feed Industry Association

The process of VFD evolved from the Food and Drug Administration's efforts to better control the use of medically-important growth promoting antibiotics in feeds since the 1960s – now called the "judicious use" policy. The VFD process evolved from an effort to create prescription medicated feeds – a process that would have caused massive disruption of medicated feed distribution due to the expected involvement of the state pharmacy boards. These boards indicated they would impose requirements on feed mills, including having either a pharmacist or veterinarian in each feed mill to dispense Rx medicated feed.

Thus, the VFD process was born in the Animal Drug Availability Act of 1996, which stated that VFDs were not prescriptions in either state or federal law. Within 70 days of enactment of ADAA, the first VFD, Tilmicosin (Pulmotil®) was approved by FDA and was the only VFD drug in the marketplace for many years. Now there are three compounds for use in swine, cattle, catfish, chickens and salmonids.

The feed industry's primary concerns involve the following: receiving from drug sponsors in a timely manner the label changes to be made and timeframe for implementation; assurances that the approved generic animal drugs will continue; that veterinarians will be trained and have knowledge of feed drugs, including approved combinations and knowledge of different approvals by different drug sponsors for the same compounds; and a reasonable plan to provide for the exhausting existing Type A premixes labeled with growth promotion claims that have up to three-year expiration dates.

Concerns shared by animal producers include the following: Receipt and storage of the VFD form (two years and not electronically), an understanding of when a VFD medicated feed can be delivered to the farm (not before receiving a VFD, including for hog supplements), having adequate supplies of old, approved drugs on-hand as the new ones come onto the marketplace and understanding that the veterinarian must complete the VFD form correctly and may not change any of the relevant information or the VFD may be invalid. Extra-label drug use (i.e. off-label use) is prohibited in medicated feed.

FDA has stated that on January 1, 2017, all drug premixes labeled at that time with growth promotion claims will be allowed to be utilized with a valid VFD for only therapeutic/preventive purposes. The feed industry has been concentrating on costs issues, realizing that the use of the currently approved drug products for therapeutic purposes will result in 2× to 10× more drug product in the feed and for likely one-time use. That makes it difficult to determine long-term use, patterns of use, stocking parameters and development and marketing of products. Business development and marketing staff are concerned about not being able to better understand the changes within only three or so months before the changes become effective.

Cooperation between the professional veterinarians, producers and the feed industry is paramount to address all major issues and ensure producers are obtaining in a timely manner the products required to prevent and treat sick animals and promote animal welfare. The feed industry is committed to that cooperation.

Practitioner experience with implemented VFD rule

Paul D. Ruen, DVM
Fairmont Veterinary Clinic, LLP, Fairmont, Minnesota

Introduction

Beginning January 1, 2017, food animal veterinarians began direct oversight of all antibiotic use in feed for prevention, control, and treatment of disease. This change occurred with the implementation of the new Veterinary Feed Directive (VFD) rule that was released by the Food and Drug Administration (FDA) in June 2015.[1] That new VFD rule followed release in December 2013 of Guidance 213,[2] which outlined a 3-year plan to transition the industry away from all growth promoter use of medically important antibiotics, eliminated over-the-counter status of these medications, and established veterinary oversight for on-farm use of therapeutic antibiotics in feed (VFD) and water (prescription).

While still early into this new oversight environment, the impact of the VFD rule is being felt by pigs, producers, feed milling industry, drug sponsors, and veterinarians in the field. In most cases, swine veterinarians have long been engaged with pork producers in the decisions related to the use of feed-grade antibiotics, and so the use of VFD's has been largely a logistical issue. The PQA Plus®[3] program has been thoroughly implemented across pork production in the US and demonstrates the close partnership between producers and their veterinarians to ensure proper care of the animals including antimicrobial therapy where appropriate. The purpose of this presentation is to share some of my practice's early experiences with and lessons learned in the field under the new VFD rule. In addition, I will discuss some of the common problems and questions that remain regarding the increased regulations.

Discussion

Swine veterinarians, unlike other food animal practitioners, were fortunate to have had some familiarity with the use of VFDs because of their required use for the feed-grade products Pulmotil® and Nuflor®. The new VFD rule added the VFD requirement for all other feed antibiotics listed as "important for humans" and approved for prevention, control, and treatment.[1] As new labels on antimicrobial products changed in December 2016, VFDs became required for the use of those products. Guidance for Industry #120,[4] released by the FDA in September 2015, provided some clarification to the details of the new VFD rule, and there has been additional information released by the FDA in 2016 to assist the food animal industry in understanding the new VFD rule. Feed medications unaffected by the rule change are bacitracin, carbadox, bambermycin, tiamulin, and ionophores.

Throughout 2016, veterinarians have been preparing to implement the new VFD rule. At Fairmont, our group determined that electronic VFD's would be the primary method used to generate VFD's for our producer clients. We identified several key issues that needed attention to be ready January 1, 2017:

1. Educating clients
2. Reaching out to suppliers – feed mills and retail
3. Veterinary clinic preparations - vets, staff, database

Client education

Our practice utilized clinic-wide producer meetings, handout materials, and individual farm strategy sessions to lead producers through the issues and the changes thrust on them by the new VFD rule. We have also made available on our website and through emails an overview of the changes created by the new VFD rule. Our clientele includes producers who may be involved in one segment of production only such as farrowing, finish, or wean/finish, and we have producers who own/manage all phases of production. Many continue to mill their own feed on-farm, while others purchase complete feed, and a few use some combination of both.

Response to our VFD education outreach varied by producer, with most desiring to streamline the feed medication component of their health plan. The new VFD rule led to some software updates in milling and production records, and many farm teams worked to adjust farm/mill processes already by mid-summer 2016. And some clients were content to let us pull them into the new era with as little engagement in the process as possible! "Oh, I'll leave that up to you," was an all-too-common response heard by our veterinary team.

Clients with multiple sow flows in the same health pyramid commonly adjusted their approach by creating one base feed medication plan across their system of production. Discussions related to the VFD rule spawned discussion and investigation into details of herd health and planning for health upgrades. *M. hyopneumoniae* elimination got a new look in those breeding herds still positive, and some additional sow farms implemented air filtration as part of their biosecurity plan.

We have many clients that own or purchase pigs from > 1 sow. The new VFD rule challenges management's planning to get the right feed to the right pig at the right time and still be efficient with stocking w/f sites that can flip between multiple flows. The list of sites under a six-month VFD can quickly become outdated when multiple pig flows stock a shared group of grower sites. The health plan and VFD control system needs to account for this challenge.

Suppliers

We are fortunate to have good working relationships with many area commercial feed mill operations, feed medication wholesalers, and medication sponsors. During 2015 and 2016 I participated in several producer education meetings held jointly by our clinic and suppliers/feed milling companies. Where we may have had prior relationships with sales and nutrition specialists, these new efforts introduced milling managers and veterinarians to each other and opened lines of communication and understanding. We obviously need to work side-by-side to manage the VFD process efficiently and accurately.

In several meetings state regulators were invited to attend and participate in question/answer sessions. Early on, it was clear the regulators differed in their own knowledge of the new VFD rule, and sometimes conflicting answers were the result. One question, in particular, led to significant disagreement among the regulators, "Who needs to register as a distributor?" For producers who manufacture feed exclusively for their own pigs the understanding is they are not a distributor. However, when the question of contract grower sites arose, many regulators held the opinion that pigs at sites not owned by the producer nor cared for directly by the producer now makes that producer a distributor. Still other regulators said that if the VFD is always written to the producer then he/she is not a distributor, but if written to the contract grower then the producer is a distributor. If consistency of oversight for antimicrobial use across the three routes of administration is a primary goal, then common sense should prevail. We write the prescriptions for water and injectable medication to the producer owner for the sites/groups/flow listed on the prescription too.

I particularly appreciated hearing the perspectives from the feed mill managers, who have been frustrated in the past with the information process of the old VFD's. Many feed mill managers have felt burdened by the responsibility to determine if a VFD is legal or not. Under the old VFD rule, it was required to record and track # tons delivered of VFD product, list veterinarian's state license, and deliver the medicated feed to only the 1 specific group/barn/site listed on the document. The new VFD rule has improved and streamlined the process to make it more efficient. Overall, however, effort has jumped dramatically due to the expanded number of medications requiring a VFD.

In some instances, companies supplying feed medications to producers with feed mills have misunderstood the new VFD rule and risked supply chain interruptions. Several producers have contacted me concerned about losing their purchasing options because suppliers communicated to producers that they will no longer be able to sell them type A feed medications. A real example via emails:

"Paul,
(Name of producer) received an Acknowledgement of Distribution form from (Name of supplier). We filled it out & went to send it back and the contact person and our representative both indicated that since we are only feeding our own pigs, come the first of the year, (Name of supplier) will no longer sell VFD medications to us. I asked them where they propose we purchase those products then and they indicated a (Name of COOP) type location. I called Sue B. at (Name of COOP) and she said that is the first she had heard of this. – Ann

Ann,
That's not right. You will need to sign an Acknowledgement of Distribution letter, but the purchasing channels are not changing. Contact Jason Gibbs, MN Dept. of Agriculture, if your supplier wants to confirm that.
(651)285-8403. – Paul

"Thanks Paul.
I did give Jason a call this morning & he agreed that is completely incorrect. (Name of Company) should still be able to sell anyone Type A and Type B medications after the first of the year as long as they have an Acknowledgement of Distribution from them."
– Ann

Fairmont Veterinary Clinic preparation

Internally, our clinic has focused on developing a plan that will generate VFD's in an accurate and timely manner for clients. To accomplish this, we needed to 1. identify and train office staff to assist with the VFD process and 2. gather and verify client information pertaining to pig sites, barn capacities, premises ID's, and other flow information. Prior to 1-1-17 our practice had just one staff person responsible for assisting our veterinary team with VFD's. In 2015 and 2016 we averaged 72 VFD's per year. Our expectation is that number will multiply by 8-10 in the 2017 calendar year. At the Fairmont Veterinary Clinic, we now have 5 office staff trained to assist with the VFD process. Our practice made the decision to work with GlobalVetLINK®'s[5] (GVL) online FeedLINK® program. GVL personnel came to our central office several times to help us in developing a plan and to conduct training sessions for our veterinarians and office staff to bring us up to speed with their web-based system, familiarize us with terminology, and work through practice VFD's so we would be prepared for the variety of production situations we see with our clients.

As a veterinary team, we have been responsible for contacting our clients to update and verify the data that must be entered onto the GVL system. Our target date for completion of this was December 1, 2016. We largely met that goal, although the site information continues to grow and change as producers' operations change and new clients come on. Information about commercial feed mills and other feed suppliers must also be included and kept current in the GVL database. While there were a few mills without email in years past, the new VFD rule has prompted change at those businesses so that electronic VFD's can be used today in virtually every situation.

When I spoke at the AASV meeting a year ago, I focused on 4 components that veterinarians needed to address to get the VFD process right: legal, accurate, efficient, and on-time. As the new VFD rule became implemented in late December 2016, there were identified some common problems and questions -

Legal

- Valid veterinary client patient relationship (VCPR)
  - How often does the veterinarian need to be on-farm, flow and site visits?

- What constitutes meaningful and real veterinary oversight?
  - Can veterinarians in a group practice help each other in signing VFD's?
- Use of VFD in compliance with laws
  - How much variation is there between regulators and states for interpretation of regulations?
- Provide client and distributor a copy of VFD
  - Who's the client and who's the distributor if the producer is a distributor?
- Retain VFD records for 2 years
  - Do some states require a different (longer) length of records storage?

Accurate

- Premise(s) where the animals are located and will be eating the medicated feed
  - How best to include multiple groups, barns, locations on the VFD form?
- VFD issuance date and VFD expiration date (maximum 6 months) based on calendar date
  - Need to have VFD's in place before 1-1-16 for feed in the feeders and bins.
- Name of drug to be used, level of drug in the feed, and duration of use to the animals
  - Clarity in descriptions to eliminate confusion in feed orders and at feed mill.
  - How do we handle intended "pulse dosing"? And what is "treatment failure"?
  - A 2nd pulse requires a 2nd VFD.
- Instructions and usage information if intended to be fed in combination with other VFD or OTC drug
  - Veterinary directions and restrictions on the VFD must match drug labels.

Efficient

- One-time notification to the FDA of intent to distribute medicated feed containing a VFD drug.[6]
  - When is a producer a distributor?
  - Will sending this notice to the FDA put you on a list for further regulatory oversight?
- Flows of pigs, and "a common treatment plan" for a farm.
  - Shared VFD for multiple groups (barns, sites) for a single producer from 1 mill system.
  - What is a reasonable definition of a production flow?

What is judicious use versus convenient use of feed antimicrobials?

On-time

- How do the 3 parties – producer, veterinarian, and distributor communicate best so the pig wins?
  - Will there be glitches in the electronic systems implemented to comply with the new VFD rule?
  - Will water antimicrobials replace some feed medication use?
- Expiring VFD's – how do we keep ahead of the process and make appropriate treatment changes?
  - How will diagnostics, medical records, and economics drive decisions?

Summary

As I write this post-election in November 2016, I am reminded that anticipating the future is a risky business – as those of us in agriculture or politics understand! In anticipating the problems that may arise with the new VFD rule, I have relied on our collective industry experiences in preparing for the new VFD rule. There will undoubtedly be unexpected problems as we hit January 1, 2017. What is certain today, however, is that veterinary oversight of feed medications must now equal that used in the other treatment routes of injectable and water medications - the consuming public and the FDA are demanding it and our veterinary profession must embrace it.

References

1. Federal Register. Final Rule: Veterinary Feed Directive. Available at: https://www.federalregister.gov/articles/2015/06/03/2015-13393/veterinary-feed-driective.

2. US Food and Drug Administration. Guidance for Industry #213, December 2013. Available at: http://www.fda.gov/downloads/AnimalVeterinary/GuidanceCompliance Enforcement/GuidanceforIndustry/UCM299624.pdf.

3. National Pork Board. PQA Plus®. www.pork.org.

4. US Food and Drug Administration. Guidance for Industry #120, September 2015. Available at: http://www.fda.gov/downloads/AnimalVeterinary/GuidanceComplianceEnforcement/GuidanceforIndustry/ucm052660.pdf.

5. GlobalVetLINK® https://www.globalvetlink.com/.

6. 21CFR558.6 http://www.gpo.gov/fdsys/pkg/CFR-2011-title21-vol6/pdf/CFR-2011-title21-vol6-part558.pdf.

GlobalVetLINK's initial experiences following the VFD and prescription changes post-January 1, 2017

J.T. Holck, DVM, MS, MBA
Feed His People, LLC, Gilbert, Iowa

Introduction

With the implementation of Guidance Document 213 on January 1, 2017, many veterinarians, producers, feed and pharmaceutical distributors are developing or changing their processes for managing Veterinary Feed Directives (VFDs), as well as veterinary prescriptions (scripts).

Many current over-the-counter (OTC) drugs have changed to VFD or script status effective January 1, 2017, so whether or not an animal health practitioner has been familiar with writing VFDs and/or scripts in the past, there is a need for understanding the changes and adjusting management strategies moving forward.

Paper and spreadsheet methods of writing VFDs and scripts involve a large amount of time, paperwork and resources to ensure clear communication, accuracy of drug label requirements, completeness of information and compliance with regulations.

Electronic software solutions from GlobalVetLINK, LLC (GVL®) reduce time spent on VFDs and scripts, reduce or eliminate paperwork, provide accurate drug label information, verify completeness of certificates and enable regulatory compliance.

GVL has provided the industry with an electronic Veterinary Feed Directive system, FeedLINK®, for over 11 years. In 2016, GVL launched an electronic solution called ScriptLINK®, which assists food animal practitioners and the supply chain in compliance with veterinary prescriptions. Together, these platforms offer a software solution for food animal practitioners to aid in compliance with the new antimicrobial regulations.

This presentation will highlight GVL's initial experiences with VFDs (FeedLINK) and prescriptions (ScriptLINK) in early 2017. This paper will briefly review the capabilities of these solutions as GVL prepared for 2017, and anticipated industry concerns and issues with implementation of the new regulations.

FeedLINK

FeedLINK VFD system is a web-based software solution, developed to assist in the streamlined fulfillment and tracking of VFDs. GVL launched FeedLINK in 2005 and continues to invest in further developments and enhancements to ensure full regulatory compliance. Another major objective of the system is streamlining the communication responsibility of veterinarians by automatically transmitting VFDs to both the feed distributor and producer. By subscribing to GVL, veterinarians, producers, and feed distributors gain access to their VFDs in a secure, 21CFR Part 11 compliant database, thus fulfilling the two year retention requirements in Guidance 120.

ScriptLINK

ScriptLINK Electronic Veterinary Prescription Certification application provides veterinarians with a secure digital Rx software system for multiple species and delivery methods. In the oversight of antibiotic usage in water solubles and injectables, this system provides online prescription certification for veterinary practices. ScriptLINK will help by optimizing prescribing, fulfillment and tracking of all prescription medication supporting the judicious use of antibiotics in food animals.

Built-in GVL SmartEngine™ Technology

The simplified process provided by GVL's platforms is driven by the patent pending GVL SmartEngine™ technology. SmartEngine technology guides the user through the labelled selections for a specific product to provide assurance to veterinarians, producers (clients) and feed mills that each VFD or script is consistent with label instructions.

VFD drugs can be searched by manufacturer, brand name, active ingredient, or indication of use by species. Once a species and product is selected, SmartEngine only allows legal options for 1) stage of production, 2) indication of use, 3) dosage, 4) duration of use, 5) combination products, and 6) expiration date. SmartEngine also auto-populates all required cautionary statements including target animal and human safety statements, special instructions, and withdrawal times.

Because extra-label use is allowed for scripts, ScriptLINK takes that into account during the creation process and provides free-text fields for the veterinarian to specify extra-label use instructions.

With the addition of more VFD and script products to manage in January 2017, SmartEngine technology is designed to help simplify the creation and management of electronic VFDs and scripts. By providing VFD expiration date reminders to all stakeholders as well as other features, the system provides compliance assurances to the veterinarian, producer (client), and feed distributor.

Integration with current feed and production software

Integrating GVL solutions with already existing software systems is a key part of GVL's business model. This provides more access to veterinarians who want to manage as much client information and records as they can through a single platform.

FeedLINK is integrated with several feed and production management software systems, including Prairie Systems, MetaFarms, Easy Automation, Inc. and PigCHAMP Inc.

In the case of Prairie Systems, FeedLINK integrates with Prairie Systems' Feed Allocation System (FAS). By integrating with Prairie Systems using the industry approved standard the actual feed order process will not change, but simply be enhanced as FAS will automatically check for a valid VFD in the FeedLINK database. If no VFD exists to enable the order to be completed, a request for a VFD will be automatically generated on the behalf of the producer and sent to FeedLINK. This submission will contain necessary information to initiate a request of the VFD to the veterinarian identified with a VCPR for that producer. The veterinarian will then review the request and generate a completed VFD or decline the request.

The integration of two third party systems working collaboratively to ensure each and every feed order is valid prior to reaching the manufacturing partner is an industry first.

Mobile friendly

Both FeedLINK and ScriptLINK systems are mobile friendly. Whether on smartphone or a tablet device, the GVL system is ready to go anywhere that has an internet connection. This mobile connectivity allows veterinarian users to create and send a VFD or script while on a farm visit, reducing the time it takes to start treatment.

Data Integrity

GlobalVetLINK's objective is to be the world's most trusted, secured and independent repository of animal health data, providing clients the confidence in the security of their data. As an independent third party company, GlobalVetLINK understands the high importance of a client's data security as well as the significance to their operation, and GVL ensures client data is secure.

Summary

GlobalVetLINK's software solutions provide food animal practitioners with a "one-stop-shop" of herd health management solutions to aid in compliance with the new antimicrobial regulations and support better decision making when prescribing interventions for clients all while helping practitioners save time.

References

1. Guidance #213 - Guidance for industry on new animal drugs and new animal drug combination products administered in or on medicated feed or drinking water of food producing animals. Available at www.fda.gov.

2. Thomas P. Countdown to Guidance 213 Practitioner View – Company. 2015 Allen D. Leman Swine Conference.

3. Main R et al. New Tools for Monitoring the Health Status of Swine Herds. 2015 ISU James McKean Swine Disease Conference.

Table 1: Key initial issues and frequently asked questions (FAQ)

Key issue	Expectation	Preparation
VFDs required for feed on inventory (bins, feeders) on January 1, 2017	Any VFD medicated feed after Jan.1 requires a VFD in order to feed to animals.	VFDs could be written in FeedLINK in December to be effective in January
Amount of paperwork	Too much to handle, not enough staff/resources to get to it all	FeedLINK eliminates paper process – stakeholders just need internet and computer, tablet or smartphone – entire process can be done digitally.
Not familiar with drug labels	Need to ensure all required VFD information is accurate and there is no off-label use	FeedLINK is powered by GVL SmartEngine technology, which enables accurate VFDs by providing specific, quality controlled, drug label information direct from the drug manufacturer.
Interpretation of regulations	There seems to be a lot of "gray areas"	GVL has been working with the FDA, state inspectors, and industry experts to align interpretations and share during webinars and customer meetings.
VFD retention and easy access in case of inspection	Veterinarian is responsible for retention of the original copy for 2 years. The feed distributor and client are responsible for retention of a VFD copy – 2 years.	FeedLINK maintains all VFDs in secure, online account for a minimum of two years. VFDs are searchable by date, owner, species, drug, etc. so certificates are easily found.
On-farm feed manufacturing – what if the client makes their own feed on-farm.	The client does not need to register as a distributor unless they plan to sell feed to someone else. They will need a VFD to purchase Type B medicated feed and will also require a VFD to feed their own animals.	GVL has been working with the FDA, state inspectors, and industry experts to align interpretations and legal processes for on-farm manufacturing of feed.
Maintaining VCPR	Follow state or federal guidelines, and provide appropriate documentation.	GVL is building a "one-stop-shop" including electronic VFD and prescription certification and the diagnostic database justification.

The VFD inspectional assignment

Center for Veterinary Medicine, Food and Drug Administration, Department of Health and Human Services

Introduction

Although FDA's VFD inspection assignment is intended to verify compliance with the requirements for the authorization, distribution, and use of VFD feeds, at this time (2016) investigators are primarily using the tool to educate producers, veterinarians, and distributors of VFD feed, including feed mills and retailers on how to achieve voluntary compliance with these requirements. The inspection tool guides our investigators through the necessary requirements of the VFD regulation during inspections with the parties responsible for retaining copies of VFD orders, letters of authorization, and whether a distributor filed a notification of intent to distribute VFD feed. These requirements come directly from the VFD regulation and are covered in the VFD brochures we have developed for producers, veterinarians, and distributors. (See references) cGMP records may also be examined if the party manufactures feed.

The inspection assignment is periodically updated. The following information is provided as an example of information that has been requested on past inspections. Ordinarily, a distributor, such as a feed mill or retail store, is the investigator's first stop in an inspection. At the distributor, the investigator will examine 3 randomly selected VFD orders for their completeness, e.g., accuracy and inclusion of all the necessary information. From these three VFD orders, the investigator will pick one VFD order to follow back to the issuing veterinarian and forward to the producer. The following elements are included in the inspection tool for the investigators to use in ensuring that the required items appear on the VFD order.

a. Veterinarian's name, address, and telephone number
b. Client's name, business or home address, and telephone number
c. Premises where the animals specified in the VFD order are located
d. Date the VFD order was issued
e. Expiration date of the VFD order
f. Is the name of the VFD drug or drugs identified on the form?
g. Species and production class of animals to be fed the VFD feed
h. Approximate number of animals to be fed the VFD feed by the expiration date of the VFD order
i. Reason the VFD order was issued (the indication)
j. Level of VFD drug in the feed and duration of use
k. Withdrawal time, special instructions, and cautionary statements necessary to use the drug according to its approved labeling.
l. The number of reorders (refills) authorized, if permitted by the drug's approval, conditional approval, or index listing
m. This required statement: "Use of feed containing this veterinary feed directive (VFD) drug in a manner other than as directed on the labeling (extralabel use), is not permitted"
n. Affirmation of intent for combination VFD drugs (see instructions)
o. Veterinarian's electronic or written signature

In addition, some elements are specific to distributors, veterinarians, or producers, and therefore the investigator will be asking the following more specific questions to individuals based on their responsibilities.

Questions specific to all distributors

1. Did the distributor notify FDA of the intent to distribute VFD feeds? (Verify VFD Distributor Notification Listing on `http://www.fda.gov/downloads/AnimalVeterinary/Products/AnimalFoodFeeds/MedicatedFeed/UCM096059.pdf`
2. Does the distributor keep copies of VFD orders for at least 2 years, if distributing VFD feed to the end user?
3. If the operation being inspected distributes VFD feed to other distributors and not to the end user, does this operation keep copies of acknowledgement letters for at least 2 years from the date of last shipment under the acknowledgement letter?

Questions specific to distributors who manufacture VFD feed

1. Does the feed label contain the VFD Caution statement?
2. Do the drug inventory or production records show the correct amount of drug added?
3. Do the labels and formulas match the VFD orders?

Questions specific to veterinarians

1. Does the veterinarian have a valid license in those states where VFD feed is being fed?
   a. If No, can the veterinarian provide any evidence that he or she is complying with applicable state veterinary licensing and practice requirements in the states in which the veterinarian does not have a license?
2. Does the veterinarian know that either the state or federal requirements for veterinary client patient relationship (VCPR) apply in each state?
   a. Can the veterinarian show any medical record(s) for the client's animals named on the VFD?
3. Does the veterinarian keep copies of VFD orders for at least 2 years?

Questions specific to producers (clients)

1. Does the client keep copies of VFD orders for at least 2 years?
2. Did the client feed the VFD feed to the authorized number of animals on the VFD order?
3. Did the client feed the VFD feed for the identified duration on the VFD order?

4. Did the client stop feeding the VFD feed prior to the expiration date on the VFD order?
5. Did the client follow the withdrawal period for the VFD feed, if any?
6. Did the client follow any special instructions or caution statements on the VFD order, if any?
7. If a combination VFD feed was fed, was its use consistent with the affirmation statement on the VFD order?
8. Does the client have labels for VFD feeds? If Yes,
   a. Does the feed label contain the VFD Caution statement?
   b. Did the drug level on the label match the drug level on the VFD form?
   c. Is the drug level and indication on the VFD form consistent with the approval?

The investigators have been asked to follow a phased enforcement approach with an emphasis on using education to obtain voluntary compliance with the VFD rule requirements, prior to pursuing more formal enforcement actions. If you have further questions, please let us know by sending an email to AskCVM@FDA.HHS.GOV.

References

The 2015 Veterinary Feed Directive rule: https://www.gpo.gov/fdsys/pkg/FR-2015-06-03/pdf/2015-13393.pdf

CVM's VFD Brochures for:

Producers:

English: http://www.fda.gov/downloads/AnimalVeterinary/DevelopmentApprovalProcess/UCM455419.pdf

Spanish: http://www.fda.gov/downloads/AnimalVeterinary/DevelopmentApprovalProcess/UCM525590.pdf

Distributors:

That manufacture feed:

English: http://www.fda.gov/downloads/AnimalVeterinary/DevelopmentApprovalProcess/UCM455425.pdf

Spanish: http://www.fda.gov/downloads/AnimalVeterinary/DevelopmentApprovalProcess/UCM525591.pdf

That do not manufacture feed:

English: http://www.fda.gov/downloads/AnimalVeterinary/DevelopmentApprovalProcess/UCM455479.pdf

Spanish: http://www.fda.gov/downloads/AnimalVeterinary/DevelopmentApprovalProcess/UCM525601.pdf

Veterinarians:

English: http://www.fda.gov/downloads/AnimalVeterinary/DevelopmentApprovalProcess/UCM455480.pdf

Spanish: http://www.fda.gov/downloads/AnimalVeterinary/DevelopmentApprovalProcess/UCM525615.pdf

Veterinary students:

English: http://www.fda.gov/downloads/AnimalVeterinary/DevelopmentApprovalProcess/UCM455481.pdf

Spanish: http://www.fda.gov/downloads/AnimalVeterinary/DevelopmentApprovalProcess/UCM525618.pdf

Other sources of information:

Webpages:

- Veterinary Feed Directive (VFD): http://www.fda.gov/AnimalVeterinary/DevelopmentApprovalProcess/ucm071807.htm
- Safe Feed: http://www.fda.gov/safefeed
- Judicious Use: http://www.fda.gov/AnimalVeterinary/SafetyHealth/AntimicrobialResistance/JudiciousUseofAntimicrobials/default.htm
- Blue Bird Labels: http://www.fda.gov/AnimalVeterinary/Products/AnimalFoodFeeds/MedicatedFeed/BlueBirdLabels/default.htm

Guidance documents:

GFI # 120: http://www.fda.gov/downloads/AnimalVeterinary/GuidanceComplianceEnforcement/GuidanceforIndustry/UCM052660.pdf

GFI # 152: http://www.fda.gov/downloads/AnimalVeterinary/GuidanceComplianceEnforcement/GuidanceforIndustry/UCM052519.pdf

GFI # 181: http://www.fda.gov/downloads/AnimalVeterinary/GuidanceComplianceEnforcement/GuidanceforIndustry/UCM052468.pdf

GFI # 209: http://www.fda.gov/downloads/AnimalVeterinary/GuidanceComplianceEnforcement/GuidanceforIndustry/UCM216936.pdf

GFI # 213: http://www.fda.gov/downloads/AnimalVeterinary/GuidanceComplianceEnforcement/GuidanceforIndustry/UCM299624.pdf

GFI # 233: http://www.fda.gov/downloads/AnimalVeterinary/GuidanceComplianceEnforcement/GuidanceforIndustry/UCM474640.pdf

Statute:

Animal Drug Availability Act of 1996: https://www.gpo.gov/fdsys/pkg/PLAW-104publ250/html/PLAW-104publ250.htm

Economic impact: Fitness traits in post-weaning pigs and sows in lieu of genetic improvement in litter size and leanness

John Mabry, PhD, Professor Emeritus
Iowa State University, Ames, Iowa

There are several different genetic technologies that have been used to improve trait performance for pig farmers. Some examples would include:

a) Terminal Cross Mating System: this is simply using a crossbred female composed of breeds that are best for reproduction as the maternal line animals. The most common examples are the use of a Landrace/Large White F1 female as the parent female. This allows the best breeds for reproduction to exert their strengths, and takes advantage of the heterosis that is expressed by the F1 female for reproductive traits.
b) Optimal Genetic System Structure: this is where the production farm uses the proper number of GGP, GP and Parent Stock females in the genetic system. Usually at least 10% of the female herd should be in the GGP and GP roles. These maternal females then are mated to a breed or line of terminal boar that best fit the needs of the producer. The most common terminal breeds are the Duroc, Pietrain, or composite terminal lines. This allows for the producer to take advantage of both maternal heterosis in the females, and heterosis for terminal traits in the market hogs.
c) BLUP based selection programs: in order to maximize the genetic progress from selection, particularly in the maternal lines, we must use the most accurate method for estimating genetic merit in the animals eligible for selection. The BLUP system of breeding value estimation is the most accurate system available, and can be used by both the genetic supplier and the commercial producer to improve the traits of economic importance.
d) Marker Assisted Selection: we can now identify individual genes that have a major influence on certain traits of economic importance, such as litter size, feed conversion and meat quality. However, the genetic assay is somewhat expensive and is usually done by the genetic supplier, but the commercial producer is the one that enjoys the benefits.
e) Mate-Select Software: it is now possible for the genetic supplier to use software that can structure the matings made at the GGP level to produce the most extreme progeny from the pure line matings, while minimizing inbreeding. This is very important in the production of the most extreme boars for AI usage in both the maternal and terminal lines. This is usually employed by the genetic supplier and the results are then relayed to the commercial producer.
f) Genomic information: it is also now possible to genotype GGP animals for a large number of gene pairs, in excess of 60,000 genes for one animal. The cost of this is still quite high, but it can result in a significant increase in the accuracy of BV estimation, but due to cost, is employed at the genetic supplier only.

It is now possible for the commercial pork producer to reap the benefits from all the above genetic technologies, either directly using them, or having their genetic supplier use them.

In terms of improvement in performance of key swine traits, over the past 30 years we have seen:

1. Litter size has increased from 10 → 14 or more pigs/litter
2. Backfat has decreased from 25mm at P2 measurement → 10mm or less
3. Lean meat percentage has increased from 45% → 57% or more
4. Growth rate has increased from a 220 lb pig @ 7 months of age → 285 lb pig @ 6 months of age
5. Feed conversion has improved from 3.50 lb feed/lb gain → 2.50 lb feed/lb gain or less

While this is an overall positive change, in making genetic improvement we have to be aware of any possible 'correlated responses' to this genetic improvement in trait performance. Pearson (1903) and Robertson (1967) opined that *"..individuals with intermediate trait values have higher fitness.."*, and, *"..extreme phenotypes cause reduced fitness.."*.

For example, as litter size has increased, it has exceeded the uterine capacity of the sow, and as a result while more pigs are farrowed and weaned per litter, the number of piglets with low birth weights has increased. The potential impacts of this include increased piglet mortality, and of those piglets that do survive to weaning have a higher percentage of pigs with sub-standard weaning weights.

Also, as backfat has decreased and lean meat percentage has increased and growth rate has increased, the percentage of animals that are more fragile and have more structural problems has increased. These negative correlated responses in fitness traits exert an economic impact with higher sow and pig mortality, higher sow culling rates, and an increased percent of sub-standard market hogs.

The overall goal of pork production is to maximize profit. So while there is a need to maximize the performance of key economically important traits in pork production, at the same time we must consider any potential correlated responses that might negatively impact profit.

Profit = return > cost of production

Profit can be improved by either increasing the return from animals sold from the farm (by increasing reproduction, survival or growth), or by lowering the cost to produce the pigs (by improving feed conversion and reducing non-productive sow days). In order to maximize the profit the pig farmer needs to know their cost of production, including the specific components. Software can be utilized to assist the pig farmer in knowing their cost of production, and to examine the overall benefits from changing any of the components. The software utilized for the results in this paper is PigProfitTracker© developed by the Iowa Pork Industry Center at Iowa State University, Ames, IA (Figure 1).

At the sow farm, the capacity is usually determined by the desired output of market hogs and level of investment of the farm owners. The correct number of gestation spaces, farrowing spaces and gilt development spaces is based on the overall desired production capacity. The primary duties are to breed the sows, farrow the sows, wean the litters, and then repeat the process. Consequently then the sow farm costs accumulate on a sow/day basis. The owner must feed the sows, pay for the facilities, labor, vet-med costs, gilt development costs, genetic costs and other variable costs. The returns from the sow farm then come from full value weaned pigs, marginal value weaned pigs, and cull sows that are sold.

There are a number of variables or 'traits' that can influence either the cost or return at the sow farm. Returns can be increased by increasing farrowing rate, increasing the number of pigs born alive (of adequate size), decreasing piglet mortality, increasing piglet weaning weight or decreasing sow mortality and cull rate. Costs can be reduced by reducing non-productive sow days and by decreasing piglet mortality. Non-productive sow days are decreased when farrowing rate is increased and wean to service interval is decreased. In order to make genetic improvement in these traits, they must have adequate genetic control to respond to selection, and the pig farmer must use the correct genetic system to get heterosis for these traits. The traits that have been shown to respond to BLUP based selection include litter size, pre-weaning mortality, litter weaning weight, and litters/sow/year (as a way to indirectly improve farrowing rate). Sow mortality can be improved by phenotypic selection in structure and soundness.

Economic impact of piglet birth weight

I have used the term 'number born alive of adequate size' as a trait to be selected for. We have seen tremendous genetic progress in litter size since the implementation of BLUP based selection. However, as we have increased litter size above 12 NBA, the number of pigs with birth weights less than 1 kg has increased, along with piglet mortality. This is due to the litter size exceeding the uterine capacity of the sow, with pigs in large litters getting less pre-natal nutrients from the sow's uterus, resulting in more pigs that are smaller at birth than desired.

People who have raised pigs have long realized that the birth weight of a piglet is associated with its potential to make a profit for the pig farmer. The lifelong pathway of commercial pigs is shown in Figure 2.

Research has shown a strong relationship between the birth weight of a piglet and its survival rate, weaning weight, market weight, and potential profitability.

See Figure 3 showing the relationship between piglet birth weight and survival rate (Dufrasne et al, JAS 2013). If the piglet has a birth weight in excess of 1.4 kg then the pre-weaning mortality is less than 10%. If the piglet has a birth weight of less than 1 kg, then the pre-weaning mortality is very high, in excess of 25%. Based on these results, it appears that a reasonable target for piglet birth weight of 1 kg and higher will keep pre-weaning mortality at profitable levels.

Research done by Smithfield Premium Genetics® and published by Dr. Justin Fix of North Carolina State University (Fix, et al 2010) examined the relationship between piglet birth weight and weaning weight at 3-4 weeks of age (Table 1).

They weighed more than 2400 piglets at birth, assigned them to birth weight classes, and then followed their weight at weaning. A linear relationship was found between piglet birth weight class and piglet weaning weight. Most noticeable was the weaning weights of less than 5 kg for piglets with birth weights less than 1 kg on average.

The authors also followed the pigs from weaning to market and tracked the cost and return of each piglet. They measured the piglets post-weaning mortality, growth rate and feed conversion, and eventual return when marketed (Table 2).

Based on these results, using input costs from the USA in 2010 (Fix et al, 2010), it is clear that as the birth weight of the piglet increased, the profit of the pig at market weight also increased significantly.

The results from Fix et al (2010) can be used to develop regression effects of piglet birth weight on economically important traits such as pre-weaning mortality, post-weaning mortality, market weight, and the percentage of 'Full Market Value Pigs'. The results showed that as pig birth weight got below 1.4kg the chance of profit got less probable. Their research suggested that for each decrease in piglet birth weight of 0.1kg (below 1.4kg) the pre-weaning mortality increased by 3%, the post-weaning mortality increased by 2%, market weight decreased by 1.63kg or 3.6 lb, and the probability of it being a 'Full Market Value' pig decreased by 2%. They defined a 'Full Market Value' pig as one that weighed more than 200 lbs (90kg), had no injuries, ruptures or other defects, and received full market value at the slaughterhouse. A pig that was not 'Full Market Value' received a discounted market price of 75%.

Figure 1: Software used to estimate profit/loss by swine producer.

Pig Profit Tracker

Breed - Finish *v.426*

Farm Scenario: *Iowa View Farms*

IOWA STATE UNIVERSITY - Iowa Pork Industry Center - (515) 294-4103, www.ipic.iastate.edu 2/26/16

Inputs

Corn price ($/bu)	3.70	Weaning wt (lb)	13	Breeding sows in herd	2400
SBM ($/ton)	265.00	Sale/marketing wt (lb)	280	Litters/sow/year	2.35
VTM ($/ton)	624.00	Wean - Finish F:G (lb, live)	2.85	Pigs weaned/litter	11.9
Additive ($/lb)	2.75	Wean - Finish mortality (%)	5.00	Replacement cost ($/gilt)	200.00
		Avg death loss wt (lb)	100	Cull sow sale weight (lb)	450
Carcass dress (%)	75	Substandard sales (%)	4	Annual replacement rate (%)	45
		Substandard wt (lb)	185	Sow mortality rate (%)	6.2

Farrowing		$ / pig weaned
Gilt Dev purchase cost/female genetics		$ 3.66
Breeding cost / semen & boars		$ 1.55
Vet / Medicine		$ 2.00
Labor		$ 3.50
Fixed (building, taxes, rent, pymt etc)		$ 7.00
Variable (ins, util, repairs, misc.)		$ 4.55
Management/ genetic/ accounting fees		$ 1.00
Manure		$ -
GMD average cost per ton ($/ton)	$ 11.00	$ 0.47
Cull sow value ($/lb)	$ 0.40	$ (2.90)
Other		$ -
Total non-feed costs		**$ 20.83**

Post-Weaning to Finish		$ / pig
Death loss		$ 2.57
Vet / Medicine		$ 2.32
Labor		$ 2.81
Fixed (building, taxes, rent, pymt etc)		$ 11.28
Variable (ins, util, repairs, misc.)		$ 8.63
Management/ genetic/ accounting fees		$ 1.50
Trucking		$ 2.00
Manure	$ (180,000)	$ (2.82)
GMD average cost per ton ($/ton)	$ 10.00	$ 3.80
P1 progeny *loss*	$ 5.00	$ 1.15
Other		$ -
Total non-feed costs		**$ 33.24**

Farrowing

	Lact. (days)	Days/yr	Daily (lb)	Total (lb)
Lactation	23	54	12.3	665
Gestation		311	5.5	1709
Total >>	Sow feed per sow per year			**2374**
	Sow feed per pig weaned			85

		%	Lb / pig	arket	$ / pig
71	Corn	71	60.3	$ 0.07	$ 3.98
25.	SBM	25.50	21.6	$ 0.13	$ 2.87
	VTM / Premix	3.30	2.8	$ 0.31	$ 0.87
	Additive	0.20	0.2	$ 2.75	$ 0.47
	Total	100.00	**84.9**	**$ 0.096**	**$ 8.19**
			Lb / pig	$ / lb	$ / pig
	Prewean Pig Feed		0.5	$ 0.50	$ 0.25
	Total SEW pig feed cost				**$ 8.44**

Post-Weaning to Finish

		%	Lb / pig	$ / lb	$ / pig
76.	Corn	76.58	582.7	$ 0.07	$ 38.50
21.	SBM	21.00	159.8	$ 0.13	$ 21.17
	VTM / Premix	2.27	17.3	$ 0.31	$ 5.39
	Additive	0.15	1.1	$ 2.75	$ 3.14
	Paylean etc. $/hd				$ -
	Total	100.00	**761.0**	**$ 0.090**	**$ 68.20**

Cost per head-- summary breakdown

	Farrowing	Wean-Fin	Total
Total feed cost	$ 8.44	$ 68.20	**$ 76.65**
Total non-feed cost	$ 20.83	$ 33.24	**$ 54.07**
Total >>	**$ 29.27**	**$ 101.44**	**$ 130.72**

Miscellaneous Outputs

Number of pigs weaned	**67116**
Number of finished hogs sold	**63760**
Avg wt sold (live)	276.2
lbs sold per sow per year (live)	7338
Farrowing corn (bu/hd)	*1.08*
Finishing corn (bu/hd)	*10.41*
	11.48
Wean-finish +deads F:G (lb)	2.84
Whole herd F:G (lb)	3.04
Wean-finish feed ($/ton)	179.26
Breed-wean feed ($/ton)	197.71
Breed-finish feed ($/ton)	180.64

Profit/loss

	$/cwt carcass	per head
Full value mkt price ($/cwt carcass)		$ 66.00
Substandard or lightweight price ($/cwt live)		$ 49.50
Income -	$ 66.24	$ 136.72
Expense - Feed	$ 37.13	$ 76.65
- Non-feed		$ 54.07
Total	$ 63.33	$ 130.72
Profit/loss per hog sold	**$ 2.91**	**$ 6.00**
Profit/loss total per year	**$**	**382,560.00**
Profit/loss attributed to farrowing		**$ 1.34**
Profit/loss attributed to finishing		**$ 4.66**

PigProfitTracker©, ISU

Figure 2: The lifelong pathway of commercial pigs.

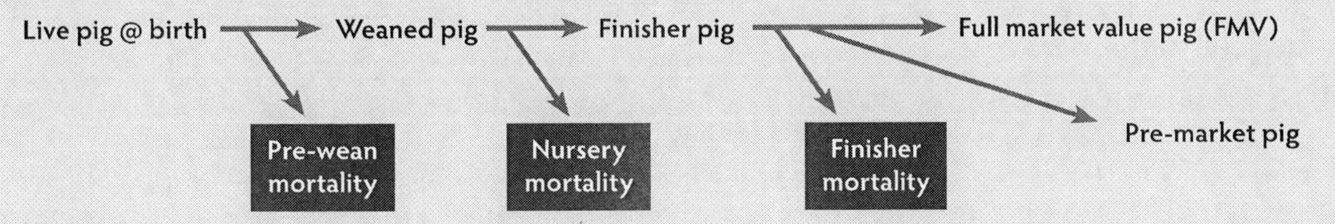

Figure 3: Relationship between piglet birth weight and survival rate.

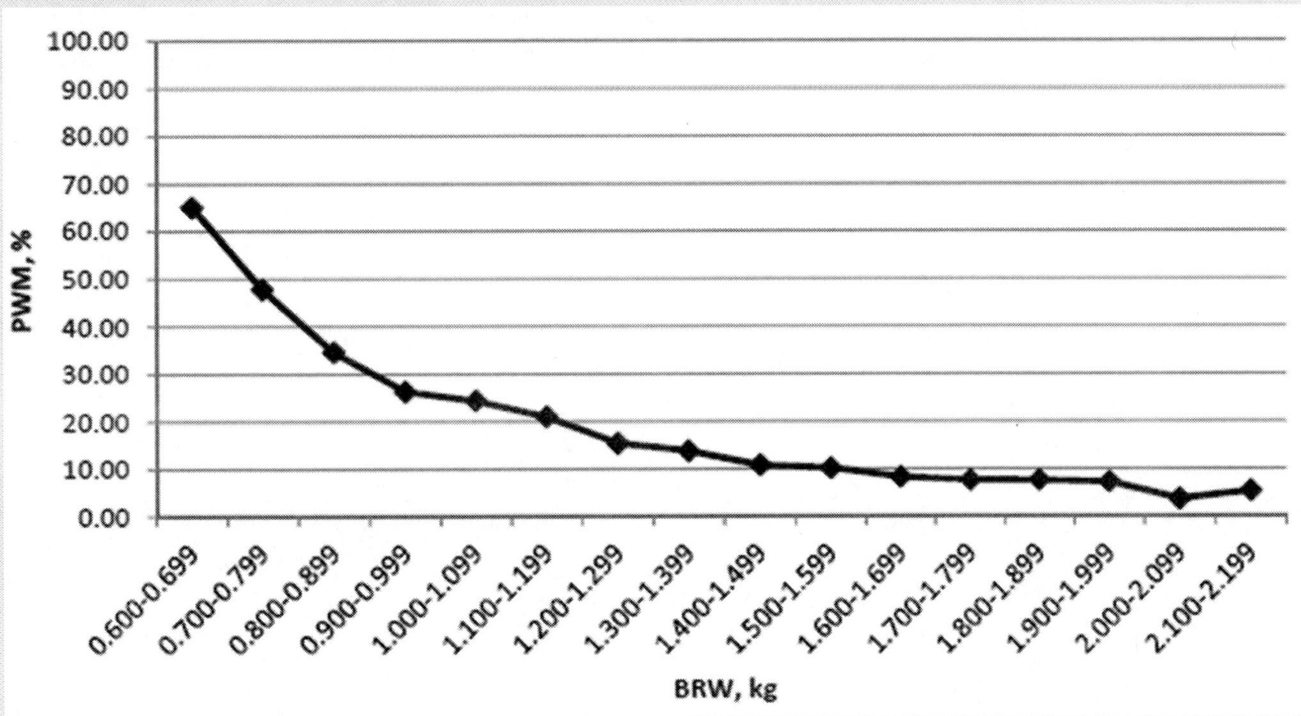

(Dufrasne et al, JAS 2013)

Using the above research based information, we can estimate the profitability of each birth weight class (1.4kg, 1.3kg, 1.2kg, 1.1kg, 1.0kg) using the PigProfitTracker® software. This software uses farm specific cost of production inputs, and user inputted performance and price information. A market carcass price of $84/cwt was assumed for FMV pigs, and $63/cwt for non-FMV pigs. A feed cost of $0.12 was used.

The relative performance and estimated profit for each birth weight class of piglet is shown in Table 3. Based on this information, as a piglet's birth weight declines from 1.4kg down to 1.0kg, the average profit per pig also declines. And based on the information and assumptions in Table 3, piglets with a birth weight of 1.0kg or less will not produce a profit for the pig farmer. How does this then impact the genetic supplier? Over the past 10 years the genetic merit for litter size has increased greatly; however, the number of piglets with low birth weights has also increased as the litter size has exceeded uterine capacity of the sow. How should the genetic supplier account for this increase in low birth weight pigs in their genetic improvement program? A logical solution is to only give born alive credit to those pigs that have an adequate birth weight to be profitable. Therefore, the move towards selection for litter size born alive with birth weight above 1kg is recommended for use in improving profitable litter size in pigs.

Table 1: Effect of birth-weight category and parity on mean weaning weight in a maternal line of barrows and gilts*.

Birth-weight category	No. of piglets (n = 2467)†	Average parity of dam	Birthweight (kg)			Weaning weight (kg)‡	SEM
			Minimum	Maximum	Mean		
1	42	2.4	0.66	.094	0.86	4.15	0.13
2	135	2.6	.095	1.10	1.03	4.65	0.07
3	243	2.9	1.11	1.26	1.19	5.03	0.05
4	387	2.7	1.27	1.42	1.35	5.38	0.04
5	491	2.8	1.43	1.58	1.51	5.76	0.04
6	467	3.2	1.59	1.74	1.67	6.08	0.04
7	319	3.4	1.75	1.90	1.82	6.39	0.05
8	238	3.7	1.91	2.06	1.97	6.64	0.05
9	145	4.0	2.07	2.85	2.24	7.15	0.07

* Pigs were weaned either at an average of 15 days of age (pigs weaned at 14, 15, or 16 days) or at an average of 20 days of age (pigs weaned at 19, 20, or 21 days). Individual piglet birt weights were partitioned into nine categories which incrementally increased or decreased by 0.5 SD (0.16 kg) from birth-weight mean (1.57 kg).

† Includes only pigs that survived until weaning.

‡ All weights differ from each other ($P < .001$; ANOVA).

Research done by Smithfield Premium Genetics® and published by Fix et al, 2010.

Table 2: Differences in predicted revenue ($) at harvest for pigs of various birth weights, compared to an average birth weight pig (1.4 kg).

			Live market hog price				
Birth weight, kg	**$43.00**	**$45.00**	**$47.00**	**$49.00**	**$51.00**	**$53.00**	**$55.00**
0.8	-$39.42	-$41.25	-$43.09	-$44.92	-$46.75	-$48.59	-$50.42
1.0	-$26.76	-$28.00	-$29.25	-$30.49	-$31.74	-$32.98	-$34.22
1.2	-$13.09	-$13.70	-$14.31	-$14.92	-$15.52	-$16.13	-$16.74
1.4	$0.00	$0.00	$0.00	$0.00	$0.00	$0.00	$0.00
1.6	$11.38	$11.91	$12.44	$12.97	$13.50	$14.03	$14.56
1.8	$20.60	$21.56	$22.52	$23.48	$24.44	$25.39	$26.35
2.0	$27.72	$29.01	$30.30	$31.59	$32.88	$34.17	$35.46
2.2	$33.03	$$34.56	$36.10	$37.63	$39.17	$40.71	$42.24
2.4	$36.84	$38.56	$40.27	$41.99	$43.70	$45.41	$47.13

Research done by Smithfield Premium Genetics® and published by Fix et al, 2010.

Economic impact of sow culling rate and sow mortality rate

When a sow goes out of service from the breeding herd, either by voluntary culling or dying, there is a cost to the farm. Even if the farm is raising their own replacement gilts, these events incur a cost to the farm. The traditional comparison when making a decision to cull a sow has been to compare the performance of a 'potential cull' sow to that of a replacement gilt. For instance, if the performance of the older sow is 10.5 born alive and 80% farrowing rate; and, the performance of the parity 1 replacement gilt is expected to be 11 born alive and 88% farrowing rate, most herds have chosen to cull the older sow.

However, there is another factor to consider in making this decision: the performance and profitability of progeny from the parity 1 sow compared to those progeny from parity 2+ sows. Klobasa et al (1986) reported on differences in immunoglobulin synthesis in first versus multi-parity sows. Deen (2002) reported that gilt progeny are more likely die in the nursery than sow progeny. Morales et al (2006) and Burkey et al (2008) showed an effect of sow parity on piglet health status. And Miller (2008)

Table 3: Relative performance and estimated profit for each birth weight class of piglet.

Variable	1.4 kg BW	1.3 kg BW	1.2 kg BW	1.1 kg BW	1.0 kg BW
Pre-wean mortality	8%	11%	14%	17%	20%
Post-weaning mortality	5%	7%	9%	11%	13%
Percent 'Full Market Value' pigs	95%	93%	91%	89%	87%
Percent 'Pre-market' pigs	5%	7%	9%	11%	13%
Avg market weight (FMV pigs)	250.0	246.4	242.8	239.2	235.6
Profit (loss) for FMV pigs	15.00	12.26	9.48	6.66	3.79
Profit (loss) for non-FMV pigs	**(-32.23)**	**(-33.82)**	**(-35.44)**	**(-37.11)**	**(-38.83)**
Net Profit (loss) per pig marketed	12.64	9.03	5.44	1.85	**(-1.75)**

Table 4: Progeny comparison

Item	Parity 1 Progeny	Parity 2+ Progeny
Weaning weight, lbs	11.7	12.5
Nursery mortality %	3.2	2.6
Nursery ADG (lbs/day)	0.90	0.96
Nursery medical costs	\$1.79	\$0.71
Finishing mortality %	4.31	2.95
Finishing ADG	1.62	1.69
Finishing medical costs	\$1.52	\$0.84
Estimated profit difference	-\$5 to -\$9 per pig	

and Miller et al (2012) examined the poor growth and survival of the progeny of gilts.

Research presented at the Leman Conference has shown that progeny from parity 1 sows have a lower weaning weight, higher mortality, lower growth rate and higher medical costs. The magnitude of this difference is shown in Table 4.

So now when the producer considers voluntarily culling a sow, they should take into consideration the salvage value of the sow, the cost of the replacement female, the reproductive performance of the replacement female, and the reduced performance of the progeny from the first parity sow compared to the progeny from the mature sow.

There is a cost to both sow mortality and sow culling rate. If we consider the factors mentioned above, assume no difference in reproduction, use a \$5 per pig reduced performance from gilt litters, and use the PigProfitTracker© software, the economic impact in a 2400 sow farm for a 1% increase in sow mortality is to lower profits by approximately \$6057 per year. And the economic impact of a 1% increase in sow culling rate is to lower profits by approximately \$1652 per year. When one considers how easy it is to see culling rate change by 5% or more, this economic impact becomes very significant.

Economic impact of post-weaning mortality and percent full market value pigs

Within a commercial pork production system, the size of the Grow-Finish farms is set by the sow farm size and pig flow. The pig farmer needs the correct number of nursery and finisher spaces (or wean-finisher spaces) to match pig flow from the sow farm. The role of these farms is to start the pigs on feed after weaning (the bigger the pig the better), then to sell the pigs when they reach the desired market weight. As a result, the growth rate, feed conversion and mortality of the pigs are very important traits. The return when the pigs are marketed are based on the number of pigs marketed with adequate size, as the packer will discount any pigs that are under a minimum market weight. There can also be some impact on price based on the meat quality of the pigs, but this is not usually a large impact.

So at the Grow-Finish farms, the fixed costs are based on the per head capacity of the farm, and variable costs are based on several factors for each pig, especially feed conversion and mortality rate and weight when mortality occurs. The return is influenced by the number of full value pigs marketed, the growth rate, and mortality rate. One important fact that bears mention here is that the pig's genetics for post-weaning traits comes equally from the sire and dam of the piglet.

The economic value of post-weaning mortality is well recognized. When the pig is weaned, then moved to the grow-finish facilities, it is expected to eat feed, convert the feed to lean gain, and then be marketed as a full value market hog. When a pig dies after weaning it has incurred much of the cost of a market hog, but gives no value in return. Using the PigProfitTracker© software and an assumption that the average weight of post-weaning mortalities is 100 lbs, the economic impact of a 1% increase in post-weaning mortality is $30,914 for a 2400 sow breed to finish farm.

The economic impact of a pig failing to meet the 'Full Market Value® standards is not as well recognized. In this case the pig does not die, but also does not reach market weight of at least 200 lbs. When this occurs the packer discounts the price given to the pig by at least 25%. Using the PigProfitTracker© software the economic impact of a 1% increase in non-Full Market Value pigs is $28,682. While these pigs do generate some return, they are very inefficient in their growth and the reduction in market return is quite severe.

Current popular genetic marketing strategy

Several genetic suppliers have focused heavily on improving litter size and pigs weaned per sow per year, in both their genetic improvement programs and in their marketing strategies. Many customers who have switched to these genetic suppliers have seen correlated responses in higher sow mortality, higher sow culling rates, higher post-weaning mortality, and an increase in the percent of non-Full Market Value Pigs. What is the economic 'trade off' in this situation?

Using the PigProfitTracker© software, with a starting assumption of a 2400 sow breed to finish herd, and an economic assumption of $6/pig profit. Reproductive and performance starting points were 28 PSY, 45% culling rate, 6.2% sow mortality rate, 5% post-weaning mortality rate, and 4% of pigs were non-Full Market Value pigs as defined earlier in this paper. We then imposed three different scenarios with an increase of reproductive performance from 28 PSY to 31 PSY. In Case 1 (no correlated loss of fitness) we kept the sow culling and mortality rates, and pig mortality and non-FMV rates the same. In Case 2, we assumed a small correlated decrease in fitness traits, and in Case 3, a moderate decrease in the fitness traits. These results are summarized in Table 5.

If we just increase PSY from 28 to 31 and assume no loss in the fitness traits, then profit from the farm increases by 23%, as shown in Case 1. However, if we assume a slight loss of fitness as shown in Case 2, the annual profit from the farm is essentially the same as our 28 PSY starting point. And if we assume a moderate loss of fitness as shown in Case 3, the annual profit from the farm is reduced by 30% from our 28 PSY starting point. It is not logical to assume we can increase PSY from 28 to 31 with no loss of fitness in the sows or pigs, so Case 1 would not be an expected outcome. The lesson here is that the producer needs to consider both the increase in reproduction from the 31 PSY sow genetics, but also the producer should consider what level of loss in the fitness traits for the sows and pigs would be expected under their farm management and environment.

At this point, let me summarize several of the 'Take Home Points' from this discussion:

- Over the past 30 years the use of genetic technologies has made significant genetic improvement in reproduction, growth rate, feed conversion, and lean meat percentage.
- At the same time we have seen an increase in the percent and number of low birth weight pigs, which eventually has lower weaning weights, poorer survival, slower growth and less profit potential.
- Associated with this movement towards more extreme phenotypes we have seen a correlated reduction in the fitness traits of both sows and piglets (mortality rates, culling rates, sub-standard weaners and finishers)
- It is important that genetic suppliers include phenotypic structure in their genetic improvement programs as a way to minimize any loss of fitness.
- It is also possible for genetic suppliers to include fitness traits such as mortality rates and culling rates in their genetic improvement program, most likely through progeny test programs.

Table 5: Profit comparison

	Average	Case 1	Case 2	Case 3
PSY	28	31	31	31
Sow culling %	45	45	48	50
Sow mortality%	6.2	6.2	8	10
PW mortality %	5	5	6	7
% Non-FMW pigs	4	4	5	6
Profit	$382,560	$471,535	$382,058	$295,070

- For the commercial pork producer it is logical to focus their genetic program towards one that will maximize long term profits at their farm rather than chase short term improvements in targeted traits of current interest.

References

1. Burkey T.E., Miller P.S., Johnson R.K., Reese D.E., Moreno R. 2008. Does dam parity affect progeny health status? Nebraska swine report. University of Nebraska, Lincoln, NE.

2. Dufrasne, M., Misztal, I., Tsuruta, S., Holl, J., Gray, K.A., Gengler, N. 2013. Estimation of genetic parameters for birth weight, preweaning mortality, and hot carcass weight of crossbred pigs.J. Anim. Sci.91:5565–5571.

3. Fix, J.S., Cassady, J.P., Herring, W.O., Holl, J.W., Culbertson, M.S., and See, M.T. 2010. Effect of piglet birth weight on body weight, growth, backfat, and longissimus muscle area of commercial market swine. Livest. Sci.127: 51–59

4. Klobasa F., Butler J.E., Werhahn E., Habe F. 1986. Maternal-neonatal immunoregulation in swine. II. Influence of multiparity on de novo immunoglobulin synthesis by piglets. *Veterinary Immunology and Immunopathology* 11, 149–159.

5. Miller Y.J. 2008. Investigating the poor growth performance and survival of the progeny of gilts. PhD Thesis, University of Sydney.

6. Miller, Y.J., A.M. Collins, D. Emery, D.J. Begg, R.J. Smits, and P.K. Holyoake. 2012. Piglet performance and immunity is determined by the parity of both the birth dam and the rearing dam. An. Prod. Sci. 53: 46-51.

7. Morales J., Mateos G.G., Manteca X. 2006. Effect of parity and rearing segregation at birth on productive performance and health status of pigs. *Journal of Animal Science* 84, 292

8. Pearson K. Mathematical contributions to the theory of evolution. XI. On the influence of natural selection on the variability and correlation of organs. 1903.Phil. Trans. R. Soc. A. 200:1–66.

9. Robertson A. The nature of quantitative genetic variation. In: Brink R.A, editor. Heritage from Mendel. University of Wisconsin Press; Madison, WI: 1967. pp. 265–280.

Semen supplier contributions to sow herd performance

GC Althouse, DVM, PhD; RL Linardi, DVM, PhD; EM Woodward, PhD
Department of Clinical Studies, New Bolton Center, School of Veterinary Medicine, University of Pennsylvania, Kennett Square, Pennsylvania

Introduction

The use of high quality extended boar semen is critical to achieving desired herd fertility goals. When using artificial insemination (AI) in a breeding program, the semen supplier de facto becomes an external driver to a customer's herd reproductive performance. Depending upon circumstance, this input can either aid in or interfere with achieving herd production goals. Abrupt increases in regular return rates leading to repeat services may serve as early indicators of semen quality issues. End-point data suggestive of semen quality problems may include low farrowing rate, decreases in average total born per litter, and/or increases in litter scatter. Expectations of fluctuations in this end-point data can be quite variable, driven not just in the etiology of the semen quality issue, but also by herd management factors which can exacerbate the problem. When reviewing the literature, earlier studies[1] have reported mean reductions of 17% and 1.2 piglets born alive due to poor AI semen quality on farms (N = 37) experiencing ongoing reproductive problems. A more recent retrospective assessment on poor reproductive performance in 600 herds (250-5,000 sows) that use AI found that semen quality issues caused decreases in reproductive performance in 120 (33.3%) of the herds.[2]

When assessing the impact of a boar stud's contributions to herd fecundity, a process control strategy should be employed to identify and eliminate special causes of variation. The diagnostician should obtain both pertinent pre-use data on the extended semen product originating from the stud, along with the post-use fertility data from the sow farm(s). Temporal overlayment of these data often aids in narrowing down the time frame of a more detailed diagnostic focus. Assurances to the accuracy and completeness of information from the stud and sow farm(s) are critical to diagnostic success. This report highlights some of the more recent issues which have been identified as leading to disturbances to semen quality and subsequent herd fertility through active field investigations.

Initial ejaculate quality

All progressive boar studs today employ some type of assessment of a freshly collected boar ejaculate. The most common ejaculate parameters assessed include ejaculate volume, color, odor, sperm motility, sperm morphology, and sperm concentration. Ejaculate color/odor, sperm motility and sperm morphology are qualitative assessments, whereas ejaculate volume and sperm concentration data are quantified in order to determine eligible dose numbers from a given ejaculate. Minimum requirements of fresh boar semen for use in AI programs have been established.[3,4] Color and odor are very quick and simple procedures which are used to identify the presence of certain spermicidal contaminants (i.e., urine, blood, diverticulum fluid). The remaining qualitative parameters of sperm motility and sperm morphology are typically assessed to verify that ejaculates meet or exceed a desired threshold level prior to inclusion in the production of an AI dose. Current recommendations suggest at least ≥ 80% progressively motile sperm in conjunction with having ≥ 75% morphologically normal sperm cells[3,4] and are typically defined in a contract between the sow farm and stud. Any investigation should confirm that threshold quality parameters are being met, along with affirming that the stud uniformly applies consistent, objective processes using validated techniques.

Stud contaminants

Bacteria are a normal contaminant in a boar ejaculate. Other important inputs from the stud, particularly water used in making semen extenders, have been identified as origins of contamination in the extended porcine semen product.[5,6] Several bacterial contaminants have been found to reduce extended semen shelf-life and, if used in an AI program, reduced herd reproductive performance.[5-8] Good hygiene and sanitation procedures should be in place at a stud in order to minimize dose contamination.[9,10] Regular monitoring of the extended semen product should also be in place in order to confirm that preventative contamination procedures are being performed. The monitoring program should be typical to that of an acceptable infectious disease surveillance program.[11] Most progressive studs today will monitor both their extended semen product and purified water system on a minimal monthly basis. Additional screening of samples is warranted when there are changes to stud personnel, protocols, new materials, and installation of equipment.

In mid- to late-2015, a field investigation was initiated due to increased variability in weekly conception rates on multiple farms receiving extended semen from a single stud. Randomly selected extended semen samples were submitted to the Reference Andrology Laboratory (University of Pennsylvania) for full semen analysis, including microbiology. Results showed multiple samples having a *Serratia marcescens* spermicidal contaminant. Subsequently, a methodical process of testing was performed throughout the stud to identify and eliminate sources of contamination. Microbiological monitoring of extended semen doses was performed in concert with implementation of any semen process management changes to confirm success.

Figure 1: Disruption to sperm plasma membrane fluidity associated with exposure to select plastics. Note variation in percentage of sperm exhibiting transitional fluidity between the control (C1), treatment 1 (T1) and 2 (T2) groups.

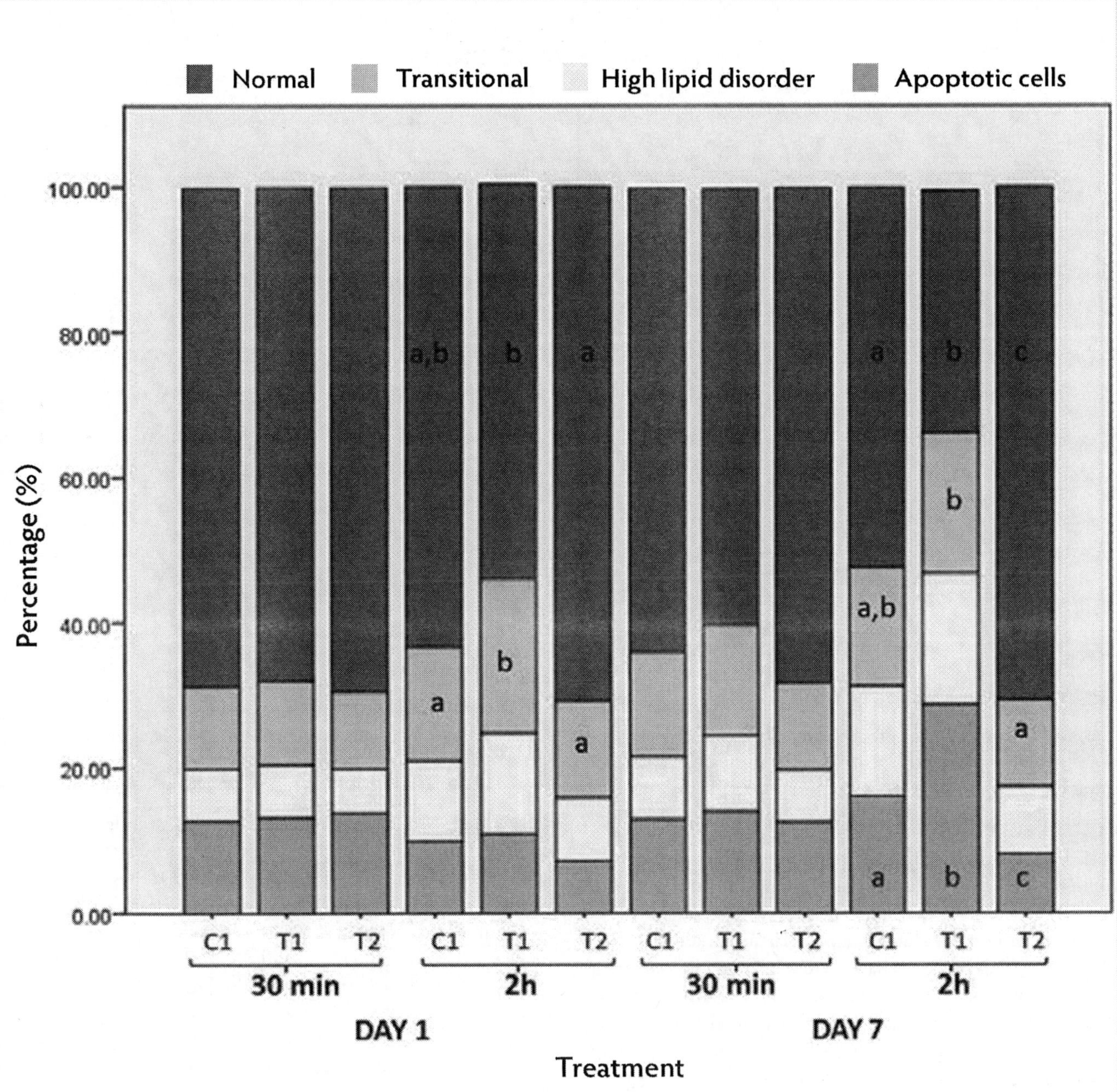

Different letters mean significant differences ($P > 0.05$) within time periods and treatment groups (ANOVA).

Figure 2: The effect of semen extender on percentages of acrosome-intact, viable (AIV) sperm over time. Under stress incubation, Extender 2 has less ($P < 0.05$) AIV sperm than either currently used extender (Control) or test extender 1.

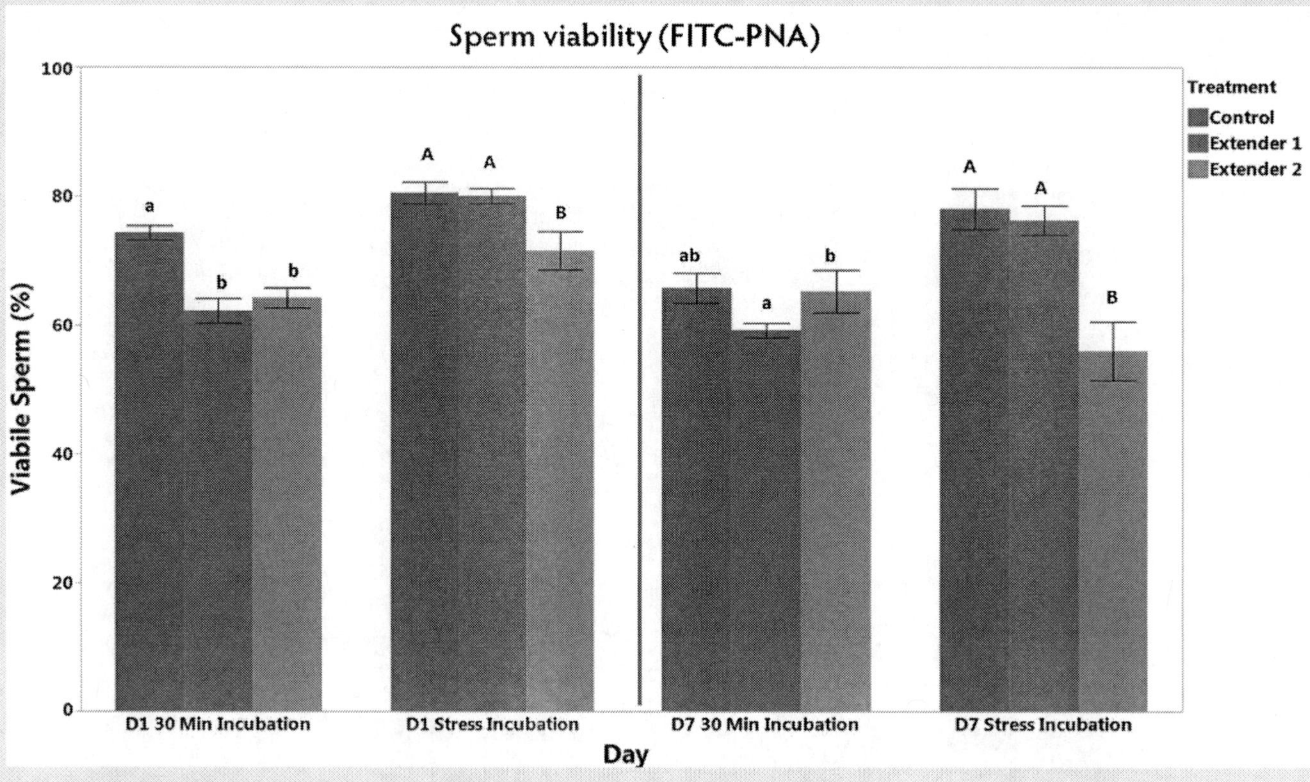

Comparisons were made between treatments at each sampling period. Different letters indicate significant differences ($P < 0.05$). Error bars indicate SEM.

Material supply issues

Adoption of AI in the swine industry has been broad, with this reproductive technology now having a global footprint. As a maturing market, the competition by the suppliers of materials for boar studs remains diverse and intense. Classic marketing strategies by AI materials suppliers have included being a low cost provider or purveyor of quality products. The former drives supplier revenue through sales volume, while the latter achieves revenue through the production of consistent, quality products. In an ideal situation, the optimal materials supplier to a boar stud would be one that combines both cost reduction and product quality improvement. Ideally, the materials supplier should be able to demonstrate to the stud acceptable supply chain practices which assure not only consistency and quality of raw ingredients, but also an appropriate level of *in vitro* and *in vivo* screening processes on the final product prior to distribution. When inadequately screened, supplies may have an unintended negative consequence on reproductive performance. Below are a few recent examples.

Example 1: Multiple sow farms began reporting reduced litter size (total born) around the same time, initiating a field investigation. Three (N = 3) boar studs sourced the extended semen used on these farms. Sow farm reproductive records were collected and assessed, amounting to ~130,000 services. Of these services, 77% were performed using extended semen in blister bags, with the remaining services using extended semen held in tubes. An initial data review showed distinct fecundity differences based upon whether blister bags or tubes were used for breeding. Retained blister bags and tubes were tested using standard spermiogram analyses (e.g., sperm motility and morphology) over a 7-day storage period, with no differences observed between the receptacles. Subsequently, additional testing was developed in order to assess sperm cell plasma membrane fluidity and applied to samples stored in blister bags and tubes. Time dependent degradation in plasma membrane fluidity was found to occur in association with samples held in the blister bags (Figure 1), suggesting a deleterious effect of the bag on the sperm. Subsequent work by others[12] identified the leaching of specific compounds (i.e., BPA, BADGE, PVC, phthalates) from the multilayer blister bags which perturbed sperm cells.

Example 2: In 2015, multiple sow farms reported a reduction in litter sizes. Further investigation using data obtained from 36 breeding farms confirmed an average decrease of 0.7 less piglets/litter, starting breeding week 13. Focusing on the time period that reduced litter sizes commenced, the stud's material supply records were assessed, identifying that new semen extender product lots were incorporated during that time period for extended semen production. Further *in vitro* testing of the extender was initiated, along with comparison to other commercial semen extenders. Boar semen samples held in the extender in question were found to be stress sensitive (thermal and temporal), resulting in lower numbers of acrosome intact, viable sperm (Figure 2). A change in semen extender type led to a return to normal reproductive performance on the sow farms.

Conclusion

Semen quality is a vital component to achieving sow herd fecundity goals. Proactively, boar studs need to have protocols in place which objectively assure that the extended semen product meets or exceeds quality parameters and consistency. Boar studs should screen AI material suppliers to assure that they abide by acceptable supply chain practices which support product safety, consistency and traceability. Not only should these practices be able to assure proper design, monitoring, and control of manufacturing processes and facilities, but they should also include an appropriate level of *in vitro* and *in vivo* testing which supports quality and consistency of the final product prior to distribution. Detailed recordkeeping of all material/supply lots used at a stud, in conjunction with timely feedback and accessibility to sow farm reproductive records, is crucial to both the stud's quality assurance program and to delineating the contributions the stud has to a sow farm's fecundity goals.

References

1. Flowers B. Successful AI programs. Proceedings of ACT/SFT/AASP Swine Reproduction Symposium; 1996; Kansas City. 15–25.

2. Flowers WL. Semen quality assurance. Proceedings 49th Annual NC Pork Conference; 16.2.2005; Greenville. 1-9.

3. Althouse, GC: Artificial insemination in swine: Boar Stud Management. In: Youngquist RS, Threlfall WR. *Current therapy in large animal theriogenology*. 2. Saint Louis:Saunders Elsevier; 2007. S. 731-738.

4. Althouse GC: Applied andrology in swine. In: Chenoweth P, Lorton S. *Animal andrology: theories and applications*. 1. Wallingford:CAB International; 2014. S. 404-417.

5. Althouse GC, Lu KG. 2005. Bacteriospermia in extended porcine semen. *Theriogenology* 63:573-584.

6. Payne, B.J., Clark, S., Maddox, C., Ness, A. 2008. Achromobacter xylosoxidans in extended semen causes reproductive failure in artificially inseminated sows and gilts. *J Swine Health Prod* 16:316-322.

7. Althouse G.C., Kuster, C.E., Clark, S.G., Weisiger, R.M. 2000. Field investigations of bacterial contaminants and their effects on extended porcine semen. *Theriogenology* 53:1167-1176.

8. Maroto Martín, L.O., Muñoz, E.C., De Cupere, F., Van Driessche, E., Echemendia-Blanco, D., Rodríguez, J.M., Beeckmans, S. 2010. Bacterial contamination of boar semen affects the litter size. *Anim Reprod Sci* 120:95-104.

9. Althouse, G.C. 2008. Sanitary procedures for the production of extended semen. *Reprod Dom Anim* 43(Suppl 2):374-378.

10. Althouse, G.C., Reicks, D.E., Spronk, G.D., Trayer, T.P. 2003. Health, hygiene and sanitation guidelines for boar studs providing semen to the domestic market. *J Swine Health Prod* 11:204-206.

11. Althouse, G.C., Galligan, D.T. 2006. Product quality monitoring and assessment of semen by veterinary laboratories. Proc. (Seminar #5) 37th Amer Assn Swine Vet Ann Mtg, Kansas City, MO, pp.21-23.

12. Nerin C., Ubeda J.L., Alfaro P., Dahmani Y., Aznar M., Canellas E., Ausejo R. 2014. Compounds from multilayer plastic bags cause reproductive failures in artificial insemination. *Sci Rep*. 9;4:4913. doi:10.1038/srep04913.

A pig's early challenges

Maria Pieters, DVM, PhD
Department of Veterinary Population Medicine, College of Veterinary Medicine,
University of Minnesota, St. Paul, Minnesota

Introduction

Doubtlessly, getting piglets off to a good start is the main goal of swine reproductive herds. However, how this is accomplished or the many aspects that can influence this process may vary from farm to farm, system to system, or simply, from time to time. What was considered state-of-the-art in the past may have included practices that currently can be considered obsolete or that may have been considerable modified. And on the other hand, new information may point at the importance of practices that in previous years were not even thought to have influence in swine health and production. And as weird as it may sound, life may not start with conception, but long before that. In this document, some examples of practices that have been evidenced to significantly affect the early and very early development of newborn pigs are presented, in order to illustrate this topic.

Early (and very early) challenges

Gilt acclimation

The process of making incoming gilts successfully adapted to a recipient farm is currently recognized as one of the most important procedures for improved health and production. Obviously, practices will vary depending on the microorganism(s) in question, on the availability of replacement animals, on the farm location, etc. Gilt acclimation is considered influential, not only in the immediate performance of the reproductive female, or the existing females in the herd, but most importantly, on the piglets that will be born to those newly introduced animals. Making every effort to guarantee that incoming gilts are properly prepared for the farm's own challenges may be seen as a demanding job with no direct result. Nevertheless, when properly done, all the dedication put into this process will certainly be paid back. The effects of poor practices for gilt acclimatization will be observed in a wide spectrum, affecting not only health, but production performance as well. Thus, special attention needs to be taken regarding this prominent process.

One significant example is the one of gilt acclimation for *Mycoplasma hyopneumoniae* (*M. hyopneumoniae*), due to the future piglets' implications. Regularly, *M. hyopneumoniae* endemically infected sow farms receive the influx of naïve gilts to become part of the reproductive herd. The fact that naïve gilts are brought into positive farms translates into the need to equalize the status of the new groups and the existing sow population. The failure to make this transition smooth has been appeared to have occurred in a high proportion of sow farms and those effects have been evidenced in the lack of disease control in downstream flows.

Briefly, active transmission of *M. hyopneumoniae* between sows and gilts appears to occur with the introduction of naïve animals, and depending on the gilts' age of introduction, the excessively persistent infectious process of *M. hyopneumoniae* may extent to farrowing, lactation, and beyond. Gilts reaching farrowing with an active infection are potential bacterial shedders, which predisposes piglets to colonization with the bacterium during the lactation period. While baby pigs will not suffer from disease right after colonization, they will become pathogen amplifiers during the nursery phase and usual clinical presentation appears during the finishing period. It has been previously suggested that the level of colonization at the end of the lactation period can be used as a predictor of downstream disease presentation.

Likewise, examples exist for other pathogens that can cause alterations in the disease dynamics in sow farms, having a profound effect on piglet's early colonization and potential disease development. Examples of these include other bacterial species, either pathogenic or of commensal nature, as well as viral agents, which can be circulating in the existing female population, or which can be exacerbated by the introduction of a sufficiently large group of animals with a different health status.

Antibiotic treatments

Treatments with antibacterial drugs can be applied at the sow or the piglet level, having a significant influence on piglets. In the case of sow antibiotic treatment, targeted therapy will modify bacterial populations, not only for the intended pathogen to be controlled, but also for other species that are susceptible to the chosen antibiotic. Effects of antibiotic treatments in sows can have effects in piglets in various ways, from the direct transfer of metabolites during lactation, to the direct transfer of microorganisms while co-housed with baby pigs.

In the same manner, antibiotic treatments applied to piglets in the lactation period will significantly modify the composition of bacterial populations in the different systems.

Recent information, specifically in swine, has suggested that alterations in the microbiome can be responsible for disease presentation in various degrees, not only affecting bacterial agents, but also clinical demonstrations associated to viral agents.

Immunizations/vaccinations

Arguably most important than other practices, immunizations or vaccinations applied at the sow level will affect the immune response in the baby pig. While this process can be affected by many various factors, depending on the infectious agent, on the animals' age, on the health status, and on previous exposure, among others, the transfer of maternal immunity is crucial for the baby pig. On the other hand, vaccinations and immunizations in baby pigs need to be carefully decided, based on the farms' status and overall vaccination/immunization program.

Cross-fostering

Usually regarded as a way to uniform litters and affect only production, research data has suggested that cross-fostering practices can have a significant influence in the piglet's immune response. Although this and other practices may have been established in the industry for many years, a careful examination of their potential benefits and disadvantages of such practices should be exercised.

Batch farrowing

Synchronizing groups of females for farrowing and lactation is making a comeback as a production system in reproductive herds. Little information regarding the effect of this production system in overall herd health has been performed, however, data is being generated on such evaluations.

Summary

New perspectives in swine health and production point at the importance of events happening early in the life of a piglet, even before birth. It is the responsibility of the swine professional to be informed of research generated looking at evaluation of current practices and to carefully evaluate the benefits of applying such practices in the field.

Considerations for batch production

C. Scanlon Daniels, DVM, MBA
Circle H Headquarters, LLC, Dalhart, Texas

Introduction

Batch farrowing systems have experienced considerable interest and increase in use by producers in the United States over the last several years. This paper will review the various types of batch production, the motivations by producers to consider batch production and some of the considerations before and during implementation.

Background

This topic has been reviewed in the popular press over that last few years.[1-5] Extensive (outdoor or transitional) swine production has long utilized batch farrowing. Farrowing twice per year allows the production of pigs to better match seasonal weather considerations compared to farrowing every week. As swine producers developed more intensive (indoor) production systems through the 70's and 80's, weekly production systems became more prevalent in order to produce more pigs and more efficiently utilize facility investments. In more recent times, improving the health of weaned pigs has led an increasing number of producers to return to batch production.[5]

Types of batch production schedules

Every two weeks, every 3 weeks, every 4 weeks and every 5 weeks are commonly considered batch production schedules.[2] The one chosen will depend on a multitude of factors, including sow farm size, growing pig facility size, labor considerations, desired weaning age, boar and/or semen availability, weaned pig transport distance, weaned pig transport capacity utilization, farrowing facility downtime and degree of all in-all out (AIAO) farrowing management desired.

With every 2 week farrowing, there are 10 groups of sows total with two farrowing groups. This system can work well for farms that have been under a weekly production schedule. With farms that have two farrowing barns, or a farrowing that is split by a central hallway, AIAO farrowing house flow by common hall can be achieved. Farrowing every three weeks with seven groups of sows also results in two groups of sows in farrowing at one time. With a six week turn on the farrowing rooms, this schedule can allow for an older weaning age, but will reduce throughput compared to a weekly production system. It allows for a longer breeding window since the maximum wean age in this system is greatly elevated. Farrowing every four weeks with five groups is a very nice fit for farms that have utlized a weekly production system. This system allows for AIAO by entire farrowing capacity but does not allow for extending breeding beyond one week without significant compromises in weaning age on the later services.

Farrowing every five weeks can accommodate a longer window of breeding (more than one week) although changing to this system from weekly production will decrease overall output due to greater farrowing house downtime and/or increases in weaning age attributable to the 5 week turn.

Applications

There are some scenarios that lend themselves to batch farrowing better than others. In the examples discussed in this section, it is assumed that farms contemplating change have been under a weekly production schedule with 20 groups of sows. Under this assumption, no changes to the physical capacities of the farm are required such as new building construction.

A common scenario for batch production conversion is a small farm that is interested in producing larger numbers of pigs at one time to better supply the market with like age and health status weaned pigs. Another is a farrow to finish herd that wants to convert to only weaned pig production. A third is when individual sow farms convert to batch farrowing as part of a system or group of farms. In this manner of production, the individual farms produce batches of pigs, but the system retains a weekly production schedule.

Small farm scenario

For this scenario, we will use a 350 sow farrowing operation. In a weekly production system weaning 10.4 pigs per litter at 25 pigs per sow per year (PSY) it should wean around 168 pigs per week from 16 to 17 sows. This farm should have approximately 72 farrowing stalls to accommodate this level of production. When the farm switches to batch farrowing using the 5 group/4 week schedule, it can farrow sows one out of every four weeks. Assuming similar efficiencies, it should wean around 672 pigs every four weeks.

Farrow to finish conversion

In this example, we will utilize a 2000 sow farrow to finish operation. After conversion, existing gestation stalls are used through pregnancy confirmation, nurseries are converted to farrowing and finishers are used for pen gestation. In conventional operation, this farm should farrow around 92 sows per week. At the same productivity used in the previous example, it would wean around 950 pigs per week. Assume the existing farrowing house has 368 farrowing stalls and we want to double the capacity of this farm. The existing nursery would have around 22,800 square feet of pig space to be converted to farrowing. If we use 48 square feet per farrowing stall (6 × 8 ft) this would allow for the addition of up to 475 more farrowing stalls if all the nursery space could be utilized. If this farm is single site, a five group/four week schedule would work well and allow for the weaning of approximately 7650 weaned pigs every 4 weeks. If the nursery is separate, a 10 group/2 week schedule can work well since both farrowing sites (existing & nursery conversion) could be managed AIAO. If operated in a 10 group/2 week fashion, approximately 3825 pigs would be weaned every two weeks.

System of farms

In this example, four 2400 sow farms are grouped with each utilizing a 5 group/4 week batch production system. In this system, the four farms rotate farrowing, weaning and breeding each week on a four week cycle. This system allows for immediate breedback of recycles and opportunity matings since all weekly breed groups are represented in the system. Sows that fail to conceive can be moved at the time of heat detection, post pregnancy check or prior to farrowing from their original group to the farm housing their new batch. Weekly weaned pig production remains constant in this example; all the weaned pigs originate from one farm each week rather than all four farms each week.

Advantages and disadvantages of batch production

Transition to batch farrowing has been documented to reduce exposure in growing pig flows on farrow to finish farms to *Lawsonia intracellularis, Mycoplasma hyopneumoniae* and *Actinobacillus pleuropneumoniae*.[6] Other pros and cons associated with the transition to batch farrowing have been reviewed.[7-14] Batch farrowing produces larger populations of pigs with less age variation. It allows for increased utilization of AIAO flow in farrowing and shorter fill time in growing pig sites. Boar capacity for heat detection, natural mating or semen production for artificial insemination must be considered. Greater labor specialization can be achieved in some cases. With fewer groups of sows in the farm there is lower risk of disease and/or immunity subpopulations within gestation and farrowing. Pharmacologic interventions such as P.G. 600® (serum gonadotropin and chorionic gonadotropin), Matrix® (altrenogest), Chronomate™ (altrenogest) and/or prostaglandin may require greater use. Batch farrowing systems inherently have more non productive days. Batch farrowing systems lack the ability to increase services the week after missing target or reduce services the week after exceeding target. The space available for gilts and wean sows needs to be larger in a farm utilizing batch farrowing. The lack of availability of nurse sows is a limitation.

Summary

Significant disease outbreaks like what the industry previously experienced Porcine Circovirus Type II and more recently with Porcine Epidemic Diarrhea Virus provide the opportunity and incentive to review production practices that influence the health of weaned pig flows. In addition, idling or depopulation, change of ownership, and repurposing of farm sites allow for the same. Batch farrowing systems reduce the impacts of endemic disease in the farrowing house and post weaning and deserve an additional look by producers and veterinarians. Transition plans and steady state management of batch farrowing systems are complex and warrant extensive study and preparation before implementation.

References

1. Rix M, Ketchem R. National Hog Farmer. How batch farrowing impacts production. September 6, 2011. Available at: http://www.nationalhogfarmer.com/genetics-reproduction/how-batch-pig-farrowing-impacts-production-0906. Accessed December 26, 2016.

2. Rix M, Ketchem R. National Hog Farmer. An in-depth look at batch farrowing. September 12, 2011. Available at: http://www.nationalhogfarmer.com/weekly-preview/0912-batch-farrowing. Accessed December 26, 2016.

3. Rix M, Ketchem R. National Hog Farmer. Should you consider batch farrowing? April 7, 2014. Available at: http://www.nationalhogfarmer.com/business/should-you-consider-batch-farrowing. Accessed December 26, 2016.

4. Rix M, Ketchem R. National Hog Farmer. Are you thinking about switching to batch farrowing? September 12, 2016. Available at: http://www.nationalhogfarmer.com/reproduction/are-you-thinking-about-switching-batch-farrowing. Accessed December 26, 2016.

5. Rix M, Ketchem R. National Hog Farmer. Batch farrowing interest grows with demand for larger groups of weaned pigs. October 10, 2016. Available at: http://www.nationalhogfarmer.com/reproduction/batch-farrowing-interest-grows-demand-larger-groups-weaned-pigs. Accessed December 26, 2016.

6. Vangroenweghe F, et al. Health advantages of transition to batch management system in farrow-to-finish pig herds. *Veterinarni Medicina*, 57, 2012 (2): 83-91. Available at: http://vri.cz/docs/vetmed/57-2-83.pdf. Accessed December 26, 2016.

7. Aarts D. Practical management of the transition to batch farrowing. London Swine Conference – Tools of the Trade. April 1-2, 2009. Available at: http://www.londonswineconference.ca/images/pdfs/2009/LSC2009_DAarts.pdf. Accessed December 26, 2016.

8. Brown P. Advantages and disadvantages of batch farrowing. *In Practice*, 2006; 28:94-96.

9. Casanovas C. WATTAgNet.com. How batches affect the labor schedule. July 1, 2009. Available at: http://www.wattagnet.com/articles/974-how-batches-affect-the-labour-schedule. Accessed December 26, 2016.

10. Casanovas C. WATTAgNet.com. Labour line-up on 3-week batches. July 1, 2009. Available at: http://www.wattagnet.com/articles/982-labour-line-up-on-3-week-batches. Accessed December 26, 2016.

11. Casanovas C. WATTAgNet.com. How many batches for your sows? July 1, 2009. Available at: http://www.wattagnet.com/articles/865-how-many-batches-for-your-sows. Accessed December 26, 2016.

12. Casanovas C. WATTAgNet.com. Rotations for the farrowing room. April 3, 2008. Available at: http://www.wattagnet.com/articles/1007-rotations-for-the-farrowing-room. Accessed December 26, 2016.

13. Barugh I. New Zealand Pork. Batch Farrowing. Available at: http://www.nzpork.co.nz/images/custom/pork_production_farrowing_batch_farrowing.pdf. Accessed December 26, 2016.

14. Gadd J. Nottingham University Press. 2011. Modern Pig Production Technology: A Practical Guide to Profit. Chapter 20: Batch Farrowing. 459-466.

Managing Danish nurseries in prolific sow herds with minimal antibiotic use

J.M.J. Agerley, DVM
SvineVet, Moltrupvej 49a, DK-6100 Haderslev, Denmark

Introduction

Since 1995 a myriad of legislations, have been passed in Denmark, aimed at reducing antimicrobial consumption. In 1995 the veterinarians lost their right to sell Antibiotics. The idea was to take away any economic incentive. In the year 1998 and 2000, AGP's (Antibiotic Growth Promoters) where banned for weaners and Finishers. In the year 2000 the national Danish drug surveillance database, VetStat; was created. VetStat records all data from prescriptions veterinarians make to a farm and has access to the national animal registry database. By knowing the amount of antimicrobials being used per animal on the farm, ADD's (Average Daily Doses) are calculated. And average daily dose is a standard dose of a drug per animal per day. In 2010 new legislations where passed in Denmark, where a maximum tolerated ADD use per animal group was set. Roughly every 2 years since then the maximum accepted consumption levels have been decreased. The challenge for Danish farms is keeping Productivity up with a low consumption of antimicrobials.

The national average of weaned pigs per sow has increased from 27.2 in 2008 to 31.4 in 2015.[1] In these years not many farms have been able to expand and as a general assumption the extra pigs weaned per sow have to be produced in the same buildings. Thus pig density has increased while antimicrobial consumption has gone down. From 2009-2015 Antimicrobial consumption in Danish pigfarms has decreased by 22%.[2]

Farrowing house

Management of the farrowing house is the foundation for a healthy and productive nursery. Danbred sows are very prolific and some farms reach up to18 liveborn in average. This has reduced weaning age and weaning weight on a national level and especially in farms, that wean more than 35 piglets/sow/year (Table 1). When scaling piglets at weaning we find piglets as light as 2 kg's in some farms.

Very rigid management rules have to be applied when making foster sows and when pigs are evened out after birth. In very high performing farms more piglets are moved in litter equalization after birth than in farms with fewer piglets weaned per sow. This is due to inequality of the sows especially in farms with floor feeding and inequality in piglet birth weight in farms with very high litter sizes. If piglets are not matched to the right size of sow and utter after birth milk production suffers and too many piglets fall behind. In many farms this scenario leads to an excess collection of piglets to foster sows, because of hunger and disease, thus increasing the pathogen transmission in the farrowing room.

Optimizing farrowing crate usage for milk production is important, but fully optimizing crates with 15-30% of the piglets being with foster sows means having older piglets in the same row/ room as younger piglets. Leaving the percentage of crates empty that will be needed for foster sows after birth, decreases piglet kilos weaned per crate, but health benefits in the nursery seem to outweigh lower weaning weight.

Managing week sizes and its effect on nursery performance is often overlooked in Denmark. We have seen farms with up to 40% difference sow numbers per week. This variation follows through to a large variation in weaning age. Weeks with many sows, cause an equal number of litters to be weaned lighter and younger.

Early milk production is positively influenced from fibers in the gestation feed and by having the right back fat thickness. Focusing on early milk production reduces mortality.

Nursery

On average Danish farms are weaning 15% more piglets per sow. As stated above not all farms have been able to expand their building mass, as a general assumption Danish nurseries have become more crowded. Decreased weaning weight, especially in farms weaning more than 35 piglets per sow, adds to the crowdedness of the nurseries. "Add on" nurseries with a capacity of 400 piglets have become part of the solution to create cheap space. They are used to handle foster pigs and piglets with low weaning weight. They can be used to optimize pig flow, which has been distorted due to variation in weaning weight forcing some farmers to move piglets backwards to keep sales weight up.

By giving the smallest piglets 2 extra weeks in the "add on" nurseries in an all in all out system, the smallest piglets in the nursery become the largest. Typically 10-15% of the piglets are weaned to these "add on" nurseries. Most of the antimicrobials in Denmark are used on treating diarrhea. A large part of the diarrhea in Danish nurseries is feed related. Feed and feeding management is important when aiming at reducing antimicrobial consumption on Danish farms.

References

1. LANDSGENNEMSNIT FOR PRODUKTIVITET I SVINEPRODUKTIONEN 2015, notat nr. 1611.

2. Danmap 2015 - Use of antimicrobial agents and occurrence of antimicrobial resistance in bacteria from food animals, food and humans in Denmark. Available at `http://www.danmap.org/`. Visited 16 November 2016.

Table 1: Production data from Danish farms from 2008 to 2015

	Best 25% 2015.0	2015	2014	2013	2012	2011	2010	2009	2008
Weaned/sow/year	33.80	31.40	30.60	30.00	28.80	28.10	27.50	27.20	27.30
Liveborn/sow	16.50	15.90	15.60	15.40	15.10	14.80	14.50	14.20	14.10
Litters/sow/year	2.33	2.27	2.26	2.25	2.26	2.26	2.26	2.25	2.24
Piglet mortality %	11.80	13.40	13.60	13.70	13.70	13.90	14.20	14.00	13.90
Weaning weight kg	6.40	6.80	6.90	7.00	7.00	7.10	7.20	7.40	7.40
ADWG Nursery g/day Ref (7-30kg)	479.00	443.00	441.00	441.00	438.00	435.00	439.00	446.00	447.00
Nursery mortality	2.80	3.10	2.90	2.90	2.90	2.90	2.80	2.60	2.60
FCR Fu/kg	1.73	1.88	1.92	1.91	1.94	1.94	1.94	1.92	1.94
Transfer weight kg	31.50	30.80	30.90	31.00	30.60	31.10	31.40	31.40	31.70
Kg's produced per nursery pig	25.10	24.0	24.0	24.0	23.6	24.0	24.2	24.0	24.3

Biosecurity: The strengths and weaknesses in our industry

R.B. Baker, DVM, MS

Some historical perspectives

A few years ago while attending a national meeting, a friend and colleague from days at Premium Standard Farms chastised me about Biosecurity. Charlie Arnott was quickly becoming the "go to" on consumer preferences, paradigms, and concerns about their meat purchase decisions and how they perceived food animal industries. His impact has certainly been significant and he continues to guide us in many of our consumer decisions. His comment shook me a bit when he lamented that we use the term "biosecurity" and no one, neither producers nor consumers had a clue what it really meant and even more it sounded "scary" to consumers. I mumbled some indignant defense and sought the security and comfort of another swine veterinarian to talk to.

But this idea didn't go away. It haunted a bit and eventually evolved into some hours of idle time thought about what the term really should mean and the effectiveness of our efforts. Charlie was right in that most of us scientific types when considering biosecurity are really thinking about exclusion of unwanted micro and in some cases macro organisms. This neglect of the other two components of Biosecurity, biocontainment, and biomanagement, and the near total focus on bioexclusion have certainly facilitated the rapid spread of those transboundary agents (exotic/foreign animal diseases) that have entered the North American swine production system since the late 1970's. Prior to that time *Actinobacillus pleuropneumonia*, *Streptococcus suis* type II, PRRSV, Influenza A H3N2, porcine circovirus type II, PEDV, deltacorona virus,, Senecavirus A and a few others were unknown. The list of new agents in our industry continues to grow. Admittedly some of these agents were likely here and later discovered with improving technologies but many initially came from other parts of the world spreading relentlessly after arrival. So, I ask myself what are we doing wrong. Is it a failure to observe and practice what we know or is it time to rethink and redesign some of our biosecurity efforts? Surly both are involved. The other thought is how should we discuss biosecurity with the vast majority of non-scientific consumers. Perhaps biosecurity should be considered "health assurance" or even better "health and well-being security" when addressing those audiences.

The breeding stock perspective

As our industry evolved from small single site production farms to the more sophisticated two and three site enterprises as introduced in 1987 by Dr. D.L. Harris,[1] likewise this drastically changed the structure of today's efforts at disease control and the continuous improvement of health assurance strategies. The development of biosecurity systems and philosophy started with the breeding stock industry. A great history of the development of multisite production in the US and those that spearheaded the movement can be found in the *Multisite Pig Production* book referenced in the endnotes. Prior to that era, a bucket of soapy water and long handled scrub brush (occasionally clean coveralls) was the only deterrent of disease transmission by the veterinary practitioner as he or she traveled from farm to farm in the course of daily activities.

Producers of those earlier times as well as producers today want disease freedom and genetic improvement from their sources of replacement stock whether internally or externally produced. As genetic breeding stock companies replaced traditional home reared or replacement stock purchased from purebred producers also did the principals of health assurance evolve. These "principals" were first introduced by Tom Alexander and first implemented by the Pig Improvement Company. To quote Dr. Alexander from his Principals of Health Assurance: **"The first and foremost priority is to supply breeding stock of such a health status and in such a way that it will not lead to outbreaks of disease in customers' herds[1]."** This remains a challenge but one the breeding stock industry has become reasonably good at. Unfortunately, we have not done as well on the commercial side of pig production.

Along with the development of large populations of pigs (North Carolina, Iowa, southern Minnesota, etc.) in close proximity to other sources of growing pigs came a whole plethora of new disease and biosecurity challenges. With these challenges came an ever expending development of biosecurity methodologies and standards. Many of these mimicked the original health assurance principals set forth by Dr. Alexander in the early 1960's. Those principals have evolved along with new technology but generally withstood the test of time. Many of these are still applicable today for commercial production and worth reviewing. The original principals are as follows with a few additional clarifying comments:[2]

1. **"Genetic Improvement is a long term business and therefore disease control in breeding companies is also a long term business."** Health stability and pathogen freedom from economically significant diseases is key for long term success not only for the breeding stock industry but also the commercial pork producer.
2. **"Every new nucleus herd should be set up with the healthiest pigs available."** The development of hysterotomy, medicated early weaning, embryo transfer, rollover elimination,

and other technologies have augmented this effort.

3. **"For insurance purposes there should always be more than one nucleus herd in a company and there should be more than one genetic pyramid, the pyramids being separate one from another."** Genetic movement across pyramids has been successful through frozen semen transfers, semen introductions, hysterotomies, embryo transfer, etc. In the future next generation sequencing and other evolving disease detection methodologies may allow movement of growing and mature stock between nucleus herds and other genetic pyramid sites.
4. **"It is both impossible and unnecessary to try to standardize the health status of all the herds throughout a breeding organization."** Much like one size of biosecurity doesn't work for all producers, the health status of a specific multiplier won't be suitable for all customers. Single sourcing with matched health has been the standard although freedom from certain agents like PRRSV virus and other costly diseases are a breeding stock industry must. This is also important in commercial systems producing genetic stock and to protect different segments of their production system.
5. **"Location is the most important factor in preventing contamination and herd breakdown."** This has generally been a paradigm for all biosecurity efforts in the North American pig industry and many other parts of the world. However, it doesn't always work if feed ingredients, feed mill, feed delivery, pig transport, and other basic exclusion methods aren't effectively practiced.
6. **"Because of the distribution and movement of pigs within a pyramid, it is important to concentrate the main disease control measures at the apex of the pyramid."** At this level all bioexclusion efforts are heightened as well as surveillance for new clinical signs and laboratory disease monitoring.
7. **"Maximum precautions must be taken on every farm to prevent contamination."** As an industry we often fail to keep our eye on the biosecurity ball. Even in filtered barns and systems there are many other opportunities for disease entry. A single bioexclusion step does not eliminate the necessity of the others.
8. **"The movement of people between herds must be controlled and one night, two night, and three night rules must be laid down for each herd."** This original principal has been adapted and expanded over the years since Dr. Alexander first introduced the down time rules. They were based on the potential economic loss due to a disease outbreak where the top of the pyramid was most valuable. Thus they deserve the longest down times between farms. According to Dr. Alexander these rules were based on information from Pirbright on the survival times of Foot and Mouth virus and paralleled their down time rules for employees and visitors before visiting livestock farms after leaving the research facility. This was a very contentious issue for practicing veterinarians of that era as commercial farms took notice. They were sensitive to the potential loss of revenue since it decreased the number of farm visits when serving a breeding stock genetic pyramid and when producer customers decided if it is good for the genetic companies then it must also apply to us.
9. **"Breeding stock companies must have total control over pig transport and all pig movement."** It took a while but today this is apparently the situation in most of the breeding stock business but on the commercial side much of pig movement is contracted or owned with little highly monitored quality control efforts. Rules don't work for transport since many of them are difficult to abide by. It is highly likely that PCV-2 spread through the North American swine industry via transport vehicles. Other agents are often spread by pig trailers as well, especially PRRS, TGE, PED, Salmonella sp., E.coli and numerous other agents of varying economic importance.
10. **"To prevent excessive dissemination of disease and because breeding stock companies cannot standardize health, companies must practice straight line distribution to their customers."** This rule was often compromised during the rapid growth in herd size and new facilities in the United States. To accommodate multiple source stocking and respond to breeding stock herd health issues in source herds the Vet to Vet was established to match health status and minimize this risk.
11. **"Breeding stock companies must monitor for pathogens over a wide spectrum of diseases and be prepared for new diseases when they arrive."** Originally this was largely done by observation of clinical signs and testing for pseudorabies and brucellosis. This is far more complex today were all shipments of replacements are frequently tested serologically and polymerase chain reaction (PCR) for a large spectrum of pathogens.
12. **"Never do any test if you cannot be sure of interpreting the results."** Our ability to interpret has certainly improved with newer testing technologies but false positives and negatives still plague us from time to time. I have suffered through many episodes when stocking systems or farms where the results were at best confusing and occasionally led to bad decisions.
13. **"Breeding stock companies should never routinely use drugs or vaccines in a herd which might mask the presence of an infection that would be damaging to a customer herd."**
14. **"Veterinary work should be predominately regional and local rather than central."** As a practitioner this is certainly preferred but requires specialization and constantly improving disease and application knowledge.

There were a few additional principals concerning stocking new multiplication from a single source after a complete depopulation. The original downtime for a depop-repopulation of a multiplier was six weeks after cleaning and disinfection. For trucks two days after clean, wash, and disinfection before returning to a multiplier etc. It is apparent that many of these "rules" are still applicable today but are often neglected in commercial production.

The strengths and failures of current efforts

In the years since the 1960's numerous perceived and documented risks associated with health breakdowns have been proposed and established. Field experience often doesn't align with experimental results and many of our practitioner observations have led to false outbreak paradigms and interventions. There are far more of these "standards" than the scope of this paper can adequately discuss but some of the recent developments and strategies are included from a disease based perspective along with mentioning a few assumptions that failed to deliver.

Recent lessons learned from the PEDV epidemic of 2013-2016

The appearance of porcine epidemic diarrhea virus in multiple unconnected sites in April and May of 2013 aroused suspicion of feed contamination. Not since the days of Pseudorabies (PRV) had feed ingredients been suspected as a significant vector of swine diseases. During the PRV elimination corn was often stored for 30 days in dry conditions before grinding. Dual purpose trucks used for grain haul and livestock transport were rejected at mills attempting to avoid transmission of the virus. Those days are long gone but the importance of feed biosecurity remains. The saga of PEDV spread through feed contamination remains unsolved but most Canadian and US practitioners agree to that likelihood. A case control study by AASV was the first indication that a feed ingredient was likely involved but not identified.[3] A follow-up investigation by the USDA/APHIS/VS considered a large number of possible scenarios of virus entry and spread in the earliest known infected farms.[4] Although no conclusion could be definitively determined recycled and contaminated feed totes were a most likely source of the virus entry into the US in their opinion but an unidentified secondary carrier was necessary for distribution. Other possibilities for introduction were also considered but none identified.

This episode in our disease history certainly demonstrated significant weaknesses in both farm and national level biosecurity. Efforts have been made to address both but it is certain that many possible routes of disease agent entry exist and the degree of difficulty to adequately address all these weaknesses is still lacking especially at the border in a free trade era. Once a new agent enters feed mill effective biosecurity which would prevent dissemination is often lacking. Like the border there are many biosecurity risks associated with our mills. A recent update on feed mill biosecurity presented at the 2016 James D. McKean Swine Disease Conference discussed many possible routes of mill contamination and potential disease dissemination.[5] There is little doubt that a great effort is required to maintain a safe feed supply for the swine industry.

The PRRSV influence on our biosecurity standards

During the late 1980's biosecurity efforts were forever changed as a new devastating disease emerged. Originally designated "Mystery Swine Disease" the new virus spread relentlessly during the great restructuring era of the industry. Originally described in the US, the syndrome appeared in Canada, Europe, and most countries with significant swine production by 1992 when the syndrome was officially named PRRS; porcine reproductive and respiratory syndrome.[6]

After the virus was identified researchers methodically tested proposed routes of transmission. The demand for large numbers of gilts had become the major route of dissemination long before virus identification and the development of testing methodologies. Thus the virus became established in much of the expanding commercial and breeding stock industries. As information on modes of transmission were established so were steps taken to prevent introduction. Biocontainment was largely ignored as the industry scrambled to find management solutions and exclusion methods. The discovery of semen transmission and post insemination infection had a major impact on stud biosecurity, deliver of semen, monitoring, and site location/isolation. Today most boar studs are PRRSV negative, filtered, daily test for PRRSV, and are completely isolated from other pig populations. Interestingly all the intensive exclusion steps taken at studs still didn't protect many of them from PEDV introduction. This was likely due to incomplete feed ingredient and feed mill biosecurity.

Once sequencing became a popular epidemiology tool the role of transport vehicles became apparent. A number of studies by Scott Dee highlighted this risk. Soon "Trailer Bakers", TADD's, and BioDry's emerged. Trailer down times, complete drying, and driver management became important. The good news, these efforts appeared to lessen the spread of other agents such as TGE and swine dysentery but did not contain PRRS expansion. Boots and coveralls were implicated fueling down times of 48 hours or more in most commercial units[7]. Shower in and out facilities became standard on most breeding herds. During this period of intensive PRRSV research, insect studies indicated a possible role of mosquitos and house flies. A lot of barns added fly screening to their barns but in retrospect this was one of the failures in PRRSV exclusion. Transfer stations or bump chutes were in vogue and some are still in use. This was a reasonable attempt to keep pig transportation off farm grounds and allowed off site cleaning of the final loadout step.

As additional biosecurity methods piled up many producers determined that the cost and disappointment of exclusion methods made PRRSV management the more attractive alternative. Vaccines had become available after 1992 and although not broadly protective they were widely used in the US. A later spin on vaccination has been area field virus inoculation of replacements followed by a "cool down" period before addition to the breeding herd. Like vaccine this method fails to cross protect when genetically diverse field strains reenter a site. Veterinarians and scientists that preferred the elimination and exclusion route lost out to the management approach philosophy although the hope of eradication remains.

The cost of a virulent PRRSV outbreak in a breeding herd is significant. Numerous economic studies have established the national and breeding herd cost of PRRSV. The annual cost to the industry has been estimated from $560 million in 2005 to $664 million in 2011. The annual cost per sow is estimated at $255 and the growing pig cost from $6.25 to $15.25.[8-10] These estimates plus actual known outbreak costs experienced by producers and production companies stimulated a closer look at filtration opportunities. Several research studies looked at this possibility and exclusion opportunity in swine dense areas.[11-15] There has been a good deal of debate about the return on investment for barn filtration. Many barns are now filtered thorugh negative pressure ventilation which creates numerous challenges in retrofit instillations with maintaining a leak free system. More recently positive pressure filtration has been developed and applied in new swine structures. Although there have been reports that these systems create their own maintenance issues they are generally more effective at near complete filtration of incoming air. To save costs some farms and systems have elected to only filter during those months of increased PRRSV activity; mid-October - February. Field experience has shown that filtration isn't the only procedure needed to prevent a new introduction in swine dense geographies but filtered farms have drastically decreased the frequency of new PRRSV introductions.[15] In addition to filtration many larger production companies have also added on site biosecurity officers whose full time job is to monitor for air leaks and all other biosecurity functions associated with the site. There is no doubt that this has also contributed to the filtration success. The continual challenge is aggressive monitoring and implementation of all bio-exclusion efforts. This will remain for many years until the swine industry develops significant biocontainment strategies on growing pig sites.

The role of swine influenza virus

During the year 1995 it was determined that Ontario swine had HI titers to a human H_3N_2 that was circulating in people at the time.[16] The virus was not isolated and assumed to be non-virulent in pigs. In 1997 an influenza virus was detected in a neonatal pig with a concurrent PRRSV infection. By 1998 H_3N_2 viruses were reported in moderate to severe influenza outbreaks in North Carolina, Iowa, Minnesota, and Texas. The North Carolina isolate contained genes from Swine, Avian, and human viruses.[17] After 80 years of H_1N_1 relative stability this new virus was particularly hard on the breeding age swine especially pregnant vaccinated gilts. This quickly precipitated H_3N_2 vaccine production. The influenza vaccines were approved and marketed for adult swine. After significant national exposure and breeding herd vaccination the new variant has become a growing pig issue in most situations and the virus is also a frequent clinical issue in suckling pigs. Antigenic drift and shift has occurred making it difficult for vaccine companies to appropriately change vaccine isolates. Autogenous and the Harris replicon vaccine dominate US sales and are frequently used in growing pigs.

A number of preventative biosecurity strategies were developed by the breeding stock and commercial industry in response to the new influenza threat. Mandatory or voluntary farm staff vaccinations have become a common recommendation. This was a completely novel bioexclusion strategy. The effectiveness of this is untested but since the pandemic of 1999 the pandemic H_1N_1 virus has become endemic in the swine industry but rarely is considered a significant risk to swine. Seasonal human influenza vaccines do contain the non-swine adapted pandemic isolate somewhat justifying farm worker vaccination. It is most likely that this program mostly benefits the workforce. Another strategy mostly implemented by breeding stock companies has been checking the oral temperature of each farm employee before they enter the farm on a daily basis. Workers are instructed to stay at home if they exhibit flu like symptoms. To my knowledge this strategy has not been highly effective in preventing outbreaks although this approach is extremely difficult to ascertain effectiveness in the field.

Additional biosecurity considerations

Through the course of industry change and expansion beginning mid-1970's our biosecurity efforts have frequently lagged behind disease developments. PRRSV clearly demonstrated that the improvements in health and biosecurity strategies made prior to its introduction in the US were immediately obsolete. A weak preventative underbelly existed across North America. The days of disinfectant foot baths and multiple farm visits in a day disappeared. During this period of unparalleled change in the industry our understanding of sampling/monitoring for diseases likewise improved. Statistical testing controls cost but with a certain risk of uncertainty in the results. However, the development of the PCR and other genomic based tests continue to expand our knowledge and interpretation of diagnostic data when looking for the presence of disease agents. The specificity and sensitivity of antibody based tests have likewise improved and the development of oral fluid testing has greatly improved population evaluation if conducted on a timely basis. The role of the diagnostic lab has changed to a real time partner in the detection of those diseases that we know. This is not the case for most emerging new agents. In the future "next generation sequencing" will likely become a main line deterrent test utilized for biocontainment. The cost of next generation sequencing is no longer prohibitive for large producers and the breeding stock industry.

Conclusion

One approach to farm biosecurity is the hazard analysis and critical control point discovery and intervention method. This analysis relies on farm interviews of all employees seeking to understand how all things enter and exit a site. Once this is established a determination of those risky farm procedures that can be controlled by cost effective means is addressed. This evaluation eventually leads to identification of critical control points for implementation. This method is rarely used today as veterinary and other "experts" believe they know all things that occur on farms. This is a failure to appropriately investigate risks objectively and prospectively.

Transport remains a significant risk for dissemination of most economically important diseases considering the number of pigs on the road daily and generally inadequate control of biosecurity standards for trailers and drivers. This is especially difficult during winter months in the Midwest. Numerous disinfectants and protocols have been evaluated and many appear highly effective for rendering numerous viruses and bacteria non-infectious.[18] In the field these products and efforts have frequently not been adequate due to weather (cold) and incomplete cleaning and disinfectant application. This has been clearly demonstrated with the spread of PRRSV, Porcine Circovirus type 2, and PEDV. Rather than continuing to work with standard procedures and the try harder management approach, perhaps it is time for a drastic change in equipment design both on the road and at the packing plant.

Risk assessment tools have been developed for long term prospective study of those events or farm location factors, and history status that lead to eventual outbreaks. The original PADRAP was developed to evaluate PRRSV outbreak risks but has evolved to look at all disease risks in a statistically valid and continuous improvement framework.[19] These tools continue to evolve along with outbreak investigation information.

As mentioned feed ingredients, feed mill, and distribution biosecurity must be improved. This may necessitate structural as well as managerial changes at the mill. Visual inspection for contamination of all feedstuffs which enter a feed mill is a necessary first step. Improved border inspection methods will hopefully develop and be implemented in the future.

In the end both biocontainment and bioexclusion must be a part of our future swine industry if we will stop the cycle of disease out breaks with those pathogens currently circulating in North American swine populations. Doing more of the same is unlikely to solve our biosecurity deficits. Rethinking many of the paradigms from the recent past is needed along with changes in the way we do business.

References

1. Harris DL. *Multi-site pig production*. Iowa State University Press/Ames. 2000.

2. Alexander TJL. *Re-evaluation of the principals that underlie PIC health programmes and their application in different countries.* 1990. PIC white paper, Nashville TN.

3. Snelson, H., 2014. PEDV-Lessons Learned, In, 2014 *World Pork Expo*, Des Moines, IA.

4. Scott A et al. Porcine epidemic diarrhea virus introduction into the United States: Root cause investigation. *Preventive Veterinary Medicine.* 2016;123;192–201.

5. Gebhardt JT et al. Feed Biosecurity Update. *ISU James D. McKean Swine Disease Conference proceedings.* 2016;37-42.

6. Christianson WT, Joo HS. Porcine Reproductive and Respiratory Syndrome: A review. *J Swine Health and Production.* 1994;2;2;10-28.

7. Otake S et al. Transmission of porcine reproductive and respiratory syndrome virus by fomites (boots and coveralls). *J Swine Health and production.* 2002;10;2;59-65.

8. Neumann E et al. Assessment of the economic impact of porcine reproductive and respiratory syndrome on swine production in the United States. *J Am. Vet. Med. Assoc.* 2005;227;3;385-392.

9. Holtkamp DJ et al. Assessment of the economic impact of porcine reproductive and respiratory syndrome virus on U.S. pork producers. *International PRRS Symposium Proceedings.* 2011;86.

10. Alonso C et al. Financial implications of installing air filtration systems to prevent PRRSV infection in large sow herds. *Prevent. Vet. Med.* 2013;111;268-277.

11. Dee S et al. Evaluation of an air-filtration system for preventing aerosol transmission of porcine reproductive and respiratory syndrome virus. *Can. J. Vet. Res.* 2005;69;293-298.

12. Dee SA et al. Evaluation of systems for reducing the transmission of porcine reproductive and respiratory syndrome virus by aerosol. *Can. J. Vet. Res.* 2006;70;28-33.

13. Dee S et al. Evaluation of alternative strategies to MERV 16-based air filtration systems for reduction of the risk of air-borne spread of porcine reproductive and respiratory syndrome virus. *Vet. Microbiol.* 2009;138;106-113.

14. Dee S et al. Further assessment of air filtration for preventing PRRSV infection in large breeding pig herds. *Vet. Rec.* 2010;167;976-977.

15. Dee SA et al. Evaluation of the long-term effect of air filtration on the occurrence of new PRRSV infections in large breeding herds in swine-dense regions. *Viruses.* 2012;4;654-662.

16. Carman S et al. H3N2 influenza A virus recovered from a neonatal pig in Onterio-1997. *Can. Vet. J.* 1999;40;889-890.

17. Zhou, NN et al. Genetic Reassortment of Avian, Swine, and Human Influenza A Viruses in American Pigs. *J. Virol.* 1999;73;8851-8856.

18. Schneider TS et al. Evaluation of disinfection protocols to reduce virus transmission via livestock transport vehicles using model trailers and experimental conditions. *J. of Swine Health and Production.* 2015;23;6;306-316.

19. Holtkamp D et al. PADRAP: Production Animal Disease Risk Assessment Program. *Annual Meeting of the American Association of Swine Veterinarians Proceedings.* 2011;521-524.

Practical PED elimination and surveillance: The Quebec experience

Julie Ménard, DVM
F. Ménard inc., Ange-Gardien, Quebec, Canada

Introduction

Porcine epidemic diarrhea virus (PEDV) was first introduced into the United States in May 2013. Many different states were affected at the same time, in different systems and associated with different vet clinics. The initial agent for introducing PEDV in America has not been confirmed but feed ingredients are the most likely suspect.

Practitioners from the US described the amazing transmission of PEDV between farms from May 2013 to fall 2013. The described clinical signs were like TGE on steroids.[1]

In Canada, because of the high traffic for culled sows and weaned piglets between the two countries, contamination was a continuous fear.

A bit of history

The first 2 cases appeared in Canada in January 2014, both at same time: one in a farrow to finish barn in Ontario and the other at an abattoir loading dock in Quebec. The threat was real. Within 20 days, 11 farms became infected with PEDV, the majority of which were in the province of Ontario and one in Prince Edward Island.

The links between these farms: plasma proteins sourced from the U.S. Investigations by consulting vets and the Canadian Food Inspection Agency (CFIA) confirmed that plasma protein used in creep feed diets was PCR positive for PEDV and the use of this feed infected all the farms at the same time.[2,4]

At F. Menard, a swine integrated company in Quebec producing more than 1 million pigs per year, preventing the introduction of PEDV was a priority. February 10, 2014 showed a first positive PCR detected at our abattoir loading dock. Through traceability measures, 17 different farms had delivered pigs to slaughter that morning. Samples collected from all these farms helped us identify a grow-finish barn as the contaminated site. The PCR positive group of pigs never showed any clinical signs. The source of the virus came from the transporter who did not follow the biosecurity rules when coming from another contaminated slaughter plant, specifically not washing his truck adequately at night during very cold weather, and coming into our farm with the same dirty boots as the ones he used at slaughter the day before. Our producer also spread the virus to his second barn by using the same boots and clothes he used in the infected barn.

A nice ending to the story

The F. Menard supervisor over-looking that specific client was also visiting more than 50 clients every two weeks. According to our records, the farm had been contaminated for a month before we discovered that it was PEDV positive. The supervisor never contaminated any of the 50 other client barns because he simply followed the basic biosecurity rules: changing boots, coveralls and gloves between each farm and keeping his truck cab clean. The lesson learned: good biosecurity can prevent PEDV transmission.

We got rid of this first case through strict biosecurity procedures and extreme sanitation. We learned that PEDV was highly resistant in the environment since we found PCR positive samples in the entrance to the barn, in cracks in the pen walls, on producer's boots and even on the outside of the barns. We kept that barn empty for 3 months before putting a new group of pigs in, which remained negative after their introduction into the barn.

What did we improve at F. Ménard?

Even before the first case, the ultimate goal of F. Ménard was to prevent contamination of breeding herds. The economic impact of classic PEDV introduction into a breeding herd is dramatic with an average loss of 2,500 piglets per 1,000 sows and 7.4 weeks before going back to 80% of the original production level.[3] The integrated structure of F. Ménard, linking all the production units through common feed delivery, trucking, manure handling and slaughter plant, can have a devastating impact. Any one nursery or finishing farm infected with PEDV can have a cascade effect throughout our system leading to bigger losses through infected breeding herds. We had to preserve our markets.

Actions taken to prevent PED

Feed

As of February 2014, we completely stopped using porcine protein in our two feed mills. No porcine plasma and no blood meal are used in our diets. We segregated our 2 feed mills: one specialized on sow and nursery diet and the other mill on the grow-finish. This way, we can manage different ingredients and reduce the risks.

Transports

We already had specialized trucks for each stage of production. We split the risk by keeping the gilts and nucleus truck, the weaned piglet trucks and grow-finish pig trucks in 3 different heated garages. They are washed, disinfected and dried every night after the driver's shift.

We also trained the drivers in regards to biosecurity rules in order to prevent PEDV introduction in barns.

In 2015, we invest in a new truck wash facility permitting better cleaning and disinfection.

Farms

We are currently operating under multiple sites production. We then have a different crew taking care of the breeding herds, nurseries and grow-finish farms. Strict rules are respected for farm visits: shower in/out, boot and coverall change, registration for visitors. All introduced materials have to be disinfected and heated for 24 hours before being allowed into the barn. Sow farms are equipped with loading docks preventing any introduction of a transporter into the barns and therefore limiting the risk of spreading the disease.

Surveillance

We think that prevention through intensive diagnostics is essential to the control of PEDV. The quicker new infections are detected, the higher the success rate in eliminating the virus. Since January 2014, samples are collected at every high risk spots: abattoir loading dock, nursery post placements, pre-slaughter grow-finish pigs, gilts pre and post transport, washing bay and trucks. This intense monitoring has given us the opportunity to intervene very rapidly and prevent the spread of the disease.

New cases in 2015

Summer time gave some relief to new cases in Canada but with the very cold weather, PEDV came back with force. Again within two weeks, three large nursery-finish sites were contaminated and the virus spread very rapidly between the different barns on each site. Introduction of the virus again came with the transport of incoming piglets and was then spread to more than 20,000 pigs through biosecurity rules not being respected. The producers and our crew responsible for vaccinations were associated with these new cases. These waves of disease spreading involving 22 different barns showed again how PEDV can spread rapidly and highly contaminate the environment. Table 1 shows every location where PCR positive results were found.

The Quebec experience

Sixteen different sites have been contaminated with PEDV in Quebec from February '14 to November '16. Of these 16 sites, 12 were F. Menard cases. Only one sow herd got infected and they eliminated the viruses within 6 months. The success for containment of the virus from this site is related to the speed of detection and intervention, strict follow up of elimination protocols and protection of surrounding herds. The detailed protocols of washing, disinfection and biosecurity put in place are ultimately part of the success. The producer and veterinarian worked hand in hand to get rid quickly of the virus.

The intervention of our special Quebec entity, the EQSP has certainly contributed to the success of this project.

Table 1: PEDV PCR positive environmental sampling

Deep pit	+++	Farm loading dock	++
Floor	+++	Producer's truck	++
Pen wall	++	Pressure washer	++
Farm office	+	Service man's truck	+
Entrance to barn	+	Abattoir loading dock	+
Outside barn	+	Truck washing bay	+
Attic	+		

What is EQSP?

EQSP or "Équipe québécoise de Santé Porcine" was founded in June 2013 in response to this emerging disease. Sponsored by producers, slaughter plant and pig industry, they have an important role in coordination. The Ontario Swine Health Advisory Board (OSHAB) and Canadian Swine Health Board have the same goal: to control and eliminate PEDV in Canada. Therefore all new cases are declared through these organizations with locations and type of production specified. These groups help producers with elimination planning and ensure follow up. They also supervise environmental swab samplings at abattoirs, assembly yards, truck washes and feed ingredient mixing, and compile results that they published regularly.

They also provide the producers with guidelines and protocols for PEDV elimination and biosecurity procedures.

The Quebec experience – Our clue for PEDV elimination

I think that awareness of the coming event was probably the first reason why our province reacted so quickly and intensively to PEDV. Sharing of information with US colleagues, rigorous implementation of elimination protocols, and team work are all factors which influenced our success.

Fear has been a tremendous actor and the fact that we never accepted to live with this disease made the difference.

Finally, I must admit that luck has been part of the ease in elimination because feed ingredients have not been involved in our first contamination.

Lessons learned at F. Ménard

Surveillance is certainly a core value in our company. For every diseases of economic relevance, preventive diagnostic testing is routinely done.

PEDV was also an excellent opportunity to heighten our biosecurity measures and put in place structure to protect ourselves against other emerging diseases.

Finally, basics respect of rules such as strict all in-all out and sanitation are all important measures to get rid of any diseases.

The threat for reintroduction

In Quebec we know that as long as there are cases in other Canadian provinces or in US, we are at risk. Poor truck sanitation, comingling in assembly yards, lack of biosecurity compliance and potentially contaminated feed ingredients are all risk factors. I think that as compared to PRRS it is much more simple to get rid of PEDV and prevent its reintroduction.

It is why as a global industry, we all have to work together at fighting this costly disease.

Acknowledgements

Thank to Dr Jean-François Doyon for its professionalism in the sow farm eradication project. A special thanks to Dr Charles Surprenant my colleague and to the F. Ménard production team. Together we are working every day to improve biosecurity measures and prevent the spread of diseases.

References

1. Dufresne, L. 2014. Field experience with porcine epidemic diarrhea. AASV annual meeting, Dallas, Texas. 613-616.

2. MacDougald, D. 2014. People and trucks – Keeping PED out of Midwest US and Canada. AASV annual meeting. Dallas, Texas.

3. Morrison, R., Goede, D. 2014. Epidemiology and economic impact of PED. AASV annual meeting. Dallas, Texas. 605-611.

4. Pasick, J., Berhane, Y., Ojkic, D., Maxie, G., Embury-Hyatt, C., Swekla, K., Handel, K., Fairles, J., Alexandersen, S. 2014. Investigation into the role of potentially contaminated feed as a source of the first detected outbreaks of porcine epidemic diarrhea in Canada. Transboundary and Emerging Disease. 61: 397-410.

US swine industry structure and disease control: A "wicked" problem

Mike Lemmon, DVM

Introduction

The definition of "wicked" problems have been characterized as problems that are somewhat unpredictable, exist in complex systems and are multi-causal. By definition they are not stable, they have emergent kinds of behaviors and rarely have clear cut solutions. These "wicked" problems often take a multitude of different disciplines to understand and to attempt solutions. Disease control in our industry is a "wicked" problem by this definition and past experience.

The purpose of this paper is to review some of the causal health control issues our industry has created, suggest some factors we should be reviewing and consider industry course adjustments that should be discussed for the overall health of our industry. Past speakers at this session have alluded to disease control problems with our current production model without calling for specific changes. I will lean on references from previous authors and presenters on similar topics to defend and support the thoughts presented here. It is not my intent to be politically correct in this paper nor overly critical of many of our decisions and practices that have gotten us to this point. With "wicked" problems, even hindsight is not always 20/20!

Some of my thoughts, comments and conclusions are heavily influenced from my veterinary and production experience not only over the last 40 years here in the US, but the last 20 years in various countries in Asia. Dealing with "wicked" disease problems is an everyday occurrence in those countries where I spend half of my time. *I also must disclose that I am associated with and may financially benefit from any reference made to genetic practices or building systems mentioned in this paper or represented at this conference.

Casual issues are "wicked" contributors

Various authors from this session last year mentioned and alluded to some of the causal issues related to the current swine industry disease control issues. Dr. Rodger Main commented that today's multisite production systems; extensive pig movement within and across broad geographic regions; a growing list of emerging and re-emerging diseases; a growing appreciation of the health status of a region, pig flow or system as it affects the performance and financial health of pork producers has created a need for more partnering, communication and collaboration in an effort to better manage and control rapidly emerging disease and disease control issues.[1] Indiana alone in 2015 imported 1,082,320 pigs from NC; 508,617 from OH; 504,737 from OK 476,612 from IL; 388,092 from MI; 175,700 from WY and 153,605 from Ontario. 92% of these pigs are feeding pigs with an average of 913 pigs per permit.[2] This continuous cross country pig movement makes regionalization efforts in the face of a disease outbreak a "wicked" problem! Not only is Indiana at risk of exposure to an emerging disease but the 22+ states those 3514 trucks pass through are at a similar risk. There are no available statistics on pigs traveling through IN to other destinations.

Dr. Craig Rowles suggested in his paper on "Lessons learned from highly pathogenic avian influenza (HPAI)" that as our industry has been driven by decisions based on continual incentive to drive costs lower, we have incurred the potential for increasing the risk of disease introduction.[3] Accounting departments are continually trying to maximize tangible production efficiencies sometimes at the unknowing expense of pig health, disease control or other value propositions. Jan Powell, ingredient procurement director for JBS United in Sheridan Indiana, suggests that this effort is out of control in relation to all of the other cost control or value propositions that are available.[4] Cheaper is the easiest decision to explain and justify to accountants and other decision makers who may not be as globally informed about the ramifications of their decisions as they should be.

The largest majority of vitamin premixes used in swine production are sourced outside the US with the vast majority coming from China.[4,5] This saves a few cents per ton of complete feed and even fewer cents per pig over domestic or European sources, but puts the entire swine industry of our country at risk of exposure to contaminants from a country that has nearly every swine disease known to mankind! Dr. Scott Dee et. al. developed a model that suggested contaminated feed ingredients could serve as a transboundary risk factor for PEDV and possibly other diseases.[6] What was the industry cost of PEDV? The 30,000 foot view of this ingredient procurement practice appears to lack good judgement! Our domestic vitamin industry has eroded over the last couple of decades, it won't come back without incentives. This is a prime example of efforts to maximize local efficiencies by compartmentalized decision makers having unintended negative global effects on our businesses, industry and consumer.

In the 80's, the advent of medicated early weaning (MEW), early weaning (EW), and 2 and 3 site systems along with all in all out (AIAO) management of these systems provided segregation of pigs in age groups to minimize the vertical spread of disease. This move from single site to multi-site systems has been embraced by the industry as a necessary move to manage system health and disease spread. Multi-site production has enabled specialization of labor, control of certain diseases and rapid expansion of contracting opportunities and industry integration. However, in his 2016 Howard Dunne Memorial Lecture, Dr. Daryl Olson stated that our multi-site model has created an industry that is perfectly designed to propagate acute viral diseases, thus contributing to our "wicked" problem![7]

In addition, the extra biosecurity, management, site approval, roads, manure management, labor, decrease in husbandry skills, transportation, animal movement and animal stresses, have added costs to this multi-site system that do not exist on single site systems. The cost advantage of the advertised and accepted control of disease from a multi-site system has been overwhelmed by the higher infrastructure, transportation and management costs as well as increased biosecurity risks to the industry caused by this system structure.

A complicating factor resulting from increased production specialization has been the change in the way individual farms access genetics. Today, few people in the industry understand genetics or want to be distracted by the genetic selection and multiplication process. Producers have been influenced to believe that genetic development and multiplication can be done more efficiently under controlled off-site genetic multiplication systems. The resulting need for periodic introduction of large numbers of potentially naïve animals into a sow production unit continually challenges the immune status of the sow farm and opens it up for unintended disease introductions. With today's technology, on-site multiplication systems are arguably equally efficient in providing competitive genetics to our production systems with much lower bio-security risks and acclimation costs.

The resulting integration of our industry has inadvertently resulted in at least 2 separate and segregated entities. The sow farm has the goal of producing as many weaned pigs/sow/year (PSY) as possible to lower overhead costs and meet production incentives. The finishing system wants good quality viable pigs that will efficiently grow to market. This competing model has resulted in the manipulation of genetic inputs to favor production at the sow farm at the expense of the finishing system without full recognition of the consequences of those genetic manipulations! The resulting focus to increase born alive has resulted in lower average birth weights, lower viability to weaning and to market, as well as poorer finishing performance.

The swine industry has basically utilized the same building technology for the last 35 years. Tunnel ventilation was an engineer's dream in simplicity and a builder's fancy. Cheap, easy to construct and simple to understand – I suggest these barns maybe too simple for the pig's good. This horizontal ventilation system encourages horizontal disease spread, requires large groups of pigs in a single airspace for efficient operation and has driven the industry to large sow farms with weekly weaning of at least 1000 pigs per week to maintain AIAO at the finishing sites. This trend in large sow farms has been further advanced with "Quad" systems (barns of 4,000 head in each barn) and the desire to run entire production sites AIAO. Enhancements with some "Pit" ventilation and better controls have been made, but horizontal ventilation systems have a basic flaw that encourages horizontal disease transmission and pig environment control challenges for the farm manager.

These tunnel ventilation systems by nature are compromise systems for 2/3 of the pigs in every barn. With the minimum of 3°C (5.5°F) temperature difference from one end of the barn to the other, many pigs are at the edge of their comfort zone much of the time. In 2002, Dr. John Baker provided us with an insight to the pigs "Comfort Zone" and consequences of falling out of that narrow range.[8] Any change or variation of their environment is often enough to push the compromised pigs at the upper or lower limits of the zone into stressful periods that weaken their native immune system and make them more susceptible to chronic diseases. These pigs exhibit themselves as chronic poor doers or non-full value pigs at market time. At the very minimum, airborne disease particles are carried by the crossflow air movement from pigs at one end of a tunnel ventilated building to pigs at the other end in a matter of minutes or seconds!

Our diseases are changing. In the days of Specific Pathogenic Free (SPF) pigs, regulations required us to keep 30 feet separation from non-SPF pigs to prevent bacterial disease spread. Simply put, that may have been a little naïve. Now we work with diseases that spread over miles and, under certain conditions, it seems that no farm is safe from an aerosol exposure to disease! Expensive air filtration systems have yet to prove themselves universally effective in keeping farms protected from an aerosol spread of disease.

All of these implemented management changes in the name of improving or at least maintaining herd health have in practice had little success. In evaluating USDA pig crop data (119,800,000), that closely supplies the pigs in the 2015 USDA slaughter numbers (109,771,000) a simple analysis reveals a wean to slaughter death loss of 8.37%. The same data from 1995 (99,600,000 Pig crop and 94,500,000 slaughter) represents a 5.1% death loss.[9,10] It appears that wean to finish survivability has actually decreased under our present production model. Add this 8.37% to the 10-15% average pre-weaning death loss experienced in our systems and we are losing 20%+/- of our pigs from birth to market in a production and management model specifically designed to keep pigs alive and healthy! A 64% increase in wean to finish death loss is hardly a resounding success story for our industry. This phenomenon occurs at a time when animal health and welfare concern are the stated priority of every production system that has arisen from swine industry integration.

The 30,000 foot view that these data and other observations suggest presents us with a truly "wicked" problem: one which we must address for long term public acceptance, viability and welfare of our animal production industry. There are alternatives to both the practices we are employing and product we are currently providing to the public. Are we brave enough as an industry to honestly assess and implement these alternate practices?

Implications are real, and scary

After traveling the globe the last 20+ years and immersing myself into the production problems of other pig producing countries, I marvel at our good fortune of not having a Foreign Animal Disease (FAD) outbreak these last several decades. PEDV was indeed a wake-up call. But it has nearly been forgotten. We currently ship 27% +/- of our pork out of our country. In the event of an FAD (you pick it—an existing, emerging or new one) the USDA must report the outbreak to the World Organization for Animal Health (OIE) within 24 hours of diagnosis.[11] End users of our product then have opportunity to determine whether or not they will accept meat from our processors. Many foreign buyers would likely stop importing our products immediately. We only have a few days of meat storage in our system. So think about what this means to our supply chain. Personally, I can only eat so many pork chops...

Most of our industry responses to a possible FAD have been reactive. The USDA in conjunction with other local, state and federal agencies has a good plan in place in the event of an FAD.[12] However, with today's industry structure and mobility, any virulent and highly infectious disease outbreak (think PEDV) will likely spread too fast to definitively diagnose, control, eradicate or even regionalize effectively in a naïve national herd. USDA and other federal agencies have embraced the One Health (OH) concept that the health of animals, people and viability of ecosystems are linked.[13] OH focuses on delivering collaborative, multidisciplinary proactive solutions to our "wicked" problem of disease control.

As animal and food producers, we no longer work in the relative vacuum we did 40 years ago. Anything negative that happens anywhere in our production or processing chain can go viral on social media in a matter of minutes either unintentionally or through honest release of information. How many of us can honestly justify to the public our 20%+/- across the industry death loss from birth to market in addition to the rising pre-weaning mortality (PWM) most farms now experience? How do we explain and justify elimination of small "non-viable" pigs at birth or weaning? Think "Charlotte's Web" for what that outcry might look like.

It is time for this "wicked" problem to get reassessed. That means we have to re-look and re-think every aspect of animal production and the systems we have created. We must commit to a long term and disciplined approach to improving our whole industry, not just the local efficiency of an individual production system or farm.

The fact that our intensified animal production practices have come under the scrutiny of the One Health initiative should be incentive enough for us to take this outside scrutiny seriously and review every aspect of our industry as it relates to sustainability and disease control. We claim to be the experts in our industry. It would behoove us to tackle our "wicked" problem before someone else does!

Wicked problems – wicked solutions

If it is time for a systematic reassessment of pig production practices in the US, and I suggest it is, these are some of the things that our industry need to consider and evaluate. Many of these actions are simple, but some will require a generation of time to accept, adapt and implement. Many of them go against some of our current practices.

Our almost blind acceptance and insistence for our multi-site systems need to be re-evaluated. It is an expensive system and creates competitive and communication challenges within each production system. The added biosecurity, multiple site approvals, added transportation costs, inherent building inefficiencies, site coordination, gilt acclimation, labor and added administrative costs become fixed overhead to multi-site systems. Multi-site systems contribute to the "leap frog effect" of disease spread that makes it nearly impossible to regionalize or contain a disease outbreak. The economic landscape of multisite systems has changed. Cost/benefit review and comparison against single site systems should be attempted.

We need to reconsider the single site production system as a viable alternative to our current primary production system. With our current understanding of disease, herd immunity and immune stimulation strategies, better production facilities and AIAO strategies, single site systems are much more viable now than they were 35 years ago. With a single site system, much of the fixed overhead associated with multi-site systems goes away. Once a single site system achieves truckload quantities of inputs and market pigs, further efficiencies of size may not be able to be justified from a production and risk standpoint. With today's access to information and risk management, single site managers are much more aware of effective business and management practices than in the past. This awareness will make these independent managers more viable than in the past.

As an industry, we need to take responsibility for the contaminants leaving our buildings. Seventy percent (70%) of new human diseases come from animals![13] Now that we know aerosol spread of disease occurs across considerable distances, we can no longer plead ignorance to the fact that we may, through our practices, jeopardize other producers' livelihoods, well-being and even human health! In the event of a new zoonotic influenza outbreak, do we want to be upwind or downwind from the nearest hospital? How do we maintain our social license to produce animals and accept the responsibility of risks to other people when we are aware of disease agents leaving our farm space and are doing nothing about it? If PEDV was a zoonotic disease and caused severe diarrhea in humans, would we be in business of producing pigs today? We have been extremely lucky that the species barrier for disease transfer between species has been the huge impediment to zoonosis that it has.

Building systems need to change to provide better environments for our animals than what exists today. Better technologies are available, For example, we need to rediscover the importance of supporting the innate immune systems of the pigs in our production units. Husbandry skills need to be reemphasized instead of reliance on digital directions. Vertical ventilation systems with a downward air flow provide opportunities for smaller rooms, more uniform and controlled environments, less horizontal spread of disease in the event of a disease challenge and more full value pigs to survive to market. Room and farm size is not dictated by vertical ventilation systems. AIAO practices are easily achieved no matter what the farm size might be. Inlet and exhaust filter systems are rather easily adapted to vertical systems.

Pig movement from one location to another needs to be minimized as much as possible. More strict controls than we currently have need to be put into place when movements of pigs occurs to sites other than the slaughter plant. Should pigs be quarantined to their birth farm until market? A recent study on truck movement among pig farms and production systems in Canada indicated that trucks used for the shipment of pigs in production networks may play a part in and increase the spread of disease.[14] Single site production systems would certainly aid in minimizing truck traffic. Our USDA has prepared a Foot and Mouth Disease (FMD) Response Plan that initially relies heavily on controlling animal movement.[12] Might it be prudent to consider adopting as common practices some of these disease outbreak control strategies in a preemptive manner in anticipation of the likely occurrence of an FAD or zoonotic disease outbreak?

New production siting and respect for current operations should be given more consideration in order to stop the possibility of diseases leap frogging from farm to farm. How close is too close between pig units? As practicing veterinarians, many of us have witnessed the "leap frog effect" of disease transfer in our practice areas. I certainly have witnessed this phenomenon in the Asian countries where I work. When any pig disease is identified in an area, we all instinctively take extra biosecurity precautions. The effectiveness of many of these precautions is not well documented, but we take them nevertheless. What is the correct safe distance between pig farms to discourage disease transfer? It is certainly more than the 30 feet required in the SPF regulations! But is one kilometer, 3 kilometers, or 5 kilometers needed to effectively mitigate disease transfer risk? What factors other than distance are involved? We need more study and understanding of this issue.

We need to close our sow herds to live genetic inputs whenever possible. This can be easily done in single site units or modified single site units. Select breeding animals can simply be diverted from the market pens to the gilt development unit (GDU). Multi-site systems would need to expand their GDUs for replacements weaned on the sow farm. Genetic inputs are more predictable than ever and genetic traits are quantified with indexes and accuracies so that any production manager can implement an internal genetic multiplication plan successfully with a minimum of guidance. This single effort would help minimize disease risk at the sow herd level and simplify acclimation programs. Today one wonders if PRRSV would have gotten the foothold in the US that it did if there were less or at least more controlled, movement of breeding stock into commercial farms.

Solutions to these issues and other innovative modifications to existing norms that might improve disease control will take time. The evolution toward our current production model has occurred over the last 35 years and will not stop on a dime. Systems are inherently self-serving and leaders will be inclined to do what is best for themselves, their system, or their immediate customer regardless of the long term effect on the industry. PrV was finally eliminated from domesticated herds only after legislation forced all producers to comply. Change should and will require a multi-disciplined approach to sorting out solutions to these "wicked" problems of disease control.

Funding for health control measures

Today pig producers are viewed more and more as food providers not farmers. Regulations and increased responsibilities are pushed down the chain to our animal production systems. The increased costs of these regulations, controls, training, documentation, compliance audits and administration increase the cost, risk and complexity of animal production. The interdisciplinary involvement of One Health movements will only add scrutiny to our industry and will be accompanied by more oversight into each aspect of our production activities. We should, at a minimum, be engaged in finding a way to apply science and ethical practices in accordance with our training and oath to safeguard what is best for the animal and our consumer.

As we have been successful at producing the least cost pork in the world, we have put ourselves and our industry at risk from inadvertent disease introduction, a more complicated human/animal relationship, and economic consequences no one envisioned a generation ago.

We should investigate innovative funding opportunities for disease management and control as the human/animal disease connection and its effect on the nation's health and food supply becomes more apparent. Should the production animal industries be entirely responsible for safeguarding world health? No one sector of our food production, processing and distribution system should carry the burdens of these costs. A check off at the processor level that would help initiate, regulate, and monitor production activities that add risk to our food supply may be in order. A check off would help fund those initiatives that would protect us all and may be a requirement for us to be reliable suppliers to our processing partners in the future. The alternative is further integration in the entire swine production and processing industry.

Final thoughts to ponder

As an industry and as a science-based profession, we have continually been surprised and fooled by the exponential power and flexibility of microbes to thrive and adapt to whatever obstacles and impediments we put in their way. It is time to review our basic premise of fighting these microbes, and return instead to promoting tools that help us provide a better understanding and attention to individual animal needs in our high density production systems. OH advocates are already concerned about our intensified production systems. More effort should be placed on supporting the pigs' innate disease fighting mechanisms at the same time we explore better ways to support their humoral responses to our flexible and resilient microbial friends. There needs to be a return to the industry training of employees that focuses on stockmanship and animal needs instead of the many current practices of promoting inflexible systems that result in our caretakers trained to be robots following digital directions. We must develop industry standard metrics to measure and benchmark the health of our industry's production units to insure that local efficiencies don't jeopardize global health and sustainability.

Certainly, in light of our growing "wicked problem," a review of our industry procurement, production and harvesting structure needs to be done to assess and strategize how to minimize the threat of geographical and zoonotic spread of disease associated with our production industry.

References

1. Main et al. Keep your feet on the ground but stick your head in the "cloud" AASV. 2015; 433,434.

2. Ash M. IN State Board of Animal Health swine import statistics.

3. Rowles. Lessons learned from highly pathogenic avian influenza (HPAI) AASV. 2015; 431,432.

4. Powell J. JBS United, Inc. personal communication October 2016.

5. Orr D. JBS United, Inc personal communication. 2014.

6. Dee et. al. Modeling the transboundary risk of feed ingredients contaminated with porcine epidemic diarrhea virus AASV. 2015; 421.

7. Olson D. Howard Dunne Memorial Lecture: The pig always wins AASV. 2014; 17.

8. Baker JE. Effective environmental temperature. J Swine Health Prod. 2004; 12(3): 140-143.

9. USDA Pig Crop Report.

10. USDA Hog Slaughter Report.

11. Brown C. Emerging disease response: Global roadmap but local drivers AASV. 2015: 419, 420.

12. USDA Foot and Mouth Disease (FMD) response ready reference guide-quarantine, movement control, continuity of business December 2015.

13. Gray G. Modern livestock production and novel influenza virus generation: Are the benefits worth the risk? Coppoc Lecture. 2015.

14. Thakur KK et. al. Analysis of swine movement in four Canadian regions: network structure and implications for disease spread Transbound Emerg Dis. 2016 Feb;63(1): e14-26. Doi: 10.1111/tbed.12225. Epub 2014 Apr 17.

Voluntary regional PRRS control: Pitfalls and progress

David J. Wright, DVM

Introduction

Porcine reproductive and respiratory syndrome (PRRS) is not a regulated disease so it is up to the swine industry to manage and control this costly disease. Voluntary regional PRRS control programs have been implemented as a tool to assist in this effort.

The "Minnesota N212 voluntary regional PRRS elimination project" originated in Stevens County in 2004. Since then this producer-driven project has expanded to include all of Minnesota north of Hwy 212. The goals are to identify all swine herds in the area, determine the PRRS virus (PRRSV) status of all participating herds, encourage elimination when feasible, and eventually eliminate PRRSV from the area. More than thirty other ARC&E (area regional control and elimination) projects have been organized. They are unique disease control programs in that they are voluntary and producer-led as compared to conventional programs that are coordinated and implemented by government agencies.

Voluntary regional control projects are ongoing social experiments that focus on collaboration and cooperation among swine producers, veterinarians, industry, and academia. This paper will use the Minnesota N212 voluntary regional PRRS elimination project to explore the pitfalls and progress of these programs.

Project overview

After swine sites have been identified and PRRSV status has been established, the project coordinator is responsible for tactfully disclosing any herd status changes—either new outbreaks or improved herd status. He or she also updates regional databases, generates maps and organizes regional meetings.

All participants are asked to sign a consent letter or participation agreement. The N212 consent letter allows sample collection, access to diagnostic results, and the ability to share those results with others on a regional map. Participants give up confidentiality in the hope that shared information will benefit the region. It is critically important to remind participants that shared information is a privilege, not a right, and that all information should be used with discretion.

Ten Commandments of regional PRRS control

The N212 project has promoted the "Ten Commandments of regional PRRS control" as a way to lay out expectations for participants:

1. **Test your herd.**
   Collect periodic blood samples or oral fluid samples if you believe your status is negative. Sequence the virus if you are PRRS positive.
2. **Share your status.**
   Disclose your status with regional participants, neighbors and coordinator.
3. **Communicate and cooperate—with veterinarians, neighbors, and regional coordinator.**
4. **Secure your farm.**
   Practice internal and external biosecurity. Insist on clean transport vehicles.
5. **Practice all-in/all-out management—in rooms, buildings, and sites.**
6. **Buy negative—boars, semen, replacements, and finishing fills.**
7. **Eliminate wild virus from your herd.**
   Implement herd closure and elimination on sow farms. Depopulate and repopulate nurseries, growers and finishers.
8. **Consult your veterinarian.**
   Use your veterinarian to develop a PRRSV control and elimination plan.
9. **Respect your neighbor.**
   Do not over-react. Social contact is not a high risk for PRRS transmission. Build trust!
10. **Protect the sow herds.**
    Protecting the sow herds reduces downstream infections.

Pitfalls and limitations of voluntary programs

Voluntary programs are limited by their voluntary nature and have led to many of the pitfalls common to the N212 project and other ARC&E projects. Unlike mandatory government disease control programs, most voluntary programs do not achieve 100% participation. Reluctance to participate originates from a number of producer attitudes: 1) Ignorance of the project, 2) Apathy, 3) Inconvenience, 4) Perception of limited value, 5) Cost of time and testing, 6) Fear of a negative response to the disclosure of a positive herd test, and 7) Hostility due to a previous bad experience. It is up to participating producers and the coordinator to strive to dismantle these areas of resistance to encourage participation.

Another pitfall is that not all participants test and report status changes regularly—particularly among independent finishers and small producers. This leads to inaccuracies in maps and frustrates outbreak investigations. There is also no way to determine the status of pigs coming into the region. The Minnesota Board of Animal Health reports that 6.5 million pigs move into Minnesota from other states and an additional 1 million pigs are imported from Canada.[1] Without status information on these pigs, it is difficult to coordinate effective control measures.

Progress report

Table 1 summarizes the PRRSV status and participation trends of N212 sites over the last several years. Table 2 offers a snapshot of the number of pigs in participating herds and their PRRSV status. Notice that while just 69% of the sites are participants, 92% of the sows and gilts in the region are from participating herds. As might be expected, larger herds and systems participate and report more than the smaller independent herds. Also, while only 12% of the herds are PRRSV positive, 34% of the sows and gilts in the region are positive. Figure 1 depicts an epidemic curve of reported new PRRSV infections in the N212 area over the past five years. Despite progress in enrolling herds and improved participation, the incidence of new PRRSV outbreaks has prompted more creative initiatives to make the program more valuable to producers.

New initiatives

In 2015 producers, veterinarians and members of the Minnesota Board of Animal Health (MN BAH) met to explore feasible options to improve PRRSV and PEDV control in Minnesota. The N212 project used this dialogue to initiate a proposal to partner with the Board of Animal Health. The proposal requested that certificates of veterinary inspection (CVI's) of all pigs entering Minnesota include a statement disclosing the PRRS and PED status of the herd of origin, if known. No movement restrictions would have been imposed and all pigs, regardless of status, could have entered the state. The purpose of this proposal was to gather information about the PRRS and PED status of pigs entering Minnesota that could lead to a better understanding of the dynamics of disease control. It was presented to producer groups, the Minnesota Pork Board Executive Committee and the Minnesota Swine Health Task Force for feedback and discussion. The MN BAH was very supportive in cooperating with the swine industry, but only if it was requested by Minnesota pork producers. The proposal ultimately failed to gain support from a majority of Minnesota pork producers out of concerns that it may not have provided adequate value to industry and the potential for added cost to producers.

This year the N212 project is placing a renewed emphasis on protecting the sow farms. All sites within a five mile radius of participating sow herds will be identified as a "mini-region" and will be encouraged to enroll. The objective is to reduce the

Table 1: N212 PRRS site status

	Jan 2014	Jan 2015	Nov 2016
Active Sites	765	781	823
Signed participants	505	523	564
Percent participants	66%	67%	69%
Negative PRRS	254 (50% of participants)	310 (59%)	326 (58%)
Positive PRRS	59 (12%)	44 (8%)	65 (12%)
Stable or stable vaccinate	87 (17%)	72 (14%)	74 (13%)
Unknown PRRS status	105 (21%)	97 (13%)	100 (18%)

Table 2: N212 PRRS pig status, Nov 2016

	Boars	Sows/gilts	Nursery/finish	Total
Pigs in N212 region	727	176,482	1,051,155	1,319,757
Pigs from participants	565	163,096	928,834	1,121,553
Percent participants	78%	92%	88%	85%
PRRS negative	564	72,645	561,248	637,074
Percent negative	100%	45%	60%	57%
PRRS positive	0	54,067	100,935	157,308
Percent positive	0	34%	11%	14%
Stable/Vaccinated	0	5,520	190,210	224,690
Pct stable/vaccinated	0	3%	20%	20%

Figure 1: Epidemic curve of N212 PRRSV outbreaks

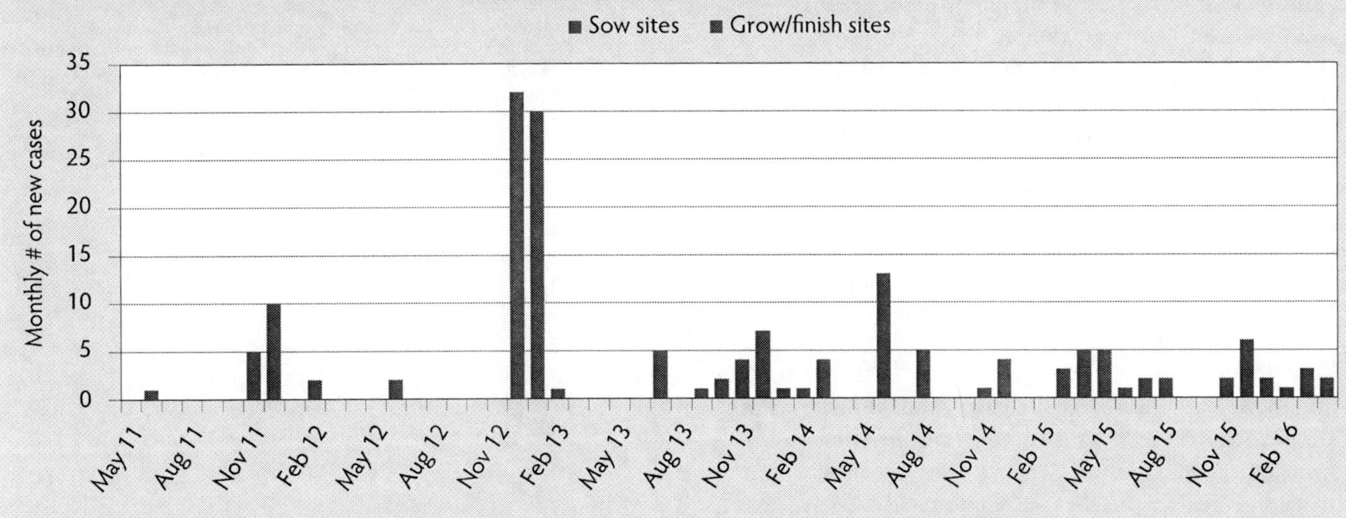

incidence of new infections in sow herds within these mini-regions. Outcomes will be measured by comparing the incidence of new infections in mini-regions with 100% participation vs. the incidence in those with less than full participation. Any sow herd with a goal of elimination with or without the use of vaccine is eligible to participate. Large sow herds are expected to test monthly; small sow herds semi-annually or if clinical signs appear; and nursery/finishing sites are asked to report the status of the herd source at each fill or test on arrival.

Financing the N212 project

When the N212 project began, major financial support came from the PRRS CAP fund, Boehringer Ingelheim, the University of Minnesota Swine Disease Eradication Center and the Leman Scholarship. The past three years the Minnesota Pork Board has provided the majority of funding. This year funding will come directly from contributions from producers, veterinarians and industry. Mapping and data management has always been provided by the University of Minnesota.

Value of regional control programs

The N212 project and other ARC&E projects offer a service to the swine industry. The most important question that must be asked is, "Is there value to the program?" Pablo Valdes-Donoso investigated this question in a research project using the N212 project as a working example. He concluded, "This study has established a systematic approach to quantify the effect of RCPs (Regional Control Programs) on PRRS control. There is evidence that RCP-N212 has attracted a growing proportion of producers to share disease status information, suggesting a rising awareness that sharing information can lead to more effective disease control. By evaluating the effect of participation on the occurrence of PRRS, the value of sharing information among producers may be demonstrated, in turn justifying the existence of RCPs. These results provide useful indicators regarding the evolution of the RCP-N212, and, ultimately, support for disease control in Minnesota. Furthermore, the methods presented here may be applied to measure progress in other RCPs."[2]

Earlier this year when the project lost its funding due to a reallocation of Minnesota Pork Board funds to other research projects, a group of participating producers and veterinarians was asked if the project was valuable enough to continue. The consensus of the group was to continue the project, to finance the program with contributions from producers, veterinarians, and industry, and to offer suggestions for improvement. Despite the limitations already mentioned, producers and veterinarians still feel there is value.

Here are a few specific examples of how the N212 project has benefited producers:

1. Pig swaps have been arranged to minimize the risk of a new infection in a neighborhood.
2. The decision whether or not to filter has been based, in part, by information from the N212 project.
3. Information distributed by the N212 project helps producers and their veterinarians decide whether or not to vaccinate for PRRS.
4. Routine and emergency feed delivery routes have been established using N212 information.
5. Biosecure routes for breeding stock delivery have been arranged using maps from the N212 project.
6. The weekly national incidence reports for the N212 area have been distributed to participating sow farms.

7. Producers are kept abreast of current research through meetings and correspondence.
8. The disclosure of a positive site eliminates the risk of disseminating inaccurate rumors.
9. Disclosures of positive sites reduce complacency in biosecurity.
10. The project fosters a spirit of collaboration and cooperation among producers and veterinarians.
11. A platform is in place to assist in dealing with other emerg ing diseases.
12. Voluntary regional disease control programs are a coordinated step toward the goal of national PRRS elimination.

Acknowledgements

Thanks to Montse Torremorell, Bob Morrison, Minnesota Pork Producers, Minnesota Pork Board, University of Minnesota-Swine Disease Eradication Center, Boehringer Ingelheim, and participating veterinarians for their support and encouragement for the N212 project.

References

1. Summary of Certificates of Veterinary Inspection in 2015 provided by the Minnesota Board of Animal Health.

2. Pablo Valdes-Donoso, Lovell, S., Jarvis2, Wright, D., Alvarez, J., Perez, A. (2016) Measuring Progress on the Control of Porcine Reproductive and Respiratory Syndrome (PRRS) at a Regional Level: The Minnesota N212 Regional Control Project (Rcp) as a Working Example. PLOS ONE | DOI:10.1371/journal.pone.0149498 February 19, 2016.